RECENT ADVANCES IN FINITE-TIME THERMODYNAMICS

RECENT ADVANCES IN FINITE-TIME THERMODYNAMICS

CHIH WU, LINGEN CHEN AND JINCAN CHEN

Nova Science Publishers, Inc.
Commack, New York

Editorial Production:	Susan Boriotti
Office Manager:	Annette Hellinger
Graphics:	Frank Grucci and Jennifer Lucas
Information Editor:	Tatiana Shohov
Book Production:	Donna Dennis, Patrick Davin, Christine Mathosian, Tammy Sauter and Lynette Van Helden
Circulation:	Maryanne Schmidt
Marketing/Sales:	Cathy DeGregory

Library of Congress Cataloging-in-Publication Data

available upon request

ISBN 1-56072-664-4

6080 Jericho Turnpike, Suite 207
Commack, New York 11725
Tele. 516-499-3103 Fax 516-499-3146
e-mail: Novascience@earthlink.net
e-mail: Novascil@aol.com
Web Site: http://www.nexusworld.com/nova

Printed in the United States of America

CONTENTS

PREFACE

Finite-time thermodynamics (FTT) is one of the newest and most challenging areas in thermodynamics. The objective of this book is to provide results from research, which continues at an impressive rate. The authors make a concentrated effort to reach out and encourage academic and industrial participation in this book and to select papers that are relevant to current problems and practice. The numerous contributions from the international community are indicative of the continuing global interest in finite-time thermodynamics. All represent the newest developments in their respective areas.

In 1824 Sadi Carnot, a French engineer, published his famous article "Reflexious sur la puissance motrice du feu" (Reflections on the motive power of fire), in which he reported the results of the first systematic study of the physical processes governing steam engines. This was the start of a new science: thermodynamics. Among the important topics in thermodynamics has been the formulation of criteria for comparing the performance of real and ideal processes. Carnot showed that any heat engine absorbing heat from a higher temperature reservoir to produce work must transfer some heat to a sink reservoir of lower temperature. He also showed that no heat engine could perform better than the Carnot heat engine. The early tradition was carried by Clausius, Kelvin and others on using thermodynamics as a tool to find limits on work, heat transfer, efficiency, coefficient of performance, energy effectiveness and energy figure of merit for energy conversion devices. The basic laws of thermodynamics were all conceived for irreversible processes. Since Gibbs, However, the subsequent development of thermodynamics has turned away from the process variables of heat and work toward state variables. The Carnot-Clausius-Kelvin view emphasizes the interaction of a dominant thermodynamic system and focuses on equilibrium states and reversible processes. The only fact concerning real processes is that these irreversible processes always produce less work and more entropy than the corresponding reversible processes.

In the real engineering world, no practicing engineer wants to design a heat engine that runs infinitely slow (reversible) without producing power. The need to develop power in real energy conversion devices is one reason why ideal, reversible performance is seldom approached.

Classical equilibrium thermodynamics can be extended to quasistatic processes. Conventional irreversible thermodynamics has become increasingly powerful, but its microscopic view does not lend itself to the macroscopic view preferred by practicing engineers.

Finite-time thermodynamics is designed as an extension to traditional thermodynamics in order to obtain more realistic limits to the performance of real processes, and to deal with processes or devices with finite time characteristics. Finite-time thermodynamics is a method for the modeling and optimization of real devices that owe their thermodynamic imperfection to heat transfer, mass transfer and fluid flow irreversibilities. In engineering, it is also known as "Thermodynamic Optimization" or "Entropy Generation Minimization". The method is based upon the most basic principles of thermodynamics, heat transfer, and fluid mechanics, and forms a new interdisciplinary branch of science.

This multi-authored book deals with the fundamental and applied research of finite-time thermodynamics. It illustrates how the gap between thermodynamics, heat transfer, and fluid mechanics is bridged.

In the paper by Lingen Chen, Fengrui Sun, Chih Wu and Jianzheng Gong, the history, development, and state-of-the-arts of finite time thermodynamics are reviewed from the point of view of both physics and engineering. Applications of finite-time thermodynamics to heat engines, refrigerators, heat pumps and other thermodynamic processes and devices with finite-size and/or finite-time constraints are discussed. Jincan Chen and B.Andresen investigate the optimal performance of an endoreversible Carnot heat engine used as a cooler. The optimal bounds of the primary performance parameters are determined. D. A. Blank and C. Wu investigate the effects of various heat transfer mode combinations and finite thermal capacity rate heat reservoirs on the power characteristics of endoreversible regenerative reciprocating Stirling and Stirling-like heat engines. G. Bisio and F. Devia review some topics of applications of FTT in heat engines. The impact of heat transfer mode on the heat engine performance is dealt with. D. A. Blank, S. Bhattacharyya and C. Wu investigate the optimal performance of an endoreversible reversed reciprocating Stirling cycle with ideal regeneration. The optimal cooling load and heating load limits are determined based on Newton's heat transfer law. S. Bhattacharyya derives the optimum operating temperatures of the electrodes of a thermionic generator with respect to maximum power output. D.A. Blank and C. Wu determine the maximum power output of an irreversible closed Joule-Brayton cycle coupled to variable temperature heat reservoirs. S. Goktun derives the coefficient of performance (COP) of an irreversible heat engine driven vapor compression and absorption refrigerator system with the irreversibilities of heat resistance and internal dissipation. V. Badescu investigates the optimum ratio between the collector and radiator area corresponding to the maximum power for the solar space power station. J. Chen and Z. Yan analyze the influence of heat resistance and regenerative losses on the cooling load vs. COP characteristic of a magnetic Stirling refrigerator. M.E.H. Assad derives the fundamental relation between cooling load and COP of an irreversible refrigerator with heat resistance and internal dissipation losses based on the heat transfer law $q \propto (\Delta T)^n$. S. Sieniutycz treats the multistage optimization of endoreversible thermal machines by Bellman's method of dynamic programming and discrete maximum principle with a constant Hamiltonian to maximize the total power output. M.E.H. Assad and C. Wu derive the power and efficiency of a MHD power plant when MHD generator is operating at constant velocity and constant Mach number conditions. A. DeVos and B. Desoete illustrate the applications of finite time thermodynamics during the thermal, chemical and electronic processes when minimum entropy generation is the optimization objective. J.B. Woodward demonstrates the relationship between the power output and the NTU (number of heat transfer units) of the heat exchanger in the Diesel waste-heat recovery systems. B. Sahin and A. Kodal investigate the exergy optimization of an irreversible combined heat and power plant. A. Kodal derives the fundamental relation between the heating load and COP of an irreversible heat pump with the irreversibilities of heat resistance and internal dissipation based on the heat transfer law of $q \propto \Delta(T^n)$. Z. Yan and L. Chen derive the fundamental relation between power output and efficiency of an irreversible heat engine with the irreversibilities of heat resistance, heat leak and internal dissipation based on the heat transfer law of $q \propto \Delta(T^{-1})$. N.E. Wijeysundera analyzes the effect of internal and external irreversibilities on the optimal performance of vapor absorption cooling system. L. B. Erbay investigates the power optimization of an irreversible heat engine with internal dissipation and combined radiation and convection heat transfer. A. M. Tsirlin and V. Kazakov discuss the optimal operating conditions for the systems with regeneration. S. Goktun and H. Yavuz derive the optimum operating temperature of a focusing solar collector-driven three-heat-reservoir heat pump with internal and external irreversibilities. B. Agnew, A. Anderson, T. H. Frost and R. M. Tozer optimize the performance of endoreversible single- and double-stage absorption refrigerators. R. Y. Nuwayhid and F. Moukalled investigate the power vs. efficiency behavior of the thermoelectric generators. C. Y. Cheng, S. C. Lee and C. K. Chen analyze the effects of various design parameters on the performance of irreversible closed Brayton cycle coupled to variable temperature heat reservoirs. M. Feidt provides a unified description of engines, refrigerators and heat pumps coupled to two or three heat reservoirs. A. Calvo Hernandez, A. Medina, J.M.M. Roco and S. Velasco investigate the optimum performance of a generalized irreversible Carnot engine based on the maximization of the power output and the per-unit-time efficiency. F.Angulo-Brown, T. D. Navarrete-Gonzalez and J. A. Rocha-Martinez calculate the power and efficiency of an irreversible Otto cycle model including irreversibilities of friction and combustion reaction. M. Dentice d'accadia, M. Sasso and S. Sibilio pay attention to the search for the optimum operation conditions of a heat engine-driven heat pump with engine waste heat recovery from both thermodynamic and thermoeconomic points of view. L. Chen, Y. Bi, and C. Wu establish a steady flow combined heat pump cycle model with heat resistance, heat leak and internal dissipation, and derive the optimal relation between heating load and COP. P. Salamon responds to some criticisms of finite time thermodynamics and highlights some of the achievements of the subject.

The readers should find the papers informative and useful for analysis and design of thermodynamic systems with improved performance. We hope that this collection of work devoted to finite-time thermodynamics will provide encouragement for further research in the area.

The editors wish to express their sincere appreciation to all the authors for their fine contribution, and all the reviewers who carefully reviewed and significantly enhanced the quality of the papers. The book would not have been possible without the dedicated effort of these individuals. We also appreciate the Nova Science Publisher personnel for their complete cooperation throughout the publication process.

PROGRESS IN FINITE-TIME THERMODYNAMICS

LINGEN CHEN, CHIH WU, FENGRUI SUN AND JIANZHENG GONG

Abstract--The historical background, research development, and the state-of-the-art of finite time thermodynamic theory and applications are reviewed from the point of view of both physics and engineering. The emphasis is on the performance optimization of thermodynamic processes and devices with finite-time and/or finite-size constraints, including heat engines, refrigerators, heat pumps, chemical reactions and some other processes, with respect to the following aspects: the study of Newton's law systems, an analysis of the effect of heat resistance and other irreversible loss model on the performance, an analysis of the effect of heat reservoir model on the performance, as well as the application for real thermodynamic processes and devices. It is pointed out that the generalized thermodynamic optimization theory is the development direction of finite time thermodynamics in the future.

Key Words: Finite time thermodynamics, heat engine, refrigerator, heat pump, irreversible thermodynamic processes, generalized thermodynamic optimization.

1. INTRODUCTION

More than 170 years ago, Sadi Carnot, a French engineer, published his famous article [1] and established a new field of science: classical thermodynamics. With the developments of theoretical physics, physical chemistry, chemical physics and engineering, the considerable progresses in the modernization of thermodynamic theory have been made. In the early 1990s, Taylor & Francis published six books within Advances in Thermodynamics Series [2-7], in which the state-of-the-arts and the future potentials of major branches of modern thermodynamics in the end 1980s were summarized and analyzed fully. They showed that the major branches, including Finite Time Thermodynamics [5], of modern thermodynamics are full of promise.

2. HISTORICAL BACKGROUND OF FINITE TIME THERMODYNAMICS

Carnot [1] showed that no heat engine operating between high-temperature heat source at temperature T_H and low-temperature heat sink at temperature T_L could perform better then the Carnot engine. The largest fraction of the heat, which can be converted into work, is

$$\eta_C=1-T_L/T_H \,. \quad (1)$$

traditionally known as the famous Carnot efficiency. This conclusion provided the upper limit of efficiency for any heat engine working between T_H and T_L. The early tradition was carried out by Clausius, Kelvin and others on using thermodynamics as a tool to find limits on work, heat transfer, efficiency, coefficient of performance (COP), energy effectiveness and energy figure of merit of energy conversion devices. The basic laws of thermodynamics were all conceived about irreversible processes. However, the subsequent development of thermodynamics has turned away from the process variables of heat and work toward state variable since Gibbs. The Carnot-Clausius-Kelvin view emphasizes the interaction of a thermodynamic system with its surroundings, while the Gibbs view makes the properties of the system dominant and focuses on equilibrium states. Contemporary classical thermodynamics gives a fairly complete description of equilibrium states and reversible processes. The only fact concerning real processes is that these irreversible processes always produce less work and more entropy than the corresponding reversible processes. Reversible processes are defined only in the limit of infinitely slow execution. The solution of optimization problem in classical thermodynamics are reversible thermodynamic processes, i.e., processes in which the system preserves internal equilibrium, the total entropy of the system and the environment does not increase, the rate of exchange between the system and the environment is infinitesimally small, and the process duration is infinitely long, and as a consequence of this is zero output rate (power output for engine, cooling load for refrigerator, and heating load for heat pump) for the duration average. Performance limits obtained with the aid of reversible processes are independent of the equation of state of the system and are limiting in the sense that they remain unattainable in all real processes.

All real thermodynamic processes are irreversible since the rate of exchange between the system and the environment is not infinitesimally small, the system does not maintain internal equilibrium, and the process duration is finite. The classical reversible bounds are greatly too high for real processes and require further refinement. The basic questions are: Are reversible limits close enough to real performance to be useful in guiding the improvement of processes? Whether it is possible to find more realistic limits to the performance of real (finite-time and /or finite-size) processes and devices? Can we use such bounds to find better criteria of merit useful in evaluating real processes and devices? Can we use such criteria of merit to optimize the performance of real processes and devices and to provide scientific and accurate guidance for solving practice engineering problems?

The remake and innovation of classical thermodynamics and the solution of thermodynamic bounds for finite-time and /or finite-size thermodynamic processes characterized by finite rate of exchange between the system and the environment are the early major subjects of finite time thermodynamics. Novikov (1957)[8], Chambadal (1957)[9], and Curzon and Ahlborn (1975)[10] noticed this problem. They treated a real Carnot engine with power output being limited by the rates of heat transfer to and/or from the working substance, and showed independently that the hot- and/or cold- end temperature of a power plant can be optimized such that the power output is maximum. The efficiency at maximum power point is

$$\eta_{CA}=1-(T_L/T_H)^{1/2} . \quad (2)$$

It provides a new performance bound, which is different from Carnot efficiency, for the heat engine characterized by finite rate, finite duration and finite-size. It is a new efficiency bound derived first in nuclear engineering literatures [8,9] and rediscovered in physics literature [10]. Since mid 1970s, the research work for searching the performance bounds of thermodynamic processes and carrying out thermodynamic optimization has achieved large progress in both physics and engineering. In the filed of physics, it was termed Finite Time Thermodynamics (FTT) by R. S. Berry, B. Andresen, P. Salamon, M. J. Ondrechen, A. M. Tsirlin and S. Sieniutycz, among others, or Endoreversible Thermodynamics by A. De Vos. In the field of engineering, it was termed Entropy Generation Minimization (EGM) by A. Bejan, R. J. Krane, D. P. Sekulic, M. Feidt and V. Radcenco, among others, or "Thermodynamic Modeling and Optimization" by A. Bejan and K. C. Ng, among others. The fundamental characters of both FTT and EGM are the same, that is, bridge the gap between thermodynamics, heat transfer, and fluid mechanics, and thermodynamically optimize performance of real finite-time and /or finite-size thermodynamic systems with the irreversibilities of heat transfer, fluid flow and mass transfer toward decreasing the irreversibility of the total system. In this review the research field is termed "Finite Time Thermodynamics".

With the support of more than 20 science foundations, a large number of research work has been performed in this new field of interdisciplinary domain. The research field concerns the conventional thermodynamic devices, such as heat engines, refrigerators and heat pumps, and quantum thermodynamic systems, direct energy conversion devices, fluid flow processes, heat transfer processes, heat exchangers, mass transfer processes, chemical reaction processes, thermal isolation systems, thermal energy storage systems as well as the operation (on/off) of time-dependent processes. Today, finite time thermodynamics is a large and active field. More than 1400 publications, including books [5,11-19], Doctor thesises [20-24] and review articles [25-41] were performed on this field between 1975 and 1998 in English, French, Russian and Chinese, among others.

3. The Connotation of Finite-Time Thermodynamics

The region covered by finite time thermodynamics is very wide. The applications of FTT include the all processes with thermal phenomena and all devices and systems operating with thermal phenomena with the constraints of finite-time and/or finite-size. The life processes of animals, the superconducting transition phenomena, and wind energy system of earth are also the object of FTT study.

The research line of FTT is as follows: making some assumptions for real process to establish thermodynamic model, given a series of constraints to define the probably time pathway of the process, solving the objective extreme for the given path or the optimal path for the specified optimization objective to obtain optimum performance criteria for the defined process. Therefore, FTT can answer some global questions, which the classical thermodynamics does not try to answer, and the conventional irreversible thermodynamics can not answer because of its micro, differential viewpoint. Examples of such questions are, as follows: What is the least energy required by a given machine, to produce a given work in a given time? What is the most work that can be produced by a given machine in given time, utilizing a given energy? What is the most efficient way to run a given thermodynamic process (optimal path) in finite time? What is the optimal time-dependent (on & off) process? What is the optimal distribution between heat exchanger heat transfer surface areas or heat conductances corresponding to the optimal performance of the thermodynamic devices for the fixed total heat exchanger heat transfer surface area or total heat conductance? What are the quantitative and qualitative features of the effects of heat resistance, internal irreversibility and heat leakage on the performance of real thermodynamic processes and devices? etc.

FTT is the further extension of conventional irreversible thermodynamics. The methods used for solving the FTT problem are usually variational principle and optimal control theory. Therefore, such problems of extremal of thermodynamic processes were first named as Finite Time Thermodynamics by Andresen et al[42] and as Optimization Thermodynamics or Optimal Control in Problems of Extremals of Irreversible Thermodynamic Processes by Orlov et al[28]. When the research object was extended from reciprocating devices characterized by finite-time to the steady state flow devices characterized by finite-size, one release that the physical property of the problems is the heat transfer owing to temperature deference. Therefore, Grazzini [43] termed it as Finite Temperature Difference Thermodynamics, and Lu [44] termed it as Finite Surface Thermodynamics. While Bejan introduce the effect of temperature difference heat transfer on the total entropy generation of the systems, taken the entropy generation minimization as the optimization objective for design thermodynamic processes are devices, and termed as "Entropy Generation Minimization" or "Thermodynamic Optimization " [11,17,18,37-41].

It is necessary to point out that classical thermodynamics is really thermostatics. While the conventional irreversible thermodynamics focus on the equations of motion, and leads naturelly to a formulation in terms of differential equations and to examination of the local, differential behavior of the systems. Therefore, FTT is a new and independent scientific discipline, this discipline can not be simply classified as classical thermodynamics and conventional irreversible thermodynamics. The relation between FTT and conventional irreversible thermodynamic was discussed recently by Verhas and De Vos [45].

4. THE PRESENT STATUS OF FTT

4.1 NEWTON'S LAW HEAT ENGINE SYSTEM COUPLED TO INFINITE THERMAL CAPACITY RESERVOIRS

Classical thermodynamics investigates the reversible cycles. Reversible cycles takes infinite time (or infinite heat exchanger surface area) to complete. Therefore, the output rate averaged for the period (or the specific output rate averaged for the total heat transfer surface area) approaches to zero. Finite time thermodynamics (FTT) emphasizes the trade-off between coefficient of performance and output rate. Endoreversible (internal reversible) cycle is the fundamental physic model adopted in FTT analysis. Hence, De Vos termed FTT as Endoreversible Thermodynamics [15,45]. In an endoreversible cycle, the working fluid undergoes quasistatic process, and the sole irreversibility is the heat resistance loss between the heat reservoirs and the working fluid. Endoreversible model was a theoretical model first applied by Novikov [8], Chambadal [9] and Curzon and Ahlborn [10], and named by Rubin [46]. When the working fluid undergoes quasistatic Carnot cycle, the cycle is called as endoreversible Carnot cycle. It is also called as Curzon-Ahlborn cycle. In the most study, it is assumed that the heat transfers between the heat reservoirs and working fluid obey Newton's (linear) Law and the heat reservoirs are infinite thermal capacity (constant-temperature) heat reservoirs. Such system is called as Newton's law system coupled to infinite thermal capacity reservoirs [47]. The studies for this type of systems involve two aspects: The first is to determine the objective function extremals and the relation between objective functions for the given thermodynamic system, i.e., to derive the finite time thermodynamic performance bounds, fundamental optimal relation and the performance optimization criteria. The second is to determine the optimal thermodynamic process (optimal path) for the given optimization objective. The optimization objectives include many performance criteria, for example, work, power, efficiency, effectiveness, entropy generation rate, profit, specific power (power averaged for total heat transfer surface area), power density (power averaged for the maximum specific volume in the cycle), and ecological objective (compromise between power output and the entropy generation).

4.1.1 THE OBJECTIVE EXTREMAL FOR GIVEN THERMODYNAMIC PROCESS (THEORETICAL CYCLE)

The theoretical cycle is referred to endorevesible Carnot cycle in order to distinguish it from the practice engineering cycles. For the given thermodynamic process, one can solve the optimal working fluid absorbing and releasing temperatures and the optimal times (or heat transfer surface areas), and obtain the performance bound, and the fundamental optimal relation between efficiency and power output. In 1957, Novikov [8], a former USSR nuclear power specialist investigated the performance of a nuclear gas turbine power plant. After deriving the optimal gas working temperature corresponding to maximum power output with the consideration of the temperature difference heat transfer in the generator, he obtained the efficiency of the plant is the same as Eq. (2). Just in 1957, Chambadal [9] obtained the same result when he analyzed the power plant driven by a stream of hot single-phase fluid. In engineering, Eq. (2) was reprinted in power engineering and engineering thermodynamics textbooks [48-51]. The efficiency formula Eq. (2) was rediscovered in 1975 by Curzon and Ahlborn [10], two Canada physicists. They treated a Carnot engine with power output being limited by the rates of heat transfer to and from the working substance. After optimizing the working temperatures corresponding to maximum power output, they derived the efficiency at maximum power, which is the same as Eq. (2). Therefore, Bejan [52] suggested that the maximum power efficiency Eq. (2) may be called the Novikov-Chambadal-Curzon-Ahlborn (NCCA) efficiency (here, we call it CA efficiency for convenient). The important significance of CA efficiency is that it provides the efficiency bound of a Carnot engine at maximum power output .It possess the universality for Newton's law system [53,54]. The CA engine model and the CA efficiency have been introduced in thermodynamics textbooks [55-60]. In physics, most researchers have studied the performance optimization of Carnot and Carnot-like cycles for simplicity and clearness.

Rubin [46] derived the maximum efficiency η_{max} for the fixed heat input. $\eta_{max} < \eta_C$ if the heat input doesn't equal to zero. The power output is obtained by taking the efficiency as the cost. This is the important inspiration of FTT [61]. Gutowicz-Krusion et al [62] derived the maximum power output P_{max} and the corresponding efficiency η_P of a Carnot engine with the volume inequality constraint. η_P is dependent on the compression ratio. η_P approaches to η_{CA} when compression ratio approaches to infinite. Spence et al [63] derived the relations between power output P, efficiency η and the operating frequency and the optimal frequency corresponding to the P_{max}. Rebhan et al [64] further derived the relations between P/P_{max}, η/η_{CA} and the operating frequency. Besides the maximum-power characteristics of heat engine [65], one also pays attention to the optimal efficiency at arbitrary power output. Yan [66] derived the relation between the optimal efficiency and the optimal power output for an endoreversible Carnot engine. The relation includes η_C at P=0 and P_{max} at η_{CA}, and is termed fundamental optimal relation. Chen et al [67] derived the fundamental optimal relation of the engine with finite compression ratio. Yan [68] taken the η^{λ} P as optimization objective to optimize η and P. Sun et al [69-71] defined two characteristic parameters to represent the degrees of deviation of working temperatures of working fluid and the power from the corresponding optimal values, obtained the holographic power versus efficiency spectrum, and formed the finite time thermodynamic criteria for the parameter selection of an endorevesible Carnot engine based on the compromise optimization between power and efficiency.

Sun et al [72] fist noticed the thermodynamic mechanism difference between steady flow and the reciprocating engines, used finite heat transfer surface area to replace the finite time constraint, taken the specific power—ratio of power to the total heat transfer surface area—as the optimization objective, and derived the minimum total heat transfer surface area principle and the heat transfer surface area characteristics [73]. Specific power optimization and the optimal heat transfer surface area analysis have been become an active engineering research job [44,74-83].

Besides the power and efficiency objectives, Salamon et al [47] investigated the optimal performance of an endoreversible Carnot engine by taking the exergy efficiency (effectiveness), exergy loss and profit as optimization objectives. On the basis of Salamon's work [47], in 1990 Chen et al provided finite time exergoeconomic analysis method using the combination of finite time thermodynamics and thermoeconomics [5,84-86], and derived the finite time exergoeconomic performance bounds, optimal relations and parameter optimization criteria [87-89]. The similar idea were provided also by Ibrahim et al [90] and De Vos [91,92]

In 1991, Angulo-Brown [93] introduced the ecological objective E—compromise optimization objective between power output and entropy generation rate. The efficiency η_E at maximum E is about $(\eta_C+\eta_{CA})$ /2, this is a general property of Newton's law system endorevesible engine, and was applied to predict the performance of superconductor materials [98].

Sahin et al [99] taken the power density–ratio of the power output to the maximum specific volume in the cycle–as optimization objective to investigate the performance optimization for a Carnot and the other engines.

Geva and Kosloff [22, 100-106] and Wu et al [24,107,108] considered the quantum mechanical characteristic of the working fluid, discussed the power, efficiency, entropy generation, ecological objective and exergoeconomic performance optimizations for heat engines and three-level amplifier consisting of spin-1/2 systems and oscillator system as working fluid, and obtained some new results different from the conventional working fluid.

4.1.2 Optimal Thermodynamic Cycles for the Given Optimization Objective

This is to solve the problems how to obtain the maximum work output, power output or efficiency for the fixed external conditions. It needs to specify the control variables and their limits, build the govern equations, seek the constraints that are imposed on the system, and determine the objective function. Gutowicz-Krusion et al [62] proved that in all acceptable cycles, endoreversible Carnot cycle with larger compression ratio can produce maximum power, i.e., CA cycle is the optimal configuration with the only Fist and Second Law constraints. Rubin [46,109] obtained the six-branch and eight-branch cycle with different constrained conditions. Salamon et al [47] proved that the optimal process corresponds to constant rate of heat exchange between working fluid and heat reservoirs for various objectives. Tsirlin and Rozonoer [13,110-112] derived the optimal thermodynamic configuration and optimal cycle region for endorevesible engine with different constrained conditions. Lampinen et al [113] optimized the heat engine configuration by using heat accumulation function provided in Mathematical Thermodynamics.

4.2 Effect of Loss Model on the Optimal Performance of Heat Engines

In practice engine, not all the heat transfers between working fluid and heat reservoirs obey Newton's law. Furthermore, besides heat resistance, there exist heat leak and internal dissipation.

4.2.1 Effect of Heat Resistance (Heat Transfer Law) Models

One can derives performance bounds and fundamental optimal relation for endoreversible heat engine with different heat transfer laws. Gutowicz-Krusion [62] first derived the efficiency bound of a Carnot engine with heat transfer law of $q \propto (\Delta T)^n$. Chen et al [114] derived the optimal relation between power output and efficacy based on this law. A great deal of literatures have studied the power versus efficiency output relation, ecological optimal performance and finite–time exergoeconomic optimal performance for Carnot engines with radiative heat transfer, $q \propto \Delta\ (T^n)$ heat transfer and mixed heat transfer law [23, 57, 60, 115-127]. Chen et al [128] and Orlov [129] analyzed the effect of heat transfer law on the optimal configuration of heat engines.

4.2.2 Effect of Other Irreversibilities

Andresen et al [130] provided the tricycle formalism, and analyzed the performance of Carnot engine with irreversibilities of heat resistance and friction. Fairen et al [131] analyzed the effect of inertia on the performance and the optimal configuration of a heat engine. Richter et al [132] investigated the heat engine performance with friction, inertia and non-equilibrium effect, and Delgodo [133] derived the optima configuration of this type of engine.

In Refs. [8,43,134-141], a constant was applied to represent the all other irreversibilities in the engine except heat resistance, including heat leak, friction and non-equilibrium. On the basis of Bejan's heat resistance and heat leak losses model [11,142], Chen et al [143-145] further derived the optimal relation between power output and efficiency, which is consistent with the real engine performance [146].

On the basis of analyzing the effects of heat resistance, heat leak [142-145,147-149] and internal dissipation on the heat engine performance, Chen et al [23,150,151] established a generalized irreversible Carnot engine model, which account of effect of the heat resistance, heat leak and the other internal dissipation, and derived its optimal performance. The effect of heat transfer law on the performance of a generalized irreversible engine was also analyzed by Chen et al [152].

4.3 Effect of Heat Reservoir Model on the Optimal Performance of Heat Engine

4.3.1 Finite Thermal Capacity Heat Reservoirs

It is often the case in practice the power is generated from heat, which is carried by a finite amount of materiel with finite heat capacity, rather than from heat extracted from an isothermal, infinite reservoir. In the reversible (infinite-time) limit, the cycle, which extracts the maximum work from a finite heat source, is qualitatively different from the Carnot cycle, and its theoretical efficiency is considerably smaller [153]. For the endoreversible cycle, the research of effect of finite heat reservoir on the performance includes two aspects: The first is to determine the optimal performance of the given finite thermal capacity cycles, such as Carnot cycle [43, 154], Rankine cycle [155], Brayton cycle [140], etc. The optimization of the first aspect may be carried out with the fixed heat input [154] or with the variable heat input [43,140,155]. The second is to determine the optimal configuration of heat engines with the given conditions. For example, the optimal configuration of endoreversible constant-temperature heat reservoir heat engine is the Curzon-Ahlborn engine, the optimal configuration of Newton's law system variable–temperature heat reservoir heat engine is a generalized endoreversible Carnot engine [156] in which the temperature of the heat reservoirs and the working fluids change exponentially with time and the ratio of the temperatures of the working fluid and the heat reservoir is a constant. The studies have shown that the heat transfer law [157] and heat leak [23,158] do affect the optimal configuration.

4.3.2 External Heating and Pumping Heat Type Reservoir

Band et al [159-164] investigated the problem of maximizing the work obtained from an ideal gas heated by a heat source at an arbitrary rate in a cylinder with a movable piston. The Newton's heat transfer law between the gas and the external heat bath was adopted. Chen et al [23, 165] determine the optimal path by assuming that the heat transfer between the gas and the external heat bath obeys the linear phenomenological law, which is often used in non-equilibrium thermodynamics.

4.4 Analysis for Practice Heat Engine Plants and Thermal Processes

4.4.1 Analysis for Combined Power Plant

Rubin [166] determined the optimal configuration of the combined heat engine cycle formed by two endoreversible heat engines with intermediate heat reservoirs. Chen and Yan [167] determined the optimal configuration of that cycle without any intermediate heat reservoir. Wu et al [167-169] analyzed the optimal power output of combined power plants, while Chen et al [170,171] derived the specific power output of the combined power plants. The effect of heat leak was analyzed by Chen et al [172]. Based on the heat resistance mode [166-171], the heat resistance and heat leak model [172], and the heat resistance and internal irreversibility model [173-175], Chen et al [23,176] established a generalized irreversible combined heat engine model and derived its optimal characteristics.

4.4.2 Analysis for Internal Combustion Engine

The piston in a conventional internal combustion moves in a sinusoidal path. In Refs. [177, 178], the piston paths of the Otto and Diesel cycles are optimized by using the optimal control theory. The efficiency and the power output of the optimal cycle could increase more than 10% than those of the conventional one with same irreversibility source. There are also a number of articles concerning the research of the effect of heat transfer, friction and combustion irreversibilities on the performance of Otto [23,174-184], Diesel [23,181,185-187] and Dual [23,188,189] cycles.

4.4.3 Analysis for Steam and Gas Turbine Cycles

Chen et al. [190] proved that the holographic spectrum and optimization criteria for endoreversible Carnot engine given in Refs. [69-72] could be extended to any heat engine by Clausius secting method for cycle analysis, and a real nuclear power plant secondary loop Rankine cycle was analyzed and optimized using the cycle-secting method by Sun et al [191]. There also exist much literature, which concern the power and specific power optimization for endoreversible and irreversible Rankine (steam turbine) cycles, including Refs. [16,192]

A number of researchers [16-18,57,97,140,143,193-199] have taken the work, power and specific power as optimization objective to investigate the performance optimization of open and closed, endoreversible, simple and regenerated Brayton (gas turbine) cycle coupled to constant and variable –temperature heat reservoirs. The analysis was performed using numerical calculation. Sahin et al [200,201] analyzed the power density optimization of Brayton cycle without accounting for the heat transfer irreversibility. References [203,204] investigated the power optimization of Brayton cycle without heat resistance. Chen et al [23,204-210] analyzed the performance of simple and regenerated closed gas turbine cycle coupled to constant- and variable-temperature heat reservoirs with the irreversibilities of heat resistance, nonisentropic compression and expansion loss and piping pressure drops. The analytical expressions between the power output and pressure ratio as well as between efficiency and pressure ratio were derived. These results include those of conventional engineering cycle analysis [202,203,211,212], and more important, discovery some new phenomena. The important new conclusion is that the regeneration does affect the power output of a closed regenerated Brayton cycle.

4.4.4 Analysis for Stirling Cycle

A lot of researchers have analyzed the power, specific power and power density optimization of endoreversible and irreversible Stirling engines, and analyzed the effects of heat transfer law and the quantum characteristics of the working fluid on the Stirling engine performance [16,20,24,213-223]. The typical work includes the followings: Radcenco and Popescu [16,20,216-218] investigated the power optimization and the optimal distribution of heat transfer surface area of endoreversible and irreversible cycle with various heat transfer law, Wu et al [24,220,221] considered the quantum characteristic of working fluid, and Ibrahim et al [222,223] introduced the finite time analysis to the dynamic performance optimization of Stirling engine.

4.4.5 Analysis for the Other Thermal Power Cycles and Plants

The finite time thermodynamic theory and analysis method have been applied to all of the possible thermal power cycles and plants including Ericsson cycle [224], Rail cycle [225], Atkison cycle [226], heat and power cogeneration cycle [227], thermoacoustic devices [24,228], thermoelectric heat engine [229], solar driven heat engine [230], heat energy storage devices [231], thermionic engine [232], MHD power plant [233], solar cell [15,234], wind energy system [235], heat exchanger [236], weather forecast [237],material melting process [238],laser device [239],light driven engine [240],and separation process [241,242],etc.

4.4.6 Analysis for Chemical Processes and Chemical Engine

Much work have been published for studying the power and efficiency optimization problems of electrochemical system, fuel synthesis, distillation process, biochemical reaction, combustion reaction, catalytic polymerization. The typical work are those performed by De Vos [15,243], Gordon et al [244,245], Andresen et al [12,19,246,247], Angulo –Brown et al [248,249], and Chen et al [23,250-252]. Gordon [244,245] solved the maximum power of an endoreversible isothermal chemical engine. Chen et al [23,250-252] derived the optimal relation between power and efficiency of the single and combined cycle chemical engine, and analyzed the effect of mass leak on the optimal performance.

4.4.7 Analysis for General Thermal Process

Salamon et al [253] constructed finite time thermodynamic potential for endoreversible system. Some improvement of this potential was performed by Angulo-Brown [254]. Andersen et al [255] defined the finite time exergy (availability) function. Salamon et al [256] used the thermodynamic state space scale tensor in the geometrical thermodynamic [257] to define the new concept of thermodynamic length. This concept has been applied to solve many problems [27] including information coding system [258], separation process [242], measures of dissipation [259], bound of dissipation [260], and variational path optimization of free energy changes [261].

Rubin [46] suggested that uses P/P_{max} and η/η_{CA} to represent the figure of merit for thermodynamic cycles, and obtained the maximum power principle. Chen et al provided the finite time exergoeconomic figures of merit [263] and the ecological figure of merit [264] in order to evaluate the energy use process more accurately.

4.4.8 The Analogous Analysis for Heat –Engine-Like Cycles

The idea and method of finite time thermodynamic optimization for endoreversible heat engines can be extended to the generalized thermodynamic processes and devices, which are termed as heat-engine-like cycles [23], with generalized potential differences and generalized displacements [16].

Sun [59] in 1986 and Chen et al [34] in 1992 provided the idea that finite time thermodynamics may be applied in the life processes. Angulo-Brown et al [248,265,266] investigated the power versus efficiency characteristics in muscle contraction processes. Bejan [60,267,268] established the Constructal Theory of Organization in Nature, and applied it to the pulsating physiologic processes including respiration, circulation and ejaculation processes, as well as fluid flow, and heat transfer processes.

The analogy between the irreversibility of fluid flow with finite pressure drops and that of heat transfer with finite temperature differences was first discovered by Bejan [11] in 1982 and then be studied further by Radcenco [16]. In1996, Bejan [17,60,269] derived the maximum power and the corresponding efficiency bound for fluid flow device with liner and nonlinear pressure drop vs. flow resistance relation. Some improvements of this result were performed by Chen et al [23,270].

The research of chemical engine performance [244,245, 250,251] is also the analogous study. The analogy between motor, generator, circuit and endoreversible heat engine was discussed by Bejan [271] and Chen et al [250]. The analogy between computer logic computing and endoreversible heat engine was discussed by De Vos [272]. Further more, much work has been performed in recent decade on the Simulated Annealing–a global optimization procedure which exploits an analogy between combinatorial optimization problems and the statistical mechanics of physical system—by Salamon, Andersen and Gordon, see the review articles [5,27,32].

4.5 Studies for Refrigeration Cycles

The idea and method described from section 4.1 to 4.4 may be extended to the research of two-heat-reservoir and three-heat-reservoir refrigerators.

4.5.1 Newton's Law Two-Heat-Reservoir Refrigerator System Coupled To Infinite Thermal Capacity Reservoirs

If the working fluid undergoes Carnot refrigeration cycle, the cycle is called as endoreversible Carnot refrigeration. Leff et al [273] are the first to consider to extend the Curzon-Ahlborn [10] analysis method to the analysis of refrigeration cycles. Because the COP (coefficient of performance) at the maximum cooling load (or rate of refrigeration, cooling power, cooling capacity) is zero for Newton's law system endorevesible Carnot refrigerator, no further step was development in that paper [273]. Rozoner et al [110] first derived the maximum COP bound for the fixed power input for the Newton's law Carnot refrigerator. Yan [274] and Goth et al [275] derived the optimal COP (ε) for the fixed cooling load (R), i.e. the fundamental optimal relation. Yan [275] also studied the optimal performance when (εR) is maximized. Sun et al [277] extended the characteristic parameters of heat engine [69-72] to the refrigerator and derived the optimization criteria of steady-flow two-heat-reservoir refrigerator [278]. Chen et al [279] derived the finite time exergoeconomic performance bounds, optimal relation and optimization criteria for Carnot refrigerator, and established the ecological optimization objective of the refrigerator [280]. Wu et al [24,281,282] discussed the cooling load, ecological objective and exergoeconomic objective optimization of Carnot refrigerator considering the quantum characteristic of the refrigeration working fluid. Klein [283] and Wu [284] also discussed the performance of the steady flow refrigerators.

4.5.2 Effect of Loss Model on the Performance of Two -Heat - Reservoir Refrigerators

In 1989, Chen et al [123,285-288] started to study the effect of heat transfer law on the optimal performance of endoreversible Carnot refrigerator. References [289-292] used a constant parameter to represent the other all irreversibilities except heat resistance in order to study the effect of internal irreversibility on the refrigerator performance. On the basis of heat leak model provided by Bejan [11,293], Chen et al [294] derived optimal relation of the model. Furthermore, Chen et al [23,295,296] established a generalized irreversible refrigerator model, which involve the irreversibilities of heat resistance, heat leak and the other internal dissipation, and studied the effect of heat transfer law on the performance of the generalized model. The effects of finite thermal capacity and heat leak on the optimal configuration of two-heat –reservoir refrigerator and on the optimal performance of a Carnot refrigerator were analyzed in Refs. [297-301].

4.5.3 Studies for Practice Two –Heat –Reservoir Refrigerator

Chen and Yan [302] investigated the optimal configuration of the combined refrigeration cycle. Chen et al [303] discussed the specific cooling load optimization of the combined refrigeration cycle, and analyzed the effect of heat leak on the optimal performance of the combined cycle [304]. On the basis of introducing internal irreversibility [305], Chen et al [23] established a generalized irreversible combined refrigeration cycle model and researched its optimal performance.

A lot of literature has been published for the performance optimization of vapor-compression refrigerators. The typical work includes Refs. [16,21,23,306-310]. Chen et al [23,300,311,312] analyzed the performance of simple and regenerated, endoreversible and irreversible air (Brayton) refrigeration cycles coupled to constant-and variable-temperature heat reservoirs. The analytical expressions for the cooling load and COP were derived.

A lot of literature has been published for the performance optimization of simple and regenerated, endoreversible and irreversible, conventional and quantum working fluid Stirling refrigerators. The optimization objective include the cooling load, COP, ecological objective and exergoeconomic objective. The typical work includes Refs [16,21,24,221,315]. The performance of magnetic refrigerator [316], thermoelectric refrigerator [16,23,317,319] and thermoacoustic refrigerator [24] also have been studied.

4.5.4 Studies for Absorption Refrigeration Cycle

Yan et al [320,321] established the theoretical model of absorption refrigerator–endoreversible three-heat-reservoir refrigerator, and derived the COP bounds corresponding to maximum cooling load and the fundamental optimal relation between the COP and cooling load for the Newton's law system. Chen et al [322,323] derived the optimal performance of linear phenomenological law system three-heat- reservoir refrigerator. The ecological optimization [324] and exergoeconomic optimization of the refrigerator was also performed by Chen et al [324,325]. Some literature including Refs. [21,326-328] have studied the specific cooling load optimization, heat transfer surface area optimal allocation, and the effects of heat leak and internal irreversibility on the performance of three-heat- reservoir refrigerator.

4.5.5 Experimental Confirmations for Refrigeration Cycle Model

Since 1994,Ng and Gordon [329-341] have provided the experimental characteristic between cooling load and COP for reciprocating chillers [330,334-338], centrifugal compressor chillers [332-334,336] and absorption refrigerators [329,331,339-341]. They also established theoretical models with heat resistance and heat leak losses, as well as heat resistance, heat leak and internal dissipation losses. The calculation results are consistent with the experimental results. Chen et al [23,342] provided the experimental results for thermoelectric refrigerator, which are consistent with the theoretical analysis results [319].

4.6 Studies for Heat Pump Cycle

4.6.1 Studies for the Two-Heat-Reservoir Heat Pump

Blanchard [343] derived the COP bounds for the fixed heating load for Newton's law endoreversible Carnot heat pump. The similar results were also obtained by Goth et al [275]. Sun et al [277] provided the performance holographic spectrum and optimization criteria [278] of endorevesible Carnot heat pump. The specific heating load optimization, ecological optimization, exergoeconomic optimization, the effect of quantum characteristic of working fluid, and the effects of heat transfer law, heat leak and internal irreversibility on the Carnot heat pump were analyzed in Refs. [23,24,292,344-348] by Wu, Chen, Sun and etc. The effects of various irreversibilities on the optimal performance of combined heat pump cycle [23,349], thermoelectric heat pump [23,350] and air (Brayton) heat pump [23,346,351] were analyzed by Chen et al [23,346,349-351].

4.6.2 Studies for Absorption Heat Pump Cycle

Yan et al [352,353] established the theoretical model of absorption heat pump and heat transformer —endoreversible three-heat–reservoir heat pump, derived the COP bounds corresponding to maximum heating load and the fundamental optimal relation between the COP and heating load for the Newton's law system, and provided unified description for Newton's law system various endoreversible cycles using the model of finite thermal capacity high-temperature source three- heat-reservoir heat pump [354]. The effects of heat transfer law and the other major irreversibility besides the heat resistance were analyzed in Refs. [23,355-359]. Chen et al [360] provided a unified description for linear phenomenological law system various endoreversible cycles, including three-heat-reservoir heat pump and refrigeration cycles, as well as two-heat-reservoir heat engine, refrigerator and heat pump cycles.

5. Conclusion

Finite time thermodynamics or entropy generation minimization is the method of modeling and optimization of various thermodynamic processes and devices that owe their thermodynamic imperfection to heat transfer, mass transfer, and fluid flow and other transport processes. It bridges the gaps not only between thermodynamic and heat transfer, mass transfer, fluid mechanics, and other transport science, but also between the physics and the engineering

Up to now, the research subject of finite time thermodynamics is mainly the conventional thermodynamic systems in which the driving force is the temperature T and the displacement is the entropy S. Radcenco [16] proved that fundamental laws of nature systems with dissipative and conservative interactions could be described on the basis of a single equation, that is YX^n = const or Y = const $\times X^n$ named the equation of conservative generalized polytrope, where $n \in (0,\infty)$ is a generalized polytropic exponent characterizing the intensity of the interaction, Y is the generalized forces and X is a generalized displacement. The intensive magnitudes acting the part of generalized force—intensity Y might be of most various kinds including mechanical force, the weight of a body with mass, mechanical moment, linear acceleration, the angular acceleration, the linear and angular velocities, linear and tangent unitary stress, pressure, surface tension, volumetric potential, intensity of electrostatic field, intensity of magnetic field, chemical potential as well as absolute temperature. The extensive magnitudes acting as the generalized displacement X included linear and angular displacements, linear impulse, kinetic moment, linear specific elongation, angular specific deformation, specific volume, electric charge, electrostatics induction, polarization vector, magnetic induction, paramagnetization vector, mass of substance, and entropy. Therefore, the idea and method for analyzing conventional thermodynamic processes can be applied to the analysis and optimization for mechanic, electric, magnetic, chemical and life systems. The further step in finite time thermodynamics may be expected to extend the heat engine FTT theory to the generalized thermodynamic systems mentioned above. This further step will be helpful for the establishment of optimal design and operation theory for the generalized thermodynamic system, that is, to carry out the generalization of endoreversibilities and to establish the theory of Generalized Thermodynamic Optimization [23,270].

REFERENCES

1. Carnot S. Reflections on the motive power of fire. Paris: Bachelier, 1824.
2. Sieniutycz S and Salamon P (ed.). Advances in Thermodynamics. Volume 1: C_{7+} Fraction Characterization. New York: Taylor & Francis, 1990.
3. Mansoori G A (ed.). Advances in Thermodynamics. Volume 2: Fluctuation Theory of Mixtures. New York: Taylor & Francis, 1990.
4. Sieniutycz S and Salamon P (ed.). Advances in Thermodynamics. Volume 3: Nonequilibrium Theory and Extremum Principles. New York: Taylor & Francis, 1990.
5. Sieniutycz S and Salamon P (ed.). Advances in Thermodynamics. Volume 4: Finite Time Thermodynamics and Thermoeconomics. New York: Taylor & Francis, 1990.
6. Hoffman E J. Advances in Thermodynamics. Volume 5: Analytic Thermodynamics. New York: Taylor & Francis, 1991.
7. Sieniutycz S and Salamon P (ed.). Advances in Thermodynamics. Volume 6: Flow, Diffusion, and Rate Processes. New York: Taylor & Francis, 1992.

8. Novikov I I. The efficiency of atomic power stations (A review). Atommaya Energiya 3, 1957(11): 409.
9. Chambadal P. Les Centrales Nucleaires. Paris: Armand Colin, 1957: 41-58.
10. Curzon F L and Ahlborn B. Efficiency of a Carnot engine at maximum power output. Am. J. Phys., 1975, 43(1): 22-24.
11. Bejan A. Entropy Generation through Heat and Fluid Flow. New York: Wiley, 19 82.
12. Andresen B. Finite-Time Thermodynamics. Physics Laboratory • , University of Copenhagen, 1983.
13. Tsirlin A M. Optimal Cycles and Cycle Regimes. Moscaw: Energomizdat, 1985 (in Russian).
14. Feidt M. Thermodynamique et Procedes. Paris: Technique et Documentation, Lavoisier, 1987 (in French).
15. De Vos A. Endoreversible Thermodynamics of Solar Energy Conversion. Oxford: Oxford University, 1992.
16. Radcenco V. Generalized Thermodynamics. Bucharest: Editura Techica, 1994 (in English).
17. Bejan A. Entropy Generation Minimization. Boca Raton FL: CRC Press, 1996.
18. Bejan A, Tsatsaronis G and Moran M. Thermal Design & Optimization. New York: Wiley, 1996.
19. Chen L and Sun F. Finite Time Thermodynamics for Energy Systems. Wuhan: Naval Academy of Engineering Publisher, 1992 (in Chinese).
20. Popescu G. Contributions dans Latude de Ioptimisation des Systems Endo-Regenerative et Exio-Irreversible de Type Stirling sur la Base de la Thermodynamique en Temps Fini. These de Doctorat, Universite Politechica, Bucarest, 1993.
21. Vargas J V C. Combined Heat Transfer and Thermodynamic Problems with Applications in Refrigeration. Ph. D. Thesis, Duke University, 1994.
22. Geva E. Finite Time Thermodynamics for Quantum Heat Engine and Heat Pump. Ph. D. Thesis, The Hebrew University, 1995.
23.Chen L. Finite Time Thermodynamic Analysis of Irreversible Processes and Cycles. Ph. D. Thesis, Naval Academy of Engineering, 1997.
24. Wu F. Study of Finite Time Thermodynamics on Stirling Machine. Ph. D. Thesis, Naval Academy of Engineering, 1998.
25. Andresen B, Berry R S, Ondrechen M J and Salamon P. Thermodynamics for processes in finite time. Acc. Chem. Res. 1984, 17(4): 266-271.
26. Andresen B, Salamon P and Berry R S. Thermodynamics in finite time. Phys. Today, 1984 (Sept.): 62-70.
27. Andresen B. Finite-time thermodynamics and thermodynamic length. Rev. Gen. Therm, 1996, 35(418/419): 647-650.
28. Orlov V N and Rudenko A V. Optimal control in problems of extremal of irreversible thermodynamic processes. Avtomatika i Telemekhanika, 1985(5): 7-41.
29. Wu C. Power optimization of a finite-time Carnot heat engine. Energy, The Int. J., 1988, 13(9): 681-687.

30. Wu C, Kiang R L, Lopardo V J and Karpouzian G N. Finite-time thermodynamics and endoreversible heat engines. Int. J. Mech. Engng. Edu., 1993, 21(4): 337-346 .
31. Chen L and Yan Z. Finite time thermodynamics: An new branch of modern thermodynamics. Nature J., 1987, 10(11): 825-829 (in Chinese).
32. Andresen B. Finite time thermodynamics and simulated annealing. in J. Shiner . Entropy and Entropy Generation. Amsterdam: Kluwer Academic Publishers,1996.
33.Chen L, Sun F and Chen W. Finite time thermodynamics for heat engines: theory and applications. J. Engng. Thermal Energy Pow., 1989,4(2): 7-14 (in Chinese).
34. Chen L, Sun F and Chen W. New developments of finite time thermodynamics. Nature J., 1992, 15(4): 249-253(in Chinese).
35. Chen L, Sun F and Chen W. The present state and trend of finite time thermodynamics. Adv. Mech., 1992, 22(4): 479-488 (in Chinese).
36. Sieniutycz S and Shiner J S. Thermodynamics of irreversible processes and its relation to chemical engineering: Second law analyses and finite time thermodynamics. J. Non-Equilib. Thermodyn., 1994, 19(4): 303-348.
37. Bejan A. Power generation and refrigeration models with heat transfer irreversibilities. J. Heat Transfer Soc. Japan, 1994, 33(128): 68-75.
38. Bejan A. Entropy generation minimization: The new thermodynamics of finite-size devices and finite-time processes. J. Appl. Phys., 1996, 79(3): 1191-1218.
39. Bejan A. Method of entropy generation minimization, or modeling and optimization based on combined heat transfer and thermodynamics. Rev. Gen. Therm., 1996,35(418/419): 637-646.
40. Bejan A. Notes on the history of the method of entropy generation minimization (finite time thermodynamics). J. Non-Equilib. Thermodyn., 1996, 21(3): 239-24 2.
41. Bejan A. Fundamental optima in thermal Science. Int. J. Mech. Engng. Edu., 1997, 25(1): 33-47.
42. Andresen B, Berry R S, Nitzan A and Salamon P. Thermodynamics in finite time: The step-Carnot cycle. Phys. Rev. A, 1977, 15(5): 2086-2093.
43. Grazzini G. Work from irreversible heat engines. Energy, The Int. J., 1991,16(4): 747-755.
44. Lu P C. Thermodynamics with finite heat-transfer area or finite surface thermodynamics. Thermodynamics and the Design, Analysis, and Improvement of Energy Systems, ASME Adv. Energy Sys. Div. Pub. AES Vol. 35, 1995, pp. 51-60.
45. Verhas J and De Vos A. How endoreversible thermodynamics relates to Onsager's nonequilibrium thermodynamics. J. Appl. Phys., 1997, 82(1): 40-42.
46. Rubin M H. Optimal configuration of a class of irreversible heat engines I. Phys. Rev. A, 1979, 19(3): 1272-1276.
47. Salamon P and Nitzan A. Finite time optimizations of a Newton's law Carnot cycle. J. Chem. Phys., 1981, 74(6): 3546-3560.
48. Vukalovich M P and Novikov I I. Thermodynamics. Moscow: Mashinostroenie, 1972 (in Russian).
49. Novikov I Iand Voskresenskii K D. Thermodynamics and Heat Transfer. Moscow: Atomizdat, 1977(in Russian).
50. El-Wakill M M. Nuclear Power Engineering. New York: McGraw-Hill, 1962.

51. Chambadal P. Evolution et Applications du Concept d'Entropie. Paris: Dunod, 1963.
52. Bejan A. Engineering advances on finite-time thermodynamics. Am. J. Phys., 1994, 62(1): 11-12.
53. Leff H S. Thermal efficiency at maximum work output: New results for old heat engines. Am. J. Phys., 1987, 55(7): 602-610.
54. Landsberg P T and Leff H S. Thermodynamic cycles with nearly universal maximum-work efficiencies. J. Phys. A: Math. Gen., 1989, 22(18): 4019-4026.
55. Callen H B. Thermodynamics and an Introduction to Thermostatics. New York: Wiley, 1985.
56. Lund K O. Thermodynamics and Introduction to Thermostatics (2nd Edn.), New York: Wiley, 1985.
57. Bejan A. Advanced Engineering Thermodynamics. New York: Wiley, 1988.
58. Zhu M and Chen H. Thermodynamic Analysis. Beijing: Advanced Edu. Pub., 1992 (in Chinese).
59. Sun F. Advanced Engineering Thermodynamics. Wuhan: Naval Academy of Engineering Publisher, 1986 (in Chinese).
60. Bejan A. Advanced Engineering Thermodynamics (2nd Edn.), New York: Wiley, 1997.
61. Ahlborn B and Barnard A J. Efficiency reduction of heat engines due to power extraction. Am. J. Phys., 1990, 58(5): 498-499.
62. Gutowicz-Krusin D, Procaccia J and Ross J. On the efficiency of rate processes: Power and efficiency of heat engines. J. Chem. Phys., 1978, 69(9):3898-3906.
63. Spence R D and Harrision M J. Speed dependence of the efficiency of heat engines. Am. J. Phys., 1985, 53(9): 890-893.
64. Rebhan E and Ahlborn B. Frequency-dependent performance of a nonideal Carnot engine. Am. J. Phys., 1987, 55(5): 423-428.
65. Gordon J M. Maximum power points characteristics of heat engines as a general thermodynamic problem. Am. J. Phys., 1989, 57(12): 1136-1142.
66. Yan Z. The relation between optimal efficiency and power output of a Cannot engine. Chinese J. Engng. Thremophys., 1985,6(1):1-6 (in Chinese).
67. Chen J and Yan Z. The relation between optimal efficiency and power output for a Cannot engine with finite compression ratio. J. Xiamen Univers., 1989, 28(3): 255-260 (in Chinese).
68. Yan Z. η and P of a Carnot engine at maximum η^{λ}P. J. Xiamen Univers., 1986, 25(3): 279-286(in Chinese) .
69. Sun F and Lai X. Finite time thermodynamic analysis of operating temperature of working fluid and cold reservoir. Proc. Sec. Conf. Univers., Res. Assoc. Engng. Thermophys. (1986), Beijing: Science Press, 1988:91-96 (in Chinese).
70. Sun F. Finite time thermodynamic analysis of heat engine with more than two heat reservoirs. Paper presented at meeting of engineering thermodynamics, 1986 (in Chinese).
71. Sun F and Lai X. Holographic spectrum of power output and efficiency for heat engine. J. Engng. Thermal Energy Pow., 1988, 3 (3): 1-9 (in Chinese).

72. Sun F, Chen L and Chen W. Finite time thermodynamic analysis and evaluation for a heat engine with steady-state energy conversion between two heat reservoirs. J. Engng. Thermal Energy Pow., 1989, 4(2): 1-6(in Chinese).
73. Chen W, Sun F and Chen L. Area performance of a heat engine between heat reservoirs under stable energy transfer. Chinese J. Engng. Thermophys., 1990, 11(4): 365-368(in Chinese).
74. Wu C. Specific power optimization of a closed-cycle OTEC plants. Ocean Engng., 1990, 17(3): 307-314.
75. Wu C. Specific output power of a dry geothermal plant. Energy, The Int. J., 1991, 16(4): 757-761.
76. Wu C. Specific power bound of real heat engines. Energy Convers. Mgmt., 1991, 32(3): 249-253.
77. Bejan A. Power and refrigeration plants for minimum heat exchanger inventory. ASME Trans. J. Energy Resource Tech., 1993, 115(2): 148-150.
78. Wu C. Heat transfer effect on the specific power availability of heat engines. Energy Convers. Mgmt., 1993, 34(12): 1239-1247.
79. Kiang R L and Wu C. Clarification of finite-time thermodynamic cycle analysis. Int. J. Power and Energy Systems, 1994, 14(2): 68-71.
80. Wu C and Schulden W H. Maximum obtainable specific power of high-temperature waste heat engines. Heat Recovery Systems & CHP, 1995, 15(5): 450-455.
81. Bejan A. Theory of heat transfer-irreversible power plants. II. The optimal allocation of heat exchange equipment. Int. J. Heat Mass Transfer, 1995, 38(3):433-444.
82. Wu C, Lopardo V J and Lee W M. Power economics of renewable energy endoreversible heat engines. Int. J. Power & Energy Systems, 1996, 16(1): 24-28.
83. Agnew B, Anderson A and Frost T H. Optimization of a steady-flow Carnot cycle with external irreversibilities for maximum specific output. Appl. Thermal Engng., 1997, 17(1): 3-16.
84. Berry R S, Salamon P and Heal G. On a relation between economic and thermodynamic optima. Resources and Energy, 1978, 1(2): 125-137.
85. Clark J A. Thermodynamic optimization: An interface with economic analysis. J. Non-Equilib. Thermodyn., 1986, 11(1/2): 85-122.
86. Tsatsaronts G. Thermoeconomic analysis and optimization of energy systems. Progress in Energy and Combustion Science, 1993, 19(3): 227-257.
87. Chen L, Sun F and Chen W. The relation between optimal efficiency and profit rate of a Cannot engine. J. Engng. Thermal Energy Pow., 1991, 6(4): 237-240 (in Chinese).
88. Sun F, Chen L and Chen W. The efficiency and profit holographic spectrum of a two-heat-reservoir heat engine. Trans. Chinese Soc. Internal Combustion Engines, 1991, 9(3): 286-287 (in Chinese).
89. Chen L, Sun F and Chen W. Finite time exergoeconomic performance bound and optimization criteria for two-heat-reservoir heat engines. Chinese Science Bull., 1991, 36(3): 233-235 (in Chinese).

90. Ibrahim O M, Klein S A and Mitchell J W. Effects of irreversibility and economics on the performance of a heat engine. ASME Trans. J. Sol. Energy Engng., 1992, 114(4): 267-271.
91. De Vos A. Endoreversible thermoeconomics. Energy Convers. Mgmt., 1995, 36(1): 1-5.
92. De Vos A. Endoreversible economics. Energy Convers. Mgmt., 1997, 38(4): 311-317.
93. Angulo-Brown F. Ecological optimization criterion for finite-time heat engines. J. Appl. Phys., 1991, 69(11): 7465-7469.
94. Arias-Hernandez L A and Angulo-Brown F. Thermodynamic optimization of endoreversible engines. Revista Mexicana de Fisica, 1994, 40(6): 866-877.
95. Arias-Hernandez L A and Angulo-Brown F. A general property of endoreversible thermal engines. J. Appl. Phys., 1997, 81(7): 2973-2979.
96. Cheng C Y and Chen C K. The ecological optimization of an irreversible Carnot heat engine. J. Phys. D: Appl. Phys., 1997, 30(11): 1602-1609.
97. Cheng C Y and Chen C K. Ecological optimization of an endoreversible Brayton cycle. Energy Convers. Mgmt., 1998, 39(1/2): 33-44.
98. Angulo-Brown F, Yepez E and Zamorano-Ulloa R. Finite-time thermodynamics approach to the superconducting transition. Phys. Lett. A, 1993, 183(5/6): 431-436.
99. Sahin B, Kodal A and Yavuz H. Maximum power density analysis of an endoreversible Carnot heat engine. Energy, The Int. J., 1996, 21(10): 1219-1225.
100. Kosloff R. A quantum mechanical open system as a model of a heat engine. J. Chem. Phys., 1984, 80(4): 1625-1631.
101. Geva E and Kosloff R. A quantum-mechanical heat engine operating in finite time. A model consisting of spin-1/2 systems as the working fluid. J. Chem. Phys., 1992, 96(4): 3054-3067.
102. Geva E and Kosloff R. On the classical limit of quantum thermodynamics in finite time. J. Chem. Phys., 1992, 97(6): 4396-4412.
103. Geva E and Kosloff R. Three-level quantum amplifier as a heat engine: A study in finite-time thermodynamics. Phys. Rev. E, 1994, 49(5): 3903-3918.
104. Geva E and Kosloff R. On the relaxation of a two-level system driven by a strong electromagnetic field. J. Chem. Phys., 1995, 102(21): 8541-8561.
105. Feldmann T, Geva E and Kosloff R. Heat engines in finite time governed by master equations. Am. J. Phys., 1996, 64(4): 485-492.
106. Geva E and Kosloff R. The quantum heat engine and heat pump: An irreversible thermodynamic analysis of the three-level amplifier. J. Chem. Phys., 1996, 104(19): 7681-7699.
107. Wu F, Sun F and Chen L. Optimum performance of a spin-1/2 irreversible quantum Carnot engine with the power in cooperation with the entropy production rate. J. Wuhan Institute Chem. Tech., 1997, 19(2): 88-91 (in Chinese).
108. Jin X, Wu F, Sun F and Chen L. Finite time thermodynamic performance bound of a quantum Carnot engine with ecological criteria. Pow. Sys. Engng., 1996, 12(5): 53-55 (in Chinese).

109. Rubin M H. Optimal configuration of an irreversible heat engine with fixed compression ratio. Phys. Rev. A, 1980, 22(4): 1741-1752.
110. Rozonoer L I and Tsirlin A M. Optimal control of thermodynamic processes.• ,• and • . Avtomat. Telemekh, 1983(1): 70-79; (2): 88-101; (3): 50-64.
111. Kuznetsov A G, Rudenko A V and Tsirlin A M. Optimal control in thermodynamic systems with sources of finite capacity. Autom. Remote Control, 1986(6): 693-7 05.
112. Orlov V N. Optimal control of an irreversible Carnot engine. Avtomat. Telemekh., 1989(4): 64-74.
113. Lampinen M J and Vuorisulo J. Heat accumulation function and optimization of heat engines. J. Appl. Phys., 1991, 69(2): 597-605.
114. Chen L, Sun F and Chen W. Influence of heat transfer law on the performance of a Carnot engine. Chinese J. Engng. Thermophys., 1990, 11(3): 241-243 (in Chinese).
115. De Vos A. Efficiency of some heat engines at maximum power conditions. Am. J. Phys., 1985, 53(6): 570-573.
116. Wu C. Power optimization of a finite-time solar radiant heat engine. Int. J. Ambient Energy, 1989, 10(3): 145-150.
117. Chen L and Yan Z. The effect of heat transfer law on the performance of a two-heat-source endoreversible cycle. J. Chem. Phys., 1989, 90(7): 3740-3743.
118. Feidt M and Philippi I. Basis of a general approach for finite time thermodynamics applied to two heat reservoir machines. ECOS'92, 1992: pp 21-25.
119. Angulo-Brown F and Paez-Hernandez R. Endoreversible thermal cycle with a nonlinear heat transfer law. J. Appl. Phys., 1993, 74(4): 2216-2219.
120. Goktun S, Ozkaynak S and Yavuz H. Design parameters of a radiative heat engine. Energy, the Int. J., 1993, 18(6): 651-655.
121. Chen J, Wu C and Kiang R L. Maximum specific power output of an irreversible radiant heat engine. Energy Convers. Mgmt., 1996, 37(1): 17-22.
122. Navarrete-Gonzalez T D, Rocha-Martinez J A and Angulo-Brown F. A Muser-Curzon-Ahlborn engine model for photothermal conversion. J. Phys. D: Appl. Phys., 1997, 30(17): 2490-2596.
123. Chen W, Sun F, Cheng S and Chen L. Study on optimal performance and working temperature of endoreversible forward and reverse Carnot cycles. Int. J. Energy Res., 1995, 19(9): 751-759.
124. Wu C, Chen L and Sun F. Effect of heat transfer law on finite-time exergoeconomic performance of heat engines. Energy, The Int. J., 1996, 21(12): 1127-1134.
125. Chen L, Sun F and Chen W. The finite time exergoeconomic performance bound of an endoreversible Carnot engine with heat transfer law $q \propto \Delta(T^{-1})$. Nature J., 1990, 13(10): 698 (in Chinese).
126. Sun F, Chen W and Chen L. The relation between optimal efficiency and power output of a Carnot engine in case of mixed heat transfer law. Nature J., 1991, 14(4): 317 (in Chinese).
127. Chen L, Sun F and Chen W. The ecological optimal performance of finite time heat engines with heat transfer $q \propto \Delta(T^{-1})$. Gas Turbine Tech., 1995, 8(1): 16-18(in Chinese).

128. Chen L, Chen S, Sun F and Chen W. The optimal configuration of endoreversible heat engines independent of the heat transfer law. Gas Turbine Tech., 1993, 6(2): 20-33 (in Chinese).

129. Orlov V N. Optimum irreversible Carnot cycle containing three isotherms. Sov. Phys. Dokl., 1985, 30(6): 506-508.

130. Andresen B, Salamon P and Berry R S. Thermodynamics in finite time: extremal for imperfect heat engines. J. Chem. Phys., 1977, 66(4): 1571-1577.

131. Fairen V and Ross J. On the efficiency of thermal engines with power output. J. Chem. Phys., 1981, 75(11): 5485-5496.

132. Richter P H and Ross J. The efficiency of engines operating around a steady state at finite frequency. J. Chem. Phys., 1978, 69(12): 5521-5531.

133. Delgado R E J. A thermal engine driven by a thermokinetic oscillator. J. Phys. Chem., 1996, 100(26): 11144-11147.

134. Howe J P. The maximum power, heat demand and efficiency of a heat engine operating in steady state at less than Carnot efficiency. Energy, The Int. J., 1982, 7(4): 401-402.

135. Li J. Relation between power and efficiency of an irreversible Cannot engine. Phys., 1989, 18(2): 89-93 (in Chinese).

136. Yan Z. η and P of an irreversible Carnot engine at maximum η^{λ}P. J. Engng. Thermal Energy Pow., 1989, 4(6): 1-6 (in Chinese).

137. Yan Z. Relation between optimal efficiency and supplying heat of an irreversible Carnot engine. J. Xiamen Univers., 1991, 30(1):25-28 (in Chinese).

138. Yan Z and Chen L. Irreversible Carnot cycle. College Phys., 1992, 11(2): 2-5 (in Chinese).

139. Lee W Y and Kim S S. Power optimization of an irreversible heat engine. Energy, The Int. J., 1991, 16(7): 1051-1058.

140. Ibrahim O M, Klein S A and Mitchell J W. Optimum heat power cycles for specified boundary conditions. Trans. ASME J. Engng. Gas Turbine Power, 1991, 113(4): 514-521.

141. Wu C and Kiang R L. Finite-time thermodynamic analysis of a Carnot engine with internal irreversibility. Energy, The Int. J., 1992, 17(12): 1173-1178.

142. Bejan A. Theory of heat transfer-irreversible power plant. Int. J. Heat Mass Transfer, 1988, 31(6): 1211-1219.

143. Chen L, Sun F and Chen W. The power vs. efficiency characteristics of an irreversible heat engine: Heat leak as an illustration. Chinese Science Bull., 1993, 38(5): 480 (in Chinese).

144. Chen L, Sun F and Chen W. Finite time thermodynamic performance of a class of irreversible two-heat-reservoir forward and reverse cycles. Sci. Tech. Bull., 1995, 11(2): 126 (in Chinese).

145. Chen L, Wu C and Sun F. The influence of internal heat leak on the power vs. efficiency characteristics of heat engines. Energy Convers. Mgmt., 1997, 38(14): 1501-1507.

146. Gordon J M and Huleihil G. General performance characteristics of real heat engine. J. Appl. Phys., 1992, 72(3): 829-837.

147. Pathria R K, Nulton J D and Salamon P. Carnot-like processes in finite-time: Applications to model cycles. Am. J. Phys., 1993, 61(10): 916-924.
148. Moukalled F, Nuwayhid R Y and Noueihed N. The efficiency of endoreversible heat engines with heat leak. Int. J. Energy Res., 1995, 19(5): 377-389.
149. Moukalled F, Nuwayhid R Y and Fattal S. A universal model for studding the performance of Carnot-like engine at maximum power conditions. Int. J. Energy Res., 1996, 20(3): 203-214.
150. Chen L and Sun F. The optimal performance of an irreversible Carnot engine. Sci. Tech.. Bull., 1995, 11(2): 128 (in Chinese).
151. Chen L, Sun F and Wu C. A Generalized of real heat engines and its performance. J. Institute Energy, 1996, 69(481): 214-222.
152. Chen L, Cao S and Sun F. The performance of an irreversible heat engine with a universal heat transfer law. Exploration Nature, 1997,16(3): 91 (in Chinese).
153. Ondrechen M J, Andresen B, Mozurkewich M and Berry R S. Maximum work from a finite reservoir by sequential Carnot cycles. Am. J. Phys., 1981, 49(7): 681-685.
154. Yan Z. Thermal efficiency of a Carnot engine at the maximum power output with finite thermal capacity heat reservoir. Chinese J. Engng. Thermophys., 1984, 5(2): 125-131 (in Chinese).
155. Lee W Y and Kim S S. An analytical formula for the estimation a Rankine cycle heat engine efficiency at maximum power. Int. J. Energy Res., 1991, 15(3): 149-159.
156. Ondrechen M J, Rubin M H and Band Y B. The generalized Carnot cycles: A working fluid operating in finite-time between finite heat sources and sinks. J. Chem. Phys., 1983, 78(7): 4721-4727.
157. Yan Z and Chen J. Optimal performance of a generalized Carnot cycle for another linear heat transfer law. J. Chem. Phys., 1990, 92(3): 1994-1998.
158. Chen L, Sun F, Xu M and Li D. Effects of heat leak and finite thermal capacity heat reservoir on the optimal configuration of two-heat–reservoir heat engine. Gas Turbine Tech., 1996, 9(1): 4-10 (in Chinese).
159. Band Y B, Kafri O and Salamon P. Finite time thermodynamics: Optimal expansion of a heated working fluid. J. Appl. Phys., 1982, 53(1): 8-28.
160. Band Y B, Kafri O and Salamon P. Optimization of a model external combustion engine. J. Appl. Phys., 1982, 53(1): 29-33.
161. Salamon P, Band Y B and Kafri O. Maximum power from a cycling working fluid. J. Appl. Phys., 1982, 53(1): 197-202.
162. Aizenbud B M and Band Y B. Power considerations in the operation of a piston fitted inside a cylinder containing a dynamically heated working fluid. J. Appl. Phys., 1981, 52(6): 3742-3744.
163. Aizenbud B M, Band Y B and Kafri O. Optimization of a model internal combustion engine. J. Appl. Phys., 1982, 53(3): 1277-1282.
164. Gordon J M and Huleihil M. On optimizing maximum power heat engines. J. Appl. Phys., 1991, 69(1): 1-7.
165. Chen L, Sun F and Wu C. Optimal expansion of a heated working fluid with phenomenological heat transfer. Energy Convers. Mgmt., 1998, 39(3/4): 149-156.

166. Rubin M H and Andresen B. Optimal staging of endoreversible heat engines. J. Appl. Phys., 1982, 53(1): 1-7.
167. Wu C. Power performance of a cascade endoreversible cycle. Energy Convers. Mgmt., 1990, 30(3): 261-266.
168. Wu C, Karpouzian A N and Kiang R L. The optimal power performance of an endoreversible combined cycle. J. Institute Energy, 1992, 65(465): 41-45.
169. Wu C. Maximum obtainable power of a Carnot combined power plant. Heat Recovery Systems & CHP, 1995, 15(4): 351-355.
170. Chen L, Sun F and Chen W. Finite-time thermodynamic optimal performance of combined power plants. Power Engng., 1994, 14(2): 9-12 (in Chinese).
171. Chen L, Sun F and Wu C. Specific power optimization for Carnot combined power plants. Int. J. Energy-Environment-Economics, 1996, 4(1): 9-16.
172. Chen L, Cao S, Wu C and Sun F. Combined power plant performance with heat leak. Proc. Int. Conf. on Energy and Environment, 12-14 Oct., 1997, Limassol, Cyprus.
173. Ozkaynak S. The theoretical efficiency limits for a combined cycle under the condition of maximum power output. J. Phys. D: Appl. Phys., 1995, 28(10): 2024-2028.
174. Sahin B and Kodal A. Steady-state thermodynamic analysis of a combined Carnot cycle with internal irreversibility. Energy, The Int. J., 1995, 20(12): 1285-1289.
175. De Vos A, Chen J and Andresen B. Analysis of combined systems of two endoreversible engines. Open Sys. & Information Dyn., 1997, 4(1): 3-13.
176. Chen L, Sun F and Ni N. Performance optimization of an irreversible steady flow combined heat engine. Proc. Seventh Conf. Univers. Res. Assoc. Engng. Thermophys. (1998), Shanghai, 1998:400-405 (in Chinese).
177. Mozurkewich M and Berry R S. Optimal paths for thermodynamic systems. The ideal Otto cycle. J. Appl. Phys., 1982, 53(1): 34-42.
178. Hoffman K H and Berry R S. Optimal paths for thermodynamic systems. The ideal Diesel cycle. J. Appl. Phys., 1985, 58(6): 2125-2134.
179. Angulo-Brown F, Fernandez-Betanzos J and Diaz-Pico C A. Compression ratio of an optimized air standard Otto-cycle model. Eur. J. Phys., 1994, 15(1):38-42.
180. Angulo-Brown F, Rocha-Martinez J A and Navarrete-Gonzalez I D. A non-endoreversible Otto cycle model: improving power output and efficiency. J. Phys. D: Appl. Phys., 1996, 29(1): 80-83.
181. Klein S A. An explantion for observed compression ratios in internal combustion engines. Trans. ASME J. Engng. Gas Turbine Pow., 1991, 113(4): 511-513.
182. Wu C and Blank D A. The effect combustion on a work-optimized endoreversible Otto cycle. J. Instit. Energy, 1992, 65(1): 86-89.
183. Blank D A and Wu C. Optimization of the endoreversible Otto cycle with respect to both power and mean effective pressure. Energy Convers. Mgmt., 1993, 34(12): 1255-1209.
184. Chen L, Wu C• Sun F and Cao S. Heat transfer effect on the net work output vs. efficiency characteristics for an air standard Otto cycle. Energy Convers. Mgmt., 1998, 39(7): 643-648.

185. Blank D A and Wu C. The effects of combustion on a power optimized endoreversible Diesel cycle. Energy Convers. Mgmt., 1993, 34(6): 493-498.

186. Chen L• Zen F, Sun F and Wu C. Heat transfer effect on net work and/or power as function of efficiency for air standard Diesel cycles. Energy, The Int.J. 1996,21(12):1201-1206.

187. Chen L, Lin J and Sun F. Effect of friction on the power vs. efficiency characteristics of air-standard Diesel cycle. Chinese J. Engng. Thermophys., 1997, 18(5): 533-535 (in Chinese).

188. Orlov V N and Berry R S. Power and efficiency limits for internal combustion engines via methods of finite-time thermodynamics. J. Appl. Phys., 1993, 74(7): 4317-4322.

189. Blank D A and Wu C. The effects of combustion on a power-optimized endoreversible Dual cycle. Int. J. Pow. Energy Systems, 1994, 14(3): 98-103.

190 Chen W, Sun F and Chen L.. Finite time thermodynamic criteria for selecting working parameters of heat engines between heat reservoirs. Chinese Sci. Bull., 1991, 36(9): 763-768 (in English).

191. Sun F, Guo L, Wang Z, Chen L and Chen W. Three methods of finite time thermodynamics for the performance analysis and optimization of nuclear power Rankine cycles. Nuclear Pow. Engng., 1996,17(15): 420-424 (in Chinese).

192. Ibrahim O M and Klein S A. High-power multi-stage Rankine cycles. ASME Trans. J. Energy Resources Tech, 1995, 117(3): 192-196.

193. Wu C. Specific power bound of a finite-time closed Brayton cycle. Int. J. Ambient, 1990, 11(2): 77-82.

194. Wu C. Power optimization of an endoreversible Brayton gas turbine heat engine. Energy Convers. Mgmt., 1991, 31(6): 561-565.

195. Woodward J B. Optimal second-law efficiency for a Brayton cycle with an internal heat source. ASME Trans. J. Energy Res. Tech., 1995, 117(4): 343-348.

196. Cheng C and Chen C. Power optimization of an endoreversible regenerative Brayton cycle. Energy, The Int. J., 1996, 21(4): 241-247.

197. Cheng C and Chen C. Power optimization of an irreversible Brayton heat engine. Energy Sources, 1997, 19(5): 461-474.

198. Feidt M. Optimization of Brayton cycle engine in contact with fluid thermal capacities. Rev. Gen. Therm., 1996, 35(418/419): 662-666.

199. Roco J M M, Velasco S, Medina A and Calvo Hernandez A. Optimum performance of a regenerative Brayton thermal cycle. J. Appl. Phys., 1997, 82(6):2735-2741.

200. Sahin B, Kodal A and Yavuz H. Efficiency of a Joule-Brayton engine at maximum power density. J. Phys. D: Appl. Phys., 1995, 28(7): 1309-1313.

201. Medina A, Roco J M M and Calvo Hernandez A. Regenerative gas turbines at maximum power density conditions. J. Phys. D: Appl. Phys., 1996, 29(11): 2802-2805.

202. Hernandez A C, Medina A and Roco J M M. Power and efficiency in a regenerative gas turbine. J. Phys. D: Appl. Phys., 1995, 28(10): 2020-2023.

203. Calvo Hernadez A, Roco J M M and Medina A. Power and efficiency in a regenerative gas-turbine cycle with multiple reheating and intercooling stages. J. Phys. D: Appl. Phys., 1996, 29(6): 1462-1468.
204. Chen L and Sun F. Maximum power output from a closed gas turbine cycle. Ship Engng., 1995 (2): 31-33 (in Chinese).
205. Chen L, Sun F and Wu C. Performance analysis of an irreversible Brayton heat engine. J. Institu. Energy, 1997, 70(482): 2-8.
206. Chen L, Sun F and Yun J. Effects of heat resistance on the performance of closed cycle regenerated gas turbine. Chinese J. Engng. Thermophys., 1995, 16(4): 401-404 (in Chinese).
207. Wu C, Chen L and Sun F. Performance of a regenerative Brayton heat engines. Energy, The Int. J., 1996, 21(2): 71-76.
208. Chen L, Sun F, Wu C and Kiang R L. Theoretical analysis of the performance of a regenerated closed Brayton cycle with internal irreversibilities. Energy Convers. Mgmt., 1997, 18(9): 871-877.
209. Chen L, Ni N, Cheng G and Sun F. FTT performance of a closed regenerated Brayton cycle coupled to variable-temperature heat reservoirs. Proc. Int. Conf. Marine Engng., 3.7.1-3.7.7, Nov. 4-8, 1996, Shanghai.
210. Chen L, Ni N and Sun F. Effects of heat resistance and internal irreversibility on the performance of real closed-cycle regenerated gas turbine. Power Sys. Engng., 1997,13(3):5-7;13(4):5-8 (in Chinese).
211. Cole G H A. Thermal Power Cycles. London: Edward Arnold Publ. Inc., 1991.
212. Frost T H, Agnew B and Anderson A. Optimization for Brayton-Joule gas turbine cycle. Proc. Inst. Mech. Engrs., 1992, 206(3): 283-288.
213. Wu F and Hu Y. Effect of finite rate process on the performance of piston-type Stirling engine. J. Engng. Thermal Energy Pow., 1991, 6(3): 320-324 (in Chinese).
214. Blank K A, Davis G W and Wu C. Power optimization of an endoreversible Stirling cycle with regeneration. Energy, The Int. J., 1994, 19(1): 125-133.
215. Blank D A and Wu C. Power optimization of an extraterrestrial solar-radiant Stirling heat engine. Energy, The Int. J., 1995, 20(6): 523-530.
216. Radcenco V, Popescu G, Apostol V and Feidt M. Thermodynamics with determined time applied to thermal motors-case studies. Vapor machine and Stirling motor. Rev. Gen. Therm., 1993, 32(382): 509-514.
217. Feidt M, Costea M and Relabbas B. Influence of the regenerator on the optimal distribution of heat transfer surface conductance in a Stirling engine. ECOS'96, pp.179-186.
218. Popescu G, Radcenco V, Costea M and Feidt M. FTT optimization of an endo-exo-irreversible Stirling motor. Rev. Gen. Therm., 1996, 35(418/419): 656-661.
219. Erbay L B and Yavuz H. Analysis of the Stirling heat engine at maximum power conditions. Energy, The Int. J., 1997, 22(7): 645-650.
220. Wu F, Chen L, Wu C and Sun F. Optimum performance of an irreversible Stirling engine with imperfect regeneration. Energy Convers. Mgmt., 1998, 39(8): 727-732.

221. Wu F, Chen L, Sun F, Wu C and Zhu Y. Performance and optimization criteria of forward and reverse quantum Stirling cycles. Energy Convers. Mgmt., 1998, 39(8): 733-739.
222. Ladas H G and Ibrahim OM. Finite-time view of the Stirling engine. Energy, The Int., 1994, 19(8): 837-847.
223. Ibrahim O M and Ladas H G. Finite-time analysis of the Stirling engine. Proc. Intersoc. Energy Convers. Engng. Conf., 1995, Vol. 3, pp. 431-436.
224. Blank D A and Wu C. Power limit of an endoreversible Ericsson cycle with regeneration. Energy Convers. Mgmt., 1996, 37(1): 59-66.
225. Wu C. Analysis of an endoreversible Rallis cycle. Energy Convers. Mgmt., 1994, 35(1): 79-85.
226. Chen L, Lin J, Sun F and Wu C. Efficiency of an Atkinson engine at maximum power density. Energy Convers. Mgmt., 1998, 39(3/4): 337-341.
227. Sahin B, Kodal A, Ekmekci I and Yilmaz T. Exergy optimization for an endoreversible cogeneration cycle. Energy, The Int. J., 1997, 22(5): 551-557.
228. Ahlborn B and Camire J. Thermoacoustic heat pumps with maximum power transfer. Am. J. Phys., 1995, 63(5): 449-451.
229. Agrawal D C and Menon V J. The thermoelectric generator as an endoreversible Carnot engine. J. Phys. D: Appl. Phys., 1997, 30(3): 357-359.
230. Goktun S. FTT optimization of a solar-driven heat engine. Sol Energy, 1996, 56(6): 617-620.
231. Conti M and Charach C H. Thermodynamics of heat storage in a PCM shell-and-tube heat exchanger in parallel or in series with a heat engine. Sol. Energy, 1996, 57(1): 59-68.
232. Erbay L B and Yavuz H. Design parameters of a thermionic energy converter. TAIES'97, Proc. Int. Conf., pp. 20-24.
233. Sahin B, Kodal A and Yavuz H. A performance analysis for MHD power cycles operating at maximum power density. J. Phys. D: Appl. Phys., 1996, 29(6): 1473-1475.
234. De Vos A. Thermodynamics of photochemical solar energy conversion. Sol. Energy Materials Sol. Cells., 1995, 38(1): 11-27.
235. Barranco-Jimenez M A and Angulo- Brown F. A nonendoreversible model for wind energy as a solar driven heat engine. J. Appl. Phys., 1996, 80(9): 4872-4876.
236. Tsirlin A M. Optimal control of irreversible thermal and mass-transfer processes. Sov. J. Comput. System. Sci., 1992, 30(3): 23-31.
237. De Vos A and Van der Wel P. The efficiency of the conversion of solar energy into wind energy by means of Hadley cells. Theor. App. Climatol., 1993, 46(2): 193-202.
238. Gordon J M, Rubinstein and Zarmi Y. On optimal heating and cooling strategies for melting and freezing. J. Appl. Phys., 1990, 67(1): 81-84.
239. Bartana A and Kosloff R. Laser cooling of internal degrees of freedom. J. Chem. Phys., 1997, 106(4): 1435-1448.
240. Watowich S J, Hoffamann K H and Berry R S. Optimal paths for a bimolecular, light-driven engine. Nuovo Cimento B, 1989, 104(2): 131-147.

241. Mironova V A, Tsirlin A M, Kazakov V A and Berry R S. Finite-time thermodynamics: Exergy and optimization of time-constrained processes. J. Appl. Phys., 1994, 76(2): 629-636.

242. Kazakov V A and Berry R S. Estimation of productivity, efficiency and entropy production for cyclic separation processes with distributed working fluid. Phys. Rev. E, 1994, 49(4): 2928-2934.

243. De Vos A. Endoreversible thermodynamics and chemical reactions. J. Phys. Chem., 1991, 95(18): 4534-4540.

244. Gordon J M. Maximum work from isothermal chemical engines. J. Appl. Phys., 1993, 73(1): 8-11.

245. Gordon J M and Orlov V N. Performance characteristics of endoreversible chemical engines. J. Appl. Phys., 1993, 74(9): 5303-5309.

246. Schon J C and Andresen B. Finite-time optimization of chemical reactions: nA⇔nB. J. Phys. Chem., 1996, 100(21): 8843-8853.

247. Badescu V and Andresen B. Probabilistic finite time thermodynamics: A chemically driven engine. J. Non-Equilib. Thermodyn., 1996, 21(4): 291-306.

248. Angulo-Brown F, Santillan M and Calleja-Quevedo E. Thermodynamic optimality in some biochemical reactions. Nuovo Cimento, 1995, D17(1): 87-90.

249. Angulo-Brown F and Arias-Hernandez L A. van't Hoff's equation for endoreversible chemical reactions. J. Phys. Chem., 1996, 100(21): 9193-9195.

250. Chen L, Sun F, Wu C and Yu J. Performance characteristics of isothermal chemical engines. Energy Convers. Mgmt., 1997, 38(18): 1884-1841.

251. Chen L, Sun F, Wu C and Gong J. Maximum power of a combined-cycle isothermal chemical engine. Appl. Thermal Engng., 1997, 17(7): 629-637.

252. Chen L, Sun F and Wu C. Performance of chemical engines with mass leak. J. Phys. D: Appl. Phys., 1998, 31(13):1595-1600.

253. Salamon P, Andresen B and Berry R S. Thermodynamic in finite time. Potentials for finite-time processes. Phys. Rev. A, 1977, 15(5): 2094-2102.

254. Angulo-Brown F, Arias-Hernandez L A and Paez-Hernundez R. On endoreversible thermodynamic potentials. Rev. Mex. Fis., 1994, 49(1): 44-54.

255. Andresen B, Rubin M H and Berry R S. Availability for finite-time processes. General theory and a model. J. Chem. Phys., 1983, 87(15): 2704-2713.

256. Salamon P and Berry R S. Thermodynamic length and dissipated availability. Phys. Rev. Lett., 1983, 51(6): 1127-1130.

257. Ruppeiner G. Riemannian geometry in thermodynamic fluctuation theory. Rev. Modern Phys., 1995, 67(3): 605-659.

258. Flick J D, Salamon P and Andresen B. Metric bounds on losses in adaptive coding. Inform. Sci., 1987, 42(2): 239-244.

259. Hoffamann K H, Andresen B and Salamon P. Measures of dissipation. Phys. Rev. A, 1989, 39(7): 3618-3621.

260. Diosi L, Kulacsy K, Lukacs B and Racz A. Thermodynamic length, time, speed, and optimum path to minimize entropy production. J. Chem. Phys., 1996, 105(24): 11220-11225.

261. Reinhardt W P and Hunterlll J E. Variational path optimization and upper and lower bounds to free energy changes via finite time minimization of external work. J. Chem. Phys., 1992, 97(2): 1599-1601.

262. Spiracle W and Rise H. Optimal finite-time endoreversible process. Phys. Rev. E, 1995, E52(4): 3485-3489.

263. Chen L, Sun F and Chen W. The maximum profits principle for thermodynamic cycles. Nature J., 1991, 14(12): 948-949 (in Chinese).

264. Chen L, Sun F and Chen W. On the ecological figures of merit for thermodynamic cycles. J. Engng. Thermal Energy Pow., 1994, 9(6): 374-376 (in Chinese).

265. Santillan M, Arias-Hernandez L A and Angulo-Brown. Some optimization criteria for biological systems in linear irreversible thermodynamics. Nuovo Cimento, 1997, 19D(1): 99-109.

266. Santillan M and Angulo-Brown F. A thermodynamic approach to the compromise between power and efficiency in muscle contraction. J. Theor. Biol., 1998, in press.

267. Bejan A. Theory of organization in nature: pulsating physiological process. Int. J. Heat Mass Transfer, 1997, 40(9): 2097-2104.

268. Bejan A. How nature takes shape. Mech. Engng., 1997(10): 90-92.

269. Bejan A. Maximum power from fluid flow. Int. J. Heat Mass Transfer, 1996, 39(6): 1175-1181.

270. Chen L, Chen B and Sun F. On maximum power and efficiency from fluid flow. Chinese J. Engng. Thermophys., 1998, 19(4): 418-420 (in Chinese).

271. Bejan A and Dan N. Maximum work from an electric battery model. Energy, The Int. J., 1997, 22(1): 93-102.

272. De Vos A. Reversible and endoreversible computing. Int. J. Theor. Phys., 1995, 34(11): 2251-2266.

273. Leff H S and Teeter W D. EER, COP, and 2nd law efficiency for air-conditioners. Am. J. Phys., 1978, 46(1): 19-22.

274. Yan Z. The optimal relation between COP and cooling load of a Carnot refrigerator. Physics. 1984, 13(12): 768-770 (in Chinese).

275. Goth Y and Feidt M. Optimum COP for endoreversible heat pump or refrigerating machine. C. R. Acad. Sc. Pairs, 1986, 303(1): 19-24.

276. Yan Z. ε and R of a Carnot refrigerator at maximum εR. Nature J., 1984, 7(1): 73-74 (in Chinese).

277. Sun F, Chen W and Chen L. An analysis of the holographic spectrum and selection of the optimum parameters for reverse endoreversible Carnot cycle. J. Naval Acad. Engng., 1990 (2): 40-45 (in Chinese).

278. Chen W, Sun F and Chen L. Finite time thermodynamic criteria for selecting parameters of refrigerating and heat pumping cycles between heat reservoirs. Chinese Sci. Bull., 1990, 35(19): 1670-1672 (in Chinese).

279. Chen L, Sun F and Chen W. Finite time exergoeconomic performance bound and optimization criteria for two-heat-reservoir refrigerators. Chinese Sci. Bull., 1991, 36(2): 156-157 (in Chinese).

280. Chen L, Sun F and Chen W. Ecological optimization criteria for a Carnot refrigerator. Nature J., 1992, 15(8): 633 (in Chinese).

281. Wu F, Chen L and Sun F. Optimal performance of quantum Carnot refrigerators. J. Appl. Sci., 1997, 15(2): 223-228 (in Chinese).
282. Wu F, Chen L and Sun F. Compromise optimization between cooling load and entropy generation rate for quantum Carnot refrigerators. Cryogenics, 1996 (1): 1-5 (in Chinese).
283. Klein S A. Design considerations for refrigeration cycles. Int. J. Refrig., 1992, 15(3):181-185.
284. Wu C. Maximum obtainable specific cooling load of a refrigerator. Energy Convers. Mgmt., 1995, 36(1): 7-10.
285. Chen L, Sun F and Chen W. Optimal performance of a Carnot refrigerator with $q \propto (\Delta T)^n$. Cryogenics, 1989 (5): 29-33 (in Chinese).
286. Chen W, Sun F and Chen L. The optimal COP and cooling load of a Carnot refrigerator in case of q• • (T^n). Chinese Sci. Bullet., 1990, 35(23): 1837.
287. Wu C, Chen L, Sun F and Chen W. General performance characteristics of a finite-speed Carnot refrigerator. Appl. Thermal Engng., 1996, 16(4): 299-303.
288. Chen L, Sun F and Chen W. Effect of heat transfer law on the finite time exergoeconomic optimal performance of Carnot refrigerators. Cryogenics. 1990 (3): 39-43 (in Chinese).
289. Yan Z. Optimal performance of irreversible refrigerators. J. Refrig., 1990 (3): 11-14 (in Chinese).
290. Grazzini G. Irreversible refrigerators with isothermal heat exchangers. Int. J. Refrig., 1993, 16(2): 101-106.
291. Ait-Ali M M. A class of internally irreversible refrigeration cycles. J. Phys. D: Appl. Phys., 1996, 29(3): 593-599.
292. Ait-Ali M M. The maximum coefficient of performance of internally irreversible refrigerators and heat pumps. J. Phys. D: Appl. Phys., 1996, 29(4): 975-980.
293. Bejan A. Theory of heat transfer-irreversible refrigeration plants. Int. J. Heat Mass Transfer, 1989, 32(9): 1631-1639.
294.Chen L, Sun F and Chen W. The influence of heat leak on the COP versus cooling load characteristics of refrigerators. Cryogenics Superconductor, 1992, 20(4): 7-14 (in Chinese).
295.Chen L and Sun F. The maximum COP bound of an irreversible Carnot refrigerator. Sci. Tech. Bull., 1995, 11(2): 128 (in Chinese).
296. Chen L, Sun F, Wu C and Kiang R L. A generalized model of a real refrigerator and its performance. Appl. Thermal Engng., 1997, 17(4): 401-412.
297.Chen J and Yan Z. Optimal performance of an endoreversible refrigerator with finite thermal capacity heat sources. Cryogenics, 1987 (4): 27-35 (in Chinese).
298. Chen L, Sun F, Ni N and Wu C. Optimal configuration of a class of two-heat-reservoir refrigeration cycles. Energy Convers. Mgmt., 1998, 39(8): 767-773.
299.Ma Y and Duang R. Finite time thermodynamic analysis for refrigerating system coupled to variable-temperature heat reservoirs. Chinese J. Engng. Thermophys., 1993,14(4): 353-356 (in Chinese).

300. Chen L, Sun F, Gong J, Chen W and Lai X. Optimization of steady flow refrigeration cycles for specified boundary conditions. Chinese J. Engng. Thermophys., 1994, 15(3): 249-252 (in Chinese).
301. Agrawal D C and Menon V J. Finite-time Carnot refrigerators with wall gain and product loads. J. Appl. Phys., 1993, 74(4): 2153-2158.
302. Chen J and Yan Z. Optimal performance of an endoreversible-combined refrigeration cycle. J. Appl. Phys., 1988, 63(10): 4795-4799.
303. Chen L, Sun F and Chen W. Optimization of the specific rate of refrigeration in combined refrigeration cycles. Energy, The Int. J., 1995, 20(10): 1049-1053 .
304. Chen L, Wu C and Sun F. Steady flow combined refrigeration cycle performance with heat leak. Appl. Thermal Engng., 1997, 17(7): 639-645.
305. Goktun S. Coefficient of performance for an irreversible combined refrigeration cycle. Energy, The Int. J., 1996, 21(7/8): 721-724.
306. Radcenco V, Vargas J V C, Bejan A and Lim J S. Two design aspects of defrosting refrigerators. Int. J. Refrig., 1995, 18(2): 76-86.
307. Vasilescu E, Radcenco V and Popescu G. Design optimization of vapour compression chillers based on FTT. ECOS'96, pp. 153-156.
308. Kim K J, Feldman K T and Razani A. Cooling and power efficiency diagrams for compressor-driven, metal-hydride slurry air conditioners. Energy, The Int. J., 1997, 22(8): 787-796.
309. Velasco S, Roco J M M, Medina A and Calvo Hernandez A. Irreversible refrigerators under per-unit-time coefficient of performance optimization. Appl. Phys. Lett., 1997, 71(8):1130-1132.
310. Sun F, Chen L and Xu M. Minimum heat transfer surface area optimization for marine vapour-compression air-conditioners. Ship Engng., 1997(3): 33-34 (in Chinese).
311. Chen L, Wu C and Sun F. Cooling load versus COP characteristics for an irreversible air refrigeration cycle. Energy Convers. Mgmt., 1998, 39(1/2): 117-126.
312. Chen L, Ni N, Cheng G, Sun F and Wu C. Analysis of heat-transfer-irreversible regenerated Brayton refrigerator. Proc. Int. Conf. Thermodynamic Analysis and Improvement of Energy Systems (TAIE'S 97), June 10-13, 1997, Beijing, pp. 8-15. Eds: Cai R and Moran M J, Word Publishing Cooperation.
313. Tie X and Yan Z. The cooling load and COP bounds for Stirling refrigerators. Sci. Tech. Bull., 1996, 12(2): 108-114 (in Chinese).
314. Chen J and Yan Z. The general performance characteristics of a Stirling refrigerator with regenerative losses. J. Phys. D: Appl. Phys., 1996, 29(4): 987-990.
315.Jin X, Wu F and Chen L. Optimization of quantum Stirling refrigerators in the classical limit. Cryogenics, 1998(1): 41-44 (in Chinese).
316. Lin G and Yan Z. Optimization analysis of magnetic Stirling refrigerators. Chinese J. Engng. Thermophys., 1994, 15(4): 357-360 (in Chinese).
317. Chen J and Yan Z. Optimal performance of a thermoelectric refrigerator. AIP Conf. Proc., No. 316, 1995, pp. 199-202.
318. Goktun S. Optimal performance of a thermoelectric refrigerator. Energy Sources, 1996, 18(5): 531-536.

319. Chen L, Wu C and Sun F. Heat transfer effect on the specific cooling load of refrigerators. Appl. Thermal Engng., 1996, 16(12): 989-997.
320. Yan Z and Chen S. Finite time thermodynamic performance bound of three-heat-reservoir refrigerators. Chinese Sci. Bull., 1986,31 (10): 718 (in Chinese)
321. Yan Z and Chen J. An optimal endoreversible three-heat-source refrigerator. J. Appl. Phys., 1989, 65(1): 1-4.
322.Chen L, Sun F and Chen W. The analysis of COP for three-heat-reservoir refrigerators. Vac. Cryogenics, 1990, 9(2): 27-30 (in Chinese).
323. Chen L, Sun F, Chen W and Wu C. Optimal performance coefficient and cooling load relationship of three-heat -reservoir endoreversible refrigerator. Int.J. Power and Energy Systems, 1997, 17(3): 206-208.
324. Chen L, Sun F and Chen W. The ecological optimal performance for a three-heat-reservoir refrigerators. Cryogenics, 1993(3): 38-40 (in Chinese).
325. Chen L, Wu C and Sun F. Maximum profit performance of an absorption refrigerator. Int. J. Energy, Environment, Economics, 1996, 4(1): 1-7.
326. Wu C. Maximum cooling load of a heat-engine driven refrigerator. Energy Convers. Mgmt., 1993, 34(8): 691-696.
327. Bejan A and Vargas J V C. Optimal allocation of a heat exchanger inventory in heat driven refrigerators. Int. J. Heat Mass Transfer, 1995, 38(16): 2997-3004.
328. Wijeysundera N E. Performance of three-heat-reservoir absorption cycles with external and internal irreversibilities. Appl. Thermal Engng., 1997, 17(12): 1151-1161.
329. Ng K C, Bong T Y, Chua H T and Bao H L. Theoretical and experimental analysis of an absorption chiller. Int. J. Refrig., 1994, 17(5): 351-358.
330. Gordon J M and Ng K C. Thermodynamic modeling of reciprocating chillers. J. Appl. Phys., 1994, 75(6): 2769-2774.
331. Gordon J M and Ng K C. A general thermodynamic model for absorption chillers: Theory and experiment. Heat Recovery Systems & CHP, 1995, 15(1): 73-83.
332. Ng K C, Bong T Y and Chua H T. Performance evaluation of centrifugal chiller in a air-conditioning plant with the building automation system (BAS). Proc. Instn. Mech. Engrs., 1994, 208(A3): 249-255.
333. Gordon J M, Ng K C and Chua H T. Centrifugal chillers: thermodynamic modeling and a diagnostic case study. Int. J. Refrig., 1995, 18(4): 253-257.
334. Gordon J M and Ng K C. Predictive and diagnostic aspects of a universal thermodynamic model for chillers. Int. J. Heat Mass Transfer, 1995, 38(5): 807-818.
335. Chua H T, Ng K C and Gordon J M. Experimental study of the fundamental properties of reciprocating chillers and their relation to thermodynamic modeling and chiller design. Int. J. Heat Mass Transfer, 1996, 39(11): 2195-2204.
336. Chua H T, Ng K C, Gordon J M and Bong T Y. On the consistency of thermodynamic models with actual chiller performance. ECOS'95, Istanbul, 1995, pp. 339-346.
337. Ng K C, Chua H T, Ong W, Lee S S and Gordon J M. Diagnostics and optimization of reciprocating chillers: theory and experiment. Appl. Thermal Engng., 1997, 17(3): 263-276.

338. Gordon J M, Ng K C and Chua H T. Optimizing chiller operation based on finite-time thermodynamics: universal modeling and experimental confirmation. Int. J. Refrig., 1997, 20(3): 191-200.
339. Chua H T, Han Q, Ng K C and Gordon J M. Thermodynamic modeling and experimental evidence for the optimization and maximum-efficiency operation of absorption chillers. ECOS'96, Stockholm, 1996, pp. 157-166.
340. Ng K C, Chua H T and Han Q. On the modeling of absorption chillers with external and internal irreversibilities. Appl. Thermal Engng., 1997, 17(5): 413-425.
341. Chua H T, Gordon J M, Ng K C and Han Q. Entropy production analysis and experimental confirmation of absorption systems. Int. J. Refrig., 1997, 20(3): 179-190.
342. Chen L, Shen L, Gong J and Sun F. Performance analysis and experimental confirmations for thermoelectric refrigerators. Paper presented at the meeting of engineering thermodynamics, 1998, paper No. 981012 (in Chinese).
343. Blanchard C H. Coefficient of performance for finite-speed heat pump. J. Appl. Phys., 1980, 51(5): 2471-2472.
344. Wu C. Specific heating load of an endoreversible Carnot heat pump. Int. J. Ambient Energy 1993, 14(1): 25-28.
345. Wu F, Chen L, Sun F, Wu C and Hua P. Optimum performance parameters for a quantum Carnot heat pump with spin-1/2. Energy Convers. Mgmt., 1998, 39(11):1161-1167.
346. Wu C, Chen L and Sun F. Optimization of steady flow heat pumps cycles. Energy Convers. Mgmt., 1998, 39(5/6): 445-453.
347. Chen L, Wu C and Sun F. Heat Pump performance with internal heat leak. Int. J. Ambient Energy, 1997, 18(3): 129-134.
348. Sun F, Chen W, Chen L and Wu C. Optimal performance of an endoreversible Carnot heat pump. Energy Convers. Mgmt., 1997, 38(14): 1439-1443.
349. Chen L, Wu C and Sun F. Optimization of the specific rate of heat pumping in combined heat pump cycles. Energy Convers. Mgmt., 1998, 39(1/2): 113-116.
350. Chen L, Wu C and Sun F. Heat transfer effect on the specific heating load of heat pumps. Appl. Thermal Engng., 1997, 17(1): 103-110.
351. Chen L, Ni N, Sun F and Wu C. Performance of real regenerated air heat pumps. Proc. IASTED Int. Conf., High Technology in the Power Industry, Oct. 27-30, 1 997, Orlando, Florida, USA. Editor: A. Domigan, Jr., IASTED/Acta Press. pp. 387-392.
352. Yan Z and Chen S. Finite time thermodynamic performance bound of three-heat-reservoir heat pumps. Chinese Sci. Bull., 1986, 31 (10): 798-799 (in Chinese).
353. Chen J and Yan Z. Equivalent combined systems of three-heat-source heat pumps. J. Chem. Phys., 1989, 90(9): 4951-4955.
354. Chen J and Yan Z. Unified description of endoreversible cycles. Phys. Rev. A, 1989, 39(8): 4140-4147.
355. Chen L, Sun F and Chen W. The relation between optimal COP and heating load of an endoreversible three-heat-reservoir heat pump with nonlinear heat transfer. Acta Energine Solaris Sinica, 1990, 11(3): 319-323 (in Chinese).

356. Chen L, Wu C and Sun F. Optimal COP and heating load relationship of a three-heat-reservoir endoreversible heat pump. Energy Convers. Mgmt., 1997, 38(8): 727-733.

357. Dentice D'Accadia M, Sasso M and Sibilio S. Optimum performance of heat engine-driven heat pumps: A finite-time approach. Energy. Convers. Mgmt., 1997, 38(4): 401-413.

358. Goktun S. The effect of irreversibilities on the performance of a three-heat -source heat pump. J. Phys. D: Appl. Phys., 1996, 29(11): 2823-2825.

359. Wu C, Chen L, Sun F and Cao S. Optimum collector temperature for solar driven heat pumps. Energy Convers. Mgmt, 1998, 39(1/2): 143-146.

360. Chen L, Sun F and Chen W. The unified description of linear phenomenological law system endorevesible cycles. J. Energy Q., 1992(1): 6-10 (in Chinese).

OPTIMAL PERFORMANCE OF AN ENDOREVERSIBLE CARNOT CYCLE USED AS A COOLER

JINCAN CHEN AND BJARNE ANDRESEN
Ørsted Laboratory, Niels Bohr Institute, University of Copenhagen,
Universitetsparken 5, DK-2100 Copenhagen Ø, Denmark

Abstract- The optimal performance of an endoreversible Carnot heat engine used as a cooler is investigated on the basis of a simple cycle model. The cooling rate of the cooler is optimized for specified power input and total heat transfer area of the cooler. The optimal ratio of the two heat transfer areas of the cooler is derived. The optimal bounds of the primary performance parameters, such as the cooling rate, the power input, and the temperatures of the working substance in the cooler, are determined. The results obtained can serve as a good guide for the understanding of the performance of a real cooler and further perfect the systematic theory of endoreversible Carnot cycles.

Keywords: Carnot heat engine, cooler, endoreversible cycle, heat transfer, cooling rate

NOMENCLATURE

A	Total heat transfer area (m^2)
A_c	Heat transfer area between cooler and heat sink (m^2)
A_h	Heat transfer area between cooler and heat source (m^2)
A_n	Heat transfer area of a natural cooling process (m^2)
C	Parameter defined in Eq. (16) ($kW\ K^{-1}$)
P	Power input (kW)
P_{min}	Minimum power input (kW)
P^*	Dimensionless power input
q_c	Rate of heat transfer from cooler to heat sink (kW)
q_h	Cooling rate (kW)
R	$= q_h$ (kW)
R_m	Cooling rate at maximum coefficient of performance (kW)
R_{max}	Maximum cooling rate (kW)
R_n	Natural cooling rate (kW)

R^*	Dimensionless cooling rate
R_m^*	Dimensionless cooling rate at maximum coefficient of performance
R_{max}^*	Dimensionless maximum cooling rate
R_n^*	Dimensionless natural cooling rate
T_1	Temperature of working substance in high-temperature isothermal process (K)
T_{1max}	Maximum value of T_1 (K)
T_2	Temperature of working substance in low-temperature isothermal process (K)
T_{2min}	Minimum value of T_2 (K)
T_c	Temperature of heat sink (K)
T_h	Temperature of heat source (K)
U_c	Heat transfer coefficient between cooler and heat sink (kW m^{-2} K^{-1})
U_h	Heat transfer coefficient between cooler and heat source (kW m^{-2} K^{-1})
U_n	Heat transfer coefficient of a natural cooling process (kW m^{-2} K^{-1})
x	$= T_2/T_1$
ε	Coefficient of performance
ε_{max}	Maximum coefficient of performance

INTRODUCTION

The natural cooling rate of a body with temperature T_h, higher than its surrounding temperature T_c, can often be described by the Newtonian heat transfer law

$$R_n = U_n A_n (T_h - T_c), \tag{1}$$

where U_n and A_n are, respectively, the heat transfer coefficient and the area between the body and its surroundings, and the subscript n indicates "natural". To obtain a cooling rate larger than R_n one has to use an active cooler, assuming that it is not possible to increase U_n e.g. by increasing coolant circulation with a fan. A common example is a high power semiconductor, where the produced heat will result in a rise in temperature of the semiconductor which may make it lose its normal function, such that a cooler shown in Fig. 1 needs to be employed to keep it operating below a certain temperature T_h.

Although the reversible Carnot heat engines, heat pumps, and refrigerators have been described in detail in classical thermodynamics, the performance of a cooler is not brought up, because it is meaningless to discuss a reversible cooler. As this point it is necessary to expound the differences between a cooler and a refrigerator or a chiller.[1-5] A cooler absorbs heat from a high-temperature heat source at temperature T_h and rejects heat to a heat sink at temperature $T_c<T_h$. The purpose of using a cooler is thus to speed up a cooling process. When a refrigerator or a chiller is operated, it absorbs heat from a low-temperature heat source at temperature T_l and rejects heat to a heat sink at temperature $T_c>T_l$. The purpose of using a refrigerator or a chiller is to obtain a cooled space with a temperature T_l which is lower than the environmental temperature.

The theory of finite-time thermodynamics, or entropy generation minimization in engineering parlance, is a useful tool for researching the performance of irreversible cycles. But almost all the investigations of finite-time thermodynamics have concentrated on predicting the optimal performance of irreversible heat engines, heat pumps, and refrigerators.[6–13] In this paper we will discuss the optimal performance of a cooler which has real applications. Earlier engineering design is carefully described in ref. 10.

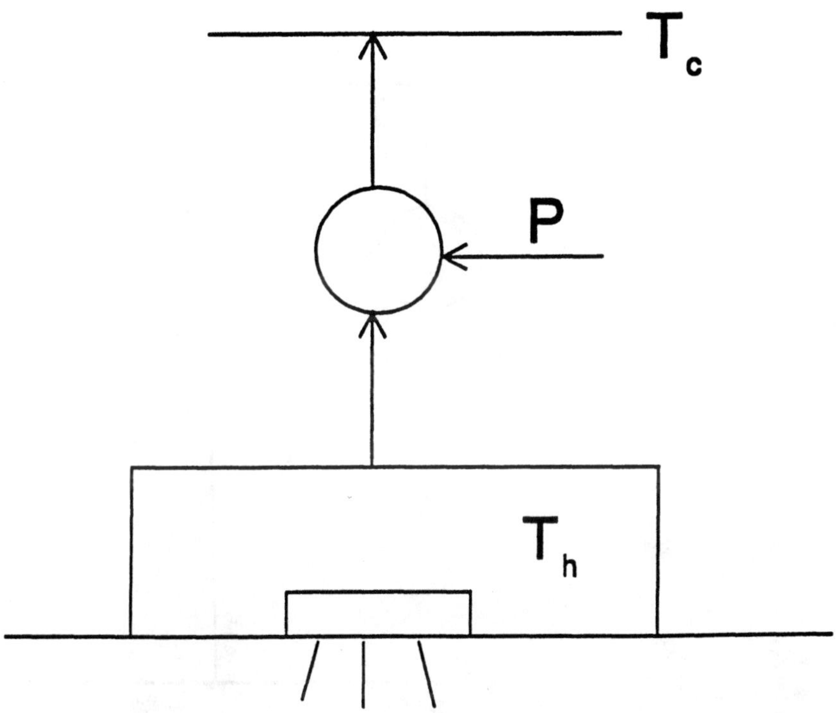

Fig. 1. Schematic diagram of a cooler.

CYCLE MODEL

If the temperature of the working substance of the cooler alternates between T_h and T_c, the rate of heat transfer from the heat source at temperature T_h to the cooler working substance and from the cooler working substance to the heat sink at temperature T_c will be equal to zero, because the heat transfer coefficients and areas between the cooler and the two heat reservoirs at temperatures T_h and T_c are always finite. Such a cooler with zero cooling rate is a reversible cooler, but it is not of realistic significance. Obviously, the temperatures of the working substance in a practical cooler are always different from the temperatures T_h and T_c of the heat source and the heat sink.

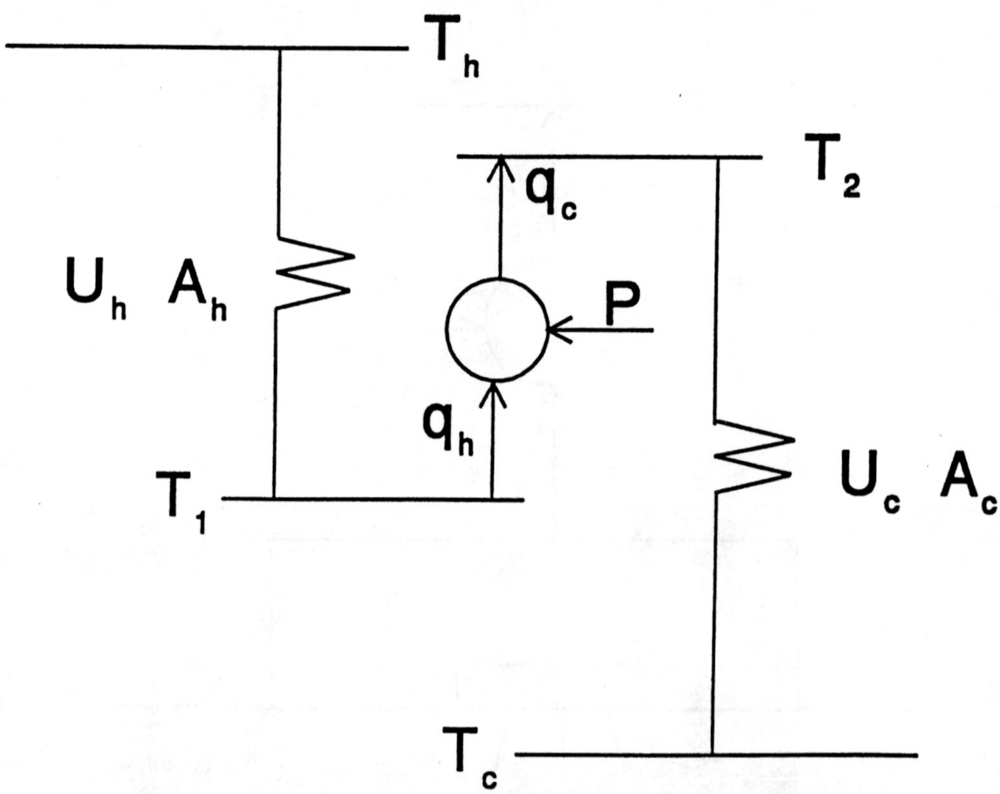

Fig. 2. The endoreversible cycle model of a cooler.

Since irreversibility of heat transfer will usually be the main irreversibility in the cooler, one can capture most of the physics of the process and thus gain primary understanding of the optimization by neglecting the other irreversibilities. It may thus be assumed that the cycle of the working substance in the cooler is an endoreversible cycle[7,8,10,13] which consists of two isothermal branches and two adiabatic branches or two other branches with perfect regeneration. A schematic diagram of the system is shown in Fig. 2, where T_1 and T_2 are, respectively, the temperatures of the working substance on the two isothermal branches, q_h is the cooling rate, q_c is the rate of heat transfer from the cooler to the heat sink, and P is the power input that the cooler needs.

When heat transfer obeys Newton's cooling law, one has

$$q_h = U_h A_h (T_h - T_1), \tag{2}$$

$$q_c = U_c A_c (T_2 - T_c), \tag{3}$$

where U_h and A_h are the heat transfer coefficient and area between the heat source at temperature T_h and the cooler; U_c and A_c are the heat transfer coefficient and area between the cooler and the heat sink at temperature T_c. The total heat transfer area of the cooler is

$$A = A_h + A_c. \tag{4}$$

According to the endoreversible cycle model mentioned above and the second law of thermodynamics one has

$$q_h/T_1 = q_c/T_2. \tag{5}$$

From Eqs. (2) – (5) we obtain the cooling rate

$$R = q_h = \frac{A}{\dfrac{1}{U_h(T_h - T_1)} + \dfrac{x}{U_c(xT_1 - T_c)}} \tag{6}$$

and the power input

$$P = q_c - q_h = (x-1)\, q_h = \frac{A(x-1)}{\dfrac{1}{U_h(T_h - T_1)} + \dfrac{x}{U_c(xT_1 - T_c)}} \tag{7}$$

where $x = T_2/T_1$.

An Optimal Relation

In the optimal design and operation of a cooler an important problem is how to allocate the total heat transfer area A between the heat transfer areas A_h and A_c of the two heat exchangers, considered in detail in refs. 3, 10, and 14. Another important problem is to determine the optimal temperatures of the working substance on the two isothermal branches of the cycle. To address these two important problems we first calculate the optimal cooling rate of the cooler for a specified power input P and total heat transfer area A of the cooler. For this purpose we introduce the Lagrangian

$$L = R + \lambda P = [1 + \lambda(x-1)] \frac{A}{\dfrac{1}{U_h(T_h - T_1)} + \dfrac{x}{U_c(xT_1 - T_c)}} \tag{8}$$

From the Euler-Lagrange equations

$$\partial L/\partial T_1 = 0, \quad \partial L/\partial x = 0 \tag{9}$$

we find that for specified P and A, when the cooling rate attains its optimal value, T_1 and x must satisfy the following equations:

$$xT_1 - T_c = \sqrt{U_h/U_c}\, x(T_h - T_1), \tag{10}$$

$$\lambda + [1 + \lambda(x-1)] \frac{q_h T_c}{U_c A(xT_1 - T_c)^2} = 0. \tag{11}$$

Solving Eqs. (6), (7), and (10), we obtain

$$x = 1 + P/R, \tag{12}$$

$$T_1 = \frac{\sqrt{U_h/U_c}\,(1 + P/R)T_h + T_c}{(1 + \sqrt{U_h/U_c})(1 + P/R)} \tag{13}$$

On the other hand, we can prove from Eqs. (2), (3), (12), and (13) that for a specified A the optimal ratio of A_h to A_c is given by

$$A_h/A_c =. \sqrt{U_c/U_h} \tag{14}$$

Solving Eqs. (4) and (14) we obtain

$$A_h =. \frac{A}{1 + \sqrt{U_h/U_c}} \tag{15}$$

Substituting Eqs. (13) and (15) into Eq. (2), we finally get

$$P = \frac{R - C(T_h - T_c)}{CT_h/R - 1} \tag{16}$$

which is a fundamental optimum relation for the endoreversible Carnot heat engine used as a cooler, with

$$C = \frac{AU_h}{(1+\sqrt{U_h/U_c})^2} \tag{17}$$

Below, Eq. (16) will be used to discuss the various optimum performances of a cooler.

OPTIMUM PERFORMANCE BOUNDS

1. MINIMUM POWER INPUT

It is proven from Eq. (16) that when the cooling rate R of a cooler is equal to the natural cooling rate R_n, the power input

$$P = \frac{R_n - C(T_h - T_c)}{CT_h/R_n - 1} = \frac{(U_n A_n - C)(T_h - T_c)}{CT_h/R_n - 1} \equiv P_{min}. \tag{18}$$

Now, Eq. (18) may be used to discuss the following different cases. (a) $U_nA_n–C<0$, $P_{min}<0$. In such a case, if $P_{min}<P<0$, we can obtain a cooling rate larger than the natural cooling rate while at the same time getting power output from the endoreversible Carnot heat engine used as a cooler. (b) $U_nA_n–C = 0$, $P_{min} = 0$. Then the net effect of using a cooler will be better than that of the natural cooling process. (c) $U_nA_n–C>0$, $P_{min}>0$. In such a case using a cooler will lose its realistic meaning if $P<P_{min}$, because the cooling rate obtained from the cooler is smaller than that of the natural cooling process. This shows clearly that P_{min} is an important parameter of coolers. Obviously, the power input of a cooler must be larger than P_{min}, i.e.,

$$P > P_{min}. \tag{19}$$

2. MAXIMUM COOLING RATE

It is also proven from Eq. (16) that as $P\to\infty$, $R\to CT_h$. Thus we can determine the cooling rate of a cooler

$$R < CT_h \equiv R_{max}. \tag{20}$$

In general, when the temperature difference between the heat source and the heat sink is small, the natural cooling rate R_n is much smaller than R_{max} so that using a cooler can provide a cooling rate R which is larger than the natural cooling rate R_n.

However, in some special cases we may find

$$R_{max} \leq R_n. \tag{21}$$

For example, a cooler with a small heat transfer coefficient U_h is operated over a large temperature difference. In such a case using a cooler is also without meaning. It is clearly seen from Eq. (21) that when the ratio of the temperatures of the two heat reservoirs,

$$T_c/T_h \leq 1 - C/U_nA_n, \tag{22}$$

the net effect of using a cooler is worse than that of the natural cooling process.

Let us consider a special case: $U_h = U_c = U_n$ and $A_h = A_c = A_n$. Then Eq. (22) may be written simply as

$$T_c/T_h \leq 1/2. \tag{23}$$

This shows clearly that it is unsuitable for a cooler to operate over a large temperature difference. This is a very natural conclusion, because the natural cooling process has a large cooling rate R_n under the influence of a large temperature difference. In fact, as long as the natural cooling rate R_n is larger or approaches R_{max}, using a cooler will be without any realistic significance.

3. Reasonable Operating Region

It is seen from Eq. (16) that the cooling rate R of a cooler is a monotonically increasing function of the power input P, as shown in Fig. 3, where $P^* = P/CT_h$ and $R^* = R/CT_h$ are dimensionless power input and dimensionless cooling rate, respectively. The thick solid line represents the case of $R=R_n$. This curve divides the (P^*,R^*)-plane into two regions A and B. When $P<P_{min}$, the cooler is operated in the region A. The cooling rate R of the cooler is smaller than R_n so that the net effect of using a cooler will be worse than that of the natural cooling process. When $P>P_{min}$, the cooler is operated in the region B. The cooler can obtain a cooling rate larger than R_n. Therefore, the region B is the only reasonable operating region for coolers.

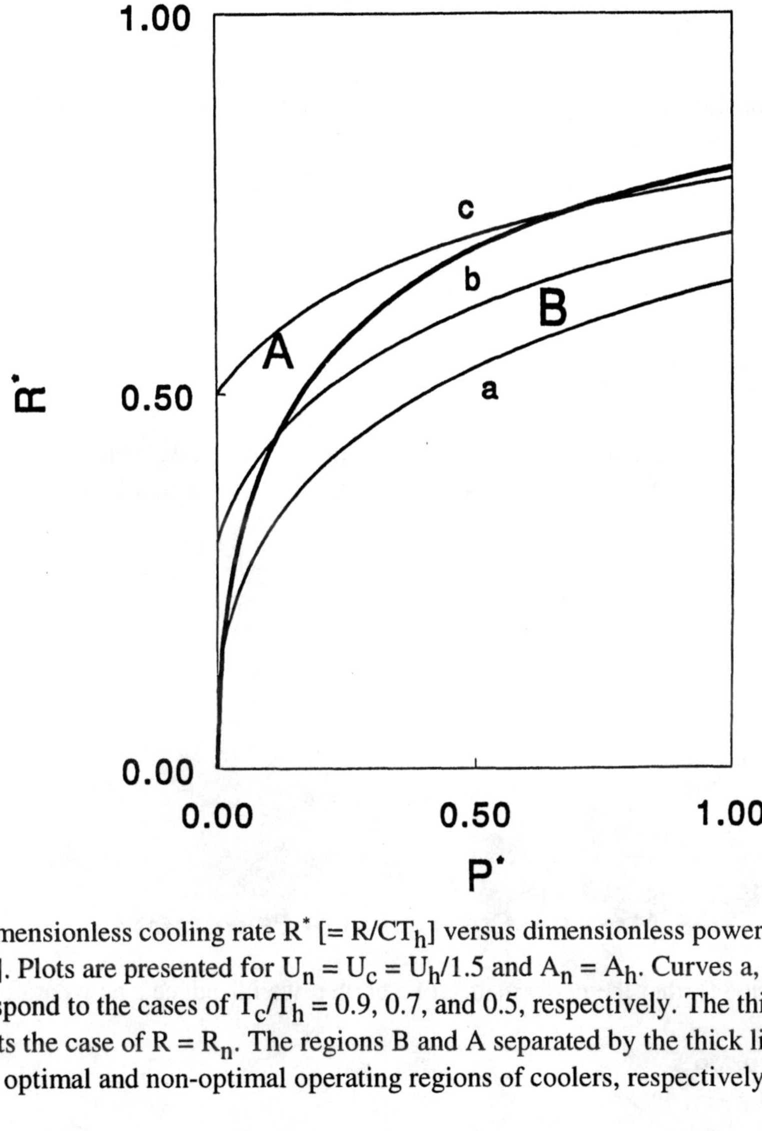

Fig.3. Dimensionless cooling rate R^* [= R/CT_h] versus dimensionless power input P^* [= P/CT_h]. Plots are presented for $U_n = U_c = U_h/1.5$ and $A_n = A_h$. Curves a, b, and c correspond to the cases of $T_c/T_h = 0.9$, 0.7, and 0.5, respectively. The thick line represents the case of $R = R_n$. The regions B and A separated by the thick line are the optimal and non-optimal operating regions of coolers, respectively.

4. Temperature Bounds of the Working Substance

From Eqs. (1), (2), (3), and (13) we find that when $R=R_n$ the temperatures of the working substance on the two isothermal branches of the cycle are, respectively, determined by

$$T_1 = T_h - \frac{U_n A_n}{U_h A_h}(T_h - T_c) \equiv T_{1max}, \tag{24}$$

$$T_2 = [T_h - \frac{U_n A_n}{U_h A_h}(T_h - T_c)]\frac{CT_c}{CT_h - R_n} \equiv T_{2min}. \tag{25}$$

In the special case of $U_h = U_c = U_n$ and $A_h = A_c = A_n$ Eqs. (24) and (25) may be written simply as

$$T_{1max} = T_c \tag{26}$$

and

$$T_{2min} = T_c^2/(2T_c - T_h) > T_h, \tag{27}$$

respectively. Because a cooler must obtain a certain cooling rate which is larger than R_n, the temperatures of the working substance on the two isothermal branches of the cycle must satisfy

$$T_1 < T_{1max} \tag{28}$$

and

$$T_2 > T_{2min}, \tag{29}$$

respectively.

5. Maximum Coefficient of Performance

To describe further the performance of a cooler, we introduce a new parameter

$$\varepsilon = (R - R_n)/P. \tag{30}$$

This parameter may be referred to as the coefficient of performance of a cooler. Substituting Eq. (16) into Eq. (30), we obtain a relation between the coefficient of performance and the cooling rate

$$\varepsilon =. \frac{(R - R_n)(CT_h - R)}{R[R - c(T_h - T_c)]} \tag{31}$$

Equation (31) shows clearly that when 0• R• $C(T_h - T_c)$, $\varepsilon > 0$. This is the operating region of an endoreversible Carnot heat engine. We will not discuss this case again.

When $C(T_h - T_c) < R < R_n$, $\varepsilon < 0$. As mentioned above, a cooler should not be allowed to operate in this region. When $R_n < R < R_{max}$, $\varepsilon > 0$. This is the reasonable operating region of a cooler. When $R=R_n$ or $R=R_{max}$, $\varepsilon=0$. This implies that the maximum coefficient of performance occurs somewhere between R_n and R_{max}, as shown in Fig. 4.

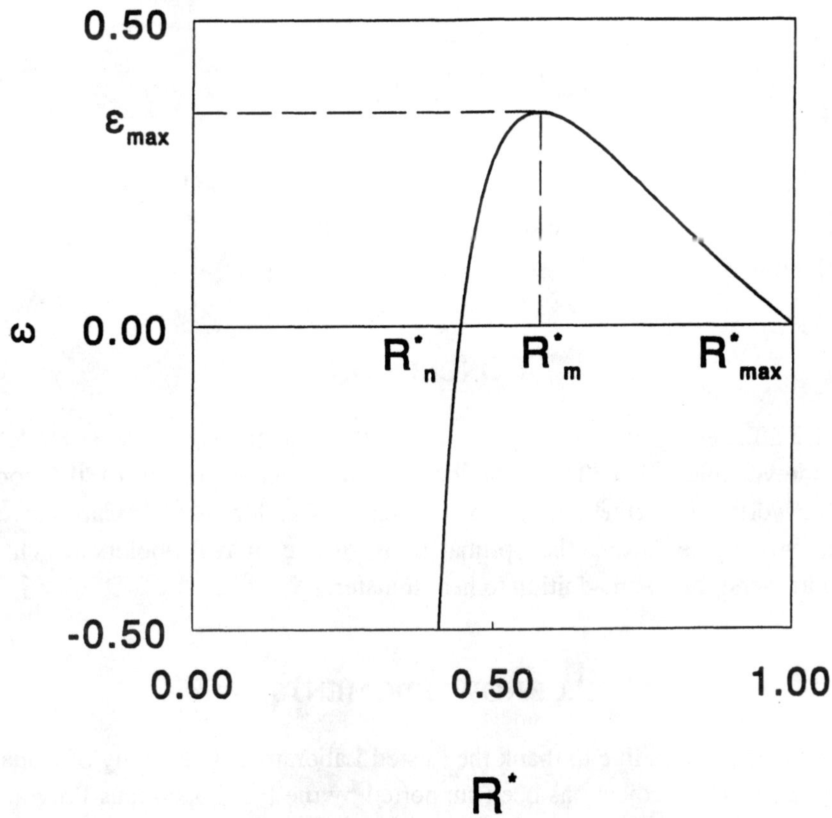

Fig.4. Coefficient of performance ε versus dimensionless cooling rate R^*. The plot is presented for $U_n = U_c = U_h/1.5$, $A_n = A_h$, and $T_c/T_h = 0.7$. R_n^*, R_m^*, and R_{max}^* are dimensionless natural cooling rate, dimensionless cooling rate at maximum coefficient of performance ε_{max}, and dimensionless maximum cooling rate, respectively.

From Eq. (31) and the extremum condition

$$\partial \varepsilon / \partial R = 0 \tag{32}$$

we obtain the cooling rate R_m at maximum coefficient of performance,

$$R_m = \frac{CT_h R_n + \sqrt{C^2 T_h T_c R_n [R_n - C(T_h - T_c)]}}{R_n + CT_c} \tag{33}$$

Substituting Eq. (33) into Eq. (31), we obtain the maximum coefficient of performance

$$\varepsilon_{max} = \frac{\sqrt{R_n}[C\sqrt{T_h T_c / R_n} - \sqrt{R_n - C(T_h - T_c)}]^2}{\sqrt{CT_c}[\sqrt{R_n T_h / T_c} + \sqrt{R_n - C(T_h - T_c)}]^2} \tag{34}$$

In general a cooler should be operated in the neighborhood of R_m so that a large coefficient of performance can be obtained. This shows that both ε_{max} and R_m are also important parameters of coolers.

CONCLUSION

A cooler affected by the influence of finite-rate heat transfer may be modeled as a simple endoreversible cycle. The main performance characteristics of the cooler are derived. The work will perfect the systematic theory of endoreversible Carnot cycles and suggests to investigate further the optimal performance of real coolers which include other main irreversibilities in addition to heat transfer.

ACKNOWLEDGMENTS

One of us (JC) would like to thank the Ørsted Laboratory, University of Copenhagen for its hospitality. This project has been supported by the EU Copernicus Program under grant no. CIPA-CT92-4026.

REFERENCES

1. Klein, S. A., Design considerations for refrigeration cycles, Int. J. Refrig. (1993) **15**, 181–185.
2. Chen, J., Yan, Z., Optimal performance of an endoreversible combined refrigeration cycle, J. Appl. Phys. (1988) **63**, 4795–4798.
3. Bejan, A., Theory of heat transfer-irreversible refrigeration plants, Int. J. Heat Mass Transfer (1989) **32**, 1631–1639.
4. Yan, Z., Chen, J., A class of irreversible Carnot refrigeration cycles with a general heat transfer law, J. Phys. D: Appl. Phys. (1989) **23**, 136–141.
5. Gordon, J. M., Ng, K. C., Predictive and diagnostic aspects of universal thermodynamic modeling for chillers, Int. J. Heat Mass Transfer (1995) **38**, 807–818.
6. Novikov, I. I., The efficiency of atomic power stations, J. Nucl. Energy II (1958) **7**, 125–128 [transl. from Atomnaya Energiya (1957) **3**, 409].
7. Curzon, F. L., Ahlborn, B., Efficiency of a Carnot engine at maximum power output, Am. J. Phys. (1975) **43**, 22–24.
8. Andresen, B., Salamon, P., Berry, R. S., Thermodynamics in finite time, Phys. Today (1984) **37** (9), 62–70.
9. Andresen, B., Finite-Time Thermodynamics, Physics Laboratory II, University of Copenhagen, Copenhagen (1983).
10. Bejan, A., Entropy Generation through Heat and Fluid Flow, John Wiley & Sons, New York (1982).
11. Chen, J., Yan, Z., Equivalent combined systems of three-heat-source heat pumps, J. Chem. Phys. (1989) **90**, 4951–4955.
12. Sieniutycz, S., Salamon, P., Finite-Time Thermodynamics and Thermoeconomics, Advances in Thermodynamics Vol.4, Taylor & Francis, New York (1990).
13. De Vos, A., Endoreversible Thermodynamics of Solar Energy Conversion, Oxford University Press, Oxford (1992).
14. Bejan, A., Power generation and refrigeration models with heat transfer irreversibilities, Thermal Sci. and Engng. (1994) **33**, 68.

GENERALIZED POWER OPTIMIZATION OF REGENERATIVE RECIPROCATING STIRLING-LIKE HEAT ENGINES WITH VARIOUS HEAT TRANSFER MODE COMBINATIONS AND FINITE NON-INTERACTIVE THERMAL RESERVOIRS

DAVID A. BLANK
Department of Mechanical Engineering
Indian Institute of Technology, Kharagpur 721302, India

and

CHIH WU
Mechanical Engineering Department
U.S. Naval Academy, Annapolis, MD 21402, USA

Abstract - An analytically compact approach (applicable to all practical modes of heat transfer) has been developed which enables the semi-decoupling of the solutions for the upper and lower operating temperatures which give internally reversible Stirling (and Stirling-like) beat engines the highest possible power potential. This highly innovative approach not only enables the analysis of the effect of various combinations of single and mixed heat transfer modes on the power optimization of these cycles, but also enables the study of trade-off between optimization with respect to "work" and optimization with respect to the "rate of work". In addition to linear modes (conduction and convection) and non-linear modes (radiation, etc.), the paper considers the effect of heat transfer between the cycle and finite reservoirs using the Log Mean Temperature Difference (LMTD) model. Also unique to this work is a new formulation which enables a specification of the allocation of the hot and cold end heat exchanger capacitance distribution which provides power optimized conditions under an combinations of modes of external heat transfer operation. To accomplish this the analysis goes beyond earlier works (in which power optimization was only based on cycle temperature bounds) and uses time symmetry to minimize cycle time, thus better optimizing overall cycle power. Because this novel procedure inadvertently results in the concurrent employment of first and second laws of thermodynamics, it not only ensures optimal allocation of thermal conductances at the hot and cold ends, but also simultaneously achieves both the minimization of internal entropy generation and maximization of specific cycle work for a given set of operating temperatures. Finally, the study looks at the work potential of different combinations of heat modes under power optimized conditions and conclusions point to the high desirability of making major use of the radiation mode in the high temperature side heat exchange and only linear modes at the low temperature side of the cycle.

INTRODUCTION

Externally reversible Stirling-like heat engine cycles produce no power. In the absence of internal irreversibilities and for the case of ideal regeneration however, these engines give rise to both a work potential and a thermal efficiency which no other real or ideal heat engine cycle can exceed when the same amount of input energy is provided and when the working substance is made up of only one constituent. Thus, this family of externally reversible heat engine cycles (which includes the Stirling [1], Ericsson [2], Rallis [3] and Carnot [4] cycles) has long been used as a yardstick against which other practical cycles have been compared (in relation to both their work potentials and efficiencies). However, in practice engineers are also concerned with the rate at which work can be generated.

This theory of irreversible thermodynamics for reciprocating externally- irreversible internally-reversible cycles gives rise to an optimum efficiency at maximum power output of $\eta = 1 - (T_L/T_H)^{0.5}$ for internally reversible Stirling-like cycles with ideal regeneration and using linear heat transfer modes [1,2,4] in contrast to the upper limit for such Stirling-like cycles of $\eta = 1-(T_L/T_H)$ obtained from classical thermodynamics. This optimum behavior is, however, only based on cycle temperature bounds.

This result varies when other modes of heat transfer (besides linear modes) are involved in the cycle. Work to quantify the efficiency of cycles operating under various combinations of heat transfer modes and with (or without) leaks has been considered by others [5-10]. However, all of these works [5-10] focus on continuous cycles only and do not give consideration to the optimum allocation of the high and low end heat capacitances. Also, all of these works consider only high and low temperature heat reservoirs having infinite thermal capacities. Other workers have considered heat engine cycles with finite thermal reservoirs [11-13], but not in mixed mode heat transfer situations.

Gyftopoulos [14] argues that for the analysis of a thermal process to be valid it must make simultaneous use of both the first and second laws of thermodynamics. Along these same lines, Bejan [15] further argues that to maximize the power of a thermal heat engine, one must minimize the internal entropy generation of the cycle. In the present work, the optimal allocation of conductances in the hot and cold ends of the cycle is carried out through a cycle time minimization procedure which has been shown to simultaneously minimize internal entropy generation (subject to operational realities) while at the same time maximizing specific work for the cycle between the optimized operating temperatures at the hot and cold ends. Additionally, this has the advantage of creating as the initial starting point (for the optimization process) an expression for power which itself represents a concurrent expression of the first and second laws of thermodynamics. Thus the present work conforms with the observations of both Gyftopoulos [14] and Bejan [15] for the case of the general Stirling-like heat engine and seeks to obtain universal expressions for the optimum performance parameters of this family of cycles subject to the most general case of operational modes. After the general formulation is given, a number of practical cases are examined. The focus of the work throughout will be power optimization. Expressions for power optimized parameters are sought which consider the proper allocation of heat conductances and which are thus independent of heat conductances.

In practice, neither optimum power nor optimum efficiency can be the sole objective of an energy conversion device. Some compromise between these two limits must be adopted. A better understanding of each of these limiting cases is thus essential to the multi-objective decision approach required to provide a better compromise between the two for a given application. The purpose of this study is to give major new insight, from a power perspective, into the general behavior of a family of practical Stirling-like cycles which include the Stirling, the Ericsson and the Rallis power cycles. As the second Emit has been well studied for some two centuries, it is now time to study the first limit for various cycles in much more detail.

THE EXTERNALLY IRREVERSIBLE, INTERNALLY REVERSIBLE FAMILY OF STIRLING CYCLE HEAT ENGINES AND ITS POWER OPTIMIZATION

The temperature-specific entropy diagram of a general reciprocating Stirling-like cycle with regeneration and finite heat reservoirs is depicted in Fig. 1. Since the external heat transfer processes 2-3 and 4-1 occur in finite time across finite temperature differences, they (2-3 and 4-1) are irreversible isothermal processes. During the heat addition process, 2-3, the finite high temperature reservoir may experience a temperature drop of T_5 - T_6. Similarly, during the heat removal process, 4-1, the low temperature reservoir may experience a temperature rise of T_8 - T_7. The cycle is thus bounded by the temperatures $T_H = T_5$ and $T_L = T_8$. On the other hand, 1-2 and 3-4 are assumed to be reversible polytropic processes. Thus, the cycle is externally irreversible, but internally reversible (endoreversible). In the analysis which follows, the working substance is assumed to be in a gaseous phase throughout the cycle.

For simplicity the regeneration process can be assumed ideal, with the heat that is rejected to the regenerator by the hot working fluid during process 3-4 (3q4) being equivalent to the heat that is supplied to the cold working fluid during process 1-2 (1q2). This assumption is most reasonable, with the efficiency of regenerators continuing to improve and with values of 98-99% already being reported [1]. Thus, the overall ideal regenerative heat transfer process can be given by :

$$\left|{}_1q_2\right| = \left|{}_3q_4\right| \quad (= 0 \text{ for Carnot}) \tag{1}$$

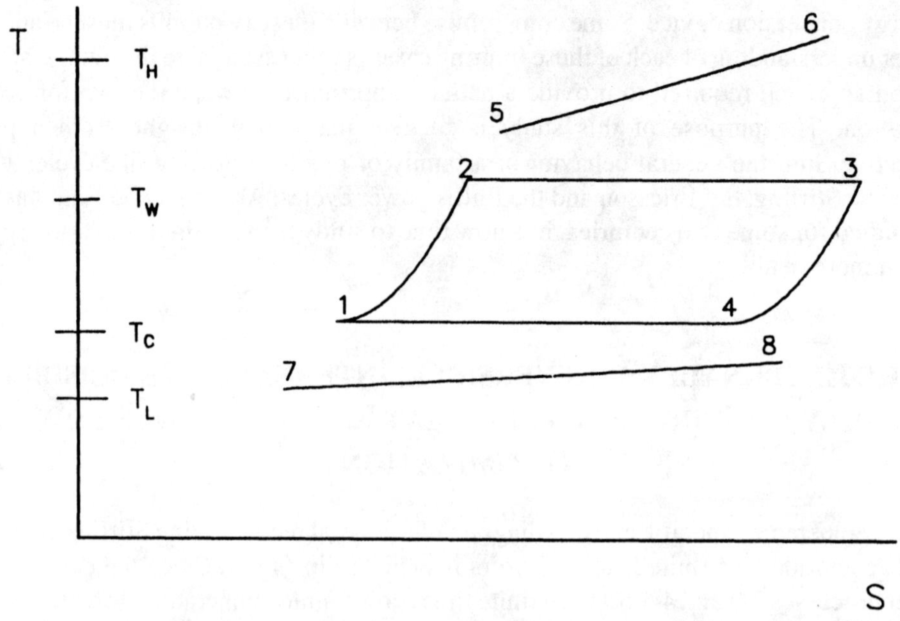

Figure 1. The General Reciprocating Stirling-like Cycle with Finite Reservoir Heat Capacities

Since the regenerator supplies the thermal energy which is transferred to the working fluid during process 1-2, the only source of external heat addition to the cycle is provided by the high temperature heat source during process 2-3. Similarly, the only assumed heat rejection from the cycle occurs during process 4-1. Thus, if the gas working fluid within the heat engine has a mass m, the magnitude of the net work for this finite-time cycle can be expressed as :

$$Wnet = m\Delta s\ (Tw - R_{\Delta s}\ Tc)(= \oint \delta Q)$$

where Δs is equal to $(s_3 - s_2)$ and where $R\Delta_s$ is the ratio $(s_4\text{-}s_1)/\Delta s$. This is the net work expression for a cycle which is externally irreversible. Note that, in contrast for the case of infinite thermal reservoirs at both the hot and cold ends of the cycle ($T_H = T_5 = T_6$ and $T_L = T_7 = T_8$) the net work output for an externally reversible cycle occurring between these temperature bounds (T_H and T_L) is given by : $(W_{net}, e_{rev})i_{irr} = m\Delta s\ (T_H - R_{\Delta s}\ T_L)$ where the subscripts 'i_{irr}' and 'e_{rev}' denotes the cycle to be externally reversible and internally irreversible, respectively. This is because of the fact that for the external heat transfer processes to be reversible, T_W must in the limit equal T_H and T_C must in the limit equal T_L.

However, as discussed in the Appendix, for Eq. (1) to bold, $R\Delta_s$ must be equal to 1.0 (for all but the Carnot cycle). Thus, for the case of ideal regeneration,

$$W_{net} = m\Delta s(T_W - T_C) \qquad (2)$$

and

$$W_{net,rev} = m\Delta s\,(T_H - T_L)\ (\equiv W_{rev},\ \text{Carnot work}) \tag{3}$$

If the polytropic coefficient for the expansion and compression process are n = k (isentropic processes) and $R\Delta_s$ equals one, the cycle reduces to the internally and externally reversible Carnot cycle and the work expression given by Eq. (3) to Carnot work (W_{rev}). Thus, for the condition of infinite thermal reservoirs, the internally and externally reversible versions of this entire family of cycles are all capable of Carnot work (W_{rev}). For such cycles the classical Carnot thermal efficiency applies. Herein, practical cycles capable attaining the Carnot efficiency have thus been termed Stirling-like. However, with power as the focus, these external heat processes must occur in finite time and thus, neither can T_W equal T_H, nor can T_C equal T_L. Moreover, practical reciprocating cycles will not normally have infinite thermal reservoirs and in general T_6, will be less than T_5 and T_8 will be greater that T_7.

A clear distinction must be made between continuous and reciprocating heat engines. In the latter, the cycle is not simultaneously engaged in the four main activities of the heat engine (heat addition, expansion, heat rejection and compression). Instead, these activities must be viewed as happening one at a time. In the case of continuous cycles, these four processes do occur simultaneously. Thus, for the general reciprocating Stirling-like power cycle with ideal regeneration, the average net cycle power can be computed by dividing the net cycle work, provided by Eq. (2), by the time for one cycle, t.

$$P = W_{net}/t = m\Delta s(T_W - T_C)/(t_{12} + t_{23} + t_{34} + t_{41}) \tag{4}$$

Where the individual constant temperature process times can "es own to be

$$t_{23} = Q_L/\dot{Q}_L = mT_C\Delta s/K_L\textstyle\sum_i a_{L,i}(\Delta T)_{L,i} \tag{5a}$$

$$t_{41} = Q_H/\dot{Q}_H = mT_W\Delta s/K_H\textstyle\sum_j a_{H,j}(\Delta T)_{H,j} \tag{5b}$$

Here, when only one mode of heat transfer is present on each of the thermal ends of the cycle, K_L and K_H are the heat conductances between the working substance and either the low temperature sink or the high temperature source, respectively, and the values of the weighting factors a_L and a_H are set a 1.0. Definitions of the weighting factors a_L and a_H and the terms $(\Delta T)_{L,i}$ and $(\Delta T)_{H,j}$ are given in Table 1 for various modes of heat transfer.

If processes 2-3 and 4-1 (between the two internal temperature limits T_W and T_C) are generalized as internally reversible polytropic processes (general polytropic processes), the process times for these can be specified as [1,2]:

$$t_{12} = t_{34} = mC_v(k-n)(T_W-T_C)/[(1-n)K_{reg}(LMTD)_{reg,ideal}] = mC_n(T_W-T_C)/K_{reg} \tag{6}$$

with K_{reg} as the heat conductance between the endoreversible cycle and the regenerator and subject to the restriction that n, the general index of expansion /compression, cannot be in the range $1 < n < k$. Here the general polytropic process specific heat is given by $C_n = C_v(k-n)/(1-n)$, where k is the ratio of the specific heat at constant pressure (C_p) and the specific heat at constant volume (C_v). In equation (6) LMTD = Log Mean Temperature Difference = $[(T_W-T_{C1})-(T_{W1}-T_C)]/\ln[(T_W-T_{C1})/(T_{W1}-T_C)]$, where T_C and T_{C1} are the temperatures of the cold working fluid inlet and outlet, respectively; T_W and T_{W1} are the temperatures of the warm working fluid inlet and outlet, respectively. It should be noted that a regenerator effectiveness of 100% implies that the outlet temperatures of the heat exchanger will equal the inlet temperatures ($T_{C1} = T_W$ and $T_{W1} = T_C$). Assuming ideal regeneration through a counter-flow heat exchanger, it can be shown [1,2] that the $(LMTD)_{reg,ideal} = 1$ K. Therefore, for the case of an ideal regenerator, the denominator in equation (6) is a constant and the time for the processes represented by this expression depends only on the quantity of heat being transferred by these processes. Note that the expression 1 K is dropped in eq. (6) and hereafter, though the temperature unit K is understood to be present.

Equations (5-6) can in turn be substituted into Eq. (4), yielding an expression for the power which is based upon reservoir and isothermal working fluid temperatures,

$$P=[T_W/[K_H(\Sigma_j a_{H,j}(\Delta T)_{H,j})(T_W-T_C)]+\beta Tc/[K_H(\Sigma_i a_{L,i}(\Delta T)_{L,i})(T_W-T_C)]+2C_n/R\ln r_v K_{reg}]^{-1} \quad (7)$$

where $b = (K_H/K_L)$,$R\ln r_v = \Delta s$, R is the gas constant and r_v the compression ratio ($v_3/v_2 = p_3/p_2$). Therefore for a given K_H value, power is a function of T_W, T_C and the capacitance ratio b and the optimization of power must be with respect to these three variables.

To obtain the optimum power for the cycle, one must minimize the time required to complete each cycle for a given amount of work output. For reciprocating Stirling-like cycles, minimum cycle time requires that the mechanical phase times of process 2-3 and process 4-1 be equal [2,4]. In other words, for minimum cycle time, there must be time symmetry for the external heat transfer processes. This also follows from the nature of the balanced cycle motion required for reciprocating engine cycles. Thus for the power optimized case, $t_{23} = t_{41}$. Also one must find the operating temperatures (T_C and T_W) which give the maximum net work rate. The first of these two steps mentioned is novel with this work and when combined with the second step provides a more complete optimum than previous works. Thus the above expression for power (Eq. (6)) can be maximized for both cycle time (t) and temperature bounds (T_C and T_W) by simultaneously demanding that:

$$\beta = K_H/K_L = T_W(\Sigma_i a_{L,i}(\Delta T)_{L,i})/T_C(\Sigma_j a_{H,j}(\Delta T)_{H,j}) \quad (8)$$

$$(\partial P/\partial T_W)_{TC,\beta} = 0 \quad (9a)$$

and

$$(\partial P/\partial T_C)_{TW,\beta} = 0 \tag{9b}$$

It should be noted here that though Eq. (8) was not derived from the second law of thermodynamics, it is in fact a statement of the rate form of the second law of thermodynamics when the entropy generation is minimum (zero). Following the arguments by Bejan [15], use of this expression must both imply and ensure the maximization of specific work for the cycle operating between two constant temperatures. The work expression used to obtain Eq. (7) is in fact optimized for specific work. Additionally, it must be noted that the application of Eq.(8) to Eq. (7) results in the concurrent employment of the first and second laws [14] as the starting point for the power optimization procedure. Thus, because of the specification for b demanded by Eq. (8), even before cycle power is optimized with respect to cycle operating temperatures, the formulation for power (Eq. (7)) is made to include both laws and to give maximum specific work for the internal temperature bounds T_W and T_C.

If we let:

$$f = K_L\,(\Sigma a_{L,i}(\Delta T)_{L,i}) \tag{10}$$

and

$$g = K_H\,(\Sigma a_{H,j}(\Delta T)_{H,j}) \tag{11}$$

we can simplify our power expression (Eq.(7)) to $P=(T_W-T_C)/[(T_W/g)+(T_C/f)+(C(T_W-T_C))]$, where $C = 2C_n/R\ln r_v\ K_{reg}$. Applying Eq. (8), along with Eq. (9a) to the aforementioned equation for P yields

$$f/g = [(T_W)_{opt}\,g' + g]/[(T_W)_{opt}\,g' - g] \equiv G(T_W)/g \tag{12}$$

where $g' = ag/aT_C$. Similarly, applying Eq. (8) along with Eq. (9b) yields

$$g/f = [(T_C)_{opt}\,f' + f]/[(T_C)_{opt}\,f' - f] \equiv F(T_C)/f \tag{13}$$

where $f' = \partial f/\partial T_C$. With these expressions (Eqs. (12-13) we can now find general isolated solutions for $(T_W)_{opt}$ and $(T_C)_{opt}$ if we in addition note that:

$$T_C = f^{-1}(T_C) \equiv T_C(G) \tag{14}$$

and

$$T_W = g^{-1}(T_W) \equiv T_W(F) \tag{15}$$

and

$$(T_C)_{opt} = \{(3+\beta)T_L + [8(\beta+1)\beta T_H T_L + (1-\beta)^2 T_L^2]^{1/2}\}/[4(\beta+1)].$$

respectively. As before, these power optimized temperature values are only coupled through Eq. (8) and can thus be used to study the effects on the cycle of using non-optimized values for one of these operating temperatures. If these expressions instead are solved simultaneously with Eq. (18) (with n = m = 1), the resultant infinite source and sink linear solution becomes

$$(T_C)_{opt} = 0.5((T_H T_L)^{1/2} + T_L)\,,$$

$$(T_W)_{opt} = 0.5((T_H T_L)^{1/2} + T_H)\,,$$

and

$$b_{opt} = 1.0.$$

Case 2

General Mode with an Infinite Source and a Linear Mode with a Finite Sink(T_H, T_7, T_8 and K_H prespecified)

For this case:

$$f = K_L(LMTD_L) = (T_8 - T_7)/\ln[(T_C - T_7)/(T_C - T_8)]\ (= G) \tag{23}$$

and

$$g = K_H(T_H^n - T_W^n)(= F) \tag{20}$$

Proceeding as in the previous case we find

$$T_C(G) = (e^y T_8 - T_7)/(e^y - 1)$$

and

$$T_W(F) = [T_H^n - ((T_C)_{opt} f' + f)/((T_C)_{opt} f' - f) f / K_H]^{1/n}$$

where

$$y = (T_8-T_7)/\{\beta[(T_H^n - (T_W)_{opt}^n][(T_W)_{opt}\,g'+g]/[(T_W)_{opt}\,g'-g]\}$$

$$g' = -nK_H T_W^{n-1},$$

and

$$f' = \partial f/\partial T_C = K_L(T_8-T_7)^2/\{(T_C-T_7)(T_C-T_8)[\ln(T_C-T_7)/(T_C-T_8)]\}^2$$

Finally, substituting these expressions into Eqs. (16) and (17) yields:

$$(T_W)_{opt} = \{[(e^y T_8-T_7)/(e^y-1)][(T_W)_{opt}\,g'-g]/[(T_W)_{opt}g'+g]\}_{opt} \tag{24}$$

and

$$(T_C)_{opt}=\{[T_H^n-[(T_C)_{opt}f'+f]/[(T_C)_{opt}f'-f](LMTD_L)/\beta]^{1/n}[(T_C)_{opt}f'-f]/[(T_C)_{opt}f'+f)\}_{opt} \tag{25}$$

If the temperature of one of these Eqs. [(24 and (25)] is pre-specified, the other can be solved in conjunction with Eq. (8). If power is optimized with both of these temperatures (neither pre-specified) then Eqs. (24) and (25) must be solved simultaneously with Eq. (18).

Case 3

Black Body Radiation and Linear Modes with an Infinite Source and a Linear Mode with a Infinite Sink(T_H, T_7, T_8 and K_H pre-specified):

For this third case:

$$f = K_L(T_C-T_L)(=G) \tag{26}$$

and

$$g = K_H[(T_H-T_W) +a_{H,rad}(T_H^4-T_W^4)] (= F) \tag{27}$$

where for black body radiation exchange between two beat exchange surfaces $a_{H,rad} = \sigma_k/U_k$. Proceeding as in the previous case we find

$$Tc\,(G) = T_L +g[(T_W)_{opt}g' +g]/[(T_W)_{opt}g'-g)K_L]$$

and

$$Tw\ (F) = \{T_H(1 + a_{H,rad}T_H^3) - (2T_C - T_L)(T_C - T_L)/[\beta T_L]/(1 + a_{H,rad}T_W^3)$$

where

$$g' = -K_H[1 + 4a_{H,rad}T_W^3]$$

Finally substituting these expressions into Eqs. (16) and (17) yields:

$$(T_W)_{opt} = \{[T_L + K_H\beta g[(T_W)_{opt}g' + g][(T_W)_{opt}g' - g]/(T_W)_{opt}g' - g)][(T_W)_{opt}g' + g])_{opt} \quad (28)$$

and

$$(T_C)_{opt} = \{\{[T_H(1 + a_{H,rad}T_H^3) - [(2T_C - T_L)(T_C - T_L)/\beta T_L][T_L/(2T_C - T_L)]\}/(1 + a_{H,rad}T_W^3)\}_{opt} \quad (29)$$

Evaluation of The Work, Power And Efficiency Potentials For Various Heat Transfer Configurations of The Family of Power Optimized Stirling Cycle Heat Engines

The above results for power optimum operating temperatures are universal for the entire family of Stirling-like cycles. Using these results, analysis of the work and efficiency potential of all members of this family can be conducted without preference to which individual family member is being considered. In other words, the Work and efficiency potential of all members will be the same for the same set of -temperature bounds, K_H and heat transfer modes. Only the power potentials of the individual cycles will vary. The temperature bounds used to compare different subcases of the cases given in the previous section are as follows: $1000K < T_H = T_5 \leq 1500K$, $T_L = T_7 = 300K$, $T_6 = T_5 - 30K$ and $T_8 = T_7 + 15K$. Calculations for power give in the second subsection for the Stirling and Ericsson cycles will all use the same A_H, m, r_v, R, K_{reg} and C_v values of 1.0 m^2 ,9.3x 10^{-3} kg, 2.0, 0.287 kJ/kg-K, 1000 kJ/K-s and 0.7165 kJ/kg-K, respectively. [1,2] A K_H value of 3.5 kJ/K-s is used whenever a linear mode is considered as one of the modes at the heat source and for comparison purposes a K_H value of $(3.5kJ/K\text{-}s)/[(T_H^2 + T_W^2)(T_H + T_W)]$ is used when considering only radiation exchange with the heat source.

POWER OPTIMIZED WORK POTENTIAL

The work per cycle at maximum power, or power optimized work, can now be determined by substituting the optimum working fluid temperature values, eqs.(16) and (19), into Eq. (2), resulting in :

$$W_{opt} = m\Delta s((T_W)_{opt}-(T_C)_{opt}) \tag{30}$$

for a variety of subcases contained within the cases given in the previous section. In general, $(W_{rev})_{eff} = m\Delta s((T_H)_{eff} - (T_L)_{eff})$ When, both the high and low temperature reservoirs are infinite, $(T_H)_{eff} = T_H$, $(T_L)_{eff} = T_L$ and $(W_{rev})_{eff} = W_{rev}$ given by Eq. (3). However, if the heat source is finite with linear heat transfer, then $(T_H)_{eff} = (T_W)_{opt} + (LMTD_H)_{opt}$. Similarly, if the heat sink is finite with linear heat transfer, $(T_L)_{eff} = (T_c)_{opt} - (LMTD_L)_{opt}$.

For Case 1 of the previous section four subcases were conducted. In Subcase 1a: m = n = 1 and $W_{opt}/(W_{rev})_{eff}$ =1/2 for all values of T_H. In Subcase 1b: m = n = 4 and and in Subcase 1c: m = n = -1. For both these subcases $W_{opt}/(W_{rev})_{eff} < ½$ for all values of T_H.. However, in Subcase 1d: m = 1 and n = 4 and the ratio $W_{opt}/(W_{rev})_{eff} > 1/2$ for all values of T_H.

Thus when the same mode of heat transfer is used at each thermal end of the cycle, the linear mode choice is capable of the highest work potential. Of all the possible combinations of using different single modes at the thermal ends of the cycle, Subcase 1d has the highest work potential. Subsequent subcase studies reveal that Subcase ld has the overall highest work potential of any case possible.

For Case 2 of the previous section only two subcases were deemed useful. In Subcase 2a: n = 1 and in Subcase2b: n = 4. Though their values are extremely close (Fig. 2) it was found that each of these has slightly less work potential than their respective infinite sink counterparts, Subcases 1a and 1d..

Case 3, which represents a merger of case 1a and 1d, is shown in Fig. 2 to be capable of exceeding the work potential of case la but not 1d. In practice this represents the most realistic of the cases formulated in the previous section. For the uniformity of Fig. 2 we have considered Case 3 to be its own subcase (Subcase 3).

However, an even more practical situation, which we call Case 4, is where the mixed mode formulation of Case 3 is present at the high temperature thermal reservoir, and where a finite sink is present at the low end. For this case we will consider 2 subcases. In Subcase 4a, as in all of the preceding subcases, both operating temperatures (T_w and T_c) are power optimum. In Subcase 4b however, a prespecified value of T_C is used which is chosen in advance to be exactly two thirds of the way between T_7 and the $(T_c)_{opt}$ found in Subcase 4a. This final and very realistic case, though not power optimized, will be shown to have the highest of the thermal efficiencies demonstrated in this study, and is seen to have a higher work potential than all subcases except Subcase ld and Subcase 2b.

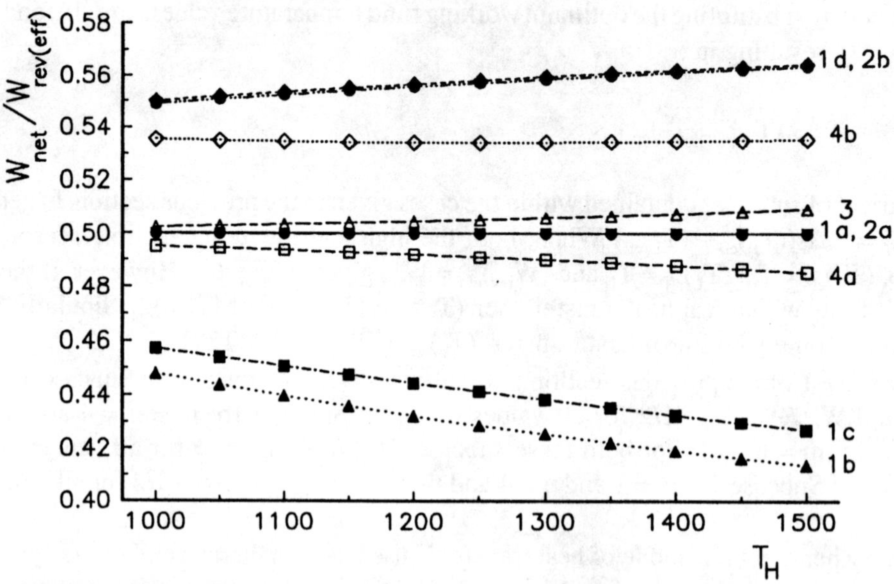

Figure 2. $\frac{W_{opt}}{(W_{rev})_{eff}}$ work ratio versus $T_H\,(K)$ for the Stirling-like cycle for various subcases

Finally, it should be noted that in subcases where only linear modes heat transfer modes were used at both ends (with either finite of infinite reservoirs) $W_{opt} < 1/2\ (W_{rev})_{eff} \leq 1/2\ W_{rev}$.

Power Optimized Efficiency and Power Potentials

Independent of the choice of heat transfer modes, the thermal efficiency for the internally reversible cycle at the condition of maximum power (optimized with respect to cycle time as well as temperature bounds) is:

$$\eta_{opt} = W_{opt}/(Q_{in})_{opt} = 1-[(T_C)_{opt}/(T_W)_{opt}] \tag{31}$$

because,

$$Q_{in} = m\Delta s T_W$$

and where $(T_c)_{opt}$ and $(T_W)_{opt}$ are given in the previous major section for the various cases. The nature of this equation is plotted in Fig. 3 versus T_H for some of the same subcases as shown in Fig. 2. For Subcase 1a (m = n = 1) (case with linear heat transfer modes and infinite thermal reservoirs at both ends of the cycle) the ratio $(T_c)_{opt}/(T_W)_{opt}$ simplifies, and

$$\eta_{opt} = 1-[T_L/T_H]^{0.5}$$

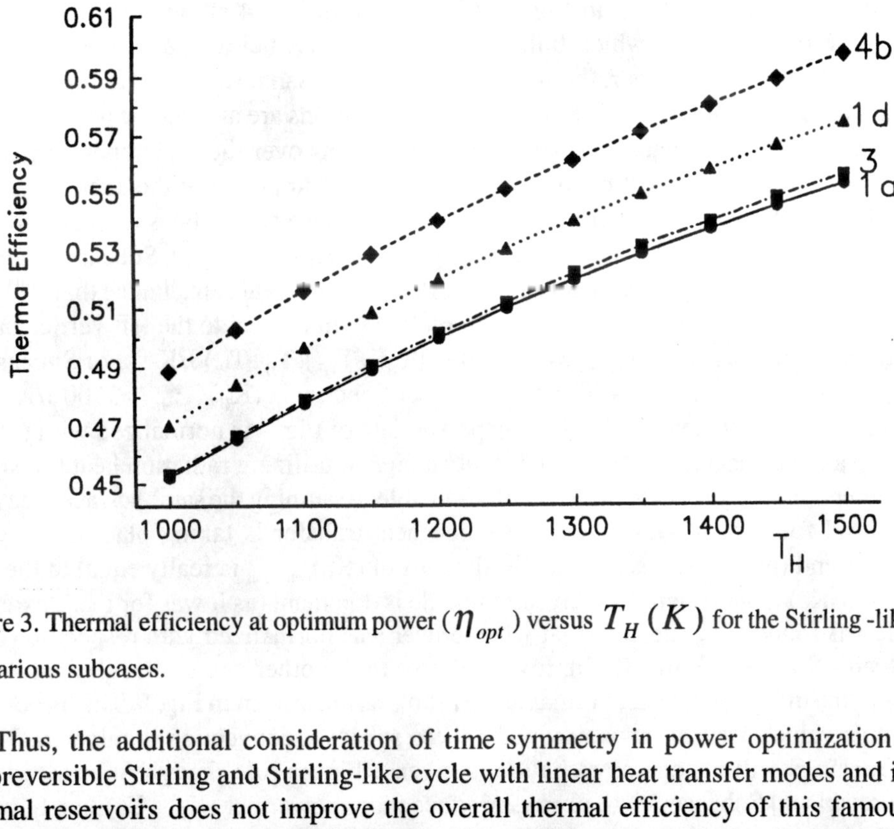

Figure 3. Thermal efficiency at optimum power (η_{opt}) versus T_H (K) for the Stirling -like cycle for various subcases.

Thus, the additional consideration of time symmetry in power optimization of the endoreversible Stirling and Stirling-like cycle with linear heat transfer modes and infinite thermal reservoirs does not improve the overall thermal efficiency of this famous case (Subcase 1a) from the Curzon and Ahlborn [4] result. As expected, the thermal efficiency of Subcase 1d, with radiation heat transfer only at the source and linear at the sink exceeds all of the power optimized cases. Subcase 4a represents the most practical and operationally feasible of the cases considered when the constant temperature operating values are power optimum at both thermal ends of the cycle. However, when Subcase 4a is modified so that a T_C is prespecified which is less than $(T_c)_{opt}$ [and $(Tw)_{opt}$ is still used], it can be seen through Subcase 4b that an efficiency which is higher than that for Subcase 1d can be obtained. As discussed later, this improved efficiency comes at the expense of the power per total capacitance ($(K_H)_{conv+rad} + K_L$) ratio.

Using the $(T_c)_{opt}$ and $(T_W)_{opt}$ values of the previous section, the expression for optimum power becomes:

$$P=\{(T_W/[K_H(\Sigma_j a_{H,j}(\Delta T)_{H,j})(T_W-T_C)]+\beta T_C/[K_H(\Sigma_i a_{L,i})(T_W-T_C)]+2C_n/[R\ln r_v K_{reg}]\}_{opt}^{-1} \quad (32)$$

where$\{\}_{opt}$ indicates that all (T_C) and (T_W) terms within any of the expressions and sub-expressions between the brackets have been replaced by $(T_C)_{opt}$ and $(T_W)_{opt}$. Close examination of Eq. (32) indicates that the optimum power potential varies with the member of the Stirling-like family considered. For representative Subcases 1a, 1d and 4b, a comparison of the power potentials of the Ericsson and Stirling power cycles have been computed and the results given in Fig. 4. Also shown in Fig. 4 is the power plot for the Stirling cycle for Subcase 3 which falls an average of 4.0% below the Stirling cycle plot shown for Subcase 4b. However, though maintained at the same K_H values for comparison, the computed K_L for Subcase 4b for Stirling cycle operations are an average approximately 51.0% higher than for Subcase 3 Stirling cycle operations over the T_H temperature range. Thus, for the same K_H the cost of improving efficiency and power at the same time is an increased K_L. Yet, when the Stirling cycle results for these two subcases (3 and 4b) are normalized with respect to total capacitance usage $[K_{tot}=(K_H)_{conv+rad}+K_L]$, Subcase 3 is found to produce an average of approximately 18% more power per total capacitance than Subcase 4b (as shown in data of Fig. 5 which has been plotted with respect to the left vertical axis), where for these two cases $(K_H)_{conv+rad}=3500+\sigma A_H(T_H^2+T_W^2)(T_H+T_W)$ J/K-s. For Subcase 1d $(K_H)_{conv+rad} =sA_H(T_H^2 + T_W^2)(T_H + T_W)$ J/K-s and for Subcase la $(K_H)_{conv+rad} =3500$ J/K-s. As can be seen in Fig. 5, it is when the Stirling power data of Fig. 4 is normalized with respect to total capacitance usage that the decisive advantage of utilizing radiation heat transfer at the high temperature side of the cycle becomes visible. Assuming the same surface area (A_H) is being used for both modes when two mode heat transfer is taking place at the high temperature end (as in Subcases 3 and 4b) the cost of $(K_H)_{conv+rad}$ is really equal to the cost of)$(K_H)_{conv}$ $(=K_H)$ alone when the convection mode is dominant (as it was for these example cases). It was found (as expected) that when power was normalized with respect to $(K_H + K_L)$ instead - Subcases 3 and 4b improved relative to the other cases, though the relative rankings of the different subcases remained the same as those seen in Fig. 5. This is because the example values used in this study make these subcases convection dominated. These relative rankings will change as radiation is given a bigger role though Subcase ld will always prevail and Subcase 3 will always produce more power per $(K_H + K_L)$than Subcase 4b. Also, as shown in Fig. 5 (from the data plotted against the right vertical axis), relative subcase rankings remain the same when the Stirling power curves of Fig. 4 are normalized with respect to Q_H instead. Thus, Subcase ld provides more power per total capacitance, more power per $(K_H + K_L)$ and more power per rate of heat addition than any of these subcases (and all the other subcases considered in this study).

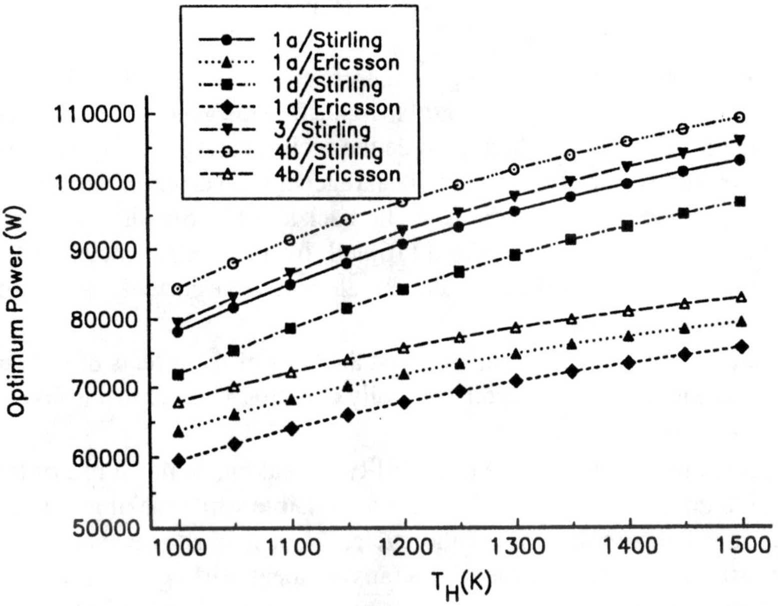

Figure 4. Comparison of the Power Potential versus TH (K) of Reciprocating Stirling and Ericsson Cycles for 4 Different Subcases.

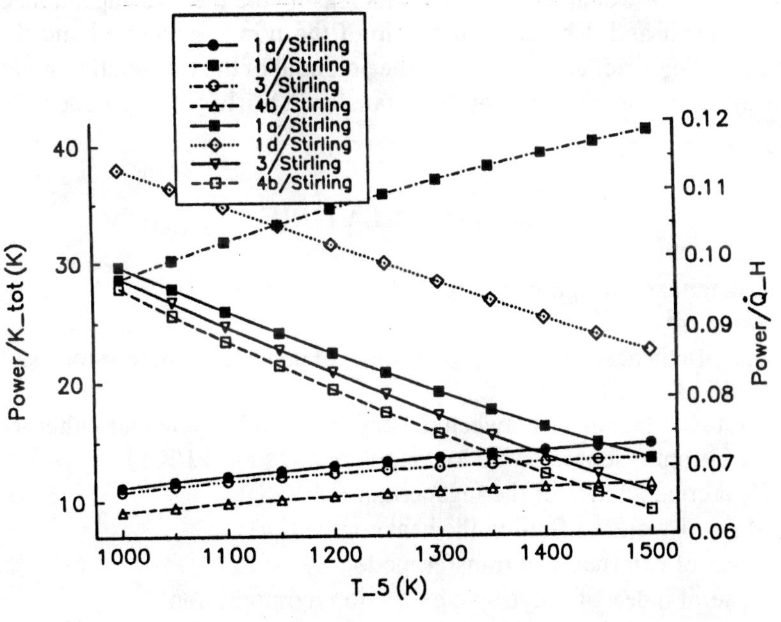

Figure 5. Comparison of the Power Potential per Total Capacitance (K_{tot})(K)(left) and per Rate of Heat Addition (Q) (right) versus TH (K) for 4 Different Subcases of Reciprocating Stirling Cycles.

CONCLUSIONS

Though the formulation and results given herein concern cycles with ideal regeneration, they do provide us with new and meaningful insights into the general behavior of the family of practical Stirling-like cycles which include the reciprocating Stirling and Ericsson heat engine cycles. Work to more closely attain ideal regeneration is one of the design objectives currently being pursued for these engines. As far back as 1972, Stirling engine regeneration efficiencies of 95 % were being reported [16]and, by 1992, efficiencies of 98-99% bad already been attained [1]. Thus, these results are relevant to the current state of technological development for these engines.

This fundamental study represents the first analysis of the effects of various practical combinations of heat transfer modes on the family of reciprocating Stirling-like cycles from the viewpoint of power optimization.

Results point conclusively to the desirability of making as much use of the radiation mode at the high temperature end of the cycle as possible while making use of only linear modes at the low temperature end of the cycle. Also, it is shown that under the same operating temperature conditions, the regeneratively ideal Stirling engine is capable of the same thermal efficiency and $W_{opt}/(W_{rev})_{eff}$ ratio as the regeneratively ideal Ericsson engine, but can give more power. Finally, it is seen for a fixed linear thermal capacitance value at the high temperature end of the cycle (K_H) that, when one is willing to consider some trade-off between work and work-rate by lowering the operating temperature of T_C to a value lower than $(T_c)_{opt}$, the $W_{opt}/(W_{rev})_{eff}$ ratio and thermal efficiency can be substantially increased along with power at the expense of increasing the thermal capacitance of the low temperature reservoir and decreasing the ratio of the power generated and the total beat capacitance used. This brief analysis shows the potential of the formulation given herein for use in the study of how to best employ beat transfer in Stirling-like engines.

NOMENCLATURE

a_i	Capacitance weighting factor
A	Area (m^2)
C_n	Specific heat for a general polytropic expansion/ compression (kJ/kg-K)
k	Ratio of specific heats, C_p/C_v
K_H, K_L	Heat Conductances between the endoreversible cycle and either the high temperature source or the low temperature sink (kJ/K-s)
K_{reg}	Heat conductance of the regenerator (kJ/K-S)
m	Mass of working fluid in the heat engine (kg)
m, n	Index of external heat transfer mode
n	General index of polytropic expansion /compression
p	pressure (N/m^2)
P, P_{max}	Net and maximum cycle power output, respectively (kJ/s)
q	Heat transfer per unit mass (kJ/kg)

Q	Heat transfer (kJ)
Q	Rate of heat transfer (kJ/s)
r_v	Compression ratio
$R_{\Delta s}$	Irreversibility factor
s, S	Specific entropy and entropy, respectively (kJ/K-kg,kJ/K)
T	Temperature (K)
T_C, T_W	Temperature of the working fluid during the isothermal external heat rejection and addition processes, respectively (K)
T_H, T_L	Temperature of the heat source and the heat sink, respectively (K)
t	Total cycle time (s)
t_{ij}	Time required to accomplish process i-j (s)
U	Overall heat transfer coefficient (kJ/K-S-m^2)
V	specific volume (m^3/kg)
W_{net}	Net work output for one cycle (kJ)
W_{rev}	Net work output per cycle for an externally and internally reversible cycle (kJ)
$(W_{rev})_{eff}$	Net work output per cycle for an externally and internally reversible cycle based on the mean temperature of the thermal reservoirs (W)
b	Heat conductance ratio
h_{opt}	Cycle thermal efficiency at the optimum power condition
s	Stefan- Boltzmann constant
g	surface emissivity

REFERENCES

1.D. A. Blank, G.W. Davis, and C. Wu, Power Optimization of an Endoreversible Stirling Cycle with Regeneration, *Energy*, 1994, 19: 125-133.

2. D. A. Blank and C. Wu, Power Limit of an Endoreversible Ericsson Cycle with Regeneration, *Energy Convers & Mgmt*, 1996, 37: 59-66.

3. C. Wu, Analysis of an Endoreversible Rallis Cycle, *Energy Convers & Mgmt*, 1994, 35: 79-811.

4. F Curzon and B. Ahlborn, Efficiency of a Carnot Engine at Maximum Power Output, *Am. J. Phys*., 1975, 43: 22-24.

5. A. Bejan, Theory of Heat Transfer - Irreversible Power Plants, *Int. J. Heat Mass Transfer*, 1988, 31, 1211-1219.

6. C. WU, Power optimization of a Finite-Time Carnot Heat Engine , *Energy*, 1988, 13: 681-687.

7. B. Andresen and J. M. Gordon, Optimal Paths for Minimizing Entroy Generation in a Common Class of Finite-Time Heating and Cooling Processes, *Int. J. Heat and Fluid Flow*, 1992, 13: 294-299.

8. G. DeMey and A. DeVos, On the Optimum Efficiency of Endoreversible Thermodynamic Processes', *J. Phys. D: Appl. Phys*., 1994, 27: 736-739.

9. A. Bejan, Theory of Heat Transfer - Irreversible Power Plants - II. The Optimal Allocation of Heat Exchange Equipment, *Int. J. Heat Mass Transfer*, 1995, 38: 433-444.
10. F. Moukalled, R.Y. Nuwayhid and S. Fattal, A Universal Model for Studying the Performance of Carnot-Like Engines a Maximum Power Conditions, *Int. J. of Energy Research*, 1996, 20: 203-214.
11. C. Wu, Power Optimization of a Finite-Time Rankine Heat Engine, *Int. J Heat Fluid Flow*, 1989, 10 N 2: 134-138.
12. W. Lee and S. Kim, Finite-Time Optimizations of a Heat Engine, *Energy*, (1990), 15: 979-985.
13. D. A. Blank and M. Maholtra, Finite Thermal Reservoirs Effects on Power Optimized Continuous Endoreversible Carnot Heat Engine Cycles, *Int. J. Power and Energy System*, being published, 1997.
14. E.P. Gyftopoulos, Fundamentals of Analyses of Processes, *ECOS'96*, Stockholm 1996.
15. A. Bejan, Models of Power Plants that Generate Minimum Entropy while Operating at Maximum Power, *Am. J. Phys.*, 1996, 64: 1054-1059.
16. R.U. Ayres and R.P. Mckenna, Alternatives to the Internal Combustion Engine, The John Hopkins Univ. Press, 1972: 180-181.

Table 1- Thermal Capacitances, Effective Temperature Heat Transfer Potentials and Mixed Mode Coefficients for Various Modes of Heat Transfer

Heat Transfer Mode	K_k	$(\Delta T)k$	a_k for mixed mode
linear with infinite reservoir	U_kA_k	L: $T_C - T_L$ H: $T_H - T_W$	1.0
linear with finite reservoir	U_kA_k	L: $LMTD_L=(T_8-T_7)/\ln[(T_C-T_7)/(T_C-T_8)]$ H: $LMTD_H=(T_5-T_6)/\ln[(T_5-T_W)/(T_6-T_W)]$	1.0
radiation with 2 black walls	σA_k	L: $(T_C)^4-(T_L)^4$ H: $(T_H)^4-(T_W)^4$	σ/U_k
radiation with 2 gray walls	$\sigma A_k/[1/\epsilon_k+1/\epsilon_l-1)$	L: $(T_C)^4-(T_L)^4$ H: $(T_H)^4-(T_W)^4$	$\sigma/\{U_k[1/\epsilon_k+1/\epsilon_l-1)\}$

Finite-time Thermodynamics and Generalized Thermodynamics Applied to Heat Engines

Giacomo Bisio and Francesco Devia

Energy and Conditioning Department - University of Genoa, Via all'Opera Pia 15 A - 16145

Key Words: Finite time process; Finite capacity reservoir; Pinch-point effect; Exergy efficiency; Expower.

ABSTRACT--The necessity of attaining a valuable power in the actual industrial processes brings forth the exclusion of reversibility in every case. This idea has led to an investigation field called finite-time thermodynamics. The aim of this paper is firstly a synthetic review of some topics of the literature regarding thermodynamics in finite time. On this basis, the impact of heat transfer mode on the performance of cyclic heat engines, in particular on their utilized fraction and exergy efficiency at maximum power is dealt with, considering various ideal models. One of these models is particularly suitable to justify the validity of some actual energy recovery plants, which were carried out in the last years. In addition, the purpose of this paper is the research of the maximum specific exergy power that can be obtained by a suitable finite temperature difference in the heat transfer processes.

1. INTRODUCTION

1. 1. Usefulness of a reversible model in the study of actual processes

Researches in thermodynamics were mainly stimulated by the aim of reduce the exergy consumption in heat engines. In this context, one can question the usefulness of comparing real processes to reversible processes which, after all, take an infinite time to complete. For example, who can be interested in works that make steel in an infinitely slow way? The basic question is: are reversible limits close enough to real performances to be useful in guiding the improvement of processes. The question leads- immediately to the consideration whether it is possible to find more realistic limits to the performance of real processes which operate in finite time and, consequently, better criteria of merit to evaluate real processes. In general terms, finite-time thermodynamics is concerned with how constraints on time or rate affect performance.

1.2. "Carnot efficiency"

The parameter ($\eta_o = 1 - T_l/T_h$) is often called "Carnot efficiency" and it is stated that the efficiency of a reversible engine is independent of the working fluid. This statement is at least of doubtful usefulness; indeed one remarks that in reality the duly defined efficiency of a cycle depends greatly on the working fluid; sometimes, the kind of working fluid is the key element.

As a matter of fact, the parameter i1o, named also more properly usable fraction, expresses the operative value of the cycle inlet energy and obviously, owing to this fact, cannot depend upon the working fluid. in a similar way, it is obvious that in reversible cycle the operative value of outlet energy should be equal to the operative value of inlet energy with unitary efficiency which is independent of the fluid.

The efficiency of an energy conversion depends greatly on both the fluid and the engine. It is useful to remind the existence of particular systems, e.g., standing wave thermoacoustic devices considered by Wheatley and others; in these devices the exergy efficiency depends greatly both on fluid proper-ties and on the geometric configuration in which the fluid works.

1.3. "Pinch-point" effect

The well known "pinch point" effect is a limitation on the exergy efficiency of the cycle owing to the following reasons:

(a) It can force a higher than otherwise necessary discharge gas temperature.

(b) The gas at the point where vaporization is complete must be at a much higher temperature than the steam and the large temperature difference results in a useless destruction of exergy.

Many authors studied the topic and various ways have been proposed and applied to overcome the above inconveniences. In next sections various analyses are considered with reference to finite-time thermodynamics.

1. 4. Exergy efficiency

The critical analysis of the concept of efficiency has led to formulate some requirements for its statement based on the two laws of thermodynamics (Bisio, 1987 and 1989) and it has been demonstrated that only exergy efficiency can fulfil all the requirements. However, not all the formulations comply with the condition that efficiency should be suitable in principle to vary between zero (for no useful effect) and one (maximum useful effect and no entropy production). This domain is comprised within the case of no entropy production (upper limit) and the one of no useful effect (lower limit).

As a reference value, s_{rf}, the entropy production corresponding to no useful effect is assumed;. multiplied by the environment temperature T_o, it gives, for the selected operation domain, the maximum loss, a parameter called losable (or reference) exergy, Ex_{rf} The actual entropy production, s_a, multiplied by T_o gives the actual exergy loss, Ex_a. The ratio of the actual entropy production to the reference one is called relative entropy production, s_r. The difference between the losable exergy and the lost one is the recovered exergy, Ex_u.

Exergy efficiency is the ratio of the recovered exergy to the losable one. One has that

$$\Phi = Ex_u/Ex_{rf} = (Ex_{rf} - Ex_a)/Ex_{rf} = (T_o\sigma_{rf} - T_o\sigma_a)/(T_o\sigma_{rf}) = (\sigma_{rf} - \sigma_a)/\sigma_{rf} = 1 - \sigma_r. \tag{1}$$

The recovered exergy must be actually usable in the real process; otherwise, by Ex_u, one should indicate the actually usable exergy and not the total recovered exergy.

1.5. Expower

Recently Radcenco et al. (1993) based themselves on the finite-time thermodynamics to define "maximum exergy power" with reference to infinite capacity thermal reservoirs.

Starting from the ideas of Radcenco et al. (1993), one purpose of this paper is the research of a definition of the "maximum exergy power", in a way slightly different (not bound to any physical model), considering thermal reservoirs both of infinite and finite capacity.

2. GENERALITIES ON FINITE-TIME THERMODYNAMICS

Researches on optimal heat exchanger temperature differences (and thus with processes evolving in finite time) were firstly reported by Chambadal (1957, 1958 and 1963) in French and by Novikov in Russian and English (1958) (see: Bejan, 1966a).

Curzon and Ahlborn (1975) examined an engine running according to an endoreversible Camot cycle, whose power depended upon the thermal flux from high- and to low- temperature reservoirs of infinite capacity, and thus, upon the actual temperature differences T_h-T_{fh} and T_{fl}-T_l, and consequent thermal irreversibilities; no other irreversibilities were considered. Curzon and Ahlborn (1975) limited themselves to the case of linear thermal transmission and no other irreversibilities were considered.

Under the above hypotheses, it was proved that the maximum power is attained with the used fraction (or thermal efficiency):

$$\eta = 1 - (T_l/T_h)^{1/2}, \tag{2}$$

and then with the exergy efficiency

$$\Phi = [1 - (T_l/T_h)^{1/2}]/[1 - (T_l/T_h)] \tag{3}$$

When the temperature differences T_h-T_{fh} and T_{fl}-T_{lh}, starting from zero, increase, the power input increases monotonically and the power output goes up till a maximum and then goes down. Consequently, the used fraction and the exergy efficiency decrease with increasing thermal power input. When inlet thermal power and outlet mechanical power tend to zero, exergy efficiency tends to one. It is to be remarked that, under the hypotheses of Curzon and Ahlborn , both η and ϕ, do not depend on the overall heat transfer coefficient, but this is very far from reality.

Subsequently many problems in the domain of finite-time thermodynamics were examined (e.g., Andresen, Salamon and Berry, 1977; Andresen, Berry, Nitzan and Salamon, 1977; Salamon, Andresen and Berry, 1980; Rubin, 1979a, 1979b and 1980; Ondrechen, Berry and Andresen, 1980; Ondrechen, Andresen and Berry, 1980; Andresen, Salamon and Berry, 1984). Many of these studies calculated the maximum power of thermal engines with several constraints and irreversibility sources.

In the case of a Camot cycle, Barrere researched for the optimal period to obtain either the maximum power or the maximum used fraction; then, this author examined more complex cycles and determined the optimal choice to divide everyone of the four phases of an original Carnot cycle in a given global number of steps.

Ondrechen, Andresen, Mozurkewich and Berry (1981) and Ondrechen, Rubin and Band (1983) dealt with cases in which only heat source or both heat reservoirs were of finite thermal capacity.

De Vos (1985) seems to be among those which first examined the consequences of a non-linear thermal transmission.

Rebhan and Ahlborn (1987) examined the influence of the relative variation of the heat absorption time and heat rejection time on the maximum power output and used fraction.

Leff (1987) considered the used fraction at the maximum power for cycles which were composed by two adiabats and two other polytropic processes different from isotherms. In addition, Leff said that the Curzon-Ahlborn thermal efficiency shares with the Carnot efficiency the useful (?) property of being independent of the working fluid and of being a function only of the extreme cycle temperatures; on the contrary, in the opinion of the present authors, this is one of the main drawback of that relation.

Various authors, e.g., Wu (1988), Lee, Kim and Won (1990) and Lee and Kim (1992), considered as a reference model an endoreversible Carnot cycle with finite capacity heat reservoirs.

Gu and Lin (1988) and Lee and Kim (1990) examined as a reference model also an endoreversible Lorentz cycle with finite capacity thermal reservoirs.

A synthesis of the main topics of finite-time thermodynamics at the date of the volume publication was reported by Sieniutycz and Salamon (1990). The physics of engineering literature on finite-time thermodynamics was reviewed most recently by Bejan (1996a).

In the next section, some models of reference ideal cycles within the field of finite-time thermodynamics are examined with their pros and cons, in a succession from the less valid to the more valid, at least in a given application domain.

3. ON VARIOUS MODELS PROPOSED WITH HEAT TRANSFER IN FINITE TIME

3. 1. Curzon and Ahlborn model with endoreversible Carnot cycle and infinite capacity heat reservoirs, characterized by heat transfers proportional to temperature differences.

The model of Curzon and Ahlborn (1975) is a remarkable improvement as regards the traditional model of a reversible cyclic process evolving in an infinite time and, thus, with zero power. However, only in very few cases the hypothesis of infinite capacity thermal reservoirs is acceptable at least approximately.

Curzon and Ahlborn reported some examples of real heat engines to demonstrate a sufficient accordance of the values found by their relation with the actual data. However, this concordance is only illusory and depends, e.g., in a coal fired steam plant from a totally arbitrary choice of the temperature of the hypothetical infinite (?) capacity source.

The result obtained by these authors, i.e. that η and ϕ do not depend upon on the heat transfer coefficients, is very far from reality and is a consequence only of the assumed hypotheses (which are not verified, not even approximately):

$$Q_h = a_h (T_h - T_{fh})\tau_h , \qquad Q_l = a_l(T_{fl}-T_l)\tau_l , \tag{4}$$

where:

Q_h heat from heat source,
Q_l heat to heat sink,
T_h temperature of the heat source,
T_l temperature of the heat sink,
T_{fh} temperature at which working fluid receives heat,
T_{fl} temperature at which working fluid gives away heat,
a_h, a_l overall heat transfer coefficients (assumed constant),
τ_h heat transfer time of the heat source,
τ_l heat transfer time of the heat sink.

For a device running according to the ideal model of Curzon and Ahlborn the diagram of Fig.1 shows the outlet mechanical power, Wm^o vs inlet thermal power, Q^o. The outlet mechanical power is maximum in correspondence of point M, whereas the used fraction is maximum in correspondence of point O, with thermal and mechanical power equal to zero. Even considering only thermal irreversibilities, it can be easily seen that these are far from being negligible and tend to zero as an infinitesimal of the same order in comparison with mechanical power, and then reversibility hypothesis is never suitable.

It is to be remarked that there are systems which are very different from the model of Curzon and Ahlborn. Standing wave thermoacoustic devices are represented in a more valid way by the diagram of Fig. 2 in which the outlet acoustical power, W_a^o, vs the inlet thermal power, Q^o, is shown. One notes that used fraction tends to zero when inlet thermal power tends to zero, contrary to what happens in Fig. I (Benvenuto and Bisio, 1989).

3.2. Gordon model with endoreversible Carnot cycle and infinite capacity heat reservoirs

Also Gordon (1990) referred to infinite capacity heat reservoirs, but assumed that heat transfers are non-linear functions of temperature differences. In this more general case, the maximum value of the used fraction is determined also by the functional temperature dependence of heat transfer, and by the ratio between heat transfer coefficients relative to heat source and heat sink.

Gordon examined in general the case in which heat transfer between the engine working fluid and the reservoirs is taken to be proportional to the difference of temperatures raised to the power "n", i.e.

$$Q_h = \beta_h (T_h^n - T_{fh}^n)\, \tau_s , \qquad Q_l = \beta_l (T_{fl}^n - T_l^n)\, \tau_l , \tag{5}$$

where:

β_h, β_l are overall heat transfer coefficients (assumed constant).

For (n = 1) the relations (5) coincide with the relations (4) of Curzon and Ahlborn.

Gordon considered in particular two cases with n very different from 1.

One is that of radiative heat transfer. In several heat engines, the primary mode of heat transfer between the engine and either or both of its heat reservoirs is radiative. Furthermore, certain systems in nature that can be analyzed effectively as heat engines, such as the generation of the earth's winds (Gordon and Zarmi, 1990; De Vos and van der Vel, 1992), are also dominated by radiative heat transfer. For a three-dimensional system, blackbody radiative power is proportional to T^4 and thus Gordon assumed (n = 4).

The second case, from linearized irreversible thermodynamics of Prigogine (1967) and others, is the heat conduction for which heat transfer is proportional to the gradient of the reciprocal of the temperature, i.e. thermal conductivity is proportional to $1/T^2$. Contrary to the statement of Gordon, this hypothesis, made only to obtain the so called principle of minimum entropy production rate at the steady state, is never verified (not even approximately) with the exception of some particular materials within very limited temperature ranges (Bisio, 1988 and 1991). Consequently, it seems useless to consider this case.

It is to be noted that the relations (5) do not allow us to state an analytic expression of the used fraction (with the exception of n = 1). Two diagrams, numerically obtained by Gordon (1990) are reported for (n = 4) (with n = 1 for a comparison).

In Fig. 3 used fraction at the maximum mechanical power, η_M, vs the parameter ($T = \beta_h/(\beta_h + \beta_l)$) is shown for ($T_l/T_h = .5$). One notes that for (n = 4) η_M is an increasing function of T, while for (n = 1) η_M remains constant (i.e. the same result of Curzon and Ahlborn).

In Fig. 4 used fraction at the maximum mechanical power, η_M, vs T_l/T_h is shown for ($\beta_h = \beta_l$). One notes that the (n = 4) curve is far from (n = 1) curve for (T_l/T_h <.5).

As regards the analysis of Gordon, we remark that the hypothesis of infinite capacity thermal reservoirs is approachable only in particular cases, e.g., in the already quoted papers (Gordon and Zarmi, 1990; De Vos and van der Vel, 1992); consequently, this model, even if represents a progress in reference to that of Curzon and Ahlborn, is not satisfying.

3.3. Lee and Kim model with endoreversible Carnot cycle andfinite capacity heat reservoirs

Various authors, e.g., Wu (1988), Lee, Kim and Won (1990) and Lee and Kim (1992), considered as a reference model an endoreversible Camot cycle with finite capacity heat reservoirs (Fig. 5).

In accordance with the concepts of Lee and Kim (1992), let us assume as constant the thermal capacity flow rates of the heat source, C_h, and of the heat sink, C_l. The inlet temperatures, T_{hl} and T_{11}, and the mass flow rates, G_h and G_l, of the heating and refrigerating fluids, as well as the minimum temperature differences (pinch-temperature differences), δT_h and δT_l, between the cycle working fluid and the heating and refrigerating fluids are chosen in an arbitrary way.

Under the chosen constraints the used fraction, η_M, and the exergy efficiency, ϕ_M, in correspondence of the maximum mechanical power are given by the relations:

$$\eta_{ml} = 1 - [(T_{11} + \delta T_l)/(T_{hl} - dT_h)]^{1/2} \quad (6)$$

$$\Phi_M = [C_l\,[(T_{11} + \delta T_l)^{1/2} - (T_{hl} - \delta T_h)^{1/2}]^2]/[(C_h + C_l\,[T_{hl} - T_{11} - \ln(T_{hl}/T_{11})]]. \quad (7)$$

It is to be remarked that the hypothesis (correspondent to reality) of finite capacity thermal reservoirs put in evidence immediately the non suitability of the Camot cycle as an ideal model.

3.4. Lee and Kim model with endoreversible Lorentz cycle andfinite capacity heat reservoirs

Gu and Lin (1988) and Lee and Kim (1992) considered as a reference model also an endoreversible Lorentz cycle with finite capacity thermal reservoirs (Fig. 6).

With constraints similar to those of subsection 3.3., the used fraction and the exergy efficiency in correspondence of the maximum mechanical power are given by the relations:

$$\eta_M = 1 - [(T_{11} + \delta T_l)/(T_{hl} - \delta T_h)^{1/2}, \quad (8)$$

$$\Phi_M = [2\,C_l\,[(T_{11} + \delta T_l)^{1/2} - (T_{hl} - \delta T_h)^{1/2}]^2]/[(C_h + C_l)\,[T_{hl} - T_{11} - \ln(T_{hl}/T_{11})]. \quad (9)$$

Considering a Carnot cycle and a Lorentz cycle having the same values of T_{11}, δT_l, T_{hl} and δT_h, one notes that the used fractions are the same in both cycles but the exergy efficiency of the Lorentz cycle is double of that of the Carnot cycle. This result, however, even if interesting, is not acceptable quantitatively, since there are practically no applications in which ($C_h = C_l$) and ($G_h = G_l$).

3.5. Roche's model with endoreversible cycle, finite capacity heat source and infinite capacity heat sink.

In the model proposed by Roche (1983) the following hypotheses are made:

(a) heating fluid is characterized by a constant thermal capacity per unit mass at constant pressure, c_p;
(b) heat sink is at the environment temperature assumed to be constant;
(c) the area, A, of the heat exchanger between the heating fluid and the cycle working fluid is fixed;
(d) the overall heat transfer coefficient per unit area, K, in the above-mentioned heat exchanger is assumed to be constant.

The last hypothesis is never exactly verified; however, this is not very far from reality when heating fluid is at a pressure near to that of atmosphere.

With a mathematical approach which utilizes the Lagrange multiplier method, the main characteristics of an ideal cycle with a given value of the area A have been determined by Roche. The obtained results are shown in a diagram with temperature, T, vs primary fluid enthalpy, I_h (Fig. 7). The heat inlet into the cycle working fluid takes place in the segment AB; this segment belongs to a right line which converges with that representing the heating fluid at the absolute zero. The extreme point A of the segment AB belongs also to the parabola which is the geometrical place of the points whose temperature is the geometrical mean between the working fluid inlet temperature and the environment temperature $(T_h\,T_o)^{1/2}$. The outlet temperature of the working fluid is given by the implicit relation:

$$[G\,c_p/(K\,A)]\,\ln(T_{hl}/T_{h2}) = 1 - (T_O/T_{h2})^{1/2}. \quad (10)$$

The mechanical power obtained by the ideal cycle is given by the relation:

$$W_m^{\,o} = G\,c_p\,[T_{hl} - T_{h2} - (T_o\,T_{h2})^{1/2}\ln(T_{hl}/T_{h2}) \quad (11)$$

and then the temperature of the point A (T_{fhl}) and B (T_{fh2}) are obtained:

$$T_{fhl} = (T_o\,T_{h2})^{1/2}, \qquad T_{fh2} = T_{fhl}\,(T_o\,T_{h2})^{1/2}. \quad (12)$$

Roche's model, even if with its limits, seems to be the more satisfying one among those till now proposed, since there is usually a remarkable temperature variation in the heat source and a very limited temperature variation in the heat sink (near to the environment temperature).

4. MAXIMUM EXERGY POWER

4.1 General considerations

To obtain the maximum exergy power in the light of finite-time thermodynamics let us assume that heat transfer takes place with a finite temperature difference and , in a way similar to that of other authors, that heat transfer is bound to this temperature difference by a linear relation. This allows us to obtain somewhat simple relations, but one must absolutely avoid to claim as valid some absurd results, which are a consequence of this hypothesis.

Contrary to the approach of other authors, we do not hypothesize any actual process, but we assume only the hypothesis of possible (in the limit) reversible processes, with exception of that relative to the heat transfer.

That is the ideal model, more acceptable than that totally reversible, to which the actual processes are to be compared.

4.2 Expower of thermal energy of infinite capacity heat reservoirs

Let us consider the thermal energy Q_1 at the temperature T_1 (i.e. an infinite capacity thermal source), its exergy value can be written as:

$$Ex_1 = Q_1(T_1-T_o)/T_1 = Q_1[1-T_o/T_1] . \quad (13)$$

As above said, on the basis of the results of Curzon and Ahlborn, one has that the maximum power is attained with the used fraction:

$$\eta = 1-(T_o/T_1)^{1/2} . \quad (14)$$

Thus, the maximum exergy power (called expower by Radcenco) is given by:

$$Ex_p = (Q_1/\tau)[1-T_o/T_1]/[1-T_o/T_1)^{1/2}] , \quad (15)$$

where: τ is time, and Q_1/τ is thermal power.

4.3 Expower of a fluid during a process from state 1 to that of the environment

Now let us consider a fluid undergoing a process from state 1 to that of the environment exchanging heat at a temperature ($T^* > T_o$).

From the first law of thermodynamics, it follows:

$$W_{x10} = Q_{1o}-(I_o-I_1) = Q_{1o}-M(i_o-i_1) , \quad (16)$$

where W_{x10} is the "external work", according to the definition of Spalding and Cole (1973), and M is the mass of the fluid.

We remember that Spalding and Cole defined (and we agree with this definition): "External work is all the work done at the control surface other than that associated with normal forces at places where material crosses the surface".

From the second law of thermodynamics, it follows:

$$Q_{1o} = T^*(S_o-S_1) = T^*M(s_o-s_1). \quad (17)$$

In addition, one has:

$$Q_{1o} = T^*M(s_o-s_1) = -KA(T^*-T_o)\tau , \quad (18)$$

where K is the overall heat transfer coefficient per unit area and A the heat transfer surface.

It follows:

$$W_{x10} = M(i_1-i_o)-T^*M(s_1-s_o) , \quad (19)$$

$$W_{x10} = [M(i_1-i_o)][KA(T^*-T_o)t]/[T^*M(s_1-s_o)]-KA(T^*-T_o)\tau , \quad (20)$$

$$W_{x10} = KA[(T^*-T_o)/T^*]\tau[(i_1-i_o)/(s_1-s_o)]-KAT_o\tau [(T^*-T_o)/T_o] . \quad (21)$$

Let us consider the parameter, named conventional mean thermodynamic temperature:

$$T_m = (i_1-i_o)/(s_1/s_o) . \quad (22)$$

The external work is given by the relation:

$$W_{x10} = KAT_m(1-T_o/T^*)\tau-KAT_o\tau (T^*/T_o-1) , \quad (23)$$

and then the external mechanical power P_{x10} is obtained:

$$P_{x10} = KAT_m(1-T_o/T^*)-KAT_o(T^*/T_o-1) . \quad (24)$$

Now the aim is of optimizing the variable T* in order that P_{x10} becomes maximum. Let us consider the parameter:

$$P_{xa} = P_{xlo}/(KAT_o) , \quad (25)$$

which is a normalized external mechanical power.

Indicating the ration T^*/T_o by x, one has:

$$P_{xa} = (T_m/T_o)(1-T_o/T^*)-(T^*/T_o-1) = (T_m/T_o)(1-1/x)-(x-1) , \quad (26)$$

$$P_{xa} = (T_m/T_o)[(x-1)/x]-(x-1) = (x-1)[T_m/(xT_o)-1] = P_{xa}(x) . \quad (27)$$

This expression is zero for (x_1 = 1) and for ($x_2 = T_m/T_o$) and then for a certain value x_{mx} the power P_{xa} presents a maximum. The following derivatives are obtained:

$$dP_{xa}/dx = (T_m/T_o)/x^2-1 \qquad d^2P_{xa}/dx^2 = -(T_m/T_o)/x^3 . \quad (28)$$

The second derivative is negative and then there is a maximum which is obtained by equating the first derivative to zero:

$$(T_m/T_o)/x_{mx}^2-1 = 0 . \quad (29)$$

Then, it follows:

$$x_{mx} = T^*_{mx}/T_o = (T_m/T_o)^{1/2} , \qquad T^*_{mx} = (T_mT_o)^{1/2} . \quad (30)$$

The maximum specific exergy power per unit mass flow rate can be written as

$$ex_{p12} = W_{x10}/\tau)/(M/\tau) = (i_1-i_o)-T^*_{mx}(s_1-s_o) , \quad (31)$$

$$ex_{p12} = (i_1-i_o)-T_o(s_1-s_o(T_m/T_o)^{1/2} < ex_{12} . \quad (32)$$

The following relation, indeed, is always verified:

$$T_m = (i_1-i_o)/(s_1-s_o) > T_o . \quad (33)$$

4.4 Expower of a fluid during a process from state 1 to state 2 (i.e. finite capacity heat reservoirs)

In a way totally similar to that of the previous case, one has:

$$W_{x12} = M(i_1-i_2)-T^*M(s_1-s_2). \tag{34}$$

Then, the maximum exergy power per unit mass flow rate follows:

$$ex_{p12} = W_{x12}/M = (i_1-i_2)-T^*_{mx}(s_1-s_2), \tag{35}$$

$$ex_{p12} = (i_1-i_2)-T_o(s_1-s_2)(T_m/T_o)^{1/2} < ex_{12}, \tag{36}$$

where ex_{12} indicates the opposite of the exergy variation during the process from state 1 to state 2.

The exergy efficiency at the maximum exergy power per unit mass flow rate (obviously, not considering mechanical irreversibilities) can be written as:

$$\Phi = [(i_1-i_2)-T_o(s_1-s_2)(T_m/T_o)^{1/2}]/[(i_1-i_2)-T_o(s_1-s_2)(T_m/T_o)], \tag{37}$$

and then the relative entropy production:

$$\sigma_r = 1-\Phi = 1-[(i_1-i_2)-T_o(s_1-s_2)(T_m/T_o)^{1/2}]/[(i_1-i_2)-T_o(s_1-s_2)(T_m/T_o)], \tag{38}$$

$$\sigma_r = T_o(s_1-s_2)[(T_m/T_o)^{1/2}-1]/[(i_1-i_2)-T_o(s_1-s_2)(T_m/T_o)]. \tag{39}$$

5. MAXIMUM POWER FROM FLUID FLOW

The heat transfer principle of power maximization in power plants with heat transfer irreversibilities was extended to fluid flow by Bejan (1996b). In that paper it is shown that when a stream flows between two pressure reservoirs ($P_1>P_2$) across linear flow resistances, a piston delivers maximum power when the pressure difference across its faces is $((p_1 - p_2)/2)$. The energy conversion efficiency at maximum power is $((1/2)(1 - P_2/P1))$, as an analog to the used fraction of relation (2).

The results of Bejan (1996b) seem meaningful and formally similar to those dealt with in previous sections considering two thermal reservoirs. However, in the opinion of the present authors two remarks are to be made.

In all practical cases the irreversibilities at the inlet and exit of a turbine or of a compressor are much lower than internal irreversibilities, contrary to the case of thermal irreversibilities of a heat engines considered by Curzon and Ahlborn.

More important, the mechanical power in a fluid flow is substantially different from the thermal power in a heat engine on the basis of the second law; then, the factor 1/2 of the Bejan's relation is not analogous, but very different from the exponent 1/2 of the relation of Curzon and Ahlborn.

The difference of entropy production rate, P*, between thermal irreversibilities and mechanical irreversibilities has been particularly dealt with in a paper accepted for publication (Bisio, 1998). For heat conduction one has:

$$P^* = (\delta Q/d\tau)\,\Delta T/[T\,(T+\Delta T)], \tag{40}$$

whereas for fluid flow one has:

$$P^* = (dV/d\tau)\,\Delta p/T.$$

The difference between the two relations, $[T(T+\Delta T)]$ vs T, in the opinion of the author, is fundamental and is connected with the fact that the operative value of thermal energy depends upon temperature.

6. APPLICATIONS

The concepts above dealt with, and in particular the necessity that the cycle working fluid does not receive heat along an isotherm have a general validity for all the systems producing mechanical energy. However, they are specifically meaningful in the range of the energy recovery systems, in which the heat source is at not very high temperatures (Bisio, 1992). The waste heat recovery systems have very different applications in the various industries, but always present the following two characteristics: (a) the primary fluid cost is practically zero; (b) the investment cost is generally high and can vary in a large domain. In economic terms, the problem is the choice of an acceptable investment cost, considering the annual exploitation of the recovery system. This investment depends upon the characteristics of the various elements of the system, everyone with a different sensitivity.

Such concepts have led to several applications in these last years; let us dealt with the main ones synthetically.

(a) The Kalina cycle (Kalina and Tribus, 1992a and 1992b; Marston and Hyre, 1995; Kalina and Brodianskii, 1997)) which utilizes a mixture of ammonia and water as the working fluid. When the liquid mixture is heated, the more volatile ammonia tends to vaporize first and at a lower temperature than does pure water. The temperature of the remaining saturated liquid rises as the ammonia concentration decreases. Thus, a better match to the temperature change of the gas is obtained and the potentiality exists for significant increases in the cycle exergy efficiency. Moreover, different parts of the cycle have different concentrations of ammonia. Thus condensation takes place at a pressure slightly above atmospheric pressure, using a relatively low concentration of ammonia, even though a relatively high-concentration is appropriate in the boiler. The Kalina cycle can be used with high exergy efficiency compared to other systems in these cases: (1) as a bottoming cycle in a combined system; (2) in the energy recovery systems, when the recovery fluid temperature is lower than 500 °C. In both cases waste gas discharge temperatures as low as 92 °C can be attained, if not limited by other constraints such as plume buoyancy or acid dew point. As the exhaust temperature is limited to higher and higher values, the Kalina cycle is disfavored.

(b) The organic Rankine cycles (ORC) of the following two types: (1) subcritical cycle with two-component (or more-component) working fluid (Bisio, 1992); (2) super-critical cycle with one-component fluid. In relation to the type of the used fluid, the application field is different, but it is always limited to very low temperatures. Mixtures of hydrocarbons have been investigated as possible ORC working fluids; it has been found, for example, that one can tailor the properties of the mixture so that its behavior is superior to that of either of its components. An important requirement of a binary mixture is that its thermodynamic characteristics, such as the critical properties and the pressure temperature relationship, can be changed by varying the composition. From the thermodynamic point of view, the superiority of the mixtures is due to their non-isothermal change of phase. A review of organic Rankine cycles for the recovery of low-grade waste heat has been recently made by Hung et al. (1997).

(c) Other more complex systems which are always characterized by the fact that the cycle working fluid receive heat in mono-phase state and then at strictly increasing temperature. Among these there is a system which uses pressurized water to recover thermal energy. Two turbines keyed on the same axis, the former fed with hot water partially vaporizing in the expansion, and the latter fed at the inlet with steam and also at the intermediate stages with flash steam, coming from a multistage flasher, allow a good energy recovery in comparisons with other systems, when the temperature of the recovered energy is not too high (Bisio and Poggi, 1991; Bisio, 1993).

In addition, it is to be remarked that also in the usual power plants the utilization of super-critical cycles, as soon as the technology allowed it, operated exactly to avoid isothermobaric vaporization.

7. CONCLUSIONS

The model of a reversible cycle which evolves in infinite time with infinite capacity thermal reservoirs is of limited applicative meaning, notwithstanding it allows simple mathematical expressions. Consequently, in these last years various authors proposed several ideal models with the aim that they could represent a better reference for the actual processes in power engines.

First of all it is necessary to give up any reference to processes which evolve in infinite time; this last model was used in the past and partly also nowadays. The models considered in this paper are all within the range of finite-time thermodynamics, even if that is not said explicitly by all the quoted authors.

A second step is the giving up of the hypothesis of linear relations in the heat transfer between the heat reservoirs and the cycle working fluid; indeed, these relations are never linear and, in addition, are very far from linearity when radiative heat transfer is prevailing.

A further step is the giving up of the reference to infinite capacity thermal reservoirs. Indeed, the hypothesis of infinite capacity is never strictly realized and is often very far from reality in actual applications.

The giving up of the hypothesis of two infinite capacity thermal reservoirs invalidates the reference to the Carnot cycle as an ideal cycle. In addition, another reason advises against the reference to the Carnot cycle as an ideal cycle: indeed, the presence of not negligible irreversibilities both in the compression and expansion strokes demands that in the cycle the ratio of net work to expansion work must have the maximum value; this requires that the compression must be in liquid phase.

The temperature variation is usually high in the heat source, whereas it is very low in the heat sink. Consequently the model proposed by Roche, in which the ideal cycle has the phase with heat inlet into the working fluid at variable temperature and the phase with heat outlet from this fluid at practically constant temperature, seems more valid than the Lorentz cycle which presents both phases with -similar temperature variations.

Such concepts have led to several applications in these last years: the paper deals with the main ones.

It must be specifically stressed that some energy recoveries at low temperatures are economically valid only by applying the concepts above-mentioned, in particular excluding heat transfer from the heat source to a cycle working fluid at constant temperature.

In order to research the maximum exergy power in the light of finite-time thermodynamics, one assumes that heat transfers occur with a finite temperature difference and that there is a linear relation between these parameters. These hypotheses allow one to obtain simple relations in the various cases; however, it is necessary to avoid any statement of validity in the actual processes for some absurd results which are the consequence only of these hypotheses. In particular, it not possible to state that the maximum exergy power does not depend on the working fluid, contrary to the claims of some authors.

On the contrary of what has been done by other authors, no kind of process is considered, but one assumes only the hypothesis of possible reversible processes with the exception of heat transfers with finite temperature difference. This is the ideal model to which the actual processes are to be compared.

Better approximation of the actual phenomenon can be reached by using more sophisticated models, which however do not allow one to have generalized mathematical expressions for exp; the values of exp are obtained only by numerical methods.

NOMENCLATURE

A	area [m^2]
C	thermal capacity flow rate [kW/K]
c_p	thermal capacity per unit mass at constant pressure [kJ/(kg K)]
Ex	exergy [kJ]
Ex_a	actually lost exergy [kJ]
Ex_{rf}	losable (or reference) exergy [kJ]
Ex_u	recovered (and actually usable) exergy [kJ]
Ex_p	maximum exergy power [kW]
ex	specific exergy [kJ/kg]
ex_p	maximum exergy power per unit mass flow rate [kW/(kg/s), kJ/kg]
G	mass flow rate [kg/s]
I	enthalpy [kJ]
i	specific enthalpy [kJ/kg]
i_o	specific enthalpy at T,, e p,, [kJ/kgl
K	overall heat transfer coefficient per unit area [kW/(m^2 K)]
M	mass [kg]
n	exponent characterizing the heat transfer dependence on temperature [dimensionless]
P_x	external power [kW]
P_{xa}	$= P_x/(K\ A\ T_o)$ normalized external power [dimensionless]
P*	entropy production rate [kW/K]
p	pressure [bar, kPa]
p_o	environment pressure [bar, kPa]
Q	heat [kJ]
Q^o	thermal power [kWI
S	entropy [kJ/K]
s	specific entropy [kJ/(kg K)]
s_o	specific entropy at T_o e p_o [kJ/(kg K)]
T	temperature [K]

T_h thermal source temperature [K]
T_{fh} temperature at which working fluid receives heat [K]
T_{fh1} inlet temperature at which working fluid receives heat [K]
T_{fh2} outlet temperature at which working fluid receives heat [K]
T_{fl} temperature at which working fluid gives away heat [KI
T_l thermal sink temperature [K]
T_m = $(i_1-i_2)/(s_1-s_2)$ thermodynamic mean temperature [K]
T_o environment temperature [K]
T^* ($T^* > T_o$) temperature of the heat transfer with the environment [K]
δT pinch point difference [K]
V volume [m^3]
W_x external work (shaft work) [kJ]
W_m^o mechanical power (kW)
W_a^o acoustical power [W]
x T^*/T_o [dimensionless]
a overall heat transfer coefficient [kW/K]
β overall heat transfer coefficient [kW/Kn]
T = $\beta_h/(\beta_h + \beta_l)$ [dimensionless]
η used fraction [dimensionless]
η_m used fraction at maximum mechanical power [dimensionless]
η_o used fraction with "zero mechanical power" (usable fraction or "Carnot efficiency") [dimensionless]
σ_r relative entropy production [dimensionless]
σ_a actual entropy production [kJ/K]
σ_{rf} reference entropy production [kJ/K]
τ time [s]
Φ exergy efficiency [dimensionless]
Φ^* maximum exergy efficiency [dimensionless]
Φ_M exergy efficiency at maximum mechanical power [dimensionless]

subscripts

h high temperature
I low temperature

REFERENCES

Andresen, B., Salamon, P., Berry R.S., 1977, "Thermodynamics in finite time: extremals for imperfect heat engines", J. Chem. Phys., **66**, 1571-1577.

Andresen, B., Berry, R.S., Nitzan, A., Salamon, P., 1977, "Thermodynamics in finite time, 1, The step-Carnot cycle", Phys. *Rev. A,* **15**, 2086-2093.

Andresen, B., Salamon, P., Berry R.S., 1984, "Thermodynamics in finite time", *Phys. Today*, September: 62-70.

Barrere, M., 1980, "Le role du temps dans l'optimisation des cycles thermodynamiques", *Rev. Gin. Therm. Fr.*, n° 228, 995-1006.

Benvenuto, G., Bisio, G., 1989, "Thermoacoustic systems, Stirling engines and pulse-tube refrigerators -Analogies and differences in the light of generalized thermodynamics", Proceedings of 24th IECEC, Washington D.C., August 6-11, **5**, 2413-2418, IEEE, New York.

Bejan, A., 1996a, *Entropy Generation Minimization,* CRC Press, Boca Raton, FL.

Bejan, A., 1996b, "Maximum power from fluid flow" , I*nt. J Heat Mass Transfer,* **39**, 1175-118 1.

Bisio, G., 1987, "On a general statement for exergy efficiency", in *"Analysis and Design ofAdvanced Energy Systems: Fundamentals",* M.J. Moran and R.A. Gaggioli, Eds., 45-50, The Winter Annual Meeting ASME, Boston, December 13-18, ASME, New York.

Bisio, G., 1988, "Thermodynamic analysis of the one-dimensional heat conduction with inhomogeneities in the heat flux direction", *Warme- und Stoffi~bertragung,* **23**, 143 -15 1.

Bisio, G., 1989, "On a general statement for efficiency", *Chemical Engineering Communications,* **81**, 177-185.

Bisio, G., 1991, "Maximum and minimum criteria to research system stability", *Proceedings of The Second World Conference on Experimental Heat Transfer, Fluid Mechanics, and Thermodynamics,* Dubrovnik, June 23-28, 1229-1236, Elsevier, New York.

Bisio, G., Poggi, S., 1991, "The possible utilization of the thermal energy discharged from sintering plants of steel works", *Proceedings of 26th IECEC,* Boston, August 4-9, **4**, 7-12, ANS, La Grange Park, IL.

Bisio, G., 1992, "Industrial waste heat recovery: the ideal thermodynamic cycle", *Proceedings of Flowers '92,* Florence, June 7-12, 503-519, Nova Science Publ., New York.

Bisio, G., 1993, "Exergy method for efficient energy resource use in the steel industry", *Energy - The International Journal,* **18**, 971-985.

Bisio, G., 1998, "Sulla formulazione della produzione entropica nell'unild di tempo nell'ambito della "termodinamica generalizzata", Accepted for publication in the *Proceedings of Accademia Ligure di Scienze e Lettere.*

Charnbadal, P., 1957, *Les Centrales Nucliaires,* pp. 41-58, Armand Colin, Paris.

Chambadal, P., 1958, "Le choix du cycle thermique dans une usine g6n6ratrice nucl6aire", *Rev. Gin. Electr.,* **67**, 332-345.

Chambadal, P., 1963, *Evolution et Applications du Concept dEntropie,* Section 30, Dunod, Paris.

Curzon, F.L., Ahlborn, B., 1975, "Efficiency of a Carnot engine at maximum power output", *Am. J Phys.,* **43**: 22-24.

De Vos, A., 1985, "Efficiency of some heat engines at maximum-power conditions", *Am. J Phys.,* **53**,70-573.

De Vos, A., van der Vel, P., 1992, "Endoreversible models for the conversion of solar energy into wind energy", J. *Non-Equilib. Thermodyn.,* **17**, 77-89.

Gordon, J.M., 1990, "The impact of heat transfer mode on the efficiency of heat engines operating at maximum power", *Proceedings of The Ninth Heat Transfer 1990,* Jerusalem, August 19-24, **4**, 329-334, Hemisphere Publishing Corporation, New York.

Gordon, J.M., Zarmi, Y., 1990, "Natural convection as an irreversible thermodynamic heat engine: predicting the Earth's average wind energy", *Proceedings of The Ninth Heat Transfer 1990,* Jerusalem, August 19-24, **5**, 485-488, Hemisphere Publishing Corporation, New York.

Gu, C.-H., Lin, L., 1988, "A heat-power cycle for electricity generation from hot water with nonazeotropic mixtures", *Energy,* **13**, 529-536.

Hung, T.C., Shai, T.Y., Wang, S.K., 1997, "A review of organic Rankine cycles (ORCs) for the recovery of low-grade waste heat", *Energy - The International Journal,* **22**, 661-667.

Kalina, A.I., Tribus, M., 1992a, "Advances in Kalina cycle technology (1980-1991): Part 1, Development of a practical cycle", *Proceedings of Flowers '92,* Florence, June 7-12, 97-110, Nova Science Publ., New York.

Kalina, A.I., Tribus, M., 1992b, "Advances in Kalina cycle technology (1980-199 1): Part 11, Iterative improvements", *Proceedings of Flowers '92* Florence ne 7- 911-919, Nova Science Publ., New York

Kalina, A.I., Brodianskii, V., 1997, ""Exergy analysis of Kalina cycle's thermodynamic efficiency", *Proceedings of Flowers '97,* Florence, July 30 - August 1, 1143-1155, SGE, Padova

Lee, W.-Y., Kim, S.-S., Won, S.H., 1990, "Finite-time optimization of a heat engine", *Energy,* **15**, 979-985.

Lee, W.-Y., Kim, S.-S., 1992, "The maximum power from a finite reservoir for a Lorentz cycle", *Energy,* **17**, 275-281.

Leff, H.S., 1987, "Thermal efficiency at maximum work output: new results for old heat engines", *Am. J Phys.,* **55**, 602-610.

Marston, C.H., Hyre, M., 1995, "Gas turbine bottoming cycles: triple-pressure steam versus Kalina", *Journal of Engineeringfor Gas Turbines and Power,* **117**, 10-15.

Novikov, 1.1., 1958, "The efficiency of atomic power stations", *J Nucl. Energy 17, 7,* 125-128, transl. from 1957, *At. Energ.,* **3** (11), 409-.

Ondrechen, M.J., Berry, R.S., Andresen, B., 1980, "Thermodynamics in finite time: A chemically driven engine", J *Chem. Phys.,* **72**, 5118-5124.

Ondrechen, M.J., Andresen, B., Berry, R.S., 1980, "Thermodynamics in finite time: Processes with temperature-dependent chemical reactions", J *Chem. Phys.,* **73**, 5838-5843.

Ondrechen, M.J., Andresen, B., Mozurkewich, M., Berry, R.S., 1981, "Maximum work from a finite reservoir by sequential Carnot cycles", *Am. J Phys.*, **49**, 681-685.

Ondrechen, M.J., Rubin, M.H., Band, Y.B., 1983, "The generalized Camot cycle: a working fluid operating in finite time between finite heat sources and sinks", J *Chem. Phys.*, **78**, 4721-4727.

Prigogine, 1., 1967, *Thermodynamics of Irreversible Processes,* 3rd ed., J. Wiley & Sons, New York.

Radcenco, Vs., Apostol, V., Radcenco, L., 1993, "Reconsideration of the exergetic method based on thermodynamics in finite time and on *expower* concept", *Proc. of ENSEC '93,* Cracow, July 5-9, J. Szargut, Z. Kolenda, G. Tsatsaronis, A. Ziebik, Eds., **1**, 289-296.

Rebhan, E., Ahlborn, B., 1987, "Frequency-dependent performance of a nonideal Carnot engine", Am. J. Phys., 55,423-428.

Roche, M., 1983, "Recuperation de chaleur - Le cycle thermodynamique ideal", Rev. Gen. Ther. Fr., n° 260-261, 561-571.

Rubin, M.H., 1979a, "Optimal configuration of a class of irreversible heat engines", I. *Phys. Rev. A,* **19**, 1272-1276.

Rubin, M.H., 1979b, "Optimal configuration of a class of irreversible heat engines, II", *Phys. Rev. A,* **19**, 1277-1289.

Rubin, M.H., 1980, "Optimal configuration of an irreversible heat engine with fixed compression ratio", Phys. *Rev. A,* **22**, 1741-1752.

Salamon, P., Andresen, B., Berry, R.S., 1977, "Thermodynamics in finite time, II, Potentials for finite-time processes", Phys. Rev. A, 15, 2094-2102.

Sieniutycz S. Salamon P., Eds., 1990, *Finite Time Thermodynamics and Thermoeconomics,* Taylor & Francis, New York (1990).

Spalding, D.B., Cole, E.H., 1973, *Engineering Thermodynamics,* 3rd ed., Edward Arnold, London.

Wu, C., 1988, "Power optimization of a finite-time Carnot heat engine", *Energy,* **13**, 681-687.

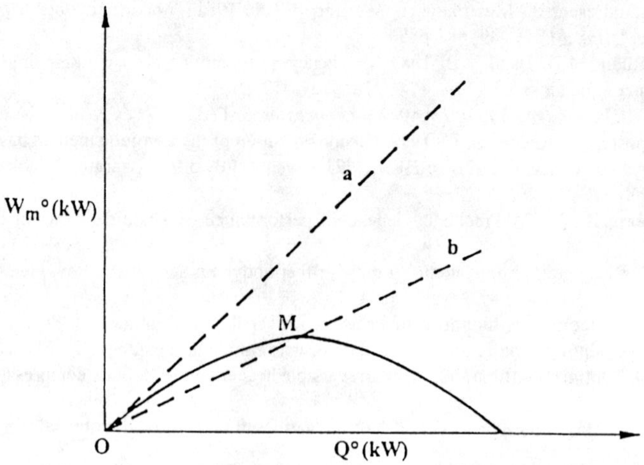

Fig. 1. Outlet mechanical power, $W_m°$, vs inlet thermal power, Q°, for a "Curzon-Ahlborn" device. Used fraction: right line (a) $\eta_o = 1 - (T_l/T_h)$; right line (b) $\eta_M = 1 - (T_l/T_h)^{1/2}$.

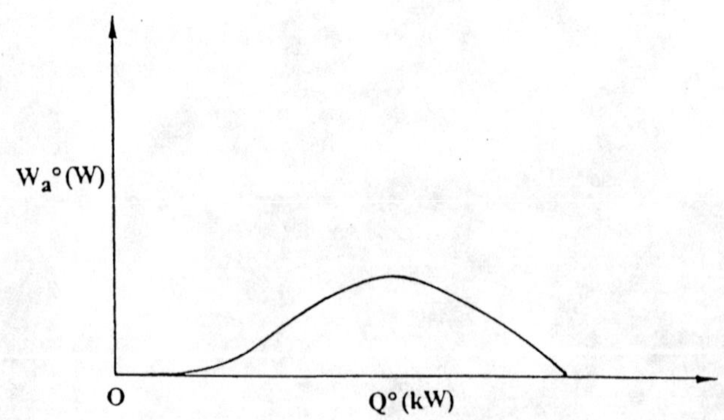

Fig. 2. Outlet acoustical power, $W_a°$, vs inlet thermal power, Q°, for a standing wave thermo-acoustical device.

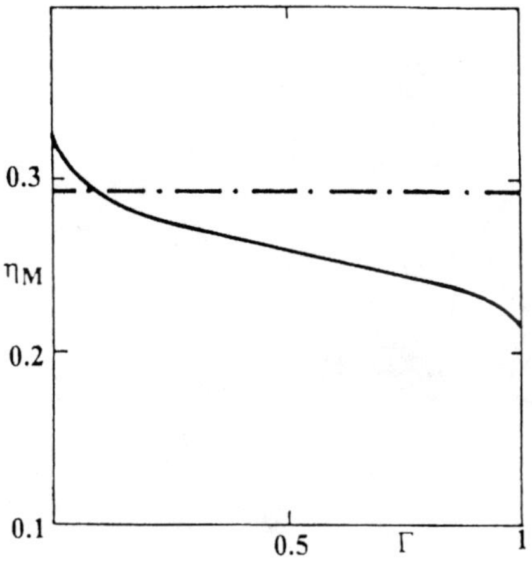

Fig. 3. Used fraction at the maximum mechanical power, η_M, vs $\Gamma=\beta_h/(\beta_h + \beta_l)$ for T_l/T_h=.5. Solid line: n=4; dashed line: n=1.

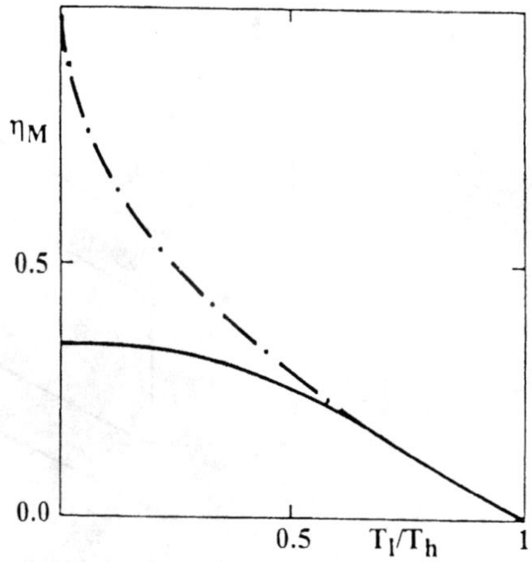

Fig. 4. Used fraction at the maximum mechanical power, η_M, vs T_l/T_h for $\beta_h=\beta_l$. Solid line: n=4; dashed line: n=1.

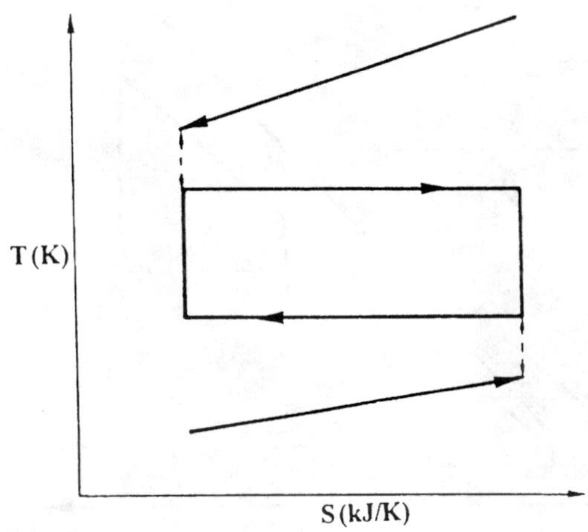

Fig. 5. Temperature, T, vs entropy, S, diagram of a Carnot cycle working between heat source and sink of finite thermal capacity.

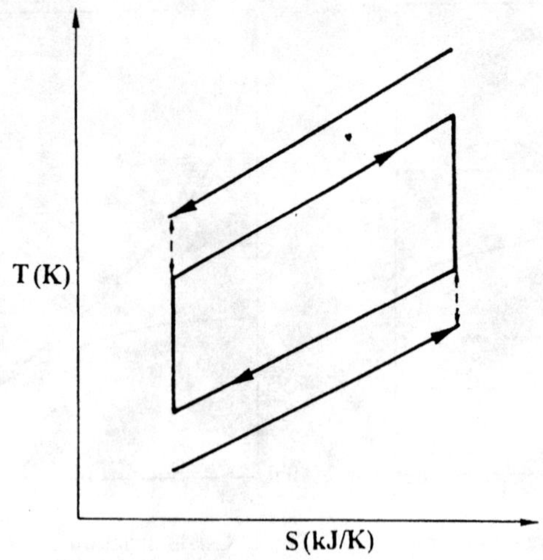

Fig. 6. Temperature, T, vs entropy, S, diagram of a Lorentz cycle working between heat source and sink of finite thermal capacity.

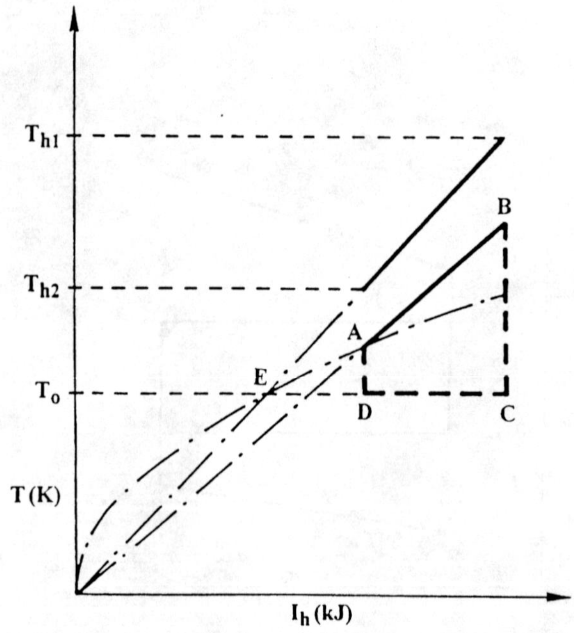

Fig. 7. Temperature, T, vs primary fluid enthalpy, I_h, diagram with a ideal thermodynamic cycle according to Roche's model.

HEATING RATE LIMIT AND COOLING RATE LIMIT OF A REVERSED RECIPROCATING STIRLING CYCLE

DAVID A. BLANK AND SOUVIK BHATTACHARYYA
Department of Mechanical Engineering
Indian Institute of Technology, Kharagpur 721302, India

and

CHIH WU
Mechanical Engineering Department
U.S. Naval Academy, Annapolis, MD 21402, USA

Abstract - An optimal heating rate and cooling rate analysis is conducted on a reversed reciprocating Stirling cycle with ideal regeneration. The maximum heating and cooling rates obtained for the cycle are based on the upper and lower temperature bounds by employing several novel irreversible thermodynamics concepts. Included among these concepts are the use of time symmetry to minimize cycle time and the use of a second law constraint to optimally allocate the thermal capacitances of the cycle and to minimize its internal entropy generation. The results obtained provide additional criteria for use in the study, design and performance evaluation of employing Stirling cycles in refrigeration, air conditioning and heat pump applications.

INTRODUCTION

Because of its potential for providing a thermal energy efficiency which theoretically approaches the Carnot efficiency, the reciprocating Stirling engine cycle is one of a handful of regenerative thermodynamic cycles that have captivated several generations of engineers and physicists. The many strong advantages of the Stirling cycle, such as its low noise and air pollution levels, its realizable potential for high thermal efficiencies and its flexibility as an external heat engine to utilize a variety of energy sources, have all worked together to propel it back into the attention of the world.

Today the world community is looking for fuel efficient and environmentally viable alternatives for many of the traditional energy conversion approaches. This development has further worked to increase technical focus on the Stirling cycle and presently the momentum of its popularity is continuing to accelerate. Today typical thermal efficiencies (overall) for reciprocating Stirling engines are about 40% compared with Otto and Diesel engines efficiencies which are about 25% and 35%, respectively[1]. Thus Stirling engines are currently being developed for applications in heat pumps using environment friendly refrigerants[2-5], energy conversion devices in both terrestrial and extra-terrestrial solar installations[6-10], submarines propulsion systems, as air-independent power plant in small unmanned autonomous underwater vehicles(AUVs)[11,12], etc. The cycle in its reversed mode is being found to be one of the viable approaches for the replacement of environmentally unacceptable halogen based refrigerants. Although cryocoolers have been the focus of Stirling engine refrigeration, recent studies[13-15] have used helium as the non-polluting working fluid in low temperature ratio refrigeration applications, e.g. residential refrigerators. Several reviews of Stirling engine development are available[16-19].

Currently, practical engineers make use of the ideal air standard Stirling cycle as a basis for analyzing trends within real versions of Stirling engines. In general, however, this type of thermodynamic analysis provides exceedingly generous estimates of the potential performance of these heat engines[18]. Thus, in order to provide a more reasonable estimate of the performance potential of a real Stirling cooler/heat pump, a reversed Stirling cycle which is externally irreversible and internally reversible is studied herein using several novel developments in finite-time thermodynamic analysis[20]. This innovative approach is based upon the inherent reality that a heat pump must extract/supply heat in a limited amount of time. The results of the analysis contained in this paper are intended for use in providing insight into the design and operating criterion necessary to attain optimum cooling and heating rates in reversed Stirling cycles with regeneration.

The reversed reciprocating Stirling cycle, depicted in Fig.1, consists of two isothermal processes separated by two isochoric processes. The working fluid is compressed isothermally at temperature T_W in process 1-2 with irreversible heat rejection to a high temperature sink maintained at T_H. The heat rejection to the regenerator is modeled using a reversible constant volume process 2-3. Isothermal expansion process, 3-4, occurs with irreversible heat addition to the working fluid, maintained at constant temperature T_C, from a low temperature source at T_L. The external heat transfer processes, 1-2 and 3-4, within real reciprocating Stirling heat pump cycles must each occur in finite time. This in turn requires that these heat transfer processes must proceed through finite-temperature differences and therefore these cycles are defined as being externally irreversible.

Assuming ideal regeneration, the working fluid gains exactly the same amount of heat in process 4-1 as it loses in process 2-3. This assumption is reasonable considering the fact that the effectiveness of regenerators continue to improve, with values of 99% currently being reported[16]. Thus, the following relation can be written for the overall ideal regenerative heat transfer processes,

$$|_4Q_1| = |_2Q_3| \tag{1}$$

If the working fluid is considered to be an ideal gas, the magnitude of the heat addition for a system with mass m during 3-4 can be found by substituting the definition for boundary work into the conservation of energy relation. Thus the heat gained by the working fluid from the low temperature source is given by:

$$Q_L = {}_3Q_4 = mRT_C \ln r_v \tag{2}$$

where R is the gas constant of the working substance and r_v is the compression ratio for the cycle. Similarly, the heat rejected by the working fluid to the high temperature sink is given by:

$$Q_H = {}_1Q_2 = mRT_W \ln r_v \tag{3}$$

The total cycle time is the sum of the individual process times:

$$t = t_{12} + t_{23} + t_{34} + t_{41} \tag{4}$$

Thus total cycle rates of heat loss and heat gain are given by:

$$\dot{Q}_{cooling} = \frac{Q_L}{t} \tag{5}$$

$$\dot{Q}_{heating} = \frac{Q_H}{t} \tag{6}$$

The individual process times can be found with the help of process rates. The rates of external heat transfer into and out of the reversed Stirling cycle may be quantified from heat transfer theory as being proportional to the temperature difference between the respective thermal reservoirs and the constant temperature states of the working fluid during each process. This implies that:

$$\dot{Q}_L = U_L A_L (T_L - T_C) \tag{7}$$

$$\dot{Q}_H = U_H A_H (T_W - T_H) \tag{8}$$

where U_L and U_H are the overall low and high temperature end heat transfer coefficients for the combined linear (conduction and convection) modes, and A_L and A_H are the surface areas of the heat exchangers between the respective low and high temperature thermal reservoirs and the system (main cycle); the product UA is called its heat conductance. In the sample calculations that follow, values will be specified only for the product of these two terms (i.e., for their heat conductance).

Dividing Eqs. (7) and (8) into Eqs. (2) and (3), respectively, yields expressions for the times associated with the external heat transfer processes,

$$\frac{mRT_C \ln r_v}{U_L A_L (T_L - T_C)} = t_{34} = t_{12} = \frac{mRT_W \ln r_v}{U_H A_H (T_W - T_H)} \tag{9}$$

The central expression of Eq. (9) ($t_{34} = t_{12}$) follows from both the objective of minimizing cycle time, and the requirement for the mechanical symmetry of the reciprocating cycle[6,8]. Total optimization of the cycle further requires the entropy generation of the internal cycle to be minimum. The second law of thermodynamics tells us that for zero entropy generation

$$\frac{U_L A_L (T_L - T_C)}{T_C} - \frac{U_H A_H (T_W - T_H)}{T_W} = 0$$

which is an expression that can also be seen to result from Eq. (9). Thus, for optimum performance (with zero entropy generation within the internal cycle), the capacitance inventory allocation for the cycle must be as follows:

$$\beta = \frac{U_H A_H}{U_L A_L} = \frac{T_W (T_L - T_C)}{T_C (T_W - T_H)}.$$

Also, modeling the regenerator as a crossflow heat exchanger and noting that at the limit of 100% effectiveness, the LMTD of the regenerator becomes unity[16] (i.e. $(LMTD)_r = 1$ K), the expression for the process times associated with heat transfer in the regenerator is,

$$t_{23} = t_{41} = \frac{mC_v (T_W - T_C)}{U_r A_r (LMTD)_r} = \frac{mC_v (T_W - T_C)}{U_r A_r (1K)} \tag{10}$$

where the expression $U_r A_r$ is the heat conductance between the cycle and its regenerator and (C_v) is the specific heat of the working substance at constant volume. For simplicity the term $1K$ will subsequently be dropped but assumed present in all the expressions containing $U_r A_r$ that follow.

Cooling Rate Analysis

Depending on the purpose at hand, the same cycle outlined above may be used for either cooling (of a source) or heating (of a sink). The first of these cases (cooling) may in turn be aimed at the air-conditioning or the refrigeration of the source.

Substituting for individual process times and simplifying, the cooling rate expression (Eq. (5)) can be rewritten as:

$$\dot{Q}_{cooling} = \left[\frac{\beta}{U_H A_H(T_L - T_C)} + \frac{T_W}{T_C U_H A_H(T_W - T_H)} + C_1\left(\frac{T_W}{T_C} - 1\right)\right]^{-1} \qquad (11)$$

where

$$C_1 = \left(\frac{2C_v}{Rln\, r_v U_r A_r}\right)$$

For given $U_H A_H$ and r_v values the cooling rate is thus a function of only three independent variables, namely T_W, T_C and β. For maximum cooling rate the following three expressions apply:

$$\left(\frac{\partial \dot{Q}_{cooling}}{\partial T_W}\right)_{T_C,\beta} = 0 \qquad (12)$$

$$\left(\frac{\partial \dot{Q}_{cooling}}{\partial T_C}\right)_{T_W,\beta} = 0 \qquad (13)$$

and

$$\beta_{opt} = \frac{(T_W)_{opt}(T_L - (T_C)_{opt})}{(T_C)_{opt}(T_W - (T_H)_{opt})} \qquad (14)$$

After making use of Eqs. (12) and (14), the temperature of the working fluid during the irreversible heat rejection process at the optimum cooing rate can be be shown to be:

$$(T_W)_{opt} = T_H + T_H{}^{0.5}\, F^{0.5} \qquad (15)$$

where

$$F = \frac{Rln\, r_v(U_r A_r)}{C_v(U_H A_H)}$$

Similarly, employing Eqs. (13) and (14), the temperature of the working fluid during the irreversible heat addition process for optimum cooling rate can be shown to be:

$$(T_C)_{opt} = \frac{T_L}{1 + G^{0.5} T_W^{-0.5}} \qquad (16)$$

where

$$G = \left[\frac{(U_r A_r)Rlnr_v}{U_L A_L)C_v}\right] = \frac{F}{\beta}$$

The expression for $(\dot{Q}_{cooling})_{opt}$ is identical to Eq. (11) with T_W, T_C and β replaced by their respective optimum values. Also, the optimized coefficient of performance is given by:

$$(COP_{cooling})_{opt} = \frac{Q_L}{Q_H - Q_L} = \frac{(T_C)_{opt}}{(T_W)_{opt} - (T_C)_{opt}} \tag{17}$$

Heating Rate Analysis

Substituting for individual process times and simplifying, the heating rate expression (Eq. (6)) can be rewritten as:

$$\dot{Q}_{heating} = \left[\frac{T_C}{T_W U_L A_L (T_L - T_C)} + \frac{1}{U_H A_H (T_W - T_H)} + C_1\left(1 - \frac{T_C}{T_W}\right)\right]^{-1} \tag{18}$$

which is also a function of the independent variables T_W, T_C and β. To obtain the optimum heat rate for a given set $U_H A_H$ and r_v values Eq. (14) must be employed together with:

$$\left(\frac{\partial \dot{Q}_{heating}}{\partial T_W}\right)_{T_C,\beta} = 0 \tag{19}$$

and

$$\left(\frac{\partial \dot{Q}_{heating}}{\partial T_C}\right)_{T_W,\beta} = 0 \tag{20}$$

Using Eqs. (14) and (19), the optimum temperature of the working fluid during the irreversible heat rejection process for maximum heating rate can be computed:

$$(T_W)_{opt} = \frac{T_H}{1 - F^{0.5} T_C^{-0.5}} \tag{21}$$

Similarly, employing Eqs. (14) and (20), the optimum temperature of the working fluid during the irreversible heat addition process for maximum heating rate can be computed:

$$(T_C)_{opt} = T_L - G^{0.5} T_L^{0.5} \tag{22}$$

The expression for $(\dot{Q}_{heating})_{opt}$ is identical to Eq. (18) with T_W, T_C and β replaced by their respective optimum values. Also, the optimized coefficient of performance is given by:

$$(COP_{heating})_{opt} = \frac{Q_H}{Q_H - Q_L} = \frac{(T_W)_{opt}}{(T_W)_{opt} - (T_C)_{opt}} \tag{23}$$

Design Conditions for Optimum Performance

A parametric study was conducted using these formulations to analyze the effect of different source temperatures, regenerator heat transfer parameter values, overall heat transfer parameter values and compression ratios on the optimum cooling load and heating load defined by Eqs. (11) and (18), respectively, under the conditions of optimum operating temperatures and β. Though the intent of this study is to demonstrate trends, an attempt has been made to provide reasonable values for use in these expressions.

Different working fluids were chosen for the various parts of the study. For the cooling device, Neon (R = 0.412 kJ/kg-K, C_v = 0.6179 kJ/kgK) was employed as the working fluid. The compression ratio, r_v, was varied between 1.5 and 3.25; the sink temperature (T_H) was chosen to be 310 K to match typical ambient conditions. The heat transfer parameter for the regenerator, $(UA)_{reg}$, was varied between 50 and 170 kW/K, while the heat transfer coefficient for the external heat exchanger (high temperature end) was set at 4 kW/K. In Fig. 2, the optimum cooling rate, $(\dot{Q})_{cooling}$, is plotted against a temperature range of 291 K to 305 K to represent typical air-conditioned space temperatures with r_v fixed at 2.5. Fig. 2 demonstrates that increases in the space temperature results in an increase in the cooling rate ($(\dot{Q})_{cooling}$). $(\dot{Q})_{cooling}$ also increases with increasing $(UA)_{reg}$ values. Fig. 3 additionally shows how $(\dot{Q})_{cooling}$ is improved by raising the compression ratio; for this figure, $(UA)_{reg}$ was set at an average value of 110 kW/K. The relative increase in $(\dot{Q})_{cooling}$ with r_v is less at higher values of r_v. Thus it is not prudent to design a Stirling cooler using high compression ratios, as this adds to the expenses of the design without having a corresponding improvement in performance. The optimum coefficient of performance, COP, at maximum cooling rate is also plotted against varying T_L, varying $(UA)_{reg}$ values and varying r_v values as shown in Figs. 4 (with r_v = 2.5) and 5 (with $(UA)_{reg}$ = 110 kW/K) respectively. Optimum COP decreases with increasing $(UA)_{reg}$ but improves with increasing design temperature of the space being cooled; COP also deteriorates with increasing r_v but the relative decrease is less at higher values of r_v. Considering the maximum cooling rates as the primary objective and the optimum COP values as secondary, a good value of r_v should be around 2.5.

For the heating device Nitrogen (R = 0.2968 kJ/kg-K, C_v = 0.7448 kJ/kg-K) was chosen as the working fluid. Figs. 6-9 shows the performance of the heating device with the same range of values for r_v and $(UA)_H$ as was used in the Neon cooling case and for a fixed T_L = 278 K. However, $(UA)_{reg}$ was varied between 25-125 kW/K instead. As can be observed from Figs. 6 (r_v = 2.5) and 7 ($(UA)_{reg} = 75\, kW/K$), maximum heating rate increases with $(UA)_{reg}$ values as well as with r_v values. However, similar to what was observed in the cooling mode, the optimum COP decreases with both increasing $(UA)_{reg}$ and r_v. $(\dot{Q})_{heating}$ is found to increase with rise in T_H.

For heat pump operation a single working fluid is desired for both the heating and cooling modes. For this purpose, Helium ($R = 2.077$ kJ/kg-K, $c_v = 3.1156$ kJ/kg-K) was chosen as the working fluid. The heating and cooling COPs for various T_H and T_L values respectively and for a fixed $r_v = 2.5$ are shown in Figs. 10 and 11. The $(UA)_H$ was varied from 4 kW/K (Fig. 10) to 6 kW/K (Fig. 11). The range of $(UA)_{reg}$ values for the cooling and heating modes were chosen differently, as indicated in these figures, for comparison. Increase in $(UA)_H$ improves the COP as well as the heat transfer rates (approximately 15% for heating and 25% for cooling) in both heating and cooling modes. This shows that $(UA)_H$ is the most influential parameter on both heating and cooling mode COPs and heat transfer rates ($\dot{Q}_{heating}$ and $\dot{Q}_{cooling}$).

Further insight into the design of regenerative reversed Stirling heat pump can be obtained using Eqs. (15)–(16) and Eqs. (21)–(22) which relate the optimum design working fluid temperatures, separately optimized for maximum cooling rate and maximum heating rate of the heat pump, to the source and sink temperatures and to the overall heat transfer parameters of the heat exchangers. Using these equations during the design process, in concert with the maximum heating/cooling rate equations, can help in the selection of heat exchangers, working fluid and operating temperatures.

CONCLUSIONS

This paper analyzes a Stirling heat pump cycle with an ideal regenerator using irreversible thermodynamics methodology to provide a more reasonable estimate of the potential performance bounds for real heat pumps. A novel second law of thermodynamics constraint imposed in the study ensures design conditions which minimize entropy generation and direct the optimum allocation of the cycles capacitance inventory ($U_LA_L+U_HA_H$). The relations thus derived are optimized to provide maximum heating and cooling rates, $\dot{Q}_{heating}$ and $\dot{Q}_{cooling}$, respectively. Parametric studies have been conducted with respect to the effect of working fluids, source and sink temperatures, regenerator heat transfer characteristics, compression ratio on optimum cooling and heating effects. The results of this analysis provide insight into the design of reversed Stirling heat pumps to achieve maximum heat rates. The choice of working fluids and compression ratio was found to be very significant design parameters, along with regenerator and source/sink heat transfer parameters. The most influential parameter, on both heating and cooling mode COPs and heat transfer rates ($\dot{Q}_{heating}$ and $\dot{Q}_{cooling}$), was found to be the conductance of the sink ($(UA)_H$).

References

1. Ayres R.U., and McKenna R.P. *Alternatives to the Internal Combustion Engine*, The Johns Hopkins University Press, Baltimore, 1972, 180-181.

2. Manaahan R. Phase 1-Supplemental Development of a Kinematic Stirling/Rankine Commercial Gas-Fired Heat Pump System. Report No. GRI-89/0133, Gas Research Institute, Chicago, 1989.

3. Rix D.H. The Potential of the Stirling Cycle Heat Pump, *Inst. of Mech. Engrs. J of Power Engg.*, 1989, 203:245-254.

4. Nakamura M., Yamaguchi K., and Ito Y. Developments of the Stirling Engine Heat Pump System, *Proc. Intersociety Energy conversion Engg. Conf.*, Boston, 1991, 488-492.

5. Fujiwara M., Sugimoto T., Mori M. and Shinozaki K. Development of Small Scale Stirling Engine Heat Pump System and its Components, *Proc. of 27th Intersociety Energy Conversion Engg. Conf.*, USA, 1992, 197.

6. Blank D.A, and Wu C. Power Potential of a Terrestrial Solar-radiant Stirling Heat Engine, *Int. Ambient Energy*, 1994, 15:131-139.

7. Shaltens R.K. *Technology 2000*, National Aeronautics and Space Administration, 1991, 1:269.

8. Blank D.A. and Wu C. Power Optimization of an Extra-Terrestrial Solar-radiant Stirling Heat Engine, *Energy: The Int. J.* , 1995, 20:523-530.

9. Noyes G.W. The Effect of Component Efficiency and Operating Conditions on the 50 kW Dish Stirling System in Riyadh, Saudi Arabia, *ASME Tran. J. Sol. Energy Engg.*, 1990, 112:244-246.

10. Badescu V. Optimal operation of a solar converter in combination with a Stirling or Ericsson heat engine, *Energy*, 1992, 17:601-607.

11. Nilsson H. Submarine Power Systems using the V4-275R Stirling Engine, *Inst. of Mech. Engrs. Power and Processes Engg.*, 1988, 202:257-267.

12. Potter I.J. and Reader G.T. Use of Stirling Engine Power Systems in AUVs, *Proc. ASME International Conf. on Offshore Mechanics and Arctic Engg.*, Alberta, 1992, 563-569.

13. Vitale N.G. and Vincent, R.J. *Recent Research in Heat Pump Design, Analysis and Application*, ASME Advanced Energy Systems Division, 1993, 28:137-141.

14. Kim S., Cho K., and Chung W. Study of Stirling cycle refrigerator, *Proc. Intersociety Energy Conversion Engg. Conf.*, Atlanta, 1993, 615-619.

15. Burt W. and Orsini R. Thermal Characterization of the BAe 80 K Stirling Cycle Cooler, *Proc. 1991 Cryogenic Engg. Conf. Advances in Cryogenic Engg*, 1991, pp. 1069.

16. Blank D.A., Davis G.W., and Wu, C. Power Optimization of an Endoreversible Stirling Cycle with Regeneration, *Energy*, 1994, 19:125-133.

17. Wu. C. Analysis of an Endoreversible Stirling Cooler, *Energy Conversion and Management*, 1997, 34:1249-1253.

18. Benson P.S. and Whitehouse N.D. *Internal Combustion Engines*, Pergamon Press, New York, 1979, 51-67.

19. Walker G., Reader G.T., Fauvel R., and Bingham E. *The Stirling Alternative*, Gordon and Breach, New York, 1993.

20. Blank D.A., Malhotra M. and Wu C. Finite Thermal Reservoir Effects on Power Optimized Continuous Endoreversible Carnot Heat Engine Cycles, *Int. J of Power and Energy Systems*, (In Press), 1997.

Nomenclature

A_H, A_L	Heat transfer surface area between the endoreversible Stirling cooler/heat pump and either the high temperature sink or the low temperature source
A_r	Heat transfer surface area of the reversed Stirling cycle regenerator
m	Mass of working fluid in the heat pump
Q_H	Isothermal heat transfer from the heat pump to the high temperature sink
Q_L	Isothermal heat transfer into the heat pump from the low temperature source
${}_1Q_2$	Isothermal heat transfer from the heat pump to the high temperature sink
${}_2Q_3$	Heat transfer captured by the regenerator from the working fluid
${}_3Q_4$	Isothermal heat transfer into the engine from the low temperature source
${}_4Q_1$	Heat transfer from the regenerator to the working fluid
$\dot{Q}_H$	Rate of heat transfer from the heat pump to the high temperature sink
$\dot{Q}_L$	Rate of heat transfer into the heat pump from the low temperature source
r_v	Compression ratio
R	Gas constant
T_C	Temperature of the working fluid during the isothermal external heat addition process
T_H, T_L	Temperature of the high temperature heat sink and the low temperature heat source, respectively
T_W	Temperature of the working fluid during the isothermal external heat rejection process
t	Total cycle time
t_{12}	Time required to accomplish the isothermal compression
t_{23}	Time required for the heat transfer process from the working fluid to the regenerator
t_{34}	Time required to accomplish the isothermal expansion stroke
t_{41}	Time required for the heat transfer process from the regenerator to the working fluid
U_H, U_L	Overall heat transfer coefficient between the heat pump and either the high temperature sink or the low temperature source, respectively
U_r	Overall heat transfer coefficient of the regenerator

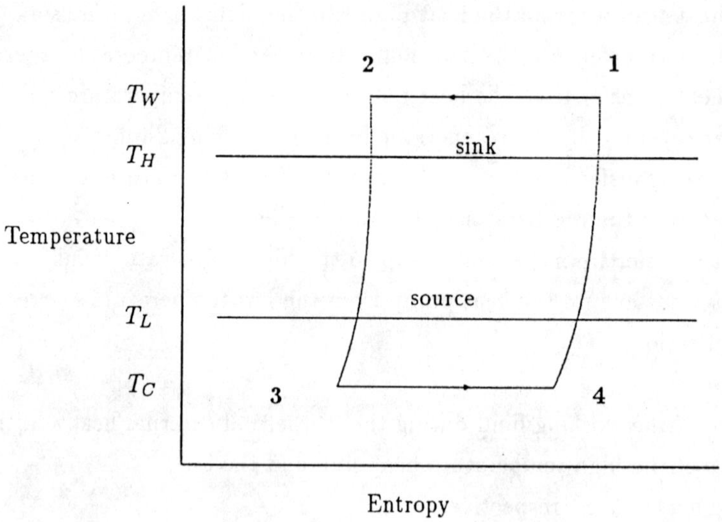

Figure 1: Reversed Stirling Cycle

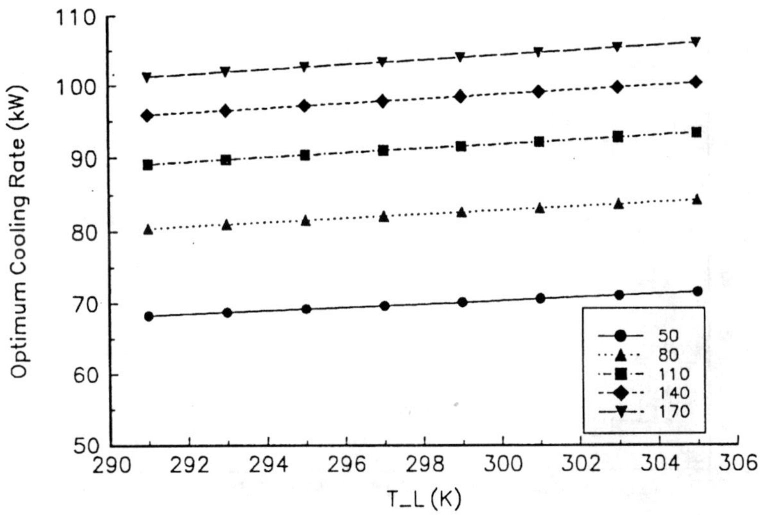

Figure 2: **Optimum Cooling Rate vs. T_L for varying U_rA_r and with r_v = 2.5 and U_HA_H = 4 kW/K**

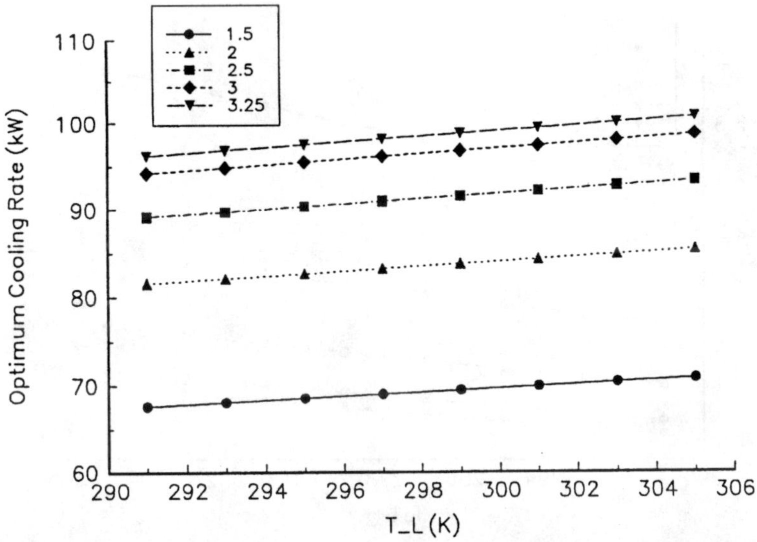

Figure 3: **Optimum Cooling Rate vs. T_L for varying r_v and with U_rA_r = 110 kW/K and U_HA_H = 4 kW/K**

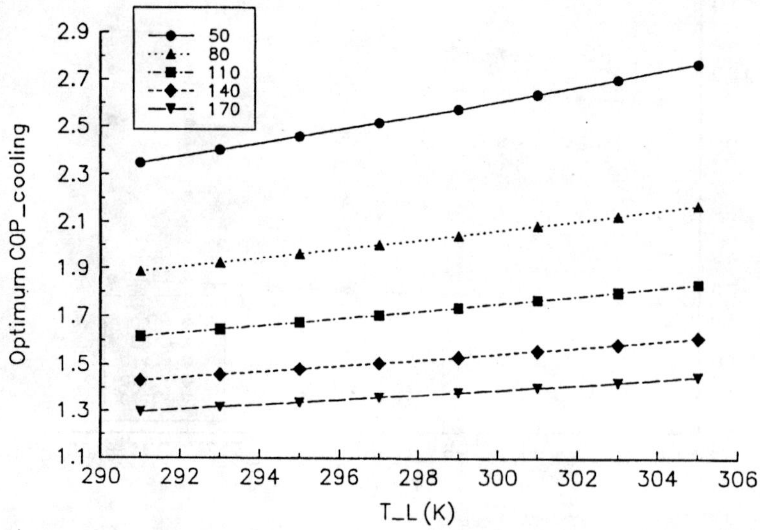

Figure 4: **Optimum COP in the cooling mode vs.** T_L **for varying** U_rA_r **and with** $r_v = 2.5$ **and** $U_HA_H = 4$ **kW/K**

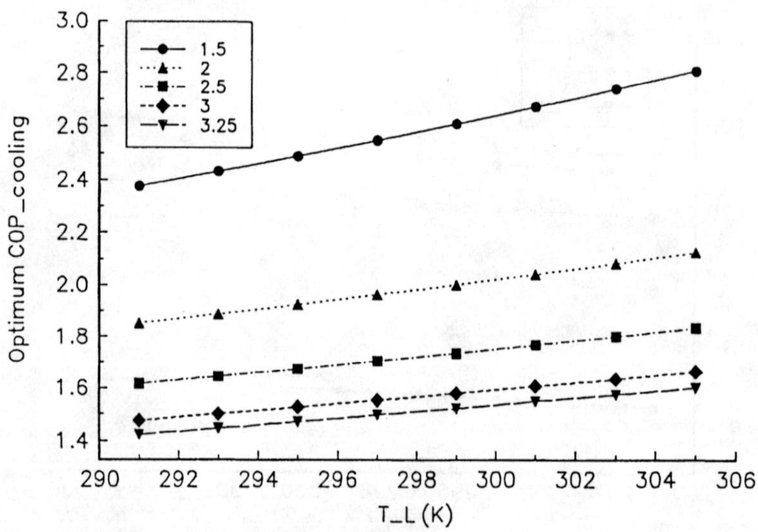

Figure 5: **Optimum COP in the cooling mode vs.** T_L **for varying** r_v **and with** $U_rA_r = 110$ **kW/K and** $U_HA_H = 4$ **kW/K**

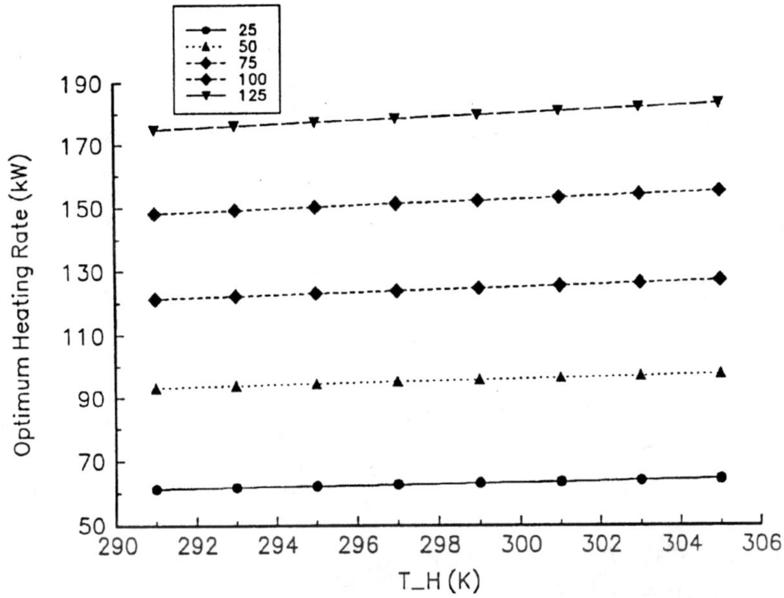

Figure 6: **Optimum Heating Rate vs.** T_H **for varying** U_rA_r **and with** $r_v = 2.5$ **and** $U_HA_H = 4$ **kW/K**

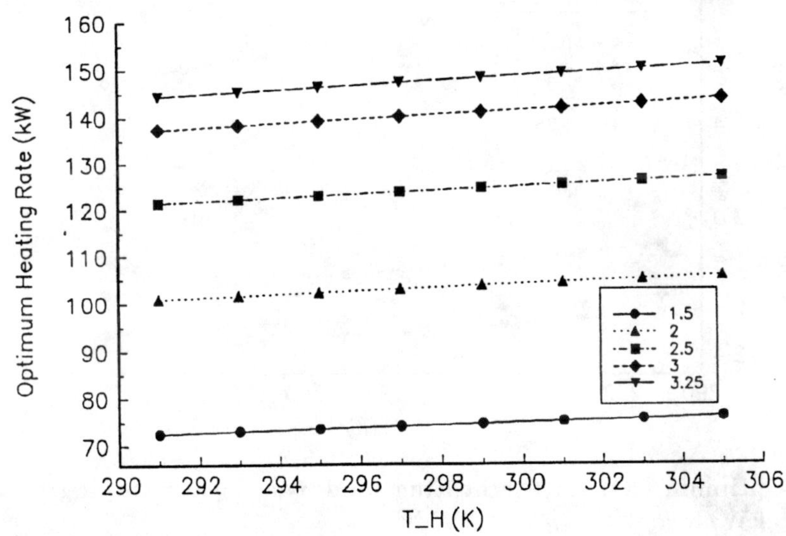

Figure 7: **Optimum Heating Rate vs.** T_H **for varying** r_v **and with** $U_rA_r = 75$ **kW/K and** $U_HA_H = 4$ **kW/K**

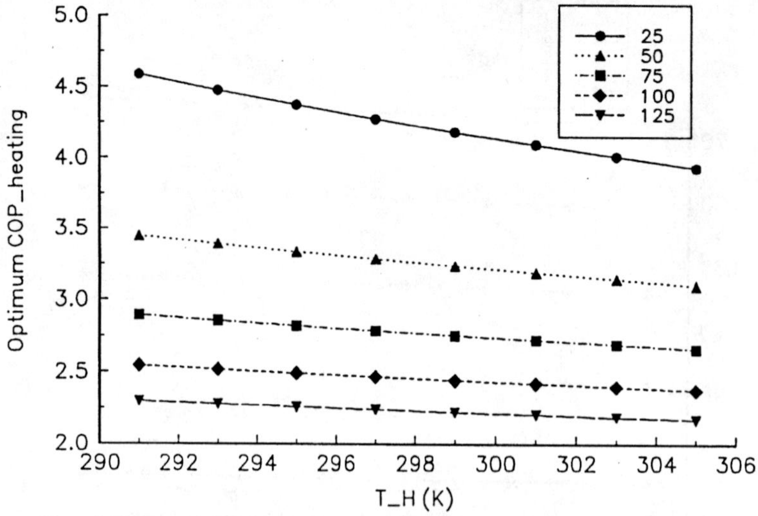

Figure 8: **Optimum COP in the heating mode vs. T_H for varying U_rA_r and with r_v = 2.5 and U_HA_H = 4 kW/K**

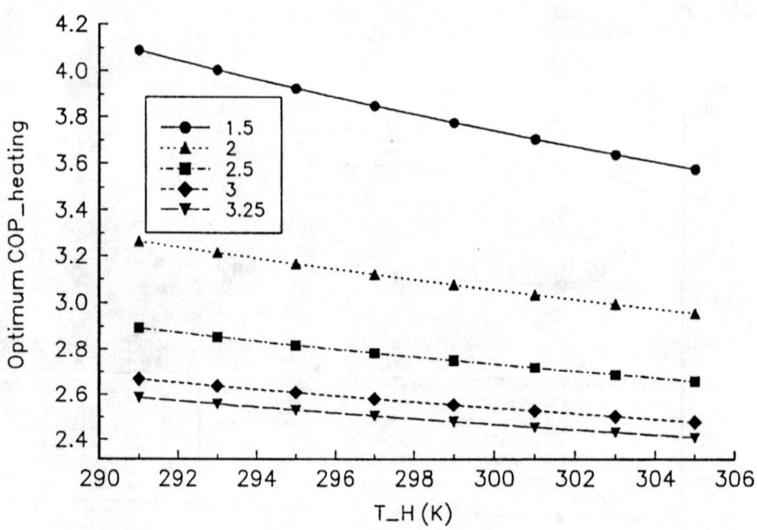

Figure 9: **Optimum COP in the heating mode vs. T_H for varying r_v and with U_rA_r = 75 kW/K and U_HA_H = 4 kW/K**

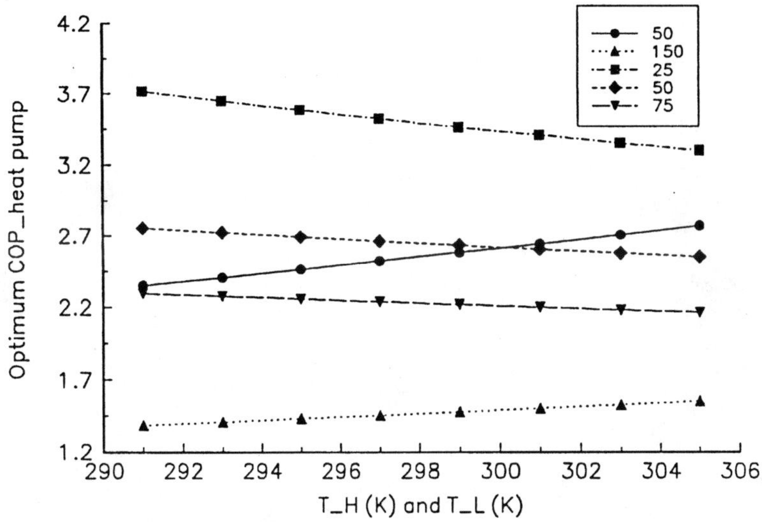

Figure 10: **Optimum COP for heat pump vs. T_H and T_L with $U_rA_r = 50$ and 150 kW/K in the cooling mode and $U_rA_r = 25$, 50 and 75 kW/K in the heating mode and with $r_v = 2.5$ and $U_HA_H = 4$ kW/K**

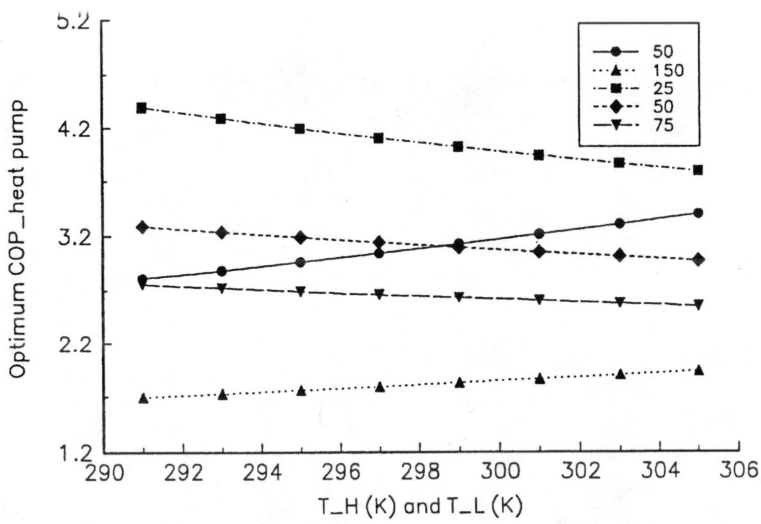

Figure 11: **Optimum COP for heat pump vs. T_H and T_L with $U_rA_r = 50$ and 150 kW/K in the cooling mode and $U_rA_r = 25$, 50 and 75 kW/K in the heating mode and with $r_v = 2.5$ and $U_HA_H = 6$ kW/K**

DESIGN OF A THERMODYNAMICALLY POWER OPTIMIZED IRREVERSIBLE THERMIONIC GENERATOR

SOUVIK BHATTACHARYYA
Department of Mechanical Engineering
Indian Institute of Technology, Kharagpur
Kharagpur 721302, India

Abstract - The operating temperatures of the electrodes in a thermionic (TI) energy generator can better be determined through thermodynamic optimization rather than through heuristic principles. Since the power output of a TI generator has high functional dependence on operating temperatures, choice of these temperatures becomes a crucial aspect in the design of any TI device.

The study reported here includes the optimisation of the operating temperatures of the electrodes of a TI generator with respect to maximum power output for a given heat source and heat sink, and for given electrode materials. Parametric studies on the variation of the optimised temperatures and power output, with varying heat source temperature, and a constant heat sink temperature have been carried out. Preliminary results show that the optimised cathode and anode temperatures increase (almost linearly) with increase in heat source temperature. Power output per unit electrode area increased exponentially with heat source temperature.

INTRODUCTION

Over the past two decades several research groups have reported[1,2] new approaches to cycle analysis employing *irreversible* or *finite time thermodynamics* concepts. The philosophy behind these developments has been an effort to blend heat transfer theory with classical thermodynamics to better define the irreversible processes of these cycles.

Although less frequently, this technique has also been employed in case of direct energy conversion cycles like the thermo-electric heat engine[3,4]. TI generators are receiving increasing attention as they promise to be direct energy conversion devices with very high power density[5]. Further, high heat rejection temperatures permit the use of light radiators. They can also be operated in series with lower temperature conversion devices to obtain improved efficiency. Such combinations are particularly attractive for use with high temperature nuclear reactors, where the diode could be part of the fuel element and a secondary coolant could transfer heat from the anode to a conventional power plant. Although cathode temperatures up to 3500 K have been considered, the limitations in the properties of available materials have so far restricted this temperature to 2200 K, the highest output being 25 W/cm^2.

MODELING THERMIONIC POWER GENERATORS

TI current density, J, is given by the *Richardson - Dushman* equation :

$$J = 4\pi \frac{mek^2}{h^3}(1-\bar{r})T^2 exp\left(-\frac{\psi}{kT}\right) = AT^2 exp\left(-\frac{\psi}{kT}\right) \tag{1}$$

where, m and e are the electron mass and charge respectively, k is Boltzmann constant, T is the absolute temperature, ψ is the TI work function above the Fermi level, $\bar{r}$ is the reflection coefficient of electrons and A is the Richardson constant. Since the electrons have to overcome the total V_c to be emitted, the current density from the cathode is given by :

$$J_c = A\,T_W^2\, exp\left(-\frac{eV_c}{kT_W}\right) \tag{2}$$

where T_W = operating temperature of the cathode. The current density from the anode is given by :

$$J_a = A\,T_C^2\, exp\left(-\frac{eV_a}{kT_C}\right) \tag{3}$$

where T_C = operating temperature of the anode. The net voltage V is the difference between the two voltages :

$$V = V_c - V_a \tag{4}$$

Thus the net electron density can be written as :

$$J = J_c - Ja = AT_W^2 exp\left(-\frac{eV_c}{kT_W}\right) - AT_C^2\, exp\left(-\frac{eV_a}{kT_C}\right) \tag{5}$$

The work output per unit electrode area is

$$\dot{W} = JV = (J_c - J_a)(V_c - V_a) \tag{6}$$

For maximum output, we need a high cathode work function and a low anode work function with the space charge being limited.

To determine the efficiency of a device, we must find the total heat input per unit area and the work output per unit area. The necessary heat input to the cathode can be determined by evaluating the energy needed for boiling off the electrons from the cathode surface i.e. the heat required to compensate the emission cooling. Each electron flowing from the cathode to the anode carries away an energy $eV_c + 2kT_W$. This is the average thermal energy of the electron as it passes over the potential of emission. Each electron emitted from the anode gives up an energy $eV_c + 2kT_C$ as it strikes the cathode. Therefore, total heat input to the cathode is given by:

$$Q_H = J_c\left(V_c + \frac{2kT_W}{e}\right) - J_a\left(V_c + \frac{2kT_C}{e}\right) \tag{7}$$

POWER OPTIMIZATION

For the purpose of optimising the operating temperatures of the cathode and the anode for maximum power, the partial derivatives of the power expression with respect to T_W and T_C respectively are equated to zero. The expression for power can be arrived at by considering the TI device as a heat engine with a heat source (T_H) and a heat sink (T_L) and operating between the two temperatures T_W and T_C. The power output is given by the product of the heat input and the Carnot efficiency, i.e $\dot{W} = \dot{Q}_h(1 - \frac{T_C}{T_W})$. But in the present case, the Carnot efficiency term has to be modified to incorporate the heat exchange between the two electrodes (by means of convection, radiation, etc.) and the other internal irreversibilities in the system. This is achieved by introducing an irreversibility factor R, ranging between 0 and 1, which gives the extent of internal irreversibilities in the system. Therefore,

$$\frac{\dot{Q}_h}{\dot{Q}_l} = R(\frac{T_W}{T_C}) \tag{8}$$

The heat transfer processes between the heat source and the cathode, and the heat sink and the anode are assumed to be linear, that is:

$$\frac{\dot{Q}_h}{\dot{Q}_l} = \frac{K_H(T_H - Tw)}{K_L(T_C - T_L)} \tag{9}$$

where K_H and K_L are the respective conductances for the heat transfer processes. Combining equations (8) and (9), we get

$$\frac{\dot{Q}_h}{\dot{Q}_l} = \frac{RT_W}{T_C} = \frac{x(T_H - T_W)}{(T_C - T_L)} \tag{10}$$

where x is the ratio of the conductances ($x = K_H/K_L$). Hence, we have

$$\dot{W} = \dot{Q}_h\left(1 - \frac{T_C}{RT_W}\right) \tag{11}$$

The power output can also be written in terms of heat rejected as shown below :

$$\dot{W} = \dot{Q}_l \left(\frac{RT_W}{T_C} - 1 \right) \tag{12}$$

Applying the first law of thermodynamics, equations (6) and (7) can be combined as :

$$\dot{Q}_L = J_c \left(V_a + \frac{2kT_W}{e} \right) - J_a \left(V_a + \frac{2kT_C}{e} \right) \tag{13}$$

Substituting for J_c and J_a from equations (2) and (3), equation (7) can be rewritten as :

$$\dot{Q}_H = AT_W^2\, exp\left(\frac{-eV_c}{kT_W}\right)\left(V_c + \frac{2kT_W}{e}\right) - AT_C^2\, exp\left(\frac{-eV_a}{kT_C}\right)\left(V_c + \frac{2kT_C}{e}\right) \tag{14}$$

Similarly equation (13) is rewritten as :

$$\dot{Q}_L = AT_W^2\, exp\left(\frac{-eV_c}{kT_W}\right)\left(V_a + \frac{2kT_W}{e}\right) - AT_C^2\, exp\left(\frac{-eV_a}{kT_C}\right)\left(V_a + \frac{2kT_C}{e}\right) \tag{15}$$

Hence equations (11) and (14) can be combined to yield :

$$\dot{W} = \left[AT_W^2 exp\left(\frac{-eV_c}{kT_W}\right)\left(V_c + \frac{2kT_W}{e}\right) - AT_C^2 exp\left(\frac{-eV_a}{kT_C}\right)\left(V_c + \frac{2kT_C}{e}\right)\right]\left[1 - \frac{T_C}{RT_W}\right] \tag{16}$$

and equations (12) and (15) can be combined to yield :

$$\dot{W} = \left[AT_W^2 exp\left(\frac{-eV_c}{kT_W}\right)\left(V_a + \frac{2kT_W}{e}\right) - AT_C^2 exp\left(\frac{-eV_a}{kT_C}\right)\left(V_a + \frac{2kT_C}{e}\right)\right]\left[\frac{RT_W}{T_C} - 1\right] \tag{17}$$

For maximum power output from the TI generator :

$$\frac{\partial \dot{W}}{\partial T_W} = 0; \quad \frac{\partial \dot{W}}{\partial T_C} = 0$$

Differentiating equation (16) partially with respect to T_W, we get

$$(T_H - T_W)(RT_W - T_C)\left[\left(V_c + \frac{2kT_W}{e}\right)\left(2T_W + \frac{eV_c}{k}\right) + \frac{2kT_W^2}{e}\right] exp\left(\frac{-eV_c}{kT_W}\right)$$

$$-T_C\left[T_W^2 exp\left(\frac{-eV_c}{kT_W}\right)(V_c + \frac{2kT_W}{e} - T_C^2 exp\left(\frac{-eV_a}{kT_C}\right)\left(V_c + \frac{2kT_C}{e}\right)\right] = 0 \tag{18}$$

Differentiating equation (17) partially with respect to T_C, we get

$$(T_C - T_L)(RT_W - Tc)\left[\left(V_a + \frac{2kT_C}{e}\right)\left(2T_C + \frac{eV_a}{k}\right) + \frac{2kT_C^2}{e}\right] exp\left(\frac{-eV_a}{kT_C}\right) +$$

$$RT_W\left[T_W^2\left(V_c+\frac{2kT_W}{e}\right)exp\left(\frac{-eV_c}{kT_W}\right)-T_C^2\left(V_c+\frac{2kT_C}{e}\right)exp\left(\frac{-eV_a}{kT_C}\right)\right]=0 \qquad (19)$$

Solving equations (18) and (19) simultaneously for T_W and T_C, the optimised working temperature of the cathode and the anode respectively can be obtained.

NUMERICAL SOLUTION

The two simultaneous non-linear algebraic equations (equations (18) and (19)) have been solved using a modified Levenberg-Marquardt algorithm. The main characteristics of this procedure involve the proper use of implicitly scaled variables and an optimal choice for the correction. The use of implicitly scaled variables achieves scale invariance of the routine and limits the size of the correction in any direction where the functions are changing rapidly. The optimal choice of the correction guarantees (under reasonable conditions) global convergence from starting points far away from the solution and a fast rate of convergence for problems with small residuals. During the course of this numerical exercise, it was found that the final solution had a dependence on the initial guess values of T_W and T_C and hence sensible initial values were assumed.

RESULTS AND DISCUSSIONS

Results have been obtained for a cathode material having a work function of 2.9 V (polycrystalline rhenium) and an anode material having a work function of 2.0 V (molybdenum). The variation of the optimized operating temperatures of the cathode and the anode with varying source temperatures and a fixed heat sink temperature of 300 K has been studied and the results plotted. The intuitive trend that the optimum values of T_W and T_C should increase with T_H has been noted. From Figure 1 it can be seen that the variation of T_W with T_H is almost linear in nature. With changing values of R (the factor of internal irreversibility) though T_W changes, the differences in the values of T_W is not very significant. This shows that R does not have a strong influence on the optimum operating temperatures. Similarly, in case of the anode temperature, the variation is linear. But the variation of T_C with changing value of R has been found to be non-uniform (Figure 2).

Figure 3 shows the variation of optimum power output with varying source temperature; it may be observed that power output is strongly dependent on the source temperature beyond a threshhold value of T_H and $\dot{W}$ exhibits an exponential rise with increasing T_H. This is because the current density function is strongly influenced by temperature and above 2000 K, an increase of temperature by 100 K, causes the current density to increase by an order of magnitude. Again, irreversibility seems to have a weak effect on power output.

Conclusions

Since the power output of a TI generator has high functional dependence on operating temperatures, choice of these temperatures becomes an important aspect in the design of any TI device. This study carries out the optimisation of the operating temperatures of TI generators with respect to maximum power output for a given heat source and heat sink, and cathode and anode materials. Parametric studies on the variation of the optimised temperatures, power output and efficiency of power generation, with the variation of temperature of heat source, and a fixed heat sink at 300 K have been carried out. Preliminary results show that the optimised cathode and anode temperatures increase (almost linearly) with an increase in heat source temperature. Power output per unit electrode area increased exponentially with heat source temperature. Internal irreversibilities were incorporated using a factor of irreversibility (R). But this might have a functional dependence on T_W and T_C. An appropriate model could be developed incorporating this feature. Additionally, the external heat transfer could be modelled for non-linear heat transfer opposed to the linear mode of heat transfer, as assumed in this study.

References

1. Bejan A. Entropy Generation Minimization: The New Thermodynamics of Finite-size Devices and Finite-time Processes, *J. Appl. Phys.*, 1996, 79:1191-1218.

2. Blank D.A. and Bhattacharyya S. A universal optimum work rate potential for continuous endoreversible Carnot heat engine cycles, *J. Appl. Phys.*,, 1996, 80:3619-3621.

3. Gordon J.M. Generalised power versus efficiency characteristics of heat engines, *Am. J. Phy.*, 1991, 59:551-554.

4. Wu C. and Schulden W. Specific heating load of thermoelectric heat engines, *Energy Conversion Management,* 1994, 35:459-464.

5. Hatsopoulos G.N. and Gyftopoulos E.P. *Thermionic Energy Conversion*, MIT Press, Massachusetts, 1979, 2:615-646.

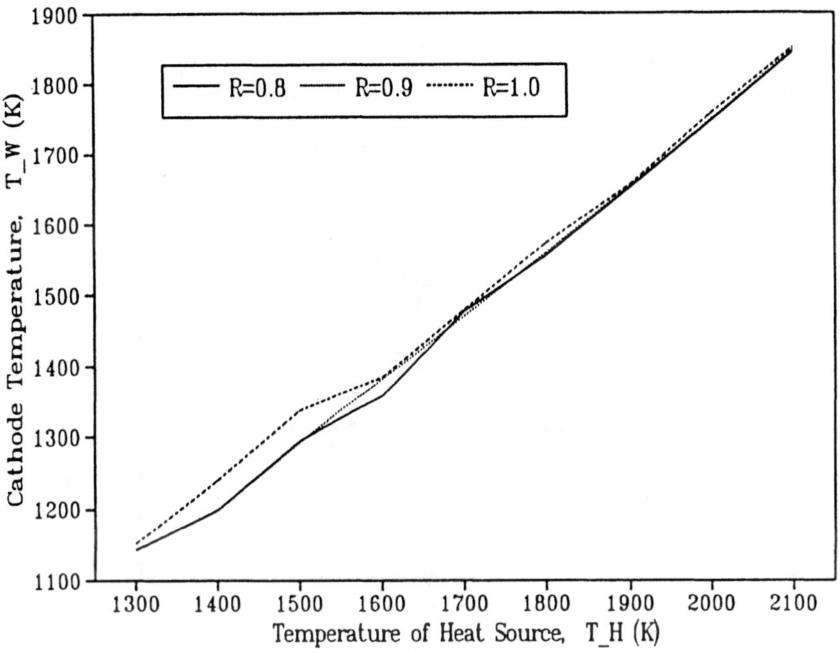

Figure 1. Variation of Optimum Cathode Temperature with Heat Source Temperature

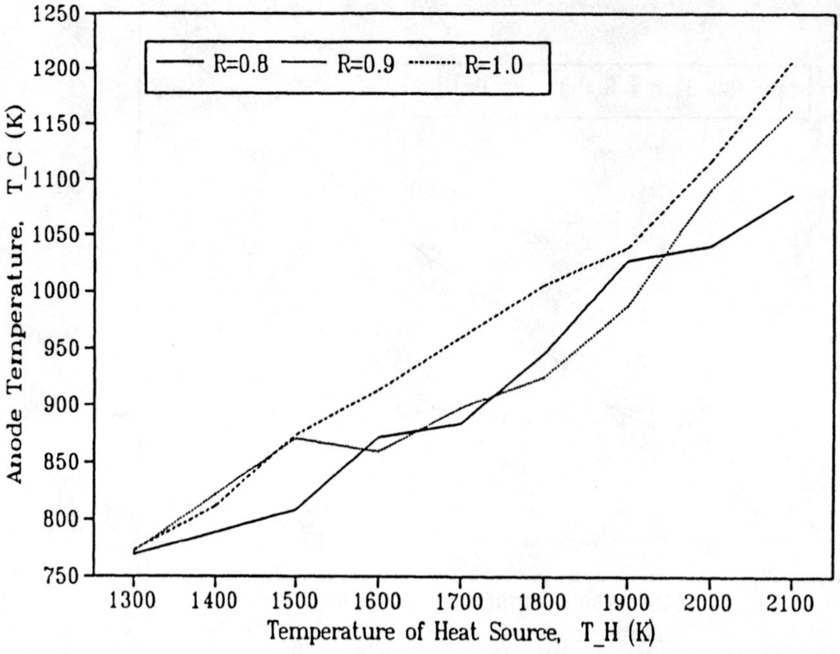

Figure 2. Variation of Optimum Anode Temperature with Heat Source Temperature

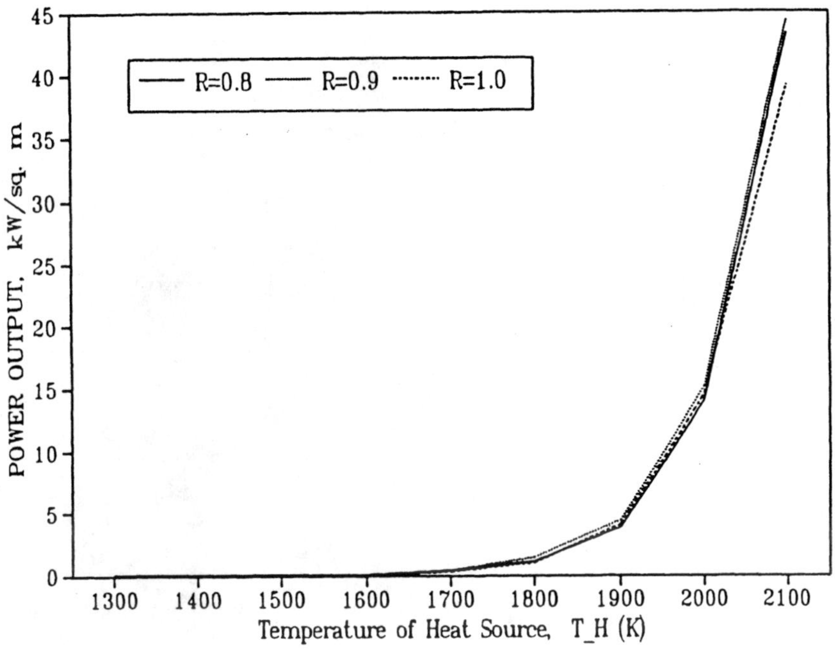

Figure 3. Variation of Power Output with Heat Source Temperature

COMBINED LAW POWER OPTIMIZATION OF CLOSED JOULE-BRAYTON HEAT ENGINE CYCLES WITH FINITE NON-INTERACTIVE THERMAL RESERVOIRS

DAVID A. BLANK
Department of Mechanical Engineering
Indian Institute of Technology, Kharagpur 721302, India

and

CHIH WU
Mechanical Engineering Department
U.S. Naval Academy, Annapolis, MD 21402, USA

Abstract- Significantly improved analytic expressions for optimum power and the operating temperatures which yield optimum power for the closed irreversible Brayton and closed Joule-Brayton heat engine cycles have been obtained through the concurrent employment of first and second laws of thermodynamics. The resulting expressions are applicable for configurations with finite reservoirs where the outlet reservoir temperature values are maintained at fixed values. In addition to the employment of this novel concurrent law procedure, the optimal allocation of the total thermal conductances at the hot and cold ends of these engine cycles is also considered. These steps enable both the minimization of internal entropy generation and the maximization of specific cycle work for a given set of power optimized operating temperatures. The analysis makes use of the ideal gas model.

INTRODUCTION

The classical air standard Brayton cycle has long been in use as a model for studying simple versions of the gas turbine heat engine cycle. Though all the thermal processes of the "Brayton" cycle are irreversible, most earlier studies idealized the various processes of the cycle, resulting in unrealistically high performance predictions. More recently, there has been renewed (and more practical) analysis of Brayton-like cycles by various researchers under the themes of power and power-density optimization. Among the first of these is Bejan [1] and Wu [2]. Bejan [1] considered a closed ideal Brayton cycle operating between an infinite beat source and infinite heat sink and demonstrated that for maximum power, the temperature extremes that mark the two constant pressure processes bear definite relationships with the heat source and heat sink temperatures. Wu [2] on the other hand looked, at a simple closed Brayton cycle used in a hybrid configuration with an indirect gas-cooled nuclear reactor in marine applications. Both the source (the nuclear reactor/cycle) and the sink (the ocean through a heat exchanger) were viewed as finite non-interacting heat reservoirs. As a follow-on Wu [3] studied a closed non-isentropic Brayton cycle. Contributing to the optimization of specific work within Brayton-like studies are the works of Leff [4], and Sahin [5,6]. Though these latter three works [4-6]do not represent irreversible thermodynamic (finite-time) analysis (as they only consider the inner cycle and not the heat transfer between the cycle with its sources), their results are useful in understanding the procedure required to minimize the internal entropy generation of Brayton cycles. Despite several flaws in the latter two references, they give valuable insights into the behavior of the open version of the Joule-Brayton cycle.

In the present Work, expressions for optimum power are sought using a procedure which can be shown in the limit of reversibility to simultaneously minimize internal entropy generation (subject to operational realities) while at the same time maximizing specific work for the cycle between the optimized operating temperatures at the hot and cold ends of the cycle. This has the advantage of creating as the initial starting point (for the power optimization process) an expression for power which itself represents (in the limit of reversibility) a concurrent expression of the first and second laws of thermodynamics. Thus the present work is made to conform with the observations of both Gyftopoulos [7] and Bejan [8](see Appendix A) for the general case of Brayton-like cycles, the Joule-Brayton heat engine cycle. Expressions for power optimized parameters are sought which consider the proper allocation of heat conductances and which are thus independent of heat conductances.

After general formulations are given for the closed versions of this important family of cycles, several practical sample calculations are analyzed. The focus of the work throughout is power optimization. In practice, neither optimum power nor optimum efficiency can be the sole objective of an energy conversion device. Some compromise between these two limits must be adopted. A better understanding of each of these limiting cases is thus essential to the multi-objective decision making approach required to provide a better compromise between the two for a given application. The purpose of this study is to give major new insight, from a power perspective, into the general behavior of a family of practical Brayton-like cycles which include the reversible Brayton, the irreversible Brayton and the Joule-Brayton power cycles. The study considers finite thermal reservoirs with pre-specified inlet and outlet temperatures. As the efficiency (work limit of these cycles has been studied extensively during the latter half of the current century, it is now time to study the power (rate of work) limit for these various cycles in much more detail.

THE CLOSED JOULE-BRAYTON CYCLE AND ITS POWER OPTIMIZATION

The temperature-specific entropy diagram of a general continuous Joule-Brayton cycle with finite heat reservoirs is depicted in Fig.1 for the general closed cycle operating between a finite source and a finite sink. Since the external heat transfer processes 2-3 and 4-1 occur across finite temperature differences, 2-3 and 4-1 are irreversible processes. During the heat addition process, 2-3, the finite high temperature reservoir may experience a temperature drop of T_5 - T_6. Similarly, during the heat removal process, 4- 1, the low temperature reservoir may experience a temperature rise of T_8 - T_7. The cycle is thus bounded by the temperatures $T_H = T_5$ and $T_L = T_7$. Also, both the compression process (1-2) and the expansion process (3-4) are assumed to be irreversible adiabatic processes. Thus, the cycle is externally and internally irreversible. In the analysis which follows, the working substance is assumed to be in a gaseous phase throughout the cycle such that the ideal gas law applies.

Since process 1-2 in the Joule-Brayton cycle is assumed to be adiabatic, the only source of external heat addition to the cycle is provided by the high temperature heat source (heat exchanger) during process 2-3. Similarly, since process 3-4 is also considered adiabatic, the only assumed heat rejection from the cycle occurs during process 4- 1. Thus, through use of the First Law of Thermodynamics, the net rate of work (power) for this continuous cycle can be expressed as :

$$P = Q_H - Q_L \ (= \oint \delta Q). \tag{1}$$

where, if the gas working fluid within the heat engine has a mass flow rate of m, the changes in the kinetic energy of the working substance during processes 2-3 and 4-1 are neglected and the heat transfer between the thermal reservoirs and the working substance in both the low and high temperature heat exchangers can be modeled using the log mean temperature difference (LMTD), then

$$Q_H = m\, c_p\, (T_3 - T_2) = K_H\, (LMTD_H), \tag{2a}$$

and

$$Q_L = m\, c_p\, (T_4 - T_1) = K_L\, (LMTD_L)\,. \tag{2b}$$

Here the log mean temperature differences at the two thermal ends are

$$LMTD_H = (T_5 - T_3 - T_6 + T_2)/\{\ln[(T_5 - T_3)/(T_6 - T_2)]\}, \tag{3a}$$

and

$$LMTD_L = (T_1\text{-}T_7\text{-}T_4\text{+}T_8)/\{\ln[(T_1\text{-}T_7)/(T_4\text{-}T_8)]\}. \quad (3b)$$

Also, c_p is the specific heat of the working substance and K_L and K_H are the heat conductances between the working substance and either the low temperature sink or the high temperature source, respectively. Note that $K_L = F_L U_L A_L$ and $K_H = F_H U_H A_H$, where F_L and F_H are factors whose values depend on the type of heat exchangers used (for a double pipe heat exchanger F = 1), U_L and U_H are the overall low and high temperature end heat transfer coefficients for the combined linear (conduction and convection) modes, and A_L and A_H are the surface areas of the heat exchangers between the respective low and high temperature thermal reservoirs and the system (main cycle).

Use can be made of pressure ratios, isentropic efficiencies and isentropic temperature relationships to replace T_2 and T_4 in the above expressions by T_1 and T_3, respectively. The pressure ratios across the compression and expansion stages of the cycle can be represented by:

$$r_L = p_2/p_1(= (p_3 + \Delta p_H)/(p_4 - \Delta p_L)) = (T_2'/T_1)^{[k/k-1)]}, \quad (4a)$$

and

$$r_H = p_3/p_4(=(p_2 - \Delta p_H)/(p_1 + \Delta p_L) = (T_3/T_4')^{[k/k-1)]} \quad (4b)$$

where $\Delta p_H(= p_2 - p_3)$ and $\Delta p_L(= p_4 - p_1)$ are the pressure drops during the heat addition and rejection processes, respectively, and k is the ratio of principal specific heat capacities cp and c, of the Joule-Brayton working substance. From Eqs.(4a-4b), it follows that:

$$T_2' = T_1 r_L^{(k-1)/k},$$

$$T_4' = T_3 r_H^{(1-k)/k},$$

and

$$r_H = ar_L - b.$$

where $a = p_1/(p_1 + \Delta p_L)$ and $b = \Delta p_H/(p_1 + \Delta p_L)$. Also, the turbine and compressor isentropic efficiencies can be expressed as:

$$\eta_c = (T_2' - T_1)/(T_2 - T_1) \quad (5a)$$

and

$$\eta_t = (T_3 - T_4)/(T_3 - T_4'). \quad (5b)$$

Using the implications of these equations [Eqs.(4-5)], the following expressions for T_2 and T_4 can be obtained:

$$T_2 = T_1[1-(1/\eta_c) + (r_L^{(k-1)/k}/\eta_c)], \quad (6a)$$

and

$$T_4 = T_3 [1 - \eta_t + \eta_t r_H^{(1-k)/k}] = T_3 [1 - \eta_t + \eta_t(ar_L - b)^{(1-k)/k}]. \quad (6b)$$

If these two temperature expressions (Eq.(6)) are substituted into Eq.(1) through Eq.(2), an expression for net cycle specific work (power per unit mass flow rate) can be derived, namely:

$$w \equiv P/m = c_p\{T_3\, \eta_t[1-(ar_L-b)^{(1-k)/k}] + (T_1/\eta_c)(1-r_L^{(k-1)/k})\}. \qquad (7)$$

For a fixed set of pressure drops during the heat addition and rejection processes, for constant specific heat values and for fixed values of turbine and compressor isentropic efficiencies, it can easily be noted that the above expression (Eq.(7)) is now only a function of the internal cycle temperature limits (T_1 and T_3) and the pressure ratio r_L.

It is through the use of Eq.(7) that entropy generation in the Joule-Brayton cycle can be minimized for a given set of internal cycle temperature limits (T_1 and T_3). It is a conclusion of the present work that this can be accomplished by maximizing specific work with respect to r_L. Bejan [8] has shown numerous examples for heat engines that point to the fact that these two processes independently lead to the same result. Taking the partial of Eq.(7) with respect to r_L (with T_1 and T_3 held constant) results in the following transcendental relationship for r_L at optimum power (maximum internal cycle specific work):

$$(r_L)_{opt} = (1/a)\{[c^k/(r_L)_{opt}]^{1/(1-2k)} + b] . \qquad (8)$$

where

$$c = T_1/(\eta_c \eta_t\, aT_3).$$

Shown in Appendix A is the fact that, in the absence of all internal cycle irreversibilities, the basis for this expression [the specific work optimization of Eq.(7)] reduces to a representation of the Second Law for the case of zero internal entropy generation. Even when irreversibilities are present, this equation [Eq.(8)] represents a Second Law constraint which ensures the minimization of entropy generation for the Joule-Brayton cycle for given internal cycle temperature bounds T_1 and T_3.

If we assume the total conductance inventory of the heat engine to be K, such that:

$$K = K_H + K_L, , \qquad (9a)$$

$$K_H = xK , \qquad (9b)$$

and

$$K_L = (1-x)\, K . \qquad (9c)$$

and if we let

$$g = (T_3 - T_2)/\, LMTD_H , \qquad (10a)$$

and

$$f = (T_4 - T_1)/LMTD_L . \qquad (10b)$$

Then the expression for net power [Eq. (1)] can be optimized with respect to the cycle's conductance inventory and cast into the following simplified expression:

$$P = K(T_3 - T_2 - T_4 + T_1)/(g + f) . \qquad (11)$$

Details of this derivation are given in Appendix B. Therefore, with T_2, T_4 and r_L themselves being functions of T_1 and T_3 only, the net power, P, of the Joule-Brayton cycle with a pre-specified total conductance inventory (K), is itself only a function of T_1 and T_3. It should be noted that because this relationship for net power (Eq.(11)) includes Eqs. (6) and (8), it represents a combined expression of the First and Second Laws of Thermodynamics. It should also be again noted that it (Eq.(11)) has already been optimized with respect to the allocation of the cycles conductance inventory. This equation then becomes the starting point for our power optimization process with respect to the external heat transfer processes to find expressions for the upper and lower temperature limits of the cycle (T_1 and T_3) which give optimum power.

Before moving on to the next phase of this power optimization process, it is valuable to describe a very time consuming dead-end experienced by the authors in attempting to numerically optimize Eq. (11) without the constraint of Eq.(8), and thus simultaneously with respect to three variables instead (T_1, T_3 and r_L). The resultant $(r_L)_{opt}$ gave T_2 and T_4 values which together with T_1 and T_3 formed a near perfect box of almost zero width (in the T ordinate) on a T-s diagram. In other words, this unconstrained solution was attempting to yield a Carnot cycle.

For closed Joule-Brayton cycles the expression for power optimized with respect to the internal cycle temperature bounds, T_1 and T_3, can be obtained from:

$$(\partial P/\partial T_1)_{T3} = 0\,, \tag{12a}$$

and

$$(\partial P/\partial T_3)_{T1} = 0\,. \tag{12b}$$

Applying these to the combined law equation for power [Eq.(11)] yields the following power optimized cycle operating temperatures limits

$$(T_1)_{opt} = \{(g+f)(1-\partial T_2/\partial T_1-\partial T_4/\partial T_1)/[\partial g/\partial T_1+\partial f/\partial T_1]\}+T_2+T_4-T_3\,, \tag{13a}$$

and

$$(T_3)_{opt} = \{(g+f)(1-\partial T_2/\partial T_3-\partial T_4/\partial T_3)/[\partial g/\partial T_3+\partial f/\partial T_3]\}+T_2+T_4-T_1\,. \tag{13b}$$

where the expressions for $(\partial T_2/\partial T_1)$, $(\partial T_4/\partial T_1)$, $(\partial g/\partial T_1)$, $(\partial f/\partial T_1)$, $(\partial LMTD_H/\partial T_1)$, $(\partial LMTD_L/\partial T_1)$, $(\partial T_2/\partial T_3)$, $(\partial T_4/\partial T_3)$, $(\partial g/\partial T_3)$, $(\partial f/\partial T_3)$, $(\partial LMTD_H/\partial T_3)$ and $(\partial LMTD_L/\partial T_3)$ needed to complete Eq.(13) are all given in Appendices B and C by Eqs. [C-1], [C-2], [C-4], [C-5], [C-6], [C-7], [D-1], [D-2], [D-4], [D-5], [D-6] and [D-7], respectively, for this closed cycle. Also needed to complete the simultaneous solution of Eqs.(13) and (8) are expressions for $\partial r_L/\partial T_1$ and $\partial r_L/\partial T_3$ which are given by Eqs. (C-3) and (D-3) of the appendix sections.

A simplified second case, the irreversible (non-isentropic) Brayton cycle, follows from the use of the results as given above, with a set equal to 1.0 and b set equal to zero ($\Delta P_H = \Delta P_L = 0$). A third and special case, the ideal Brayton cycle ($\Delta P_H = \Delta P_L = 0$ and $\eta_t = \eta_c = 1$) also follows from the above development. For this ideal case (ideal Brayton cycle) r_{opt} reduces to

$$r_{opt} = (T_3/T_1)^{k/[2(k-1)]}\,. \tag{14}$$

Thus for the ideal Brayton cycle T_2 and T_4 can be expressed as:

$$T_2 = T_4 = (T_1T_3)^{1/2}\,. \tag{15}$$

and hence

$$\partial T_2/\partial T_1 = \partial T_4/\partial T_1 = (T_3/T_1)^{1/2}/2\,, \tag{16}$$

and

$$\partial T_2/\partial T_3 = \partial T_4/\partial T_3 = (T_3/T_1)^{1/2}/2\,. \tag{17}$$

Other than these changes, all remaining partials for the ideal Brayton cycle are expressed in a manner identical to those of the more general (Joule-Brayton) case, as given in Appendices C and D.

SAMPLE CALCULATIONS AND DISCUSSION OF RESULTS

For each of the three cases described in the previous section, calculations were performed to evaluate the variation of P_{opt}/K, $(T_3)_{opt}$, $(r_L)_{opt}$ and η_{opt} (the cycle thermal efficiency at optimum power) with changes in the inlet temperature of the finite source (T_5). For all these cases k, $\Delta T_H(= T_5 - T_6)$ and $\Delta T_L(= T_8 - T_7)$ were fixed at 1.4, 265 K and 30 K, respectively, and T_7 was set at 300 K to correspond to ambient conditions. For all but the ideal Brayton cycle case η_c and η_t were both fixed at 0.9. Finally, by definition the irreversible Brayton and ideal Brayton cases specified $\Delta P_H = \Delta P_L = 0$. For the Joule-Brayton sample calculation values of $\Delta P_H = \Delta P_L = 20$ psi were used instead. The value of P_1 was set at 110 psi. The choice of the above input values for these calculations is aimed at closely simulating the combined gas turbine-nuclear reactor cycle of Daniels [2,10]. The specification of none SI units for pressure here bears no consequence, as both a and b are ratios of pressures.

Values for optimized power (P_{opt}) were obtained by substituting the power optimized operating temperatures ($(T_1)_{opt}$ and $(T_3)_{opt}$) and $(r_L)_{opt}$ into Eqs. (6) and (10) to complete Eq. (11). Similarly, values for η_{opt} were obtained from:

$$\eta_{opt} = 1-[T_4-(T_1)_{opt}]/[(T_3)_{opt}-T_2]. \tag{18}$$

Figure 2 shows the variation P_{opt}/K over a realistic range of values for T5 for the Joule-Brayton ($\Delta P_H + \Delta P_L = 20$psi and $\eta_c = \eta_t = 0.9$) and ideal Brayton ($\Delta P_H = \Delta P_L = 0$ and $\eta_c = \eta_t = 1.0$) cycle cases. Figures 3, 4 and 5 examine instead the variations of $(T_3)_{opt}$ (and $(T_1)_{opt}$) $(r_L)_{opt}$, η_{opt} respectively, with T_5 over the same temperature range and for the same three cases as Fig. 2. In all of these figures a steady rise in the vertical ordinate parameters (under consideration) with increasing source inlet temperatures is observed for the irreversible-Brayton and Joule-Brayton cycles. Figure 4 shows that the effect of turbine and compressor losses alone (the irreversible-Brayton cycle) is a drop in $(r_L)_{opt}$ (from the values of the ideal-Brayton cycle), while the effect of the further addition of pressure drops (the Joule-Brayton cycle) is a rise in $(T_L)_{opt}$ (from the values of the irreversible-Brayton cycle). The effects of the fact that the $(T_3)_{opt}$ values for the ideal-Brayton cycle plot (with respect to T_5) do not start out with as rapid a rise rate are seen in the early dips of this cycle's $(r_L)_{opt}$ and η_{opt} vs T_5 plots.

From Figs.2 and 5 it is evident that both cycle power and cycle thermal efficiency are degraded by increases in internal cycle irreversibility. Irreversibility in the internal cycle is caused by only turbine and compressor losses in the irreversible- Brayton cycle and by pressure losses in addition to turbine and compressor losses in the Joule-Brayton cycle.

CONCLUDING REMARKS

The novel analytical formulation given herein along with its sample computational results provide us with new and meaningful insights into the general behavior of the family of practical Brayton-like cycles which include the ideal Brayton, the irreversible Brayton and the Joule-Brayton heat engine cycles.

This fundamental study represents a combined First and Second Law analysis of optimized power generation using linear heat transfer modes at both thermal ends of a family of continuous Brayton-like cycles for which the thermal conductance allocation has also been optimized and internal entropy generation minimized. The work provides the pressure ratios and operating temperature limits necessary to enable a more complete power optimization of this very important family of heat engine cycles - including the very practical Joule-Brayton cycle.

NOMENCLATURE

A_H, A_L	Surface area of the, heat exchanger between the system and the heat source and sink, respectively (m^2)
F_H, F_L	Factor determining the type of heat exchanger used at the high and low temperature ends, respectively
c_p, c_v	Principal specific heat capacities of the working fluid with respect to pressure and specific volume, respectively (kJ/kg-K)
k	Ratio of specific heats, c_p/c_v
K	Total conductance inventory (kJ/K-s)
K_H, K_L	Heat Conductances between the endoreversible cycle and either the high temperature source or the low temperature sink (kJ/K-s)
P	pressure (N/M^2)
P, P_{opt}	Net and maximum cycle power output, respectively (kJ/s)
Q	Rate of heat transfer (kJ/s)
Q_L, Q_H	Rate of heat transfer between the working substance and the low and high temperature reservoirs, respectively (kJ/s)
r_H	pressure ratio across expansion stage
r_L	pressure ratio across compression stage
s	Specific entropy (kJ/K-kg)
T	Temperature (K)
T_1, T_2	Temperature of the working fluid before and after compression (K)
T_3, T_4	Temperature of the working fluid before and after expansion (K)
T_5, T_6	Inlet and outlet temperature of the heat source, respectively M
T_7, T_8	Inlet and outlet temperature of the heat sink, respectively (K)
U_H, U_L	Overall heat transfer coefficient (for linear modes) for the thermal source and sink, respectively (kJ/K-s-m^2)
η	cycle efficiency
η_{opt}	Cycle thermal efficiency at the optimum power condition
η_c	isentropic compressor efficiency
η_t	isentropic turbine efficiency
Δ	Operator indicating a change or difference in the parameter operated upon

REFERENCES

1.A. Bejan, Theory of Heat Transfer- Irreversible Power Plants, Int. J. Heat Mass Transfer, 1988, **31**: 1211-19.

2. C. Wu, Power Optimization of an Indirect Gas-Cooled Nuclear Reactor Plant, Nuclear Technology, 1989, **84**: 123-127.

3. C. Wu and R. L. Kiang, Power Performance of a Nonisentropic Brayton Cycle, J. of Engineering for Gas Turbines and Power, 1991, **113**: 501-504.

4. H. S. Leff, Thermal Efficiency a Maximum Work Output: New Results for Old Heat Engines, Am. J. Phys. 1987, **55(7)**: 602-610.

5. B. Sahin, A. Kodal, and H. Yavuz, Efficiency of a Joule-Brayton Engine at Maximum Power Density, J. Phys. D: Appl Phys., 1995, **28**: 1309-1313.

6. B. Sahin, A. Kodal, T.Yilmaz and H. Yavuz, Maximum Power Density Analysis of an Irreversible Joule-Brayton Engine, J. Phys. D: Appl Phys., 1996, **29**: 1162-1167.

7. E.P. Gyftopoulos, Fundamentals of Analyses of Processes, ECOS'96, Stockholm, 1996.

8. A. Bejan, Models of Power Plants that Generate Minimum Entropy while Operating at Maximum Power, Am. J. Phys., 1996, **64**, 1054-1059.

9. A. Bejan, Theory of Heat Transfer - Irreversible Power Plants - II. The Optimal Allocation of Heat Exchange Equipment, Int. J. Heat Mass Transfer, 1995, **38**: 433-444.

10. F. Daniels, Small Gas-Cycle Reactor Offers Economic Promise, Nucleonics, 1956, **20**: 203-214.

APPENDIX A

Gyftopoulos [7] argues that for an analysis of a thermal process to be fully meaningful it must make simultaneous use of both the first and second laws of thermodynamics. Along these same lines, Bejan [8] further argues that to maximize the power of a thermal heat engine, one must minimize the internal entropy generation of the cycle.

An expression of the Second Law of Thermodynamics for the internal cycle (i.e., cycle [system] not including its surroundings [which in this case consist of the finite thermal reservoirs]) of the overall Joule-Brayton cycle (neglecting the changes in kinetic energy of the working substance during process 2-3 and 4-1) is:

$$\oint(\delta q/T) = m \int\oint(c_p dT/T) = c_p[\ln(T_3/T_2)-\ln(T_4/T_1)] \leq 0 .$$

Thus,

$$T_1T_3 < T_2T_4 .$$

This expression is true irregardless of the nature of the external heat transfer processes.

In the ideal Brayton cycle, $\Delta P_H = \Delta P_L = 0$ and $\eta_t = \eta_c = 1$ and there is no internal entropy generation. Thus, for this ideal cycle $T_1T_3 = T_2T_4$, which agrees with Eq.(15) [though it contains less information than Eq.(15)]. When one notes that Eq.(15) results from Eq. (8), which in turn is a result of the process of maximizing cycle specific work, it becomes clear that in the absence of all internal irreversibilities (internal entropy generation), Eq. (8) reduces to a statement which satisfies the Second Law of Thermodynamics. In the more general case of the Joule-Brayton cycle there is internal entropy generation which is kept to its minimum level during power optimization through the closure constraint of Eq. (8). When the cycle specific work is maximum the internal specific entropy production is minimum. Again, for emphasis, this is true irregardless of the nature of the external heat transfer processes.

APPENDIX B

Starting with Eq.(l) and using the definitions for f and g given in Eq.(10), it is easy to show that

$$P = (K_H/g)(T_3-T_2-T_4+T_1) = (xK/g)(T_3-T_2-T_4+T_1). \qquad \text{[B-1a]}$$

and

$$P = (K_L/f)(T_3-T_2-T_4+T_1) = [(1-x)k)/f](T_3-T_2-T_4+T_1). \qquad \text{[B-1b]}$$

For optimum allocation of the conductance inventory these expressions must be equivalent. Thus,

$$x = g/(f+g) . \qquad \text{[B-2]}$$

Substituting this result [Eq. (B-2)] back into Eq.(B-1a) results in Eq. (11), power optimized with respect to the allocation of its conductance inventory.

APPENDIX C

$$(\partial T_2/\partial T_1) = [1-(1/\eta_c) +(r_L^{(k-1)/k}/\eta_c] +(T_1/\eta_c)[(k-1)/k]r_L^{-1/k})(\partial r_L/\partial T_1). \qquad \text{[C-1]}$$

$$(\partial T_4/\partial T_1) = T_3\eta_t a[(1-k)/k](ar_L-b)^{(1-2k)/k}(\partial r_L/\partial T_1). \qquad \text{[C-2]}$$

$(\partial r_L/\partial T_1) = \{1/[a(1-2k)](c^k/r_L)^{2k/(1-2k)}\{[r_L\, kc^{(k-1)}/(\eta_c\eta_t aT_3)]-c^k(\partial r_L/\partial T_1)\}/r_L^2.$ [C-3]

$\partial g/\partial T_1 = \{LMTD_H(-\partial T_2/\partial T_1)-(T_3-T_2)(\partial LMTD_H/\partial T_1)\}/LMTD_H^2.$ [C-4]

$\partial f/\partial T_1 = \{LMTD_L[\partial T_4/\partial T_1-1]-(T_4-T_1)(\partial LMTD_L/\partial T_1)\}/LMTD_L^2.$ [C-5]

$\partial LMTD_H/\partial T1=\{\ln[(T_5-T_3)/(T_6-T_2)](\partial T_2/\partial T_1)-(T_5-T_3-T_6+T_2)(\partial T_2/\partial T_1)/(T_6-T_2)\}/\{\ln[(T_5-T_3)/(T_6-T_2)]\}^2.$ [C-6]

$\partial LMTD_L/\partial T_1=\{\ln[(T_1-T_7)/(T_4-T_8)](1-\partial T_4/\partial T_1)-(T_1-T_7-T_4+T_8)[(T_4-T_8)-(T_1-T_7)(\partial T_4/\partial T_1)]/[(T_1-T_7)(T_4-T_8)]\}/[\ln((T_1-T_7)/(T_4-T_8))]^2$ [C-7]

APPENDIX D

$\partial T_2/\partial T_3 = (T_1/\eta_c)[(k-1)/k]r_L^{-1/k}(\partial r_L/\partial T_3).$ [D-1]

$\partial T_4/\partial T_3 = T_3\eta_t a[(1-k)/k](ar_L-b)^{(1-2k)/k}(\partial r_L/\partial T_3) +[1-\eta_t +\eta_t(ar_L-b)^{(1-k)/k}].$ [D-2]

$\partial r_L/\partial T_3 = \{(c^k/r_L)^{2k/(1-2k)}/[a(1-2k)]\}[-(r_L kT_1 c^{k-1})/(\eta_c\eta_t aT_3^2)-c^k(\partial r_L/\partial T_3)]/r_L^2.$ [D-3]

$\partial g/\partial T_3 = \{LMTD_H[1-(\partial T_2/\partial T_3)]-(T_3-T_2)(\partial LMTD_H/\partial T_3)/LMTD_H^2.$ [D-4]

$\partial f/\partial T_3 = \{LMTD_L(\partial T_4/\partial T_3)-(T_4-T_1)(\partial LMTD_L/\partial T_3)\}/LMTD_L^2.$ [D-5]

$\partial LMTD_H/\partial T_3 = \{\ln[(T_5-T_3)/(T_6-T_2)][(\partial T_2/\partial T_3)-1]-(T_5-T_3-T_6+T_2)[(T_2-T_6)-(T_3-T_5)(\partial T_2/\partial T_3)]/[(T_6-T_2)(T_5-T_3)]\}/\{\ln[(T_5-T_3)/(T_6-T_2)]\}^2.$ [D-6]

$\partial LMTD_L/\partial T_3 = \{\ln[(T_1-T_7)/(T_4-T_8)](-\partial T_4/\partial T_3)-(T_1-T_7-T_4+T_8)[-(T_1-T_7)(\partial T_4/\partial T_3)]/[(T_1-T_7)(T_4-T_8)]\}/\{\ln[(T_1-T_7)/(T_4-T_8)]\}^2.$ [D-7]

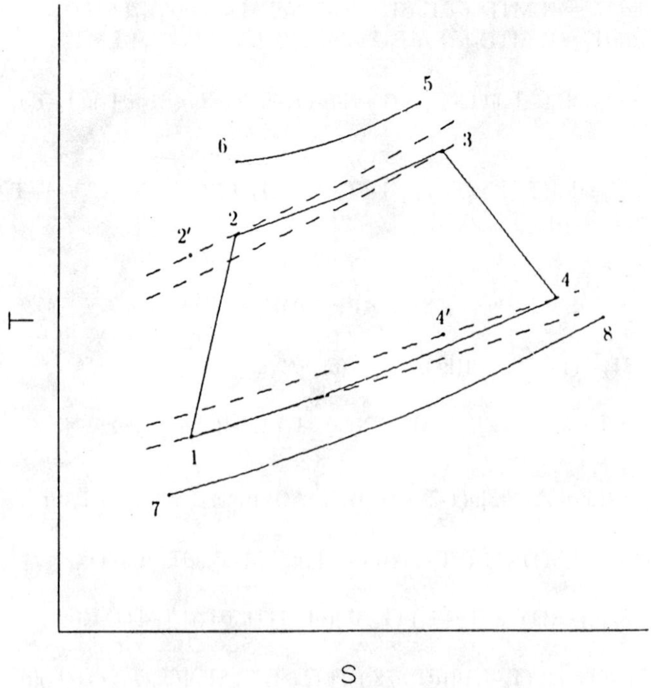

Figure 1. Temperature versus Specific Entropy Diagram for The General Closed Joule-Brayton Cycle with Finite Reservoir Heat Capacities

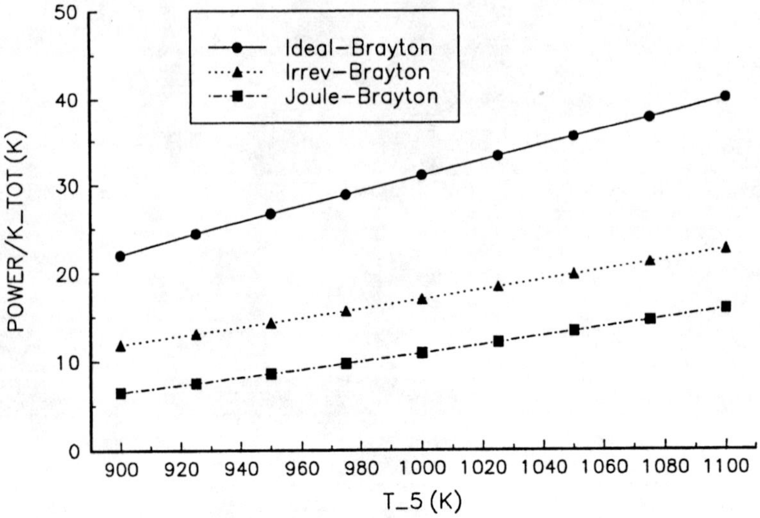

Figure 2. Power per total conductance at optimum power ($\frac{P_{opt}}{K}$) (K) versus T_5 (K) for various Brayton-like cycles.

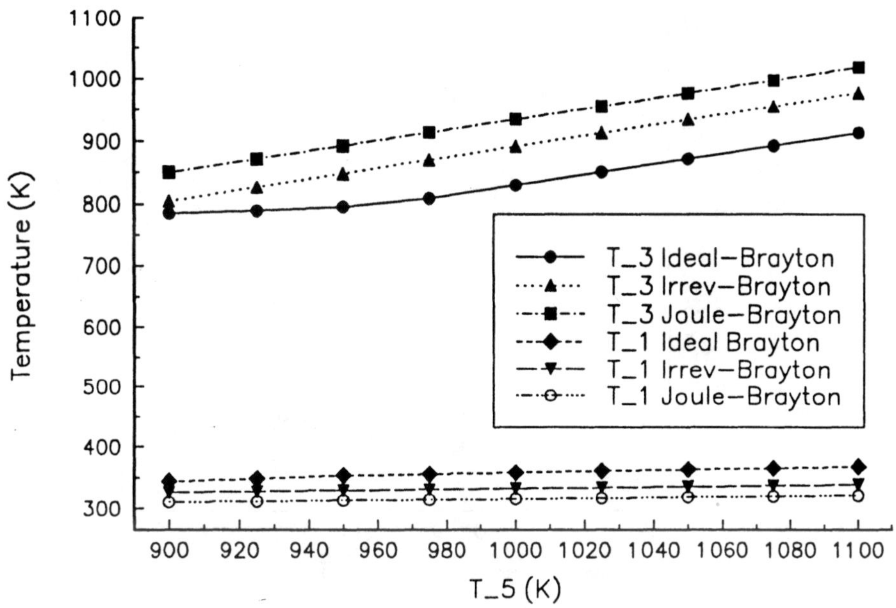

Figure 3. T_3 (K) at optimum power ($(T_3)_{opt}$) and T_1 (K) at optimum power ($(T_1)_{opt}$) versus T_5 (K) for various Brayton-like cycles.

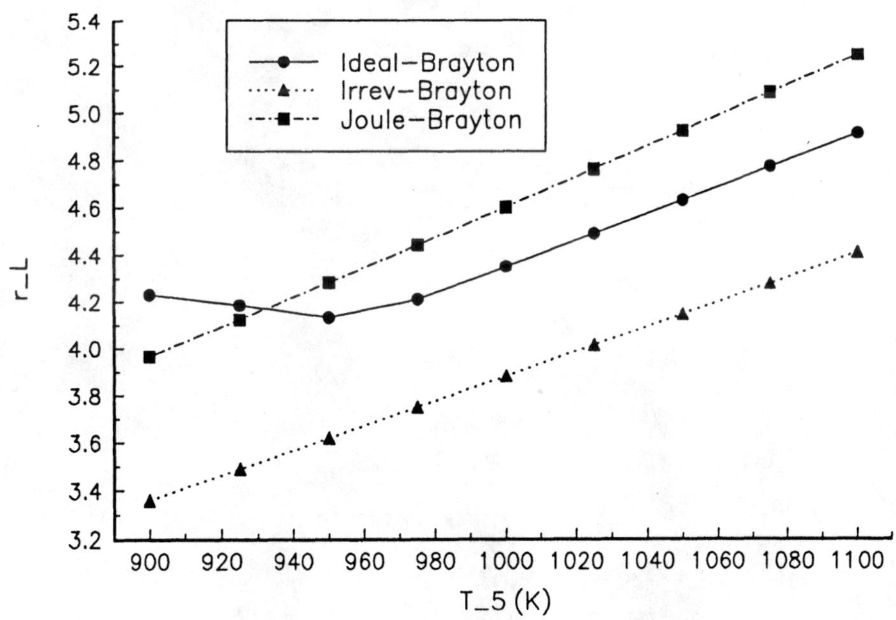

Figure 4. r_L at optimum power ($(r_L)_{opt}$) versus T_H (K) for various Brayton-like cycles.

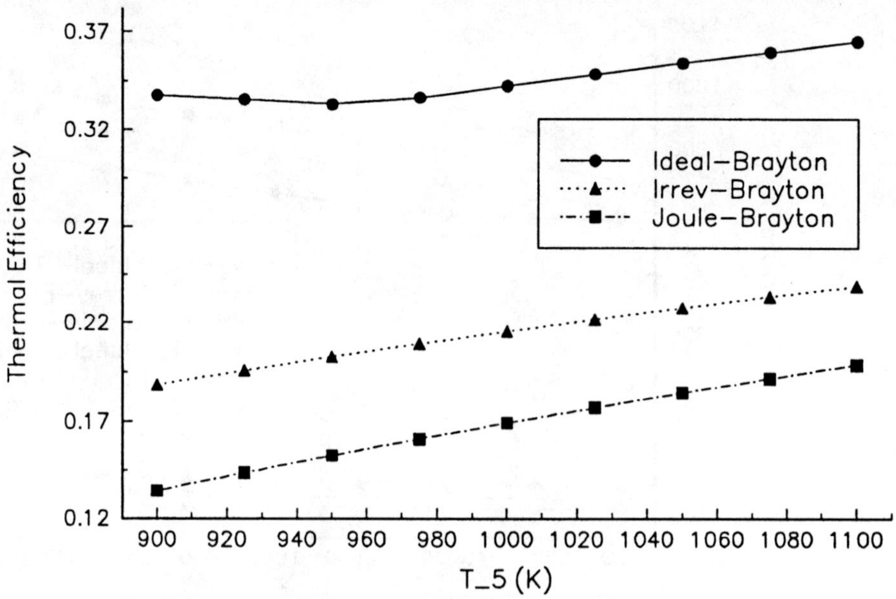

Figure 5. Thermal efficiency at optimum power (η_{opt}) versus T_H (K) for various Brayton-like cycles.

FINITE-TIME THERMODYNAMIC ANALYSIS OF AN IRREVERSIBLE COGENERATION REFRIGERATOR

SELAHATTIN GÖKTUN

Ý.T.Ü. Maritime Faculty, Department of Marine Engineering
81716 Tuzla-Ýstanbul, Turkey

Abstract- An irreversible heat engine-driven vapour compression and absorption refrigerator system is considered as a cogeneration refrigerator. The combined effects of finite-rate heat transfer and working fluid internal dissipation on the optimal performance of cogeneration refrigerator is investigated by using a finite-time thermodynamic approach. An improved equation for the coefficient of performance (COP) of the system under consideration was obtained.

Keywords - Irreversible, cogeneration, heat engine, refrigerator

INTRODUCTION

In order to raise availability of energy sources and decrease the environmental pollution of high-temperature waste heat, cogeneration cycles have been adopted by HVAC&R industry. Cogeneration is an engineering concept involving the production of both power and useful thermal energy from the same source thereby utilizing fuel more effectively than if the desired products were produced separately. The heart of a cogeneration system is a prime mover with waste heat at a useful temperature. Heat engine (HE)-driven refrigerator systems are an example for cogeneration applications. These systems typically use the work produced by the heat engine to drive a vapour compression refrigerator (VCR). At the same time the waste heat rejected by the HE is used for heating purposes or to drive an absorption or an ejector refrigerator. The above models have been studied by many authors [1-5] for investigate the performance. None of the previous investigations concerned a cogeneration refrigerator including internal irreversibilities. It is the purpose of this study to incorporate internal irreversibilities in to the heat engine-driven vapour compression and absorption refrigerator (AR) model and use it for optimization of performance with finite-time thermodynamics.

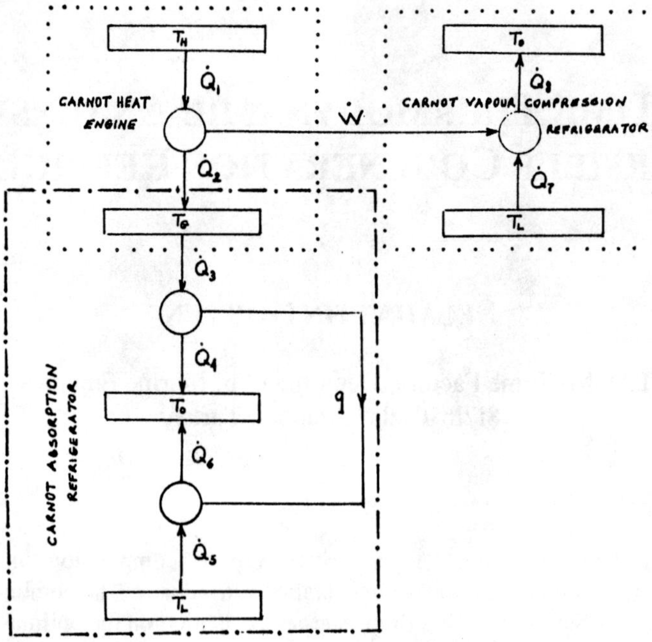

Fig. 1. A schematic diagram of an reversible cogeneration refrigerator

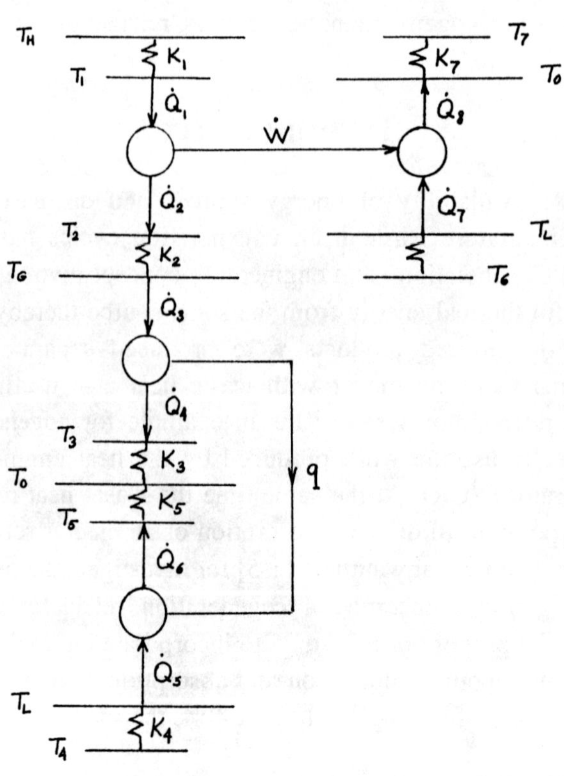

Fig. 2. A schematic diagram of an irreversible cogeneration refrigerator

THEORETICAL MODEL

A reversible HE-driven VCR and AR system operating between reservoirs at temperature T_H, T_o, T_L, consist of a Carnot heat engine (CHE), a Carnot vapour compression refrigerator (CVCR) and a Carnot absorption refrigerator (CAR) as shown in Fig. 1. CAR operates among the source (generator) at temperature T_G, the sink (ambient) at temperature T_o and the source (refrigerated space) at temperature T_L . CHE drives both CVCR and CAR units and operates between the high-temperature heat source at temperature T_H and the sink at temperature T_G. CVCR operates between the heat source at temperature T_L and the heat sink at temperature T_0. According to Fig. 1, we obtain the overall COP of an reversible cogeneration refrigerator

$$COP_{rev.} = \left[\ (\dot{Q}_5 + \dot{Q}_7) / \dot{Q}_1\ \right] = \varepsilon_{1r}(1-\eta_r) + \varepsilon_{2r}\eta_r \tag{1}$$

where ε_{1r} and ε_{2r} are the $COPs$ of the CAR and CVCR, respectively. η_r is the thermal efficience of the CHE. $\dot{Q}_5$ and $\dot{Q}_7$ are the heats absorbed from the sources by the working fluids in the CAR and CVCR, respectively. $\dot{Q}_1$ is the heat absorbed from the high-temperature reservoir by the working fluid in the CHE per unit time.

The first and second laws of thermodynamics are used to analyse the reversible system in Fig.1. The governing equations are

$$\dot{W} = \dot{Q}_1 - \dot{Q}_2 \tag{2}$$

$$\dot{W} = \dot{Q}_8 - \dot{Q}_7 \tag{3}$$

$$q = \dot{Q}_3 - \dot{Q}_4 \tag{4}$$

$$q = \dot{Q}_6 - \dot{Q}_5 \tag{5}$$

$$(\dot{Q}_1 / T_H) = \dot{Q}_2 / T_G \tag{6}$$

$$(\dot{Q}_3 / T_G) = \dot{Q}_4 / T_0 \tag{7}$$

$$(\dot{Q}_5 / T_L) = \dot{Q}_6 / T_0 \tag{8}$$

$$(\dot{Q}_7 / T_L) = \dot{Q}_8 / T_0 \tag{9}$$

Using Eqs.(2)-(9) yield

$$\varepsilon_{1r} = (T_L / T_G)[(T_G - T_0)/(T_0 - T_L)] \tag{10}$$

$$\eta_r = (T_H - T_G)/T_H \tag{11}$$

$$\varepsilon_{2r} = T_L / (T_0 - T_L) \tag{12}$$

Substituting Eqs.(10)-(12) into Eq.(1) yields

$$COP_{rev.} = (T_L / T_H)[(T_H - T_0)/(T_0 - T_L)] \,. \tag{13}$$

Although a CHE-driven CVCR and CAR system displays the highest COP, it does not produce any cooling load. Irreversible systems produce a finite amount of cooling load. Therefore, the COP given by Eq.(13) is not applicable to real cogeneration refrigerator under consideration. Irreversible HE-driven VCR and AR system exhibit $COPs$ which are significantly lower than those of the reversible system. A more realistic upper bound is needed for the COP of operating HE-driven VCR and AR system. For this reason, we consider a steady flow irreversible HE-driven VCR and AR system operating between reservoirs at temperatures T_H, T_0 and T_L as shown in Fig. 2. This system includes a finite-sized irreversible HE, an irreversible VCR and an irreversible AR. The irreversible HE provides the power $\dot{W}$ and the heating rate $\dot{Q}_2$ to the VCR and AR, respectively. Thus the VCR and AR units produce cooling loads $\dot{Q}_7$ and $\dot{Q}_5$, respectively, as shown in Fig.2. The irreversible finite-size HE operates between the source at temperature T_H and the sink at temperature T_G and the temperatures of the working fluid in the two isothermal processes are T_1 and T_2. The irreversible VCR operates between a heat source at temperature T_L and a sink at temperature T_0, the temperatures of the working fluid in the two isothermal processes are T_6 and T_7. The irreversible AR operates between the three-heat-sources at temperatures T_2, T_0 and T_L and evaporator and condenser, respectively. q is the heat-power output of the power sub-cycle of the AR unit.

According to Fig.2, we obtain the overall COP of an irreversible cogeneration refrigerator from the relation

$$COP_{irr.} = [\ (\dot{Q}_5 + \dot{Q}_7) / \dot{Q}_1\ ] = \varepsilon_1 (1 - \eta\ ) + \varepsilon_2 \eta \tag{14}$$

where ε_1 and ε_2 are the $COPs$ of the irreversible AR and VCR, respectively. η is the thermal efficiency of the finite-sized irreversible HE.

Fourier's law and the first and second laws of thermodynamics are used to analyse this system. The governing equations are:

$$\dot{Q}_1 = K_1 (T_H - T_1) \tag{15}$$

$$\dot{Q}_2 = K_2 (T_2 - T_G) \tag{16}$$

$$\dot{Q}_3 = \dot{Q}_2 \tag{17}$$

$$\dot{Q}_4 = K_3 (T_3 - T_o) \tag{18}$$

$$\dot{Q}_5 = K_4 (T_L - T_4) \tag{19}$$

$$\dot{Q}_6 = K_5 (T_5 - T_o) \tag{20}$$

$$\dot{Q}_7 = K_6 (T_L - T_6) \tag{21}$$

$$\dot{Q}_8 = K_7 (T_7 - T_o) \tag{22}$$

$$\dot{Q}_3 - \dot{Q}_4 + \dot{Q}_5 - \dot{Q}_6 = 0 \tag{23}$$

$$\dot{W} = \dot{Q}_1 - \dot{Q}_2 \tag{24}$$

$$\dot{W} = \dot{Q}_8 - \dot{Q}_7 \tag{25}$$

$$(\dot{Q}_1 / T_1) = R_1 (\dot{Q}_2 / T_2) \tag{26}$$

$$(\dot{Q}_3 / T_G) = R_2(\dot{Q}_4 / T_3) \tag{27}$$

$$(\dot{Q}_5 / T_4 = R_2(\dot{Q}_6 / T_5) \tag{28}$$

$$(\dot{Q}_7 / T_6) = R_3(\dot{Q}_8 / T_7) \tag{29}$$

$$K = K_1 + K_2 \tag{30}$$

The variables used in the preceding equations are defined as $K_i (i = 1,2,..)$ is the heat conductance, $R_j (j = 1,2,..)$ is the cycle irreversibility parameter [6]. $\dot{Q}$ is the heat transfer and K is the total size of the heat-exchanger equipment in the irreversible HE [7]. Using Eqs. (15), (16), (24), (26) and (30) and the definitions $\alpha = T_2 / T_1$, $\beta = K_1 / K_2$ and $\gamma = T_G / T_H$, the dimensionless rate of heat transfer to the finite-size irreversible HE $\varnothing$ and the dimensionless power output P of the HE are

$$\varnothing = (\dot{Q}_1 / KT_H) = \beta R_1(\alpha - \gamma) / (\beta + 1)(\alpha\beta + \alpha R_1) \tag{31}$$

$$P = (\dot{W} / KT_H) = \varnothing[1 - (\alpha / R_1)] \tag{32}$$

Using the extremal conditions

$$(\partial P / \partial \alpha) = 0 \quad \text{and} \quad (\partial P / \partial \beta) = 0 \tag{33}$$

which yield

$$\alpha_{opt} = (\gamma R_1)^{0.5} \tag{34}$$

$$\beta_{opt} = R_1^{\ 0.5} \tag{35}$$

Substituting Eqs. (34) and (35) into Eqs. (31) and (32) yields

$$\varnothing_{opt} = R_1^{\ 0.5}(R_1^{\ 0.5} - \gamma^{0.5}) / (1 + R_1^{\ 0.5})^2 \tag{36}$$

$$P_{\max} = (R_1^{\ 0.5} - \gamma^{0.5})^2 / (1 + R_1^{\ 0.5})^2 \tag{37}$$

Using Eqs. (15), (16), (24), (26), (34) and (35), the optimum working-fluid temperatures of the finite-sized irreversible HE can be written as

$$T_{1(opt)} = T_H(1+\gamma^{0.5})/(1+R_1^{0.5}) \tag{38}$$

$$T_{2(opt)} = T_H(\gamma R_1)^{0.5}(1+\gamma^{0.5})/(1+R_1^{0.5}) \tag{39}$$

The thermal efficiency of the finite-sized irreversible HE at its maximum power output is

$$\eta = (P_{max}/\varnothing_{opt}) = 1-(\gamma/R_1)^{0.5} \text{ for } \gamma < R_1 . \tag{40}$$

Using Eqs. (23), (27) and (28) the COP of the irrreversible AR can be written as

$$\varepsilon_1 = (Q_5/Q_3) = (T_4/T_G)[(R_2T_G - T_3)/(T_5 - R_2T_4)] \tag{41}$$

Solving Eqs. (21), (22), (25) and (29) under the condition that the irreversible VCR is driven by the finite-sized irreversible HE at its maximum power output yields

$$T_7 = T_0[1+(P_{max}KT_H)/(1-\psi R_3)K_7T_0] \tag{42}$$

$$T_6 = T_L[1-(P_{max}KT_HR_3\psi)/(1-\psi R_3)K_6T_L] \tag{43}$$

$$\dot{Q}_{7m} = K_6(T_L - T_6) \tag{44}$$

$$\varepsilon_2 = \dot{Q}_{7m}/(P_{max}KT_H) = \psi R_3/(1-\psi R_3) \tag{45}$$

where $\psi = T_6/T_7$ is the temperature ratio of the irreversible VCR, $\dot{Q}_{7m}$ is the cooling load produced by the VCR which is driven by the finite-sized irreversible HE at its maximum power and ε_2 is the corresponding COP. The temperature ratio ψ of the irreversible VCR is a function of the parameters $R_1, R_3, K, K_6, K_7, T_G, T_H, T_L$ and T_0 . When these parameters are given, ψ can be determined from the Eqs. (37), (42) and (43).

On substituting Eqs. (40), (41) and (45) into Eq. (14), the overall COP of the finite-sized irreversible HE-driven VCR and AR system can be obtained as

$$COP_{irr} = (\gamma / R_1)^{0.5} \{(T_4 / T_G)[(R_2 T_G - T_3)/(T_5 - R_2 T_4)] - (\psi R_3)/(1-\psi R_3)\} + \\ +(\psi R_3)/(1-\psi R_3) \quad (46)$$

NUMERICAL EXAMPLE

An irreversible HE-driven VCR and AR system operating between reservoirs at temperatures $T_H = 480K$, $T_0 = 295K$ and $T_L = 268K$ is considered. For simplicity we write $K_i = 1kW/K$. Using Eqs. (38) and (39) for $R_1 = 0.9$ and $T_G = 350K$ yields $T_1 = 457K$ and $T_2 = 370K$. Then applying Eqs. (16)-(18) and (27) yields $T_3 = 314K$ for $R_2 = 0.95$. By using Eqs. (37), (42) and (43) for $R_3 = 0.9$, the temperature ratio ψ of the irreversible VCR is found to be 0.85. Then applying Eqs. (37), (42) and (43) yield $T_7 = 300K$ and $T_6 = 255K$. Using Eqs. (19), (20) and (28) for $T_5 = 305K$ yields $T_4 = 260K$. The overall COP for the completely reversible cogeneration refrigerator under consideration was calculated by using Eq.(13) and was found to be 3.83. Calculating the overall COP using Eq. (46) for the irreversible cogeneration refrigerator yields a value of 0.54. This overall COP was lower than that found for the reversible system; however it will provide a cooling load of 17 kW and represents a more realistic COP in terms of that which is found in the real cogeneration applications.

REFERENCES

1. Åhlby, L., Hodgett, D. and Berntsson, T. 1991. Optimization study of the compression /absorption cycle. International Journal of Refrigeration 14:16-23.
2. Dentice, M., Sasso, M. and Sibilio, S. 1996. Proc.ECOS'96 (Stockholm, 25-7 June 1996) ed. P.Alvfors, L.Eidensten, G.Svedberg and J.Yan (Stockholm:Royal Institute of Technology) 135-140.
3. Howe, L.A., Radermacher, R. and Herold, K.E. 1989. Combined cycles for engine-driven heat pumps. International Journal of Refrigeration 12:21-28.
4. Þahin, B., Kodal, A., Ekmekçi, Ý. and Yýlmaz, T. 1997. Exergy optimization for an endoreversible cogeneration cycle. Energy - The International Journal 22:551-557.
5. Göktun, S. and Yavuz, H. 1997. The performance of irreversible combined heat-pump cycles. Journal of Physics D:Applied Physics. 30:2834-2840.
6. Wu, C., and Kiang, R.L. 1992. Finite-time thermodynamic analysis of a Carnot engine with internal irreversibilities. Energy - The International Journal 17:1173-1178.
7. Bejan, A. 1988. Advanced Engineering Thermodynamics. (New York: Wiley).

DYNAMIC SOLAR SPACE POWER SYSTEMS: FINITE-TIME THERMODYNAMIC MODELS

VIOREL BADESCU
Candida Oancea Institute of Solar Energy,
Faculty of Mechanical Engineering,
Polytechnic University of Bucharest,
Spl Independentei 313, Bucharest 79590, Romania

Abstract- The paper focuses on two applications of finite-time thermodynamics. First, one presents a simple model of solar space power station. Universal graphs are drawn for the optimum ratio between the collector and radiator area and for a non-dimensional parameter related to the maximum power. A brief comparison between three usual types of mirrors (conical, parabolic cylinder and paraboloidal) is performed. The differences between the performance obtained by using a Curzon-Ahlborn and an ideal Carnot engine, respectively, are outlined. Second, a model of Dyson's sphere (DS) is developed. It was proved that no optimum DS radius exists but there is a minimum DS radius, whose value is a function of the ambient temperature only. Reasonable DS radii are characterized by efficiencies of solar power generation less than 10 %.

Key words: Solar concentrator, Carnot engine, Curzon-Ahlborn engine, Optimum radiator area, Maximum power, Dyson's sphere

1. INTRODUCTION

The oil crisis of 1973 made it clear that energy sources are neither inexhaustible nor free. The criteria of merit provided by traditional reversible thermodynamics were, with few exceptions, far too unrealistic for evaluating the potential for conservation in real processes. Irreversible thermodynamics could in principle evaluate the losses in rate processes, but it requires extensive knowledge of the microscopic mechanism of the process. These observations prompted some researchers to formulate a macroscopic theory of loss processes in the tradition of reversible thermodynamics and requiring as little additional information about the system as possible. This theory is known as finite-time thermodynamics (for a good early review see [1]; new results are reviewed in [2][3]).

This paper deals with finite-time thermodynamic models of solar space power systems. Non-spectral models will be considered only. The reader interested in spectral models is directed to [4,5]. Section 2 presents two models of dynamic solar space power plant. The first model is based on classical thermodynamics while the second one proposes a finite-time thermodynamic approach. An application makes clear the differences between the results of the two models. Section 3 refers to a futuristic solar space application: the Dyson sphere. Finite-time thermodynamics allow to put limits for the sphere's radius and the efficiency of converting solar energy into work. A short conclusion ends the paper.

2. Dynamic Solar Space Power Plant

A sound work have been performed in the last years to improve the performance and economic aspects of space power systems. Different aspects concerning the design and architectural solutions, terrestrial or non-terrestrial materials, working fluids and cycles, energy storage and the operational system as well as other related topics as the launch system or the ecological aspects and the problems experienced by man when constructing the space structures have been analyzed (see e.g. [6-15]).

Presently , there are two major types of solar power systems. The first type uses photovoltaic cells to directly convert the solar radiation into electrical energy, in combination with electrochemical storage. The second one is the so called "dynamic" system. Solar concentrators reflect the flux of solar radiation towards a collector where a working fluid is heated up to drive conventional thermal engines. Electrical energy could be generated by alternators coupled to these thermal engines. This last solar power system is considered here.

2.1. System Configuration

The main components of a solar space power plant are shown in Fig. 1. They are: the solar concentrator, the collector (receiver), the thermal engine and the heat radiator. Other components of the power station as energy accumulators, transformers etc, are not relevant at this stage of the analysis. However, their weight will be taken into consideration subsequently. The solar concentrator has an opening area A_{open}. Its mirror has the optical reflectance ρ_m, the surface A_m and a superficial mass density d_m. The collector area is denoted A_c. Its temperature, emittance and absorptance, all of them supposed to be uniform, are denoted T_c , e_c and a_c, respectively. Also, the radiator superficial mass density is d_r.

Fig. 1. The main elements of a solar space power station.

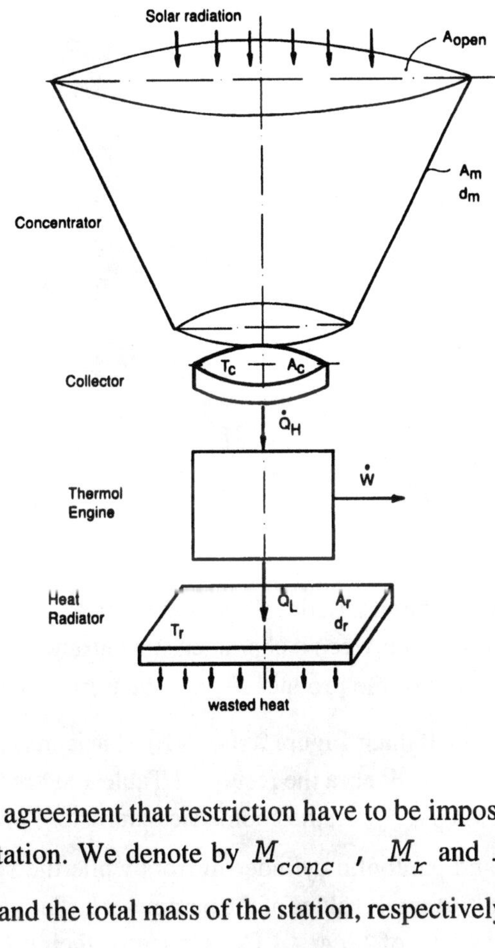

There is a general agreement that restriction have to be imposed on the total mass of a (solar) space power station. We denote by M_{conc} , M_r and M_{tot} the masses of the concentrator, radiator and the total mass of the station, respectively. The following relation applies:

$$M_{conc} + M_r = \kappa M_{tot} \tag{1}$$

with $0 < \kappa < 1$.

The concentrator and radiator masses are given by:

$$M_{conc} = A_m d_m \qquad M_r = A_r d_r \tag{2}$$

while the geometric concentration ratio C_n is usually defined as:

$$C_n = \frac{A_{open}}{A_c} \tag{3}$$

By replacing (2) and (3) in (1) we obtain:

$$C_n f A_c + s A_r = A_e \tag{4}$$

where the following notations were used:

$$f \equiv \frac{A_m}{A_{open}} \qquad s \equiv \frac{d_r}{d_m} \qquad A_e \equiv K \frac{M_{tot}}{d_m} \tag{5}$$

The shape factor f is a characteristic of the concentrating mirror. It depends on a parameter k defined as the ratio between a characteristic entrance size r and the depth h of the mirror. For the same value of the product $A_c C_n$ the mirrors with a smaller value of f are preferred because they are lighter. Figure 2 shows five basic mirrors used for solar space power applications. After some algebra the results of Table 1 are obtained. Figure 3 shows the dependence of the ratio A_m / A_{open} on the geometric parameter k=h/r. The best results are provided by the conical and parabolic cylinder mirrors while the spherical concentrators could be recommended only at small values of the parameter k. The paraboloidal mirrors are not characterized by small values of A_m / A_{open}. Consequently, their use is not always justified for those concentration ratios which are accessible to the conical or parabolic cylinder mirrors ($C_n < 250$).

2.2. Classical Thermodynamics Model

A simple classical model of solar space power station based on thermodynamic cycles was proposed by Bejan [16]. Here we follow the approach from [17] where an improved version of Bejan's model was developed. The main assumption is the existence of an ideal reversible (Carnot) thermal engine. The energy balance equations for collector and radiator, respectively, are:

$$\dot{Q}_H = A_c(\phi_{pc} - \phi_{cT}) \tag{6}$$

$$\dot{Q}_L = A_r(\phi_{rT} - \phi_{\infty r}) \tag{7}$$

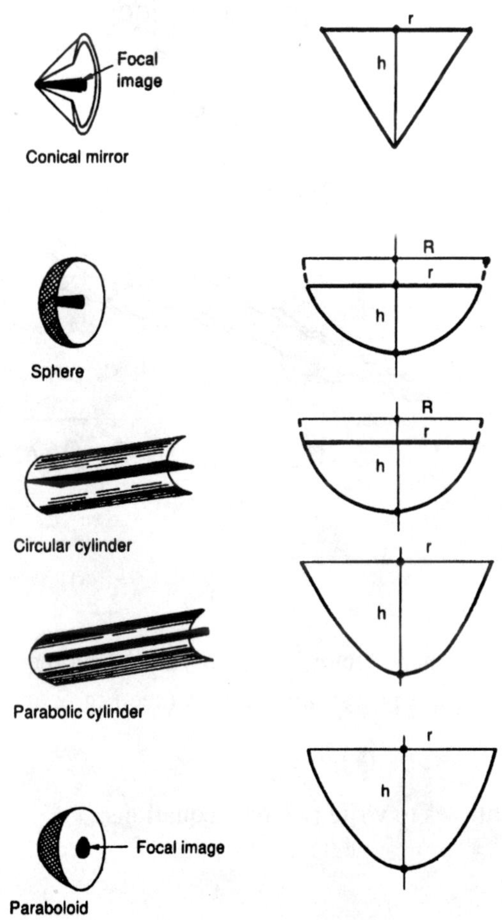

Fig. 2. Five mirror types used for solar space power applications.

Here $\dot{Q}_H$ and $\dot{Q}_L$ are the heat fluxes entering and leaving the thermal engine, respectively. Also, ϕ_{pc} and ϕ_{cT} are the energy fluxes of the radiation incident and emitted by the collector, respectively, while ϕ_{rT} and $\phi_{\infty r}$ are the energy fluxes of the radiation emitted and received by the radiator, respectively.

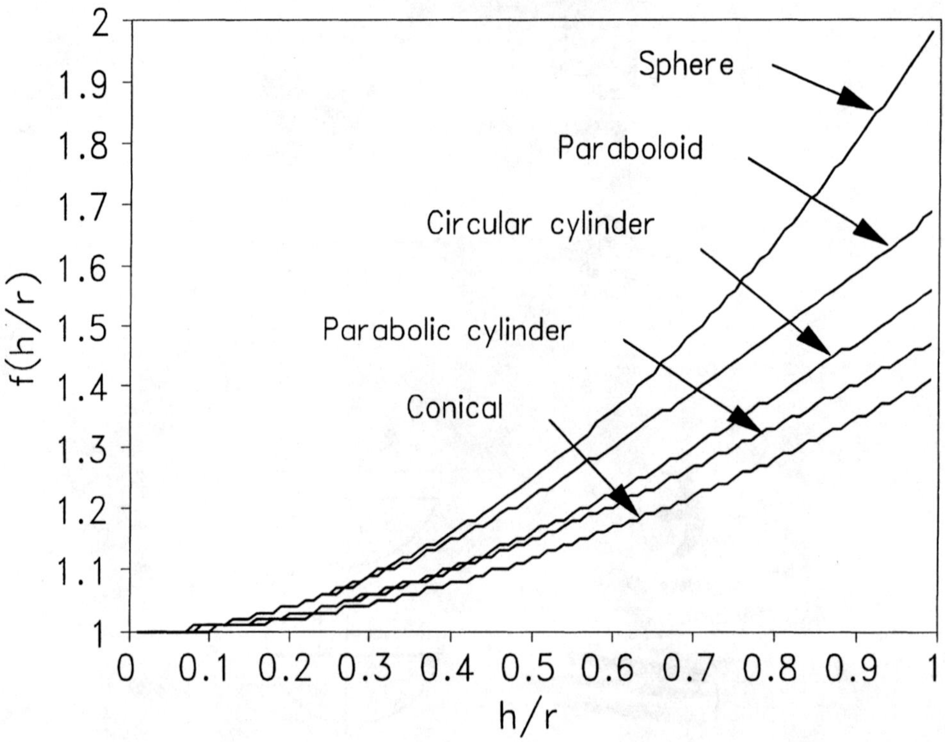

Fig. 3. The dependence of the shape factor f (Eqs (5)) on the geometric parameter k=h/r for some mirrors usual in solar space power applications (see Fig. 2 and Table 1).

The "Carnot" hypothesis allows to write two new equations:

$$\dot{W}=\dot{Q}_H\left(1-\frac{T_r}{T_c}\right) \tag{8}$$

$$\frac{\dot{Q}_H}{T_c}+\frac{\dot{Q}_L}{T_r}=0 \tag{9}$$

Equation (8) gives the power supply while (9) is a consequence of the reversible thermal engine we used.

Some details follow. The energy flux of the concentrated radiation incident on the collector is given by [17]:

$$\phi_{pc}=\rho_m a_c \frac{B(\Omega_c)}{\pi}\sigma T_s^4 \tag{10}$$

where a_c is the collector absorptance, T_s is Sun's temperature while the geometric factor B is a function of the enlarged solid angle Ω_c subtended by the Sun when viewed from the concentrator mirror, given by:

$$B(\Omega_c)=\Omega_c(1-\frac{\Omega_c}{4\pi})\cos\theta_0 \tag{11}$$

where θ_0 is the zenith angle of the Sun as reported to the normal at the concentrator aperture. The geometric factor $B(\Omega_c)$ can be computed as a function of the geometrical concentration ratio if we take into consideration that [17]:

$$C_n=\frac{\Omega_c(4\pi-\Omega_c)}{\Omega_s(4\pi-\Omega_s)} \tag{12}$$

where Ω_s is the solid angle subtended by the Sun ($\Omega_s=6.835\cdot 10^{-5}$ sr at the mean distance Earth - Sun).

The energy fluxes emitted by the collector and radiator are given by:

$$\phi_{cT}=e_c\sigma T_c^4 \tag{13}$$

$$\phi_{rT}=e_r\sigma T_r^4 \tag{14}$$

where e_c and e_r are collector and radiator emittance, respectively.

The energy flux $\phi_{\infty r}$ incident on the radiator generally depends on the temperature and/or albedo of the neighboring celestial bodies. When the case of an interplanetary station is considered, this flux is mainly due to the ambient temperature ($\approx 2-4\ K$). Consequently, it could be neglected. Moreover, the hypothesis $\phi_{\infty r} \ll \phi_{rT}$ is a very good approximation in all the practical cases.

The following notations will be used:

$$x=\frac{a_c\rho_m}{e_c}\frac{B(C_n)}{\pi}\left(\frac{T_s}{T_c}\right)^4 \tag{15}$$

By using (7) and (9-15) we obtain:

$$\frac{T_r}{T_c}=\left(\frac{e_cA_c}{e_rA_r}\right)^{1/3}(x-1)^{1/3} \tag{16}$$

Now, the ratio T_r/T_c can be replaced in (8) and the power $\dot{W}$ becomes a function of the parameter x (i.e. of the temperature T_c). The overall thermal efficiency of the power station is defined as:

$$\eta = \frac{\dot{W}}{A_c \frac{B(C_n)}{\pi} \sigma T_s^4} \quad (17)$$

The maximum efficiency η_{max} is given by:

$$\eta_{max} = a_c \rho_m \frac{x_{opt} - 1}{x_{opt} + 3} \quad (18)$$

where x_{opt} is the solution of the equation $d\eta / dx = 0$,i.e.

$$p^3 (x_{opt} - 1)(1 + \frac{x_{opt}}{3})^3 = 1 \quad (19)$$

where

$$p = (\frac{e_c A_c}{e_r A_r})^{1/3} \quad (20)$$

It is easy to see that η_{max} is a function of A_c when the constraint (4) is taken into account. Simple computations shows that a second maximization of η_{max} in respect to A_c has no local solution. However, if the maximum power $\dot{W}_{max}\ (=\frac{B(C_n)}{\pi}\sigma T_s^4 A_c \eta_{max})$ is considered, such a second maximization is possible. To this aim we have to solve the equation $d\dot{W}_{max}/dA_c=0$ by taking into account the constraint (4). The optimum value of x is obtained from:

$$(x_{opt,opt}-1)^2(1+\frac{x_{opt,opt}}{3})^3=t\frac{e_r}{e_c} \tag{21}$$

where

$$t=\frac{C_n f}{s} \tag{22}$$

The optimum values of the ratios A_r/A_c and A_e/A_c are given by:

$$(\frac{A_r}{A_c})_{opt}=\frac{t}{x_{opt,opt}-1} \tag{23}$$

$$(\frac{A_e}{A_c})_{opt}=st\frac{x_{opt,opt}}{x_{opt,opt}-1} \tag{24}$$

The non-dimensional maximum power $\dot{W}'_{max,max}$ and the optimum maximum efficiency $\eta_{max,opt}$ will be defined as:

$$\text{ax} \equiv \frac{\dot{W}_{max,max}}{a_c \rho_m A_{c,opt} \dfrac{B(C_n)}{\Pi} \sigma T_s^4} = \frac{\eta_{max,opt}}{a_c \rho_m} \equiv \frac{x}{x} \tag{25}$$

where $A_{c,opt}$ can be evaluated with (24).

Finally, an usual indicator of space power stations is obtained by using (5),(24) and (25) and taking into account that $B(C_n) = C_n B(C_n = 1)$:

$$\frac{t}{\text{max}} = \frac{d_m f}{\kappa a_c \rho_m} \frac{1}{\dfrac{B(C_n=1)}{\Pi} \sigma T_s^4} \frac{x_{opt,opt}(x_{opt,}}{(x_{opt,opt} -} \tag{26}$$

The value of this indicator may be used as a criteria to choose among different design solutions.

2.3. Finite-time Thermodynamics Model

Now, we suppose the thermal engine is of Chambadal-Novikov-Curzon-Ahlborn (shortly Curzon-Ahlborn) type. This kind of engine is a particular case of the so called "endoreversible engines" [18], which are supposed to consist of three parts :

(a) a reversible part (i.e. a Carnot part) working between two heat reservoirs (one at the high temperature, say t_1, and one at the low temperature, say t_2; usually, t_1 and t_2 are the temperatures of the working fluid during its isothermal expansion and compression, respectively).

(b) two irreversible parts containing temperature drops (i.e. the temperature fall $T_c - t_1$ accompanying $\dot{Q}_H$ and the temperature fall $t_2 - T_r$ accompanying $\dot{Q}_L$).

The Curzon-Ahlborn engine is a special case of endoreversible engine characterized by linear relationships between the heat flows and the temperature gradients (i.e. $\dot{Q}_H \sim (T_c - t_1)$ and $\dot{Q}_L \sim (t_2 - T_r)$). Details on the endoreversible and Curzon-Ahlborn engines can be found in the articles of Rubin [18] and Curzon and Ahlborn [19], respectively, or in the papers of DeVos [20,21] and Bejan [3].

Under the above assumptions two new equations can be written [20]:

$$\dot{W}=\dot{Q}_H[1-(\frac{T_r}{T_c})^{\frac{1}{2}}] \tag{27}$$

$$\frac{\dot{Q}_H}{T_c^{1/2}}+\frac{\dot{Q}_L}{T_r^{1/2}}=0 \tag{28}$$

Equation (27) gives the power supply while (28) is a consequence of the endoreversible thermal engine we used. After computations similar to those given in Section 2.2 the results shown in Table 2 are obtained.

2.4. Results and Discussions

In order to show some general results in terms of the dimensionless parameter te_r/e_c, let us look first to the usual values of material properties. The present technologies allow values close to the unity for collector absorptance and radiator emittance [22]. We'll accept here $a_c=e_r=0.9$, in order to cover the negative influences of the chemical degradation. Two types of receivers will be analyzed, i.e. a selective and a non-selective collector, with $e_c=0.1$ and 0.9, respectively. Good optics mirror allows values of normal reflectance bigger than 0.9 [22]. Here we'll consider that ρ_m varies between 0.7 and 0.9, to take into account both the time degradation and the effect of the incidence angle, which normally diminish the mirror reflectance. Figure 3 shows that usually the geometric shape function f varies between 1 and 2, whatever the mirror type is. The concentration ratio C_n varies between 5 and 100 for parabolic cylinder concentrators and between 10 and 250 in case of conical mirrors. However, the paraboloidal concentrators allows values of C_n of the order of hundreds or even thousands [23]. Almost all the above information refer to terrestrial solar power systems, too. Consequently, a relative long term experience is available and, to a certain extent, these values can be used for space power applications. A different situation occurs in case of the mass densities per unit area of mirror and radiator, d_m and d_r, respectively, where limited experience is available. So, in case of d_m early studies suppose a value of about $0.01\ kg/m^2$ and $0.025\ kg/m^2$ for plastic and metal reflectors, respectively. A value $d_m \approx 0.1\ kg/m^2$ is accepted when the structure weight is taken into account [24]. More recent works consider $d_m \approx 0.01\ kg/m^2$ [25]. Also, values of d_m between $0.1\ kg/m^2$ in case of plate reflectors and $0.033\ kg/m^2$ for inflatable mirrors were accepted by Compte et al [10]. Wallace and Bussard [26] considered the reflector mass per unit area $d_m \approx 0.16\ kg/m^2$. Some Russian studies accepted that d_m is around $2\ kg/m^2$, when the structure weight is taken into account [12]. A few data refer to the mass density per radiator area. Thus, Angelo and Buden [27] accepted a value $d_r \approx 2.3\ kg/m^2$ in case of the radiator of a nuclear power satellite. This is close to the value $d_r \approx 3\ kg/m^2$ accepted in Mozjorine et al [12] for a solar power station but is only a half of $d_r \approx 5\ kg/m^2$ which was accepted by Fay et al [28].

Figure 4 refers to the case when power is generated by using a classical (Carnot) thermal engine and covers almost all of the above situations. The large range of the parameter te_r/e_c is justified by the values which the concentration ratio C_n has in case of different types of mirrors. The maximum efficiency increases by increasing te_r/e_c, as expected. The selective collectors (say $e_c \sim 0.1$) have always better performances than the non-selective ones ($e_c \sim 0.95$). Note that for large values of te_r/e_c the maximum efficiency has a more important dependence on the mirror quality ($\eta_{max,opt}$ is linear in ρ_m; see (28)) than on the selective or non-selective collector coatings (i.e. on e_c). The optimum ratio $(A_c/A_r)_{opt}$ decreases when te_r/e_c increases (see the variation of p_{opt} and Eq (20)). An important influence on $(A_c/A_r)_{opt}$ has the collector type (namely selective or non-selective). The optimum ratio decreases when selective coatings are used. Note that when $te_r/e_c<1$ the product $(e_c/e_r)(A_c/A_r)$ is over unitary. Similar qualitative features were emphasized in case of the solar power stations equipped with a Curzon-Ahlborn engine (Fig. 5). However, important quantitative differences have to be mentioned. Thus, the values of $x_{opt,opt}$ and $\dot{W}_{\max}$ decrease, while p_{opt} and $(T_r/T_c)_{opt}$ increase when the ideal Carnot engine is replaced with the more realistic Curzon-Ahlborn engine.

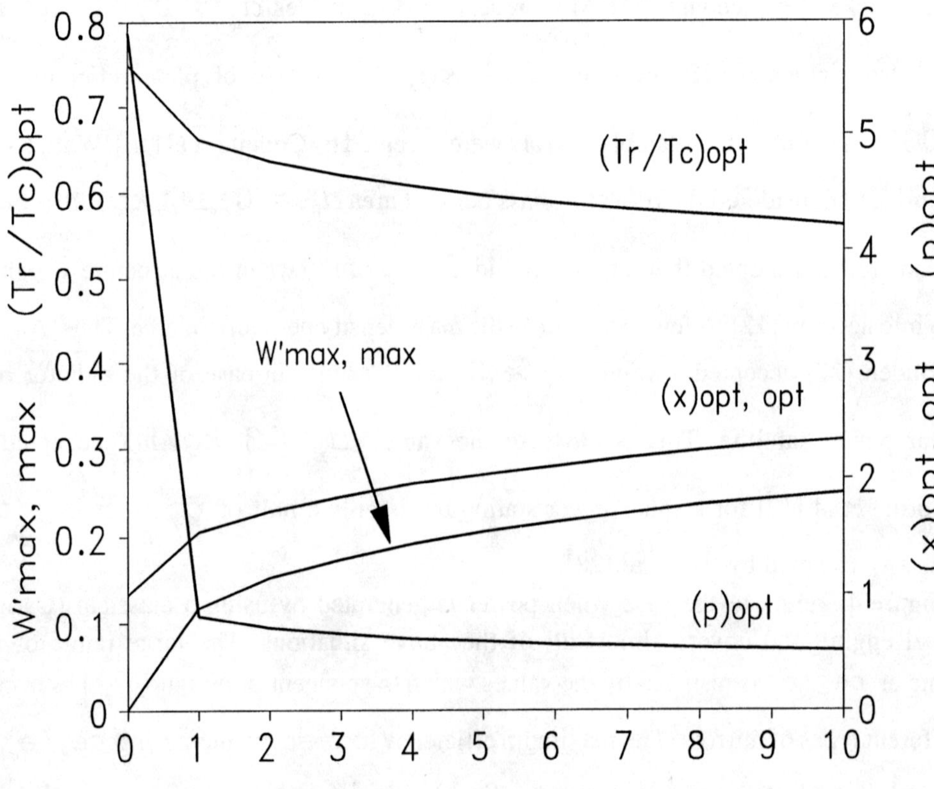

Figure 4a

Figure 4b

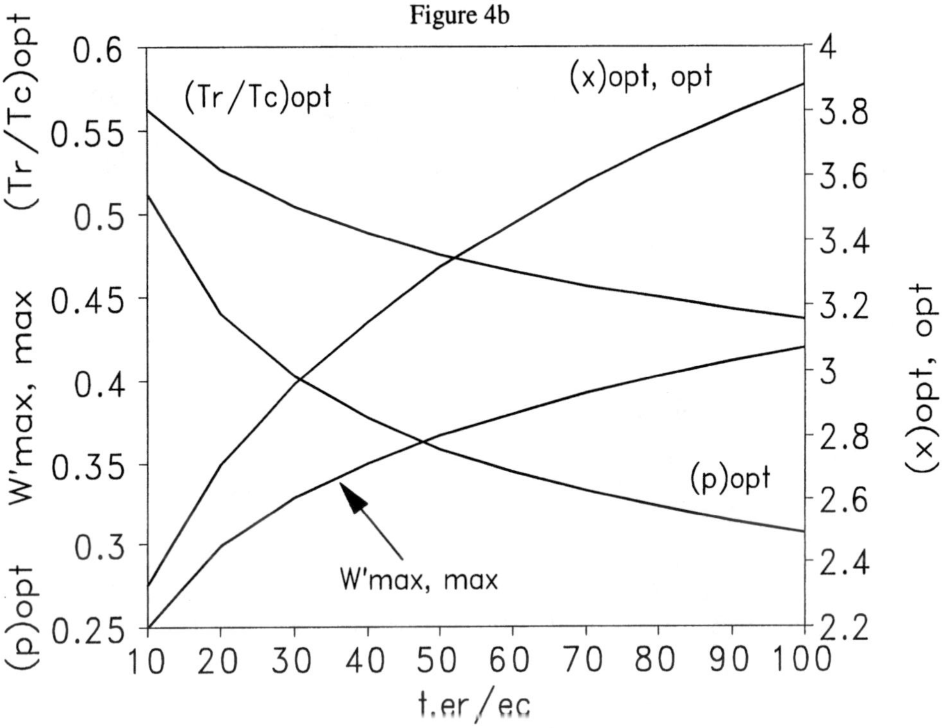

Figure 4c

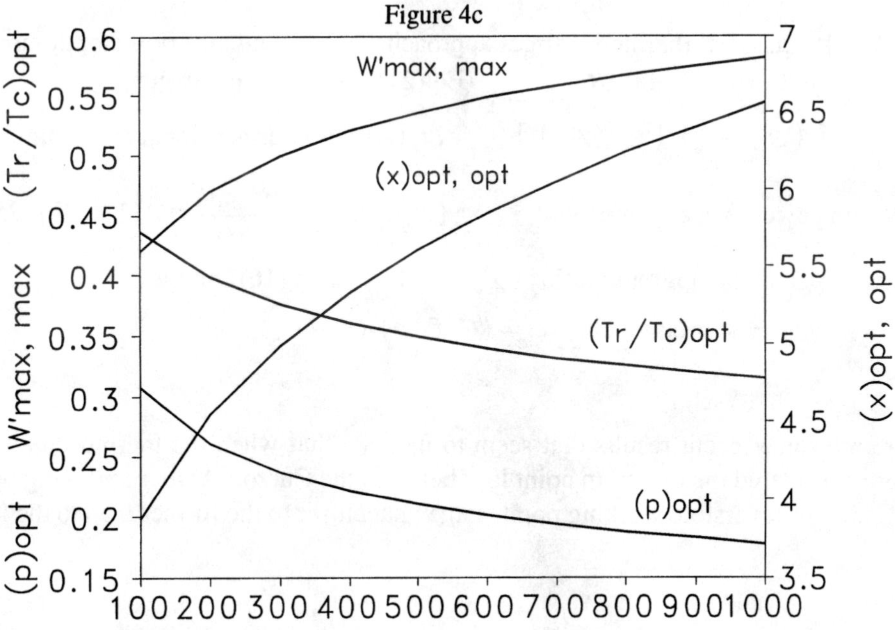

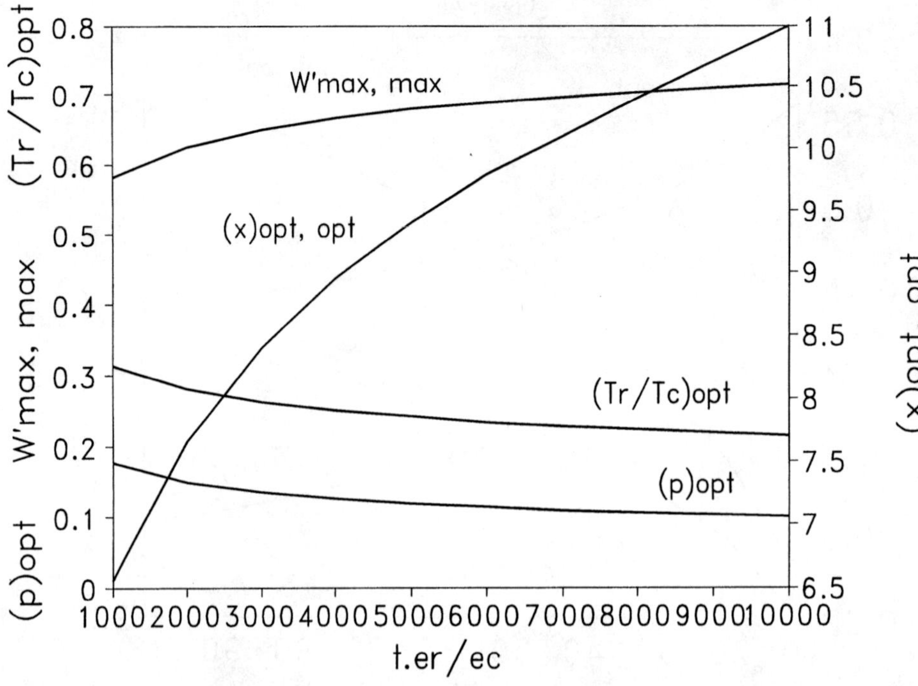

Figure 4d

Fig. 4(a-d). Classical thermodynamics approach. The dependence of the optimum optimorum value $x_{opt,opt}$ (Eq (21)), optimum parameter $p_{opt} = [(e_c/e_r)(A_c/A_r)]_{opt}$ (Eq (20)), non-dimensional maximum maximorum power $\dot{W}'max,max = \dot{W}_{max,max}[a_c \rho_m A_{c,opt} \frac{B(C_n)}{\pi} \sigma T_s^4]^{-1}$ (Eq (25)) and the optimum ratio $(T_r/T_c)_{opt}$ (Eq (16)) on the product $te_r/e_c = (C_n f/s)(e_r/e_c)$.

There are some recent results that seem to indicate that when the thermoeconomics criteria are considered the optimum point lays between the Curzon-Ahlborn working point and the Carnot or reversible working point, somewhat closer to the former than to the later [29].

Figure 5a

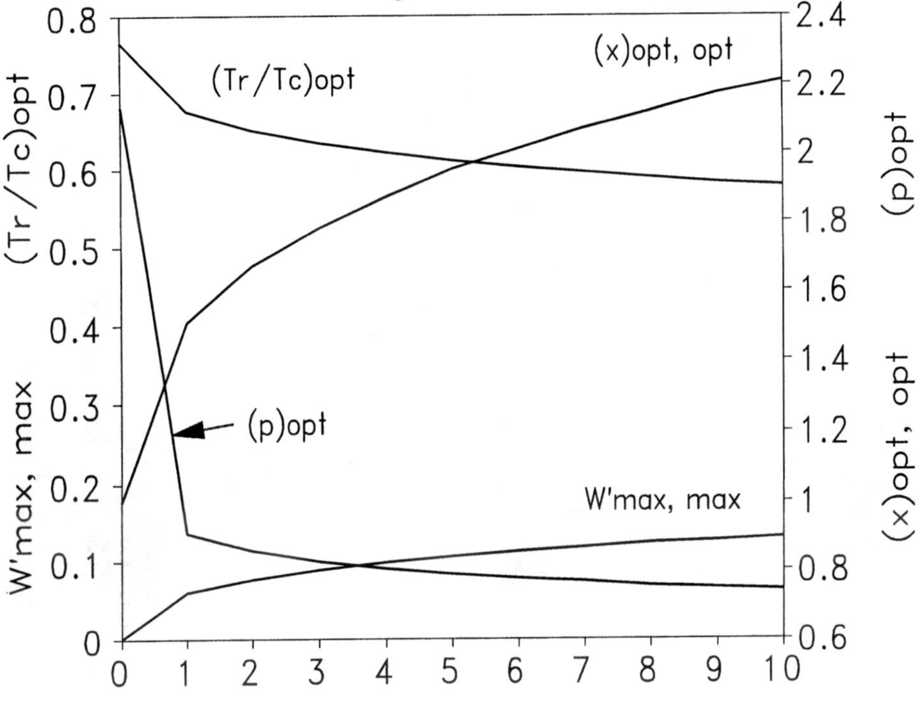

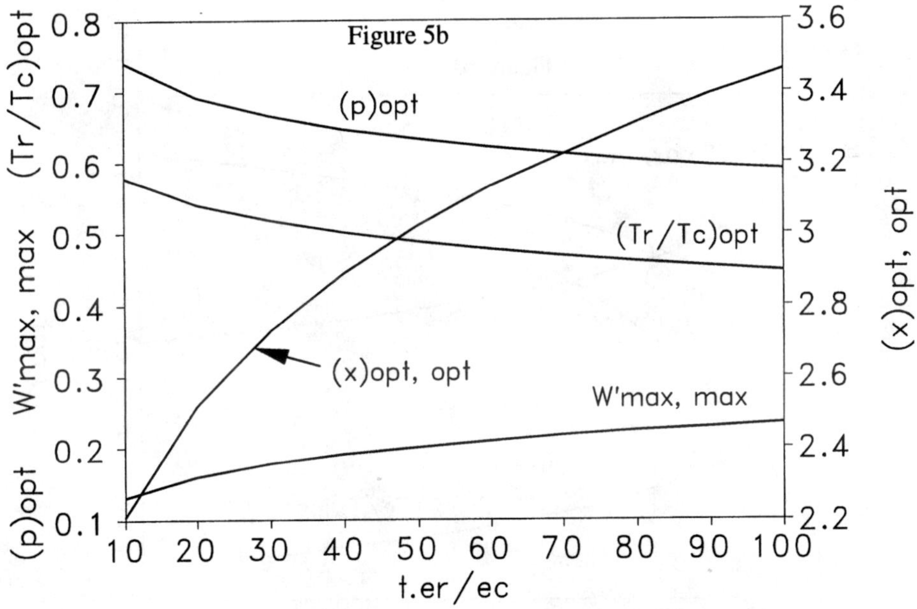

Figure 5b

Figure 5c

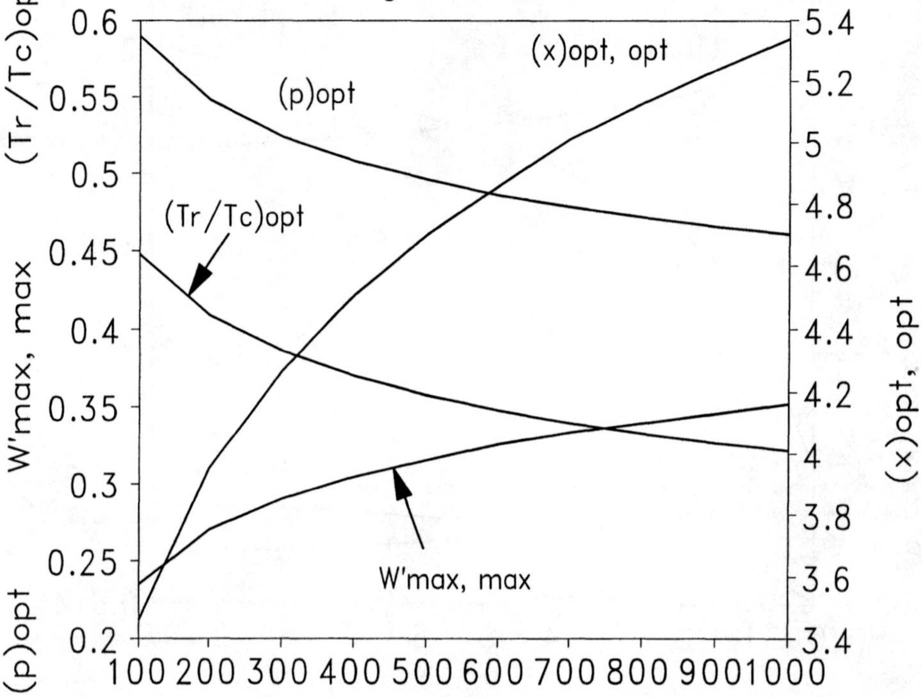

Figure 5d

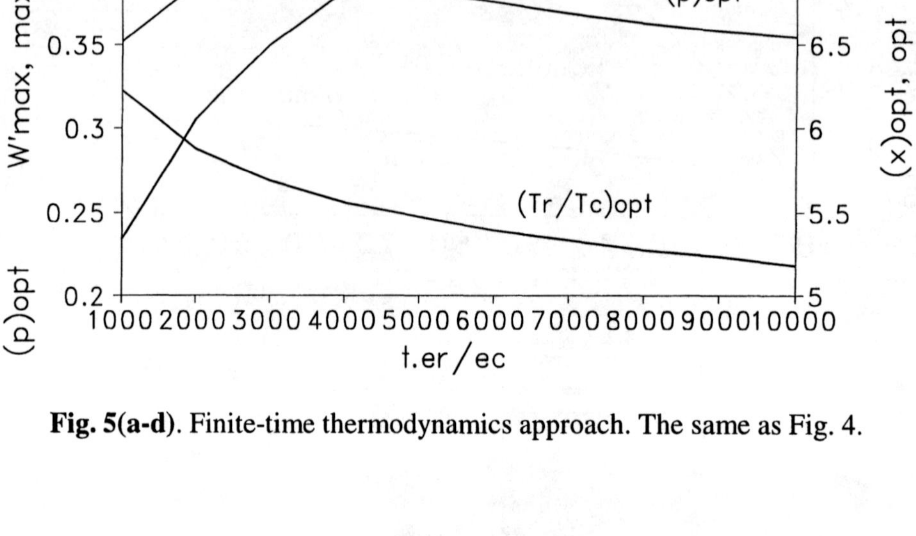

Fig. 5(a-d). Finite-time thermodynamics approach. The same as Fig. 4.

2.5. APPLICATION

The classical and finite-time thermodynamics models were applied to analyze a small Russian solar space power station. The system has an electric output power of 3.5 kW and uses an engine whose thermal output is 5 kW [6,14]. The specific weight of the system is 40 kg/kW and the total mass is $M_{tot} = 200\ kg$.

Three different concentrators were analyzed, namely a parabolic cylinder, a conical and a paraboloidal one. These mirrors have a shape factor f equal to 1.47 , 1.41 and 1.69, respectively [17]. The following numerical values were accepted throughout the application:

$$d_m = 2\ kg/m^2\ ,\ d_r = 3\ kg/m^2\ , a_c = e_c = e_r = 0.9, \rho_m = 0.9.$$

Table 2 shows the results in case of both an ideal Carnot engine and a Curzon-Ahlborn engine. The second model, which is more accurate because it takes into account the losses of available work during the engine heat transfer processes, leads to a decrease in efficiency of about fifteen percent. Also, there is a general increase of the coefficient K, which gives the ratio between the weight of the mirror and radiator and the total mass of the system. As expected, the paraboloidal concentrator has the best performance. However, the conical mirror is also an attractive solution, if we take into account that it is easier to manufacture.

3. FAR IN THE FUTURE: THE DYSON'S SPHERE

A classification of the cosmic civilizations was previously proposed by Kardashev [30]. A technological civilization is of type I, II or III as it has under its control the materials and energy resources of a planet, a star or a galaxy, respectively. An interesting model for the habitat of a civilization of type II was proposed by Dyson (see e.g. [31]). For a better usage of the star energy such a civilization will move on a sphere surrounding the aster, the so called Dyson sphere (DS). This concept was subsequently used and developed by some authors [32-35]. All of them agree that the DS will be built up by using material from asteroids and some planets. However, different architectures were proposed. Thus, some authors considered the DS in its stricto sensu while the others thought to a large number of biospheres turning around the Sun on the same circumference. Here the Dyson's sphere will be considered further on as a continuous structure, allowing a better collection of the Sun's energy. Two main hypotheses will be accepted here. First, the DS thermal ambient must be close enough to the physiological optimum of the human beings. Second, a proper DS "design" should allow the highest possible efficiency of power generation.

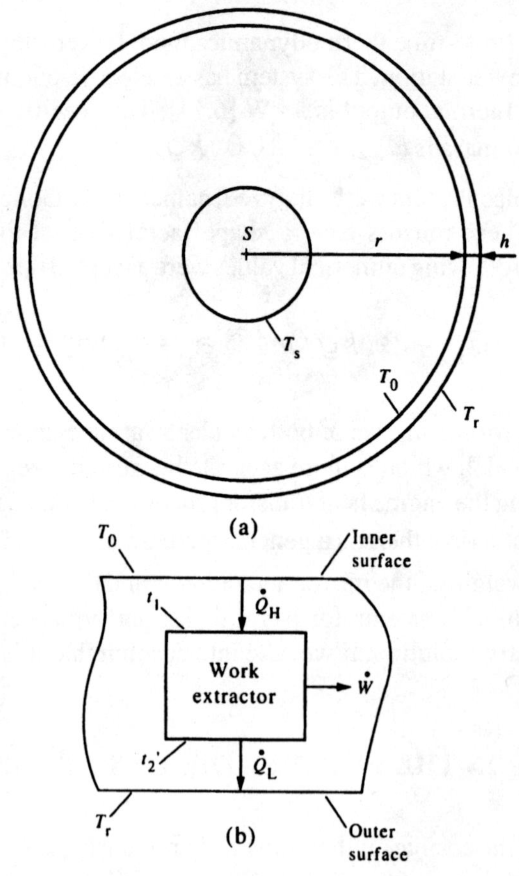

Fig.6 (a). Details of the Dyson's sphere. S-the Sun, r-DS radius, h-DS thickness. T_s-solar temperature. T_0, T_r-temperatures on the inner and outer surface of DS, respectively.

(b). Power generation within DS. $\dot{Q}_H$, $\dot{Q}_L$-heat fluxes entering and leaving the work extractor, respectively. In case the work extractor is a thermal engine, t_1 and t_2 are the absolute temperatures of the working fluid in contact with the two heat reservoirs. Otherwise t_1 and t_2 equal T_0 and T_r, respectively.

3.1. Classical and Finite-time Thermodynamics Models

Figure 6 shows details of Dyson's sphere. The inner DS surface of temperature T_0 constitutes the habitat of mankind. The difference between the temperature of the DS inner surface and the lower temperature T_r of the outer surface assures a heat flux which is used for power generation in a "work extractor". In the classical general case the work extractor is modeled as a Carnot engine. This usually provides too higher upper bounds for the efficiencies. A better approximation is possible in the particular case of thermal power generation, where differences exist between the extreme temperatures t_1 and t_2 of the working fluid and the temperatures of the two heat reservoirs (namely $T_0>t_1>t_2>T_r$). Then, the work extractor can be modeled as a Curzon-Ahlborn engine. Both cases will be considered here. Notice that in the "Carnot" case the temperatures t_1 and t_2 are not necessarily related to a working fluid and they can be simply put as $t_1=T_0$ and $t_2=T_r$.

The steady state heat flux $\dot{Q}_H$ received by the work extractor can be obtained by the energy balance of the DS inner surface:

$$\dot{Q}_H=4\pi r^2\left[\frac{R_S^2}{r^2}\sigma T_S^4+\frac{B_{Ds}}{\pi}\sigma T_0^4-\sigma T_0^4\right] \quad (29)$$

Here r and R_S are the DS and Sun radii, respectively, T_S is Sun's temperature and B_{Ds} is a geometrical factor. In the right member of (29) the first term is the energy flux received from the Sun, the second term is the energy flux received from the "sky" of temperature T_0 while the last term is the energy flux emitted by the DS inner surface. The geometrical factor B_{Ds} can be computed with:

$$B_{Ds}=\int_{\Omega_{Ds}}\cos\theta\, d\Omega \quad (30)$$

where Ω_{Ds} is the solid angle subtended by the "sky" when viewed from a particular point M on the DS inner surface. Also, θ is the angle between the radiation direction and the surface normal in M. It is easy to see that [36]

$$B_{Ds}=\pi-B_s \quad (31)$$

where B_s is the geometrical factor of the Sun. Different formula were previously proposed for B_s. Here we'll use a simplified expression which applies in case of normal incidence [37]

$$B_s = \pi \sin^2 \frac{\delta}{2} \tag{32}$$

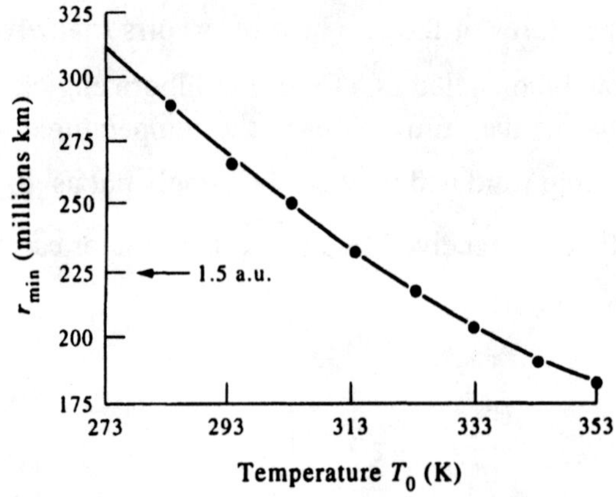

Fig. 7. The dependence of the minimum DS radius on the ambient temperature T_0. The following values of the temperature and radius of the Sun were considered: T_s=5760 K and R_s = 679 000 km, respectively.

where δ is the half-angle of the cone subtended by the Sun when viewed from the point M. The following relationship also applies:

$$\sin \frac{\delta}{2} = \frac{R_s}{r} \equiv x^{1/2} \tag{33}$$

The energy balance for the DS outer surface gives:

$$\dot{Q}_L = 4\pi (r+h)^2 \sigma T_r^4 \approx 4\pi r^2 \sigma T_r^4 \tag{34}$$

Writing the above equations we considered the Sun as an isotropic source of un-polarized black-body radiation.

Now, we shall use energy and entropy balance equations at the level of the work extractor. In case of a Carnot engine we have to use the Eqs. (8) and (9). When a Curzon-

Ahlborn engine is considered the equations to be used are (27) and (28). In both case T_c have to be replaced by T_0. We shall introduce the following notations:

$$\theta_0 \equiv \frac{T_0}{T_s} \qquad \theta_r \equiv \frac{T_r}{T_s} \tag{35}$$

By using the Eqs (29), (31-33), (8,9) and (35) we obtain the efficiency in the Carnot case:

$$\eta_{Carnot} = \frac{\dot{W}_{Carnot}}{4\pi R_s^2 \sigma T_s^4} = 1 - \theta_0^4 - x^{1/3} \frac{(1-\theta_0^4)^{4/3}}{\theta_0^{4/3}} \tag{36}$$

Similarly, by using the Eqs (29), (31-33), (27,28) and (35) we obtain the efficiency in the Curzon-Alhborn case:

$$\eta_{CA} = \frac{\dot{W}_{CA}}{4\pi R_s^2 \sigma T_s^4} = 1 - \theta_0^4 - x^{1/7} \frac{(1-\theta_0^4)^{8/7}}{\theta_0^{4/7}} \tag{37}$$

In both Eqs. (36) and (37) the only parameter dependent on DS radius is x. Simple algebra shows that η_{Carnot} and η_{CA} decrease when x increases. It is interesting to note that both efficiencies vanish for the same value of x, namely:

$$x_{max} = \frac{\theta_0^4}{1-\theta_0^4} \tag{38}$$

According to Eq (33) this means that there is a minimum DS radius given by:

$$r_{min} = \frac{(1-\theta_0^4)^{1/2}}{\theta_0^2} R_s \tag{39}$$

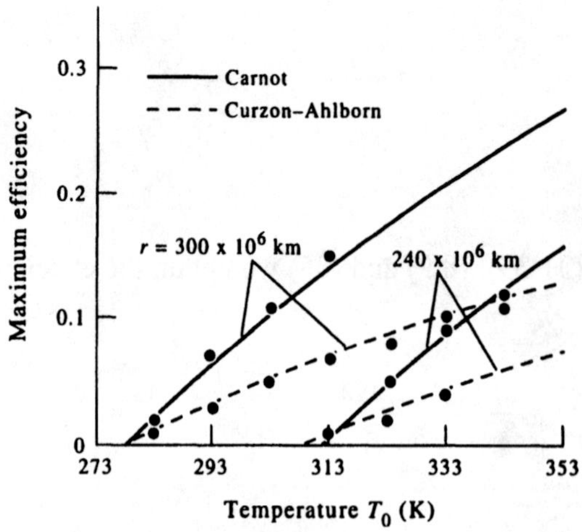

Fig. 8. The dependence of the efficiencies η_{Carnot} and η_{CA} on the ambient temperature T_0 for two values of the DS radius.

On a DS with $r<r_{min}$ power generation is not possible. Fig 7 shows the dependence of r_{min} on the ambient temperature T_0. In the previous works [38] a DS radius between 0.5 and 1.5 astronomic units was proposed. As we see, this range of values is smaller than the minimum DS radius in case a reasonable lower value of the ambient temperature T_0 is accepted.

Of course, by increasing the DS radius an increase of the power generation efficiency is obtained. However, too large values of the DS radius seem not to be acceptable since they require an enormous amount of material (which is proportional with r^2). Consequently, a realistic value of DS radius seems to be slightly superior to r_{min}. Figure 8 shows the dependence of η_{Carnot} and η_{CA} on the ambient temperature for two DS radii, namely 240 and 300 millions km. As we see, a DS radius of $300 \cdot 10^6$ km seems to be reasonable for an ambient temperature around 293 K. In this case η_{Carnot} is 7 percent but the more realistic indicator η_{CA} is 3 percent. Of course, the efficiency could increases at larger values of the ambient temperature T_0.

4. CONCLUSION

In the first part of this paper a simple model of solar space power station was developed. First we assumed a standard configuration of the system and then usual geometrical and material properties were introduced. A special attention was paid to describe appropriately the concentrating mirror. To this aim an unique non-dimensional parameter κ was found able to describe five mirror types usual in solar power applications. This parameter is defined as the ratio between a characteristic entrance size and the depth of the mirror. Values of the ratio A_m/A_{open} between the mirror surface area and its opening area were computed as a function of the parameter κ for all the five mirror types. The conical and parabolic cylinder mirrors have the smallest value of the ratio A_m/A_{open} and they should be used whenever the concentration ratio is smaller than 250. First, an ideal Carnot type engine was analyzed. Afterwards the more realistic case of a Curzon-Ahlborn engine was considered. Hence, the exergy losses during the heat transfer processes were taken into account and more accurate upper bounds are obtained for the power station efficiency. The maximum power supply $\dot{W}'_{max,max}$ is obtained when the ratio between the collector and radiator areas has an optimum value $(A_c/A_r)_{opt}$ which depends on the geometrical and optical parameters of the system. This optimum value decreases when the concentration ratio increases. The values of $\dot{W}'_{max,max}$ and $(A_c/A_r)_{opt}$ are smaller and bigger, respectively, than the values obtained when an ideal Carnot engine was considered.

In the second part of the paper a thermodynamic analysis of the Dyson's sphere (DS) was performed by using both classical and finite-time thermodynamics. No optimum DS radius exists if the criterion is the maximum efficiency of power generation. However, a minimum DS radius does exist. Its value is a function of the ambient temperature T_0 only. The minimum DS radius doesn't depend on the sort of thermodynamic approach. Differences between the two sort of thermodynamic approaches occur when solar radiation conversion efficiency is considered, as expected. Reasonable radii are characterized by efficiencies less than 10 %.

ACKNOWLEDGMENTS

Part of this work was performed with support from the European Commission (Contracts CIPA-CT92-4026 and ICOP-DISS-2168-96).

Table 1. The shape factor A_m/A_{open} for the five mirrors of Fig. 2. Here k=h/r, where h and r are the depth and the characteristic entrance size of the mirror (see Fig. 2).

Mirror	Shape factor A_m/A_{open}
Conical	$(1+k^2)^{1/2}$
Sphere	$1+k^2$
Circular cylinder	$\frac{1}{2}\frac{k^2+1}{k}\,arctg\,\frac{2k}{1-k^2}$
Parabolic cylinder	$\frac{1}{2}(1+4k^2)^{1/2}+\frac{1}{4k}\ln[2k+(1+4k^2)^{1/2}]$
Paraboloid	$\frac{2}{3}\{[(2k)^{2/3}+(\frac{1}{2k})^{4/3}]^{3/2}-\frac{1}{4k^2}\}$

Table 2. Results obtained in case of a finite-time thermodynamic model. Numbers in square brackets denote the corresponding equations of Section 2.2. For definitions see text.

$$x=\frac{a_c\rho_m}{e_c}\frac{B(C_n)}{\pi}(\frac{T_s}{T_c})^4 \qquad [15]$$

$$\frac{T_r}{T_c}=(\frac{e_cA_c}{e_rA_r})^{2/7}(x-1)^{2/7} \qquad [16]$$

$$\eta_{\max} = a_c \rho_m \frac{x_{opt} - 1}{x_{opt} + 7} \qquad [18]$$

$$p^7 (x_{opt} - 1)(1 + \frac{x_{opt}}{7})^7 = 1 \qquad [19]$$

$$(x_{opt,opt} - 1)^2 (1 + \frac{x_{opt,opt}}{7})^7 = t \frac{e_r}{e_c} \qquad [21]$$

$$(\frac{A_r}{A_c})_{opt} = \frac{t}{x_{opt,opt} - 1} \qquad [23]$$

$$(\frac{A_e}{A_c})_{opt} = st \frac{x_{opt,opt}}{x_{opt,opt} - 1} \qquad [24]$$

$$\dot{W}' \max,\max = \frac{x_{opt,opt} - 1}{x_{opt,opt} + 7} \qquad [25]$$

$$\frac{M_{tot}}{\dot{W}_{\max,\max}} = \frac{d_m f}{\kappa a_c \rho_m} \frac{1}{\frac{B(C_n = 1)}{\pi} \sigma T_s^4} \frac{x_{opt,opt}(x_{opt,opt} + 7)}{(x_{opt,opt} - 1)^2} \qquad [26]$$

Table 3. Optimum design parameters for a solar space power station in case of an ideal Carnot and a Curzon-Ahlborn thermal engine. For details, see text.

Mirror	Parabolic cylinder		Conical		Paraboloid	
C_n	50		200		1000	
	Carnot	Curzon-Ahlborn	Carnot	Curzon-Ahlborn	Carnot	Curzon-Ahlborn
$A_{open}(m^2)$	13.0	22.18	9.76	16.87	7.73	12.55
$A_m(m^2)$	17.1	32.60	13.76	23.78	13.06	21.20
$A_c(m^2)$	0.2498	0.446	0.0488	0.0844	0.0077	0.0125
$A_r(m^2)$	4.93	11.32	2.479	5.54	1.325	3.15
$M_m(kg)$	38.22	65.20	27.52	47.56	26.12	42.40
$M_r(kg)$	14.79	33.96	7.44	16.62	3.97	9.45
$\kappa\kappa$	0.26	0.49	0.18	0.32	0.15	0.26
$T_{c,opt}(K)$	763.2	772.1	993.7	1023.6	1343.4	1409.4
$T_{r,opt}(K)$	371.8	380.6	403.5	420.7	426.1	436.9
η_{max}	0.29	0.16	0.37	0.22	0.47	0.29

REFERENCES

1. B Andresen, *Finite-time thermodynamics*, Physics Laboratory II, University of Copenhagen, 1983
2. V Radcenco, *Generalized Thermodynamics*, Editura Tehnica, Bucharest, 1994
3. A Bejan, Entropy generation minimization: The new thermodynamics of finite-size devices and finite-time processes, *J Appl Phys* 79(3), 1191-1218, 1996
4. V Badescu and C Dinu, Maximum performance of omnicolor photothermal and photovoltaic converters in our planetary system, *Renewable Energy*, 6, 765-777, 1995

5. V Badescu and C Dinu, Optimization of multicolor photothermal power plants in the Solar system: a finite-time thermodynamic approach, *J Phys III France*, 6, 143-1631, 1996
6. V Prisnyakov, SPS interest and studies in USSR, *Proc of SPS 91 Symposium*, Paris, p 36-44, 1991.
7. G A Landis, *Space Power* 9(4), 365-371, 1990. P Glaser, The solar power satellites option re-examined, *Proc of SPS 91 Symposium*, Paris, p 59-64 , 1991. L Deschamps, *Space Power* 11(2), 175-182, 1992.
8. G E Maryniak and B Tillotson, *Space Power*, 7, 27-36, 1988. D Bashford, A review of advanced metallic and ceramic materials suitable for high temperature use in space structures, *IAF Paper* No 89-315, October 1989. G E Maryniak ,*Space Power*, 9(1), 39-48, 1990. G E Maryniak, Nonterrestrial resources for solar power satellite construction, *Proc of SPS 91 Symposium*, Paris, p 146-153, 1991.
9. A R Martin and V K Thompson, The impact of technology advances upon Satellite Power Systems, *Proc of SPS 91 Symposium*, Paris, p 217-223, 1991.
10. P Compte and M Ferronniere, C Marchal, Application of the concept of inflatable and rigidifiable structures to large space power stations, *Proc of SPS 91 Symposium*, Paris, p 253-259, 1991.
11. R A Akiba et al, *Acta Astronautica*, 15(11), 893-900, 1987. I Kurkin, D Sevruk, D Sidorov, M Kukolev and V Gorchakov, Modeling of the development and infrastructure of solar electric power stations, *Proc of SPS 91 Symposium*, Paris, p 260-262, 1991.
12. Y A Mozjorine, V P Senkevich, A D Koval and E A Narimanov, Small-scale space power stations : feasibility and usage prospects, *Proc of SPS 91 Symposium*, Paris, p 381-392, 1991.
13. K Eguchi, S Ogiwara and T Fujiwara, *Space Power* 9(2/3), 131-148, 1990. V Badescu, *Space Power*, 11(2), 99-106, 1992.
14. V F Prisnjakov, I N Statsenko, A I Kondratjev, V L Markov, B E Petrov and V A Gabrinets, Developing space power Brayton system with solar heat input - Research of working process of high temperature latent heat storage system, *Proc of SPS 91 Symposium*, Paris, p 465-470, 1991.
15. R J Hannigan, SPS transportation requirements : which launch system ?, *Proc of SPS 91 Symposium*, Paris, p 484-491, 1991. B A Osadin, *Space Power*, 11(2), 141-148, 1992.
16. A Bejan, *Advanced Engineering Thermodynamics*, Wiley, New York, 1988.
17. V Badescu, Optimization of a solar space power system based on thermodynamic cycles, *Int J Solar Energy*, 16, 263-275, 1995
18. M Rubin, Optimal configurations of a class of irreversible engines, *Phys Rev*, 19, 1272-1276, 1979.
19. F Curzon and B. Ahlborn, Efficiency of a Carnot engine at maximum power output, *Am J Phys*, 43, 22-24, 1975.

20. A De Vos, Efficiency of some heat engines at maximum-power conditions, *Am J Phys*, 53(6), 570-573, 1985.
21. A De Vos, Reflections on the power delivered by endoreversible engines, *J Phys D*, 20, 232 -236, 1987.
22. C. J. Hoogendorn, Optical properties of selective layers, in *Solar energy utilization. Fundamentals and applications*. Eds H Yuncu, B Kilkis, ASI Proceedings, EIEI Printing Shop, Ankara, p 26-61, 1986
23. J. A. Duffie, W. A. Beckman, *Solar Energy Thermal Processes*, Wiley, New York, 1980
24. W. R. Menetrey, Space applications of solar energy, in : *Introduction to the utilization of solar energy*, Eds A M Zarem, D D Erway, Mc Graw-Hill, New York, p 326, 1963
25. D. R. Criswell, R. D. Waldron, Lunar system to supply solar electric power to Earth, *AIAA Series : Progress in Astronautics and Aeronautics*, 10, 491-509, 1988
26. T. Wallace, R. W. Bussard, A lightweight focusing reflector concept for space power applications, *Proc of SPS 91 Symposium*, Paris, p 327-331, 1991
27. J. A. Angelo Jr, D. Buden, The nuclear power satellite (NPS) - Key to a sustainable global energy economy and solar system civilization, *Proc of SPS 91 Symposium*, Paris, p 117-124, 1991
28. E. H. Fay, M. Stavnes, R. C. Cull, Beam power options for the Moon, *Proc of SPS 91 Symposium,* Paris, p 238-247, 1991
29. A De Vos, Endoreversible thermoeconomics, *Energy Convers Mgmt*, 36, 1-5, 1995.
30. N S Kardashev, Transmission of information by extraterrestrial civilisations, *Sov Astronomy A J*, 8, 217, 1964.
31. F Dyson, *Disturbing the Universe*, Harper, New York, 1979.
32. G K O'Neill, *The high frontier - human colonies in space*, W Marrow and Co, New York, 1977.
33. M D Papagiannis, Could we be the only advanced technological civilisation in our galaxy ? in: *Origin of life*, Ed H Noda, Cent Acad Publ, Tokyo, p 583-595, 1978.
34. M D Papagiannis, Are we all alone, or could they be in the asteroid belt ?, *Q J R Astr Soc*, 19, 277-281, 1978 .
35. T S Trefil, *Living in space*, C Scribner's Sons, New York, 1981.
36. V Badescu, Thermodynamics of the conversion of partially polarised black-body radiation, *J Phys III*, 2, 1925-1941, 1992.
37. P T Landsberg and G Tonge, Thermodynamic energy conversion efficiencies, *J Appl Phys* 51(7), R1-R20, 1980.
38. V S Troitski, Searching extraterrestrial civilisation : a new strategy, *Zemlia i Vselennaia*, 6, 1983.

THE MAXIMUM COOLING RATE OF AN IRREVERSIBLE MAGNETIC STIRLING REFRIGERATION CYCLE FOR A GIVEN POWER INPUT

JINCAN CHEN AND ZIJUN YAN
Department of Physics, Xiamen University, Xiamen 361005, The People's Republic of China

CHIH WU
Department of Mechanical Engineering, U. S. Naval Academy, Annapolis, MD 21402-5042
U. S. A.

The influence of thermal resistances and regenerative losses on the performance of a magnetic Stirling refrigeration cycle is investigated, based on a general expression of the magnetization intensity of ferromagnetic materials and a linear heat-transfer law. Some fundamental optimum relations and general characteristic curves of the cycle are obtained. The maximum cooling rate is determined and other relevant performance parameters are optimized. The optimal performance of a magnetic Carnot refrigeration cycle is derived for specific choices of some parameters. Moreover, it is expounded that the calculated results are also suitable for a magnetic Stirling refrigeration cycle using a simple paramagnetic salt as the working substance.

I. INTRODUCTION

Interest in the application of magnetic refrigerators to the production of low temperatures has grown rapidly in recent years. One of the main reasons is that magnetic refrigeration has many distinctive advantages. It not only reduces environmental pollution but also provides higher efficiency and reliability than conventional gas refrigerators. This has led to a lot of investigation on the optimal design and the theoretical analysis of magnetic refrigeration cycles.[1-4]

The magnetic Stirling refrigeration cycle is one of the typal refrigeration cycles with regeneration. It can be expected to realize larger temperature spans than that of a magnetic Carnot refrigeration cycle so that it has attracted the attention of many investigators and become an animate subject in the research of magnetic refrigerators. Usually, the method of classical thermodynamics is used in most of the theoretical analysis of the magnetic Stirling refrigerator. The magnetic Stirling refrigerator is considered to consist of two reversible isothermal and two reversible isomagnetization processes, while the influence of irreversibilities existing in real systems is studied rarely. In the present paper the influence of irreversibilities due to finite-rate heat transfer and regenerative loss will be taken into account. The results obtained here are expected to be of more real significance.

II. THERMODYNAMIC PROPERTIES OF FERROMAGNETIC MATERIALS

According to the phenomenological theory of molecular fields, the relation between the magnetic field intensity H and the magnetization intensity M of a ferromagnetic material is given by[5]

$$=ng\mu_B J B_J(x) = ng\mu_B J B_J[g\mu_B J(H+\lambda M)/(kT)] \tag{1}$$

where

$$B_J(x) = \frac{2J+1}{2J}\coth\left(\frac{2J+1}{2J}x\right) - \frac{1}{2J}\coth\frac{x}{2J} \tag{2}$$

is the Brillouin function,

$$\lambda = \frac{3kT_c}{ng^2\mu_B^2 J(J+1)} \tag{3}$$

is the molecular field constant, μ_B is referred to as the Bohr magneton, g is Landé's g factor, J is the quantum number of angular momentum, n is the number of magnetic moments per unit volume, k is the Boltzmann constant, T is the absolute temperature, and T_c is the Curie temperature.

When the volume change of ferromagnetic materials is negligible, the fundamental equation of thermodynamics of ferromagnetic materials per unit volume may be expressed as

$$du = Tds + \mu_0 (H+\lambda M)\, dM, \tag{4}$$

where u and s are, respectively, the internal energy and entropy of ferromagnetic materials per unit volume, and μ_0 is the permeability of the vacuum.

Using Eqs. (1) and (4), one can prove

$$u=u(T), \tag{5}$$

$$C_M=(\partial u/\partial T)_M=C_M(T), \tag{6}$$

and

$$s=s_0(T)+\mu_0 nk[\ln \sinh(\frac{2J+1}{2J}x)-\ln \sinh\frac{x}{2J}-xB_J(x)], \tag{7}$$

where C_M is the heat capacity at isomagnetization and $s_0(T)$ is the entropy of ferromagnetic materials per unit volume when M=0. Both $C_M(T)$ and $s_0(T)$ are only a function of the temperature, but independent of the magnetization intensity M. Their values are dependent on the specific properties of ferromagnetic materials.

It is seen from Eq. (1) that the change of the magnetic field intensity H with the temperature T must satisfy the following relation:

$$H=Tf(M)-\lambda M, \tag{8}$$

where f(M) is only a function of the magnetization intensity M and determined by Eq. (1). For the process of a given magnetization intensity M, f(M) is a constant so that

$$\frac{H+\lambda M}{T}=C_1 \tag{9}$$

which is also a constant. Thus, only if the change of the magnetic field intensity with the temperature is controlled properly , can the isomagnetization process be realized.

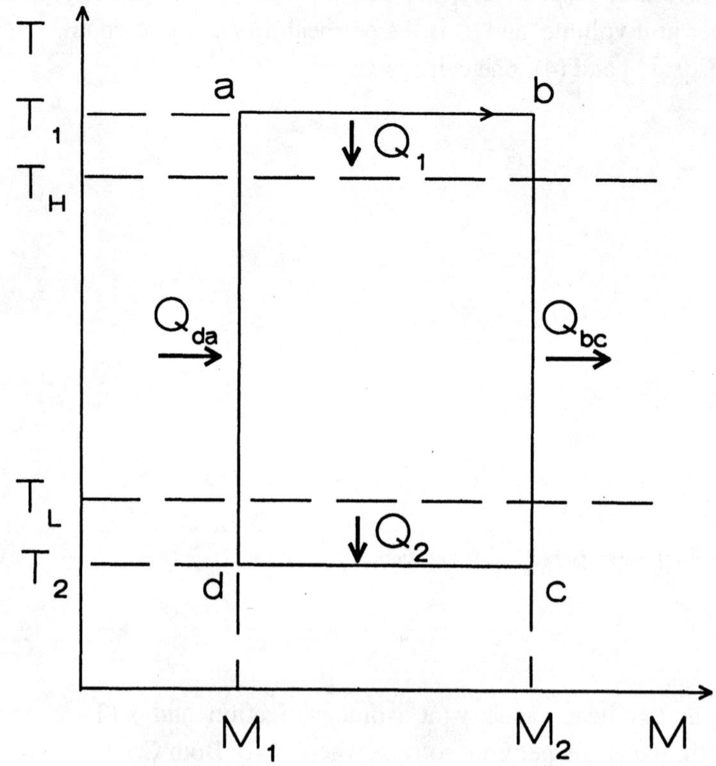

Fig.1. The T-M diagram of a magnetic Stirling refrigeration cycle.

III. CYCLE MODEL

The magnetic Stirling refrigeration cycle consists of two isothermal and two isomagnetization processes, as shown in Fig. 1, where T_H and T_L are the temperatures of the heat sink and the cooled space, T_1 and T_2 are the temperatures of the working substance in the high- and low-temperature isothermal processes, Q_1 and Q_2 are the amounts of heat exchange between the working substance and the heat reservoirs during two isothermal processes, Q_{bc} and Q_{da} are the amounts of heat exchange between the working substance and the regenerator during two isomagnetization processes.

First, we consider the cycle processes which are not restricted by time. Using Eq. (7), one can calculate that the amounts of heat released from and absorbed by the working substance during the high- and low-temperature isothermal processes are, respectively, given by

$$Q_1 = T_1 (s_a - s_b) = T_1 \Delta s_1 \tag{10}$$

and

$$Q_2 = T_2 (s_d - s_c) = T_2 \Delta s_2, \tag{11}$$

where

$$\Delta s_1 = \mu_0 nk[\ln \frac{\sinh(\frac{2J+1}{2J}x_a)\sinh\frac{x_b}{2J}}{\sinh\frac{x_a}{2J}\sinh(\frac{2J+1}{2J}x_b)} - x_a B_J(x_a) + x_b B_J(x_b)], \tag{12}$$

$$\Delta s_2 = \mu_0 nk[\ln \frac{\sinh(\frac{2J+1}{2J}x_d)\sinh\frac{x_c}{2J}}{\sinh\frac{x_d}{2J}\sinh(\frac{2J+1}{2J}x_c)} - x_d B_J(x_d) + x_c B_J(x_c)], \tag{13}$$

and

$$x_a = g\mu_B J(H_a + \lambda M_1)/(kT_1), \quad x_b = g\mu_B J(H_b + \lambda M_2)/(kT_1),$$
$$x_c = g\mu_B J(H_c + \lambda M_2)/(kT_2), \quad x_d = g\mu_B J(H_d + \lambda M_1)/(kT_2). \tag{14}$$

It is seen from Eq. (9) without difficulty that $x_a = x_d$ and $x_b = x_c$. Thus, $\Delta s_1 = \Delta s_2 = \Delta s$ is a constant which is independent of the magnetic field intensity H and the temperature T. Using the conditions mentioned above, one has

$$\frac{Q_1}{Q_2} = \frac{T_1}{T_2}. \tag{15}$$

This shows clearly that a magnetic Stirling refrigeration cycle possesses the condition of perfect regeneration[6] and can attain perfect regeneration by using a perfect regenerator. Thus, for a magnetic Stirling refrigeration cycle, one has

$$Q_{bc}+Q_{da}=\int_{T_1}^{T_2} C_M dT+\int_{T_2}^{T_1} C_M dT=0, \tag{16}$$

which can also be derived from Eq. (6).

Now, we consider the cycle processes which are restricted by time. The temperatures T_1 and T_2 of the working substance in the high- and low-temperature isothermal processes are not equal to the temperatures T_H and T_L of the heat sink and the cooled space, respectively. Equations (10) and (11) are still true. Under a linear heat-transfer law,[7-10] Q_1 and Q_2 may be expressed as

$$Q_1=k_1(T_1-T_H)t_1 \tag{17}$$

and

$$Q_2=k_2(T_L-T_2)t_2, \tag{18}$$

where k_1 and k_2 are the thermal conductances between the working substance and the heat reservoirs at temperatures T_H and T_L, and t_1 and t_2 are the times of two isothermal processes at temperatures T_1 and T_2.

When the cycle time is taken into account, not only the isothermal processes but also the regenerative processes are affected by thermal resistances, so that the cycle cannot attain the perfect regeneration. Usually, it may be assumed that the regenerative losses Q_r per cycle is

$$Q_r=(1-\eta_r)\int_{T_2}^{T_1} C_M dT, \tag{19}$$

where η_r is the efficiency of the regenerator. For a perfect regenerator, $\eta_r=1$. In general, C_M is closely dependent on the temperature.

However, in order to analyze the main characteristics of regenerative influence, C_M may be substituted by the average value C_{MA} of the heat capacities at isomagnetization between T_H and T_L so that Eq. (19) may be written as

$$Q_r=(1-\eta_r)\,C_{MA}(T_1-T_2)\,. \tag{20}$$

For a magnetic Stirling refrigeration cycle operating only between two heat reservoirs at temperatures T_H and T_L, the regenerative losses Q_r must be compensated from the heat sink during one isomagnetization process and released to the cold reservoir during the other isomagnetization process. If not, the regenerator cannot work normally. Thus, the net amounts of heat released to the heat sink and absorbed from the cold reservoir per cycle are

$$Q_H=Q_1-Q_r \tag{21}$$

and

$$Q_L=Q_2-Q_r\,. \tag{22}$$

The regenerative processes need a non-negligible time compared with the time of the isothermal processes. It may be assumed that the time of the regenerative processes is directly proportional to the temperature difference of the working substance in the two isothermal processes,[11,12] i.e.,

$$t_3=t_4=\gamma(T_1-T_2)\,, \tag{23}$$

where γ is a proportional constant which is independent of temperature. Thus, the cycle period is given by

Such a cycle model includes not only the influence of thermal resistances between

$$t=t_1+t_2+2t_3\,. \tag{24}$$

the working substance and the heat reservoirs but also the regenerative losses in the regenerative processes. It may be used to discuss the influence of both thermal resistances and regenerative losses on the performance of a magnetic Stirling cycle and obtain some optimal results which are different from those of the magnetic Carnot cycle.

IV. MAXIMUM COOLING RATE

It is well known that the coefficient of performance, cooling rate, and power input are three of the important performance parameters of refrigerators. Using Eqs. (10), (11), (17),(18), (20)-(22), and (24), we find that the coefficient of performance, cooling rate, and power input of a magnetic Stirling refrigeration cycle may be, respectively, expressed as

$$\varepsilon=\frac{Q_L}{Q_H-Q_L}=\frac{Q_L}{Q_1-Q_2}=\frac{T_2}{T_1-T_2}-a, \qquad (26)$$

$$R=\frac{Q_L}{t}=\frac{T_2-a(T_1-T_2)}{\dfrac{T_1}{k_1(T_1-T_H)}+\dfrac{T_2}{k_2(T_L-T_2)}+b(T_1-T_2)} \qquad (26)$$

and

$$P=\frac{Q_H-Q_L}{t}=\frac{T_1-T_2}{\dfrac{T_1}{k_1(T_1-T_H)}+\dfrac{T_2}{k_2(T_L-T_2)}+b(T_1-T_2)}, \qquad (27)$$

where $a=(1-\eta_r)C_{MA}/\Delta s$ and $b=2\gamma/\Delta s$.

Using a refrigerator, one always want to obtain a cooling rate as large as possible for a given power input. For this purpose, we introduce the Lagrangian

$$L=R+\nu P=\frac{1=(\nu-a)(y-1)}{\dfrac{y}{k_1(yT_2-T_H)}+\dfrac{1}{k_2(T_L-T_2)}+b(y-1)}$$

where ν is a Lagrange multiplier and $y=T_1/T_2$. From Eq. (28) and the extremal condition

$$\frac{\partial L}{\partial T_2} = 0, \tag{28}$$

we can find an optimal relation

$$y(T_L - T_2) = \sqrt{k_1/k_2}\ (yT_2 - T_H). \tag{29}$$

Substituting Eq. (30) into Eqs. (26) and (27) and eliminating T_2 yields

$$R = kT_L \frac{1 - a(y-1)}{\frac{y}{y - T_H/T_L} + bkT_L(y-1)} \tag{30}$$

and

$$P = kT_L \frac{y-1}{\frac{y}{y - T_H/T_L} + bkT_L(y-1)}, \tag{31}$$

where $k = 1/(1/\sqrt{k_1} + 1/\sqrt{k_2})^2$.

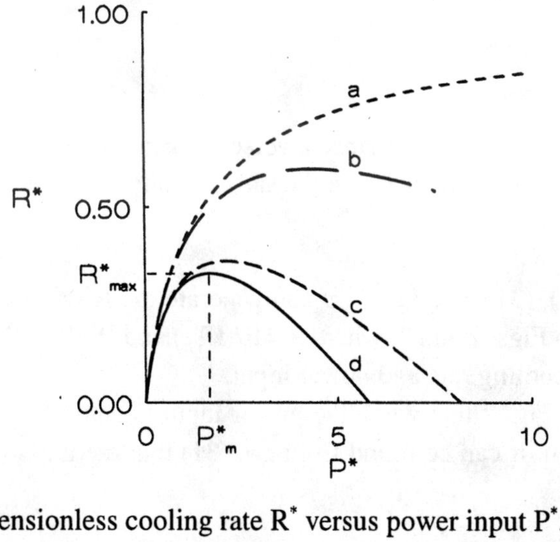

Fig.2. The Dimensionless cooling rate R^* versus power input P^*. Curves a($a=0$, $bkT_L=0$), b($a=0$, $bkT_L=0.05$), c($a=0.1$, $bkT_L=0$),and d($a=0.1$, $bkT_L=0.05$) are plotted for $T_H/T_L=2$.

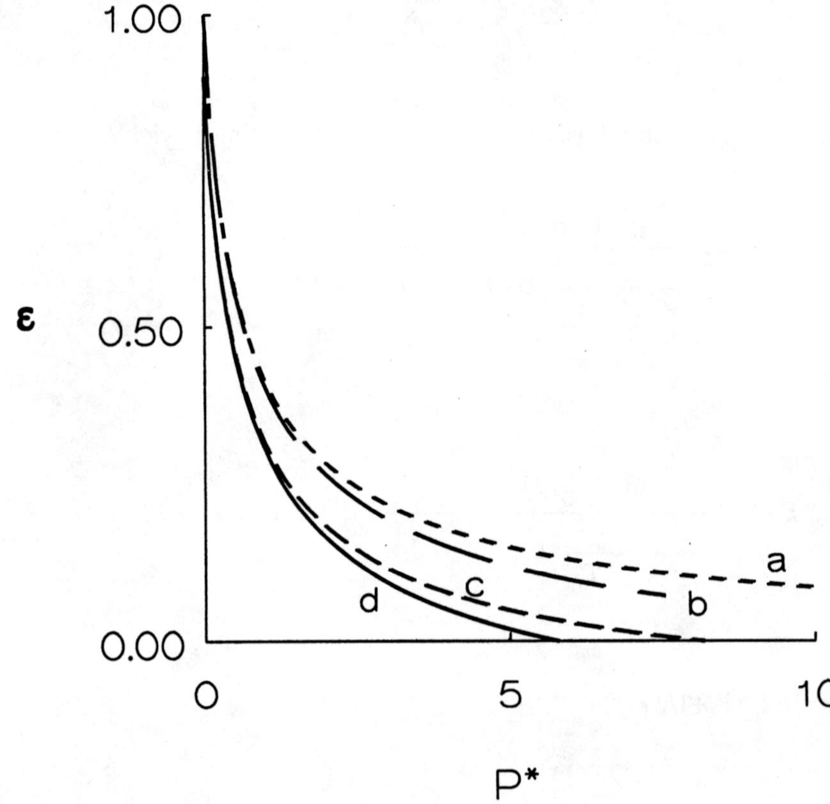

Fig.3. The coefficient of performance ε versus dimensionless power input P^*. The values of a and bkT_L are the same as those used in Fig.2.

Using Eqs. (32), (31), and (25), we can generate the R^*-P^* and ε-P^* characteristic curves, as shown in Figs. 2 and 3, where $R^*=R/(kT_L)$ and $P^*=P/(kT_L)$ are, respectively, the dimensionless cooling rate and power input.

It is seen from Fig. 2 that when the power input is equal to P_m, the cooling rate attains its maximum. It can be found from Eq. (31) that the maximum cooling rate is given by

$$R_{max} = kT_L \frac{\dfrac{(a+bkT_L)T_L}{(bkT_L+d)T_H-(a+bkT_L)T_L} - a}{\dfrac{(a+bkT_L)(bkT_L+d)T_L}{(d-a)[(bkT_L+d)T_H-(a+bkT_L)T_L]} + bkT_L}, \tag{32}$$

where

$$d = \sqrt{(1+a)(a+bkT_L)T_L/T_H - abkT_L}. \tag{33}$$

In such a case the power input

$${}_m = \frac{kT_L}{\dfrac{(a+bkT_L)(bkT_L+d)T_L}{(d-a)[(bkT_L+d)T_H-(a+bkT_L)T_L]} + bkT_L} \tag{34}$$

and the coefficient of performance

$$\varepsilon = \frac{(a+bkT_L)T_L}{(bkT_L+d)T_H-(a+bkT_L)T_L} - a. \tag{35}$$

It is also seen from Fig. 2 that the rational region of the power input should be

$$P \le P_m. \tag{36}$$

As viewed from the cooling rate, P_m determines the allowable value of the upper bound of the power input, because the cooling rate decreases as the power input increases when $P>P_m$.

Figure 3 shows clearly that the coefficient of performance is a monotonically decreasing function of the power input. From the point of view of the coefficient of performance, the power input will be required as small as possible.

Both Figs. 2 and 3 show that the power input should not be larger than P_m. If not, the magnetic Stirling refrigerator will not be operated in the optimal working regions.

According to Eq. (37), one can determine that rational ranges of the temperatures T_1 and T_2 should be

$$T_1 \le T_{1m}, \qquad T_2 \ge T_{2m}, \tag{37}$$

where

$$T_{1m} = T_H \frac{\sqrt{k_1/k_2} + (bkT_L + d)/(a + bkT_L)}{1 + \sqrt{k_1/k_2}} \tag{38}$$

and

$$T_{2m} = T_L \frac{1 + \sqrt{k_1/k_2}\ (a + bkT_L)/(bkT_L + d)}{1 + \sqrt{k_1/k_2}} \tag{39}$$

are, respectively, the temperatures of the working substance in two isothermal processes when the refrigeration cycle is operated in the state of the maximum cooling rate.

V. Discussion

1) When γ=0, this implies that the time of two regenerative processes is negligible compared with that of two isothermal processes, although the regenerative losses is taken into account. Then, Eqs. ((33), (35), and (36) may be simplified as

$$R_{\max} = kT_L (\sqrt{1+a} - \sqrt{aT_H/T_L})^2, \tag{40}$$

$$_m = kT_L \frac{(\sqrt{1+a} - \sqrt{aT_H/T_L})(\sqrt{(1+a)T_H/T_L} - \sqrt{a})}{\sqrt{(1+a)a}} \tag{41}$$

and

$$\varepsilon=\frac{\sqrt{aT_L}}{\sqrt{(1+a)T_H}-\sqrt{aT_L}}-a. \tag{42}$$

The R^*-P^* and ε-P^* characteristic curves are, respectively, shown in curve c in Figs. 2 and 3.

2) When $\eta_r=1$, the magnetic Stirling refrigeration cycle can realize the perfect regeneration. In such a case, the R^*-P^* and ε-P^* characteristic curves are, respectively, shown in curve b in Figs. 2 and 3, and Eqs. ((33), (35), and (36) may be simplified as

$$R_{max}=\frac{kT_L}{1+2\sqrt{bkT_H}+bk(T_H-T_L)}, \tag{43}$$

$$P_m=\frac{kT_L}{\dfrac{(1+\sqrt{bkT_H})T_L}{T_H-T_L+\sqrt{T_H/(bk)}}+bkT_L}, \tag{44}$$

and

$$\varepsilon=\frac{T_L}{T_H-T_L+\sqrt{T_H/(bk)}}. \tag{45}$$

It should be pointed out that such a magnetic Stirling refrigeration cycle with the perfect regeneration has the same performance characteristics as the endoreversible magnetic Carnot refrigeration cycle in which the time of the adiabatic processes is also determined by Eq.(23). Thus, the optimal performance of such an endoreversible magnetic Carnot refrigeration cycle can be directly derived from the above results.

3) When $\gamma=0$ and $\eta_r=1$, the cooling rate becomes a monotonically increasing function of the power input. The R^*-P^* and ε-P^* characteristic curves are, respectively, shown in curve a in Figs. 2 and 3. Equations (44)-(46) may be further simplified as[13]

$$R_{max}=kT_L, \tag{46}$$

and

$$P \to \infty, \tag{47}$$

$$\varepsilon = 0. \tag{48}$$

4) When $\lambda=0$, $T_c=0$. Eq. (1) becomes the expression of the relation between the magnetic field intensity H and the magnetization intensity M of a simple paramagnetic salt[14-16]

$$M = ng\mu_B J B_J(x) = ng\mu_B J B_J[g\mu_B JH/(kT)]. \tag{49}$$

When x<<1, Eqs. (1) and (50) become the Curie-Weiss law[5]

$$M = C\frac{H}{T-T_c} \tag{50}$$

and the Curie Law[15]

$$M = C\frac{H}{T}, \tag{51}$$

respectively, where

$$C = ng^2\mu_B^2 J(J+1)/(3k) \tag{52}$$

is the Curie constant.

It is clearly seen from the relation between Eqs. (50)-(52) and (1) that under the different limit conditions, the optimal performance of a stirling refrigeration cycle using paramagnetic salts as the working substance, whose properties are described by Eq. (50),

(51), or (52), may be directly derived from the above results. This show that the results obtained in this paper are of general significance.

ACKNOWLEDGMENTS

This project has been supported by the foundation of the State Education Commission, China.

ENDNOTES

[1]Z. Yan and J. Chen, J. Appl. Phys. **72**, 1 (1992).
[2]T. Numazawa, H. Kimura, M. Sato, and H. Maeda, Cryogenics **33**, 547 (1993).
[3]C. Hagmann, D. J. Benford, and P. L. Richards, Cryogenics **34**, 213 (1994).
[4]L. Chen and Z. Yan, J. Appl. Phys. **75**, 1249 (1994).
[5]S. V. Vonsovskii, Magnetism, Vol. II (Wiley, New York, 1974).
[6]J. Chen and Z. Yan, Cryogenics **33**, 863 (1993).
[7]H. M. Rubin and B. Andresen, J. Appl. Phys. **53**, 1 (1982).
[8]B. Andresen, Finite-Time Thermodynamics (Phys. Lab. II, University of Copenhagen, Copenhagen, 1983).
[9]J. Chen and Z. Yan, J. Appl. Phys. **63**, 4795 (1988).
[10]A. De Vos, Endoreversible Thermodynamics of Solar Energy Conversion (Oxford University Press, Oxford, 1992).
[11]F. A. Brown and G. R. Madrigal, Rev. Méxicana Fís. **36**, 363 (1990).
[12]J. Chen and Z. Yan, J. Phys. D: Appl. Phys. **29**, 987 (1996).
[13]Z. Yan and J. Chen, J. Phys. D: Appl. Phys. **23**, 136 (1990)
[14]R. Kubo, Statistical Mechanics (North-Holland, Amsterdam, 1965).
[15]M. W. Zemansky, Heat and Thermodynamics, 5th ed. (McGraw-Hill, New York, 1968).
[16]S. V. Vonsovskii, Magnetism, Vol. I (Wiley, New York, 1974).

PERFORMANCE CHARACTERISTICS OF AN IRREVERSIBLE REFRIGERATOR

MAMDOUH EL HAJ ASSAD
Helsinki University of Technology
Laboratory of Applied Thermodynamics

Abstract- By using finite-time thermodynamics and nonlinear heat transfer law, the general performance characteristics of an irreversible refrigerator are investigated. A general expression which relates the cooling load and coefficient of performance (COP) of a completely irreversible refrigerator is derived. The maximum cooling rate for a given COP is determined.

Keywords- coefficient of performance, cooling rate, irreversible refrigerator

INTRODUCTION

It is not practical to design a heat engine or a refrigerator that runs infinitely slowly without producing power or providing cooling. The reversible performance limits of a heat engine to produce power and a refrigerator to provide cooling are not approached in reality and these limits have been considered as the upper performance limits of the real heat engines and refrigerators.

The upper limits of a heat engine were firstly set by Carnot who showed that any heat engine absorbing heat from a high temperature heat source must transfer some heat to a low temperature heat sink in order to produce work. He also provided that no heat engine could be better than the Carnot heat engine and no refrigerator could be better than the Carnot refrigerator.

The analysis of the reversible heat engine to evaluate the engine efficiency with considering the finite heat transfer irreversibilities to and from the heat engine has been introduced [1]. The engine studied in Ref. [1] produces not only work but also power and its efficiency (endoreversible efficiency) is less than the reversible Carnot efficiency.

In the last couple of decades, finite-time thermodynamics (FTT) has been developed [2-6] to analyze endoreversible thermal cycles. The FTT models consider irreversible processes at the couplings between the system and its surroundings. The most common irreversiblity considered by FTT is the heat flux between the system and its corresponding heat reservoirs. Such flux is commonly described by a linear-heat transfer law usually called Newton's law of cooling [7, 8].

Even though FTT has been given great interest, most of the work has been concentrated on heat engines. The direct use of FTT proposed in Ref. [1] will not work when applying it to a heat pump because there in no natural maximum [9]. An expression for the COP of a heat pump running at finite speed when operating at minimum power input for a given heat power output has been derived [10]. An extensive study has been proposed [11] on the theory of heat transfer irreversible refrigeration plants. The FTT has been applied to investigate the wall gain and product loads on a refrigerator based on an endoreversible Carnot cycle for which explicit formulas for the COP and cooling rate have been derived [12]. A study has been carried out to determine the maximum obtainable specific cooling load of a refrigerator [13].

Almost all authors mentioned above have assumed linear heat transfer law (Newton's law) in their analyses.

In the present paper our aim is to determine the general performance characteristics of an irreversible refrigerator, which will serve in a realistic design criterion of a real refrigerator.

FORMULATION

The refrigerator operates between two infinitely large thermal reservoirs named heat source (low temperature) and heat sink (high temperature). The refrigerator removes heat from the low temperature heat source and transfers it to the high temperature heat sink by introducing some work into the refrigerator. The heat addition and heat rejection processes are isothermal. To have a reversible heat transfer, infinitesimal temperature differences between the refrigerator and its thermal reservoirs are required. These infinitesimal temperature differences result in zero heat transfer for finite exchanger surface areas. Hence the refrigerator will not provide any cooling load. However cooling load can be provided when the refrigerator is designed with infinitely large heat exchangers but this design is not practical. So a model of an irreversible refrigerator is needed to provide a reasonable performance characteristics which could not be achieved in reality by Carnot refrigerator.

The refrigerator in this study considers the existence of finite temperature differences between the refrigerator and its thermal reservoirs and also considers the internal irreversiblities which are caused by internal dissipations in the refrigerator. The only external irreversiblities of the refrigerator are due to the finite temperature differences mentioned above. Thus in the refrigerator, heat flows from the heat source at T_L to the cold working fluid at T_C through an evaporator and heat flows from the warm working fluid at T_W to the heat sink at T_H through a condenser. The T-S diagram of the irreversible refrigerator is shown in Fig.1.

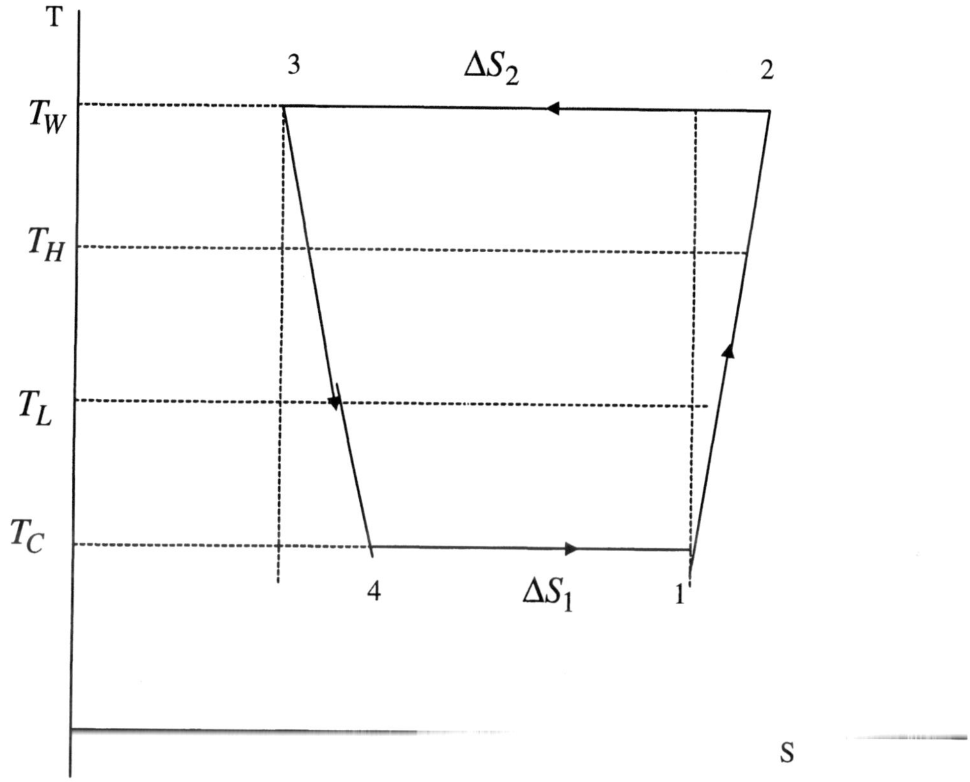

Fig.1. T-S diagram of an irreversible refrigerator.

The heat flow (Q_L) from the low temperature heat source is described in a general form by using nonlinear heat transfer relation as

$$Q_L = \alpha_n (T_L - T_C)^n t_L, \tag{1}$$

where α_n is the heat conductance and t_L is the required residence time of the cold working fluid in the heat exchanger to transfer a heat amount of Q_L.

Similarly the heat flow (Q_H) from the warm working fluid to the high temperature heat sink is

$$Q_H = \gamma_n (T_W - T_H)^n t_H, \tag{2}$$

where γ_n is the heat conductance and t_H is the required residence time of the warm working fluid in the heat exchanger to transfer a heat amount of Q_H.

The total time required for the whole cycle to be completed can be written as

$$\tau = t_H + t_L. \tag{3}$$

In Eq. (3), we neglected the time of the processes 1-2 (compressor) and 3-4 (expansion valve) which are shown in Fig.1, due to the fact that these processes occur quickly as compared to the heat addition (4-1) and heat rejection (2-3) processes.

By using Eqs. (1)-(3), the total time is obtained as

$$\tau = \frac{Q_L}{\alpha_n (T_L - T_C)^n} + \frac{Q_H}{\gamma_n (T_W - T_H)^n}. \tag{4}$$

The work input (W) to the refrigerator is obtained from the first law of thermodynamics as

$$W = Q_H - Q_L. \tag{5}$$

For an internally irreversible refrigerator, the second law of thermodynamics yields

$$\oint \frac{\delta Q}{T} = \frac{Q_L}{T_C} - \frac{Q_H}{T_W} < 0. \tag{6}$$

In order to describe quantitatively the effect of internal irreversiblities caused by the working fluid on the refrigerator performance, we introduce the cycle irreversiblity parameter R as

$$R = \frac{\Delta S_1}{\Delta S_2} = \frac{S_1 - S_4}{S_2 - S_3}, \tag{7}$$

where $\Delta S_1 = S_1 - S_4$ and $\Delta S_2 = S_2 - S_3$ are the entropy production terms of the working fluid in the two isothermal branches at T_C and T_W.

The inequality in equation (6) can be eliminated by writing

$$\frac{Q_L}{T_C} = R\frac{Q_H}{T_W}. \tag{8}$$

When $R < 1$, the refrigerator is an internally irreversible; when $R = 1$, the refrigerator is internally reversible but still externally irreversible (endoreversible refrigerator).

The power input (P) to the refrigerator, the cooling rate (q) and the COP (β) of the irreversible refrigerator are defined as

$$P = \frac{W}{\tau}, \tag{9}$$

$$q = \frac{Q_L}{\tau} \tag{10}$$

and

$$\beta = \frac{q}{P} = \frac{RT_C}{T_W - RT_C}. \tag{11}$$

After defining the parameters $z = \frac{\beta + 1}{\beta}$ and $x = T_L - T_C$, Eq.(8) becomes

$$q = \frac{\alpha_n \gamma_n R x^n \left[(T_L - x)Rz - T_H\right]^n}{\gamma_n \left[(T_L - x)Rz - T_H\right]^n + \alpha_n R z x^n}. \tag{12}$$

Equation (12) is the general relationship between the COP and the cooling rate and it describes the general performance characteristics of the irreversible refrigerator.

The cooling rate can be maximized when the derivative of Eq. (12) with respect to x is equated to zero. The following optimum parameters for a fix β can be found as

$$x_{opt} = \frac{T_L Rz - T_H}{Rz + \left(\frac{R^2 z^2 \alpha_n}{\gamma_n}\right)^{\frac{1}{n+1}}}, \tag{13}$$

$$y_{opt} = \frac{T_L Rz - T_H}{1 + Rz\left(\frac{R^2 z^2 \alpha_n}{\gamma_n}\right)^{-\frac{1}{n+1}}} \tag{14}$$

and

$$q_{opt} = \frac{\alpha_n \gamma_n (T_L Rz - T_H)^n}{\alpha_n Rz\left[1 + Rz\left(\frac{R^2 z^2 \alpha_n}{\gamma_n}\right)^{-\frac{1}{n+1}}\right]^n + \gamma_n\left[Rz + \left(\frac{R^2 z^2 \alpha_n}{\gamma_n}\right)^{\frac{1}{n+1}}\right]^n} \tag{15}$$

where the parameter y is defined as $y = T_W - T_H$.

The value of R should be greater than $\frac{T_H}{zT_L}$ to have an optimum cooling rate and that was taken into consideration during the derivation of Eq. (15), so we can write $\frac{T_H}{zT_L} < R \leq 1$.

At any given COP value and with known values of T_L, T_H, α_n and γ_n, and choosing R in the range discussed above, the optimum temperatures of the working fluid $(T_C)_{opt}$ and $(T_W)_{opt}$, and the maximum cooling q_{opt} can be obtained from Eqs. (13), (14) and (15), respectively.

NUMERICAL WORK

Consider a refrigerator that is running between $T_L = 245K$ and $T_H = 290K$. The refrigerator is set to run at $\beta = 4$. We assume the heat fluxes follow linear heat transfer law , i.e. $n = 1$. Consider the thermal conductances of the heat exchangers of the cold working fluid and the warm working fluid to be $\alpha_n = 15 {kW}/{K}$ and $\gamma_n = 20 {kW}/{K}$, respectively. The internal irreversiblity parameter is taken as $R = 0.98$ according to the interval $0.947 < R \leq 1$. Applying equations (13), (14) and (15), yields $(T_C)_{opt} = 240.57K$, $(T_W)_{opt} = 294.7K$ and $q_{opt} = 35kW$. For

$R = 1$ and with the same operating conditions, the optimum cooling rate from equation (15) is $56kW$. The Carnot COP can be obtained by using $\beta_{Carnot} = \dfrac{T_L}{T_H - T_L}$ as $\beta_{Carnot} = 5.44$ which is higher than that for the irreversible refrigerator, and at this Carnot COP, the refrigerator will provide no cooling.

CONCLUSIONS

An analysis has been carried out on an irreversible refrigerator. The nonlinear heat transfer law along with the finite-time thermodynamics has been applied to obtain the general performance characteristics of the refrigerator. The internal irreversibilities has been described by a single parameter which is introduced into the cooling rate and the COP equations. The cooling rate of the irreversible refrigerator has been maximized for a given COP. As can be seen from the results of the numerical example that the irreversible refrigerator gives more realistic performance characteristics than that of the Carnot refrigerator, since the former considers the external irreversiblities between the system and its surroundings in addition to the internal irreversiblities caused by the working fluid.

At the same operating conditions, it is shown that the cooling rate of the irreversible refrigerator is greater than that of the Carnot refrigerator. The endoreversible refrigerator gives higher cooling rate than the irreversible refrigerator, because the endoreversible refrigerator is considered to be internally reversible. So an increase in the internal irreversibility parameter will improve the performance characteristics of the refrigerator.

The model presented in this paper, provides a reasonable design criterion for use in the evaluation of the general performance characteristics of a real refrigerator.

We may conclude that the finite-time thermodynamics has been a successful extension of classical thermodynamics in the domain of determining the effect of external irreversiblities in addition to the internal irreversibilities.

NOMENCLATURE

Q = Heat flow
P = Power
R = Internal irreversiblity parameter
S = Entropy
t = Isothermal process time
T = Temperature
W = Work
α = Heat conductance in cold side heat exchanger
β = Coefficient of performance

γ = Heat conductance in warm side heat exchanger

τ = Total cycle time

REFERENCES

1. Curzon F. L. and Ahlborn B., *American Journal of Physics,* 1975, **43**, 22 .
2. Andresen B., Salamon P. and Berry R. S., *Physics Today,* 1984, **37**, 62.
3. Assad M., *International Journal of Mechanical Engineering Education,* 1997 (in press).
4. De Vos A., *American Journal of Physics*, 1985, **53**, 570.
5. Gordon J. M., *American Journal of Physics*, 1990, **58**, 370.
6. Wu C., *Energy Conversion and Management*, 1993, **34**, 1239.
7. Salamon P. and Nitzan A., *Journal of Chemical Physics*, 1981, **74**, 3546.
8. Bohren C. G., *American Journal of Physics*, 1991, **59**, 1044.
9. Blanchard C. H., *Journal of Applied Physics*, 1980, **51**, 2471.
10. Leff H. S. and Teeters W. O., *American Journal of Physics*, 1978, **46**, 19
11. Bejan A., *International Journal of Heat and Mass Transfer*, 1989, 32, 1631.
12. Agrawal D. C. and Menon V. J., *Journal of Applied Physics*, 1993, **74**, 2153.
13. Wu C., *Energy Conversion and Management*, 1995, **36**, 7.

Endoreversible Modeling and Optimization of Multistage Thermal Machines by Dynamic Programming

Stanislaw Sieniutycz
Faculty of Chemical Engineering, Warsaw University of Technology,
00-645 Warsaw, 1 Warynskiego Street, Poland

Abstract.

Nonlinear multistage processes which yield mechanical work are optimized by Bellman's method of dynamic programming to determine best discrete controls for maximum of work delivered or minimum of work consumed. Suitability of control based on the so-called driving temperature is shown for a cascade of endoreversible thermal machines. Abstract theory of recurrence equations is developed simultaneously with their application to the cascade whose dynamics is described by equations of work production and temperature change of a resource fluid. A finite-rate framework admits irreducible losses of reversible work caused by thermal resistances represented by fluid boundary layers. Efficiency changes are analyzed in terms of the driving temperature or finite heat flux. Optimal performance functions which describe extremum work are defined and governing recurrence equations are derived from Bellman's optimality principle. Analogies between optimal continuous and discrete processes are shown which help predict behavior of the latter and extend optimization techniques known for continuous systems to multistage systems. Discrete equations of state, recurrence equations and a discrete Hamilton-Jacobi equation are characteristics which exhibit strong analogy with their continuous conterparts at the limit of an infinite number of stages. Analytical solution is found for optimal temperatures and optimal work functions in terms of end states, number of stages and a finite duration. Enhanced bounds on production or consumption of mechanical energy are discussed.

PACS number(s): 05.70.-a. KEY WORDS: Carnot cycle, finite-time thermodynamics, endoreversible engines, exergy (available energy), thermodynamic limits, second law.

1. Introduction: basics of multistage endoreversible processes

We treat here a multistage process of work production from two fluids with different temperatures, the first is a finite-resource fluid and the second is the bath fluid (fluid in an infinite reservoir). Multistage processes are important because a multistage process may be regarded as a generalization of a continuous one; under special conditions allowing a continuous limit for an infinite number od stages a continuous process be obtained from a multistage one, but not inversely. Some extra information about the stage dynamics is necessary to arrive at a multistage model, even if it comes out from a discretization of known equations of a continuous process. However, the are processes which are discrete by nature, and those ones are of our main interest. The difference links between stages render mathematics difficult; this is, perhaps, why there are only a few works to the date related to the multistage optimization of the work production [1-3]. However, no general optimization theory for multistage work-production systems was obtained to the date. The purpose of this work is to present such a theory using Bellman's method of dynamic programming [4, 5]. It generalizes our recent work on optimization of continuous systems [6, 7]. The purpose of the generalization is to include controlled holdup times, mixing phenomena and effect of a finite number of stages.

We focus on a steady-state process undergoing in the multistage system presented in Fig. 1. The fluids are of finite thermal conductivity and possess their own boundary layers represented by nonvanishing thermal resistances. The multistage process is, in fact, a sequence of the Novikov-Curzon-Ahlborn processes (NCA processes) which are repeated in the stages of the cascade. The multistage NCA system contains: the resource fluid of a finite mass flow G, the envinronment or bath of the constant temperature T^e, the boundary layers as the thermal conductances, and the set of the Carnot engines, $C^1 ... C^N$, which release the mechanical work at each stage n. When the system is in the engine mode and temperatures of the resource fluid are higher than T^e then the fluid flows in direction of gradually descreasing temperatures $T^1 ... T^n$. At each stage the fluid is well mixed, and a mixing scale is assigned to the fluid flow. The fluxes of the produced work are summed, so that a cumulative power output $\mathbb{W}$ is obtained at the last stage. The power output per unit flux of the driving fluid, $W = \mathbb{W}/G$, has the dimension of work per unit mass and describes the specific work associated with the steady-flow process of power production. For the purpose of optimization by dynamic programming a suitable representation of this multistage process as a discrete optimal control is schematized in Fig. 2. Each stage n, which is represented in Fig. 2 by a distinct unit of the black box type, has nonetheless its own internal structure shown in Fig. 1 for the stage n. This structure is composed of the following elements: a well-mixed fluid cell (a shadow upper part of the pipe with the input temperature T^{n-1} and output temperature T^n), two thermal resistances of each boundary layer, a Carnot engine C^n, and a contacting part of an isothermal bath (a shadow lower part of the stage). Our main goals are: to develop an optimization model of the system, to derive optimality conditions and to find an optimization solution. By definition, the work released in engine mode is positive. An optimization problem for a cascade of engines can be stated as the maximum of work $W = \dot{W}/G$, consistent with maximum of power $\dot{W}$. For a cascade of heat pumps minimizing of the negative of the work, (-W) can be required.

2. An individual stage with a driving temperature T' as control variable

In any multistage optimization knowledge of the so-called equations of state transformation is essential. Their conventional form describes state outputs in terms of state imputs and controls at a stage. For the analyzed process, however, a modified form of these relations is suitable in which state inputs are expressed in terms of state outputs and controls. This is associated with use of the so-called forward algorithm of dynamic programming method. In any case analysis of the single stage should proceed the cascade optimization. The left-hand side of Fig. 1 (the region surrounding the second Carnot engine) shows the principle of designations used for the description of a single stage.

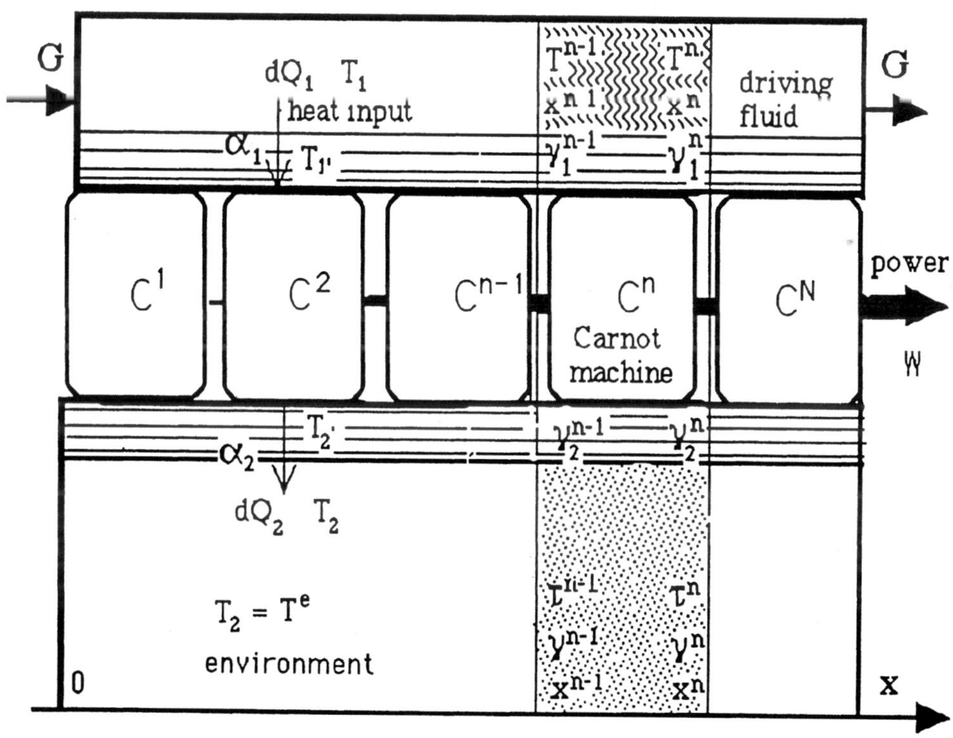

Fig. 1: Scheme of power production in a sequence of Novikov-Curzon-Ahlborn (NCA) engines and basic designations.

Two finite thermal resistances are represented by two boundary layers, each characterized by its own heat transfer coefficient, α_1 and α_2. They are contained between the upper and lower fluids and the working fluid circulating in the Carnot part of the stage. The resistances are, of course, reciprocals of the corresponding conductances $g_1 = \alpha_1 a_1$ and $g_2 = \alpha_2 a_2$, where a_1 and a_2 are the exchange surface areas at the stage. When the system is in the engine mode, the hot fluid releases the heat at a high temperature T_1, which reaches the engine part at $T_{1'}$. The low-grade heat is released by the engine fluid at the low temperature $T_{2'}$ to the environmental fluid, and reaches the bath at a low temperature $T_2 = T^e$. (In a more general case of two fluids each flowing with a finite mass flux, both fluid temperatures T_1 and T_2 vary along the path, however we restrict here ourselves to the case $T_2 = T^e$.) Clearly, for an engine mode, the efficiency of the power production at a stage, $\eta = w/q_1$, is lower than the efficiency η_C of the Carnot cycle operating between T_1 *and* T_2.

The entropy balance of the Carnot machine in a single NCA unit (an arbitrary process stage)

$$\frac{q_1}{T_{1'}} = \frac{q_2}{T_{2'}} \tag{1}$$

and the energy balance combined with the definition of the first-law efficiency,

$$q_2 = q_1 - w = q_1(1 - \eta) \tag{2}$$

yield the stage efficiency given by the Carnot formula

$$\eta = 1 - \frac{T_{2'}}{T_{1'}}. \tag{3}$$

For an engine stage this efficiency is lower than that of the Carnot engine working between the temperatures T_1 and T_2 as it applies to the "intermediate" temperatures $T_{1'}$ and $T_{2'}$. By assumption they satisfy Newtonian law of heat exchange

$$q_1 = g_1(T_1 - T_{1'}); \qquad q_2 = g_2(T_{2'} - T_2). \tag{4}$$

The quantities $T_{1'}$ and $T_{2'}$ are unknown, but they may be expressed in terms of the temperatures T_1 and T_2 and a control variable at the stage. The choice of the control variable is in principle arbitrary; for example, the controls can be: the driving heat flux q_1, the entropy flux, the efficiency, η, and the others. Here we demonstrate the suitability to descibe one-stage and multistage NCA processes by the so-called driving temperature, T' as the control variable. This concept was not systematically used to date although it has its roots in the already recognized structure of equations describing NCA processes.

The reasoning leading to the driving temperature is as follows: Since the temperatures $T_{1'}$ and $T_{2'}$ are not independent but connected by the reversible entropy balance,

$$\frac{g_1(T_1 - T_{1'})}{T_{1'}} = \frac{g_2(T_{2'} - T_2)}{T_{2'}}, \tag{5}$$

i.e. equation (1) for assumed Newtonian model of heat transfer, the efficiency (3) may be always written in terms of a single controlling temperature, for example

$$\eta = 1 - \frac{T_2}{T'}. \tag{6}$$

At this stage of reasoning, equation (6) does not mean that T' has any physical sense. However, as we proceed, we will show that it is a positive define quantity of physical type. With this property and the role it plays in the second law, T' determines the direction of heat transfer between the fluids and the Carnot engine. Using T' one can also accomplish optimization of complicated multistage processes.

Equations (3) and (6) yield the ratio of temperatures $T_{1'}$ and $T_{2'}$ in terms of T' as

$$\frac{T_{2'}}{T_{1'}} = \frac{T_2}{T'}. \tag{7}$$

Thus one can substitute $T_{2'} = T_{1'} T_2/T'$ into equation (5) to obtain the temperatures $T_{1'}$ and $T_{2'}$ in terms of T'

$$T_{1'} = \frac{g_1 T_1 + g_2 T'}{g_1 + g_2} \tag{8}$$

and

$$T_{2'} = \frac{T_2(g_1 T_1 + g_2 T')}{T'(g_1 + g_2)}. \tag{9}$$

Hence the heat flux q_1 satisfies in terms of the control T' a simple traditional formula

$$q_1 = g_1(T_1 - T_{1'}) = \frac{g_1 g_2}{g_1 + g_2}(T_1 - T') \equiv g(T_1 - T'). \tag{10}$$

Consistently with equation (6), the second heat flux is $q_2 = q_1(T_2/T')$, as may verified by

$$q_2 = g_2(T_{2'} - T_2) = g(T_1 - T')(\frac{T_2}{T'}) = q_1\frac{T_2}{T'} . \tag{11}$$

The resulting expressions in equations (10) and (11) contain the overall conductance of an inactive heat transfer, defined traditionally as

$$g = (\frac{1}{g_1} + \frac{1}{g_2})^{-1} = \frac{g_1 g_2}{g_1 + g_2} . \tag{12}$$

From equation (10) we conclude that *the driving temperature T' is the absolute temperature of an external fluid whose cooling or heating effect exerted on the fluid 1 replaces the joint effect of the Carnot machine and the second fluid under condition of the same overall conductance g.*

However, it may easily be shown that another driving temperature, T'', could be defined by the expression $\eta = 1 - T''/T_1$, in which T'' replaces T_2 in the Carnot-like expression. In this representation the ratio $T_{2'}/T_{1'}$, equation (7), would be equal T''/T_1 and a simple linear formula would be obtained for the second heat, $q_2 = g(T'' - T_2)$, whereas the first heat would satisfy $q_1 = q_2 T_1/T''$. Thus, analogously as for T', we conclude that *the driving temperature T'' is the absolute temperature of an external fluid whose cooling or heating effect exerted on the fluid 2 replaces the joint effect of the Carnot machine and the first fluid under condition of the same overall conductance g.*

The suitability of driving temperatures is caused by the fact that they make analysis of thermal effects in NCA systems simpler than other controls. This property follows from the fact that with a driving temperature the analysis of the work production process becomes broken down to a simple equivalent analysis of a heat exchange process. We shall display advantages of use of the temperature T' for multistage NCA cascades, where the efficiency-based analysis is hampered in vicinity of the environmental temperature T^e by the fact that quantities such as work and efficiency can change their signs along the process path. In most cases it is sufficient to work with one of the driving temperatues, thus we will regularly deal with T', except for special case when a symmetry is observed when replacing T' by T'', which is the case of the maximum power point. This conclusion is based on equations (15) and (16) below, which hold for both T' and T'' as then $T' = T''$. For a quick passage from the process representation in terms of T' to the representation in terms of T'' it is easiest to use the equality $T'T'' = T_1T_2$. This results from comparison of the efficiency equation (6) with its counterpart in terms of the temperature T'', $\eta = 1 - T''/T_1$. While T' and T'' are equal at the maximum power point they differ at the so-called *short circuit* point, where both w and $\eta = 0$; in this case $T' = T_2$ and $T'' = T_1$, thus the heats q_1 and q_2 are the same and equal $g(T_1 - T_2)$. On the other hand at the *Carnot point*, where $\eta = \eta_C$, $T' = T_1$ and $T'' = T_2$. These two basic points divide each characteristics into three ranges: the range of a normal work of the engine, the range of a refrigerator and the range of a heat pump. De Vos [8] discusses behavior of NCA units in various ranges of η.

The mechanical power equation may be writen in two alternative forms

$$w = q_1 - q_2 = g(T_1 - T')(1 - \frac{T_2}{T'}) = -g(T'' - T_2)(1 - \frac{T_1}{T''}) \tag{13}$$

consistent the efficiency in terms of T' *and* T''. Both representations are included in equation (13) for comparison. The stationary maximum of the power w with respect to T' occurrs at the point satisfying $\partial w/\partial T' = 0$ or

$$\frac{\partial w}{\partial T'} = g\frac{(T_1 - T')T_2}{(T')^2} - g(1 - \frac{T_2}{T'}) = g\frac{T_1T_2 - (T')^2}{(T')^2} = 0 \tag{14}$$

(analogous formula can be written down in terms of T''). Each optimal driving temperature is the geometric mean of the bulk temperatures of participating fluids

$$T' = \sqrt{T_1T_2} = T'' . \tag{15}$$

Thus T' and T'' are equal at the optimal power point. The second derivative inequality

$$\frac{\partial^2 w}{\partial (T')^2} = -2g\frac{T_1T_2}{(T')^3} < 0 \tag{16}$$

proves that the positive T' and g are sufficient for maximum. Thus, at least around the maximum power point (15) the positeveness of T' is assured; this conclusion holds also for the temperature T''. However, we will show below that T' and T'' are positive anywhere, and that this important property is caused by the positiveness of the entropy production, σ_s. Indeed, by computing the σ_s as the difference between the outlet and inlet entropy fluxes

$$\begin{aligned} \sigma_s &= \frac{q_2}{T_2} - \frac{q_1}{T_1} = \frac{q_1}{T'} - \frac{q_1}{T_1} = q_1(\frac{1}{T'} - \frac{1}{T_1}) \\ &= g(T_1 - T')(\frac{1}{T'} - \frac{1}{T_1}) = \frac{g(T_1 - T')^2}{T_1T'} \end{aligned} \tag{17}$$

the property $T' > 0$ follows for g > 0. The equality $q_2 = q_1(T_2/T')$, equation (11), was used in the transformations. The positiveness of T' as well as the simple heat formula, equation (10), make the quantity T' very useful. Moreover, because of the Carnot representation of T', equation (6) can be transformed as follows

$$\eta = 1 - \frac{T_2}{T'} = 1 - \frac{T_2}{T_1} - T_2\frac{(T_1 - T')}{T_1 T'} = \eta_C(T_1, T_2) - \frac{T_2}{gT_1(T_1 - q_1/g)}q_1 \tag{18}$$

This is a simple and suitable expression showing how a finite flux q_1 influences the process efficiency. Whenever T' is less than T, meaning that a positive amount of heat q_1 flows to the Carnot engine, the first law efficiency η drops below that of Carnot, else it raises above η_C. This conclusion is indepenent of the mode at which the stage works (engine, refrigerator or heat pump) which depends on the relations between T_1 and T_2 and their relation to the temperature of the environment, T^e. The use of T' in efficiency analysis is especially whenever the use of power w changes its sign in the way different than q_1. Yet all well-known results of an optimal NCA stage are recovered. For example, when the optimal T', equation (15), is substituted into the efficiency formula, equation (6), the well-known NCA efficiency, $\eta = 1 - (T_2/T_1)^{1/2}$, is obtained. Substituting equation (15) into equation (13) yields the maximum power in the form $w = g\sqrt{T_1} - \sqrt{T_2})^2$ which is the expression familiar from the standard treatment of NCA process [8 -11].

There are still other virtues of T'. These become explicit when dynamical problems are considered, i.e. when the temperature T_1 or T_2 change in time. Let us admit changes of T_1 assuming that it is the temperature of a finite resource fluid (of a finite amount or flow), whereas the second fluid forms an infinite isothermal reservoir, i.e. $T_2 \equiv T^e$. The superscript e stresses that T^e is both the temperature of an infinite environment and the equilibrium temperature of the first fluid with this environment.

3. Imbedding single-stage formulae into a model of multistage process

In the multistage cascade formalism the superscript n is added to all symbols, to point out that the quantities refer to the stage n. On the other hand, since $T_2 = T^e$ is now only a parameter and not a state variable, we can simplify our designations for the state variable T_1^n and the heat q_1^n by rejecting the unnecessary subscript 1; thus we will use the symbols T^n and q^n for the resource fluid temperature and driving heat flux at the stage n. We shall also take the constant temperature T^e instead of T_2^n. With these modifications, the basic equations (6) and (10), yield

$$q^n = g^n(T^n - \frac{T^e}{1-\eta^n}) . \tag{19}$$

This is an equation of thermal characteristics of the n-th stage [8]. The formula inverse with respect to the above

$$\eta^n = 1 - \frac{T^e}{T^n - g^{-1}q^n} \tag{20}$$

is also useful. Equations (18) - (20) are equivalent; each proves that the power production efficiency at an engine stage decreases with the intensity of the heat flux. However, when furnishing the cascade description, each of these is not equally suitable, as it is generally easier to incorporate equation (19) into a discrete mathematical formalism. This happens because, in the cascade formalism, some extensive quantities play essential role, which are characteristic of the cascade as a whole, not a single stage. They are usually computed as cumulative quantities over stages, and may play role of state coordinates in the cascade optimization. In general, a cumulative quantity may be a state variable whenever there are limits for its consumption or production.

Multiplying equation (19) by the stage efficiency η^n, a state equation for the cumulative work is obtained in terms of η^n. We introduce the cumulative power per unit mass flow, $W^n \equiv w^n/G$, the quantity whose units are those of the specific work

$$W^n - W^{n-1} = c\theta^n \eta^n (T^n - \frac{T^e}{1-\eta^n}) , \tag{21}$$

where $\theta^n \equiv g^n/Gc$ is a nondimensional quantity that represents the so-called number of the heat transfer units for the stage n. In terms of the driving temperature control, T', this equation becomes

$$\begin{aligned} W^n - W^{n-1} &= - c\theta^n (1 - \frac{T^e}{T'^n})(T'^n - T^n) = \\ &= - c(1 - \frac{T^e}{T^n})(T^n - T^{n-1}) - cT^e \frac{(T'^n - T^n)^2}{T^n T'^n} \theta^n , \end{aligned} \tag{21'}$$

where equation (20) has been used. The second line expression shows the split of the work production rate into a classical-type term and a dissipative term. The total work is the sum of the above expressions over the stages. Since the optimization problem is that of maximum W^n for n = N, it is convenient to treat the variable W as the zero-th state

One may note that the nondimensional variable of the time interval type, $\theta^n \equiv g^n/(Gc)$, appears linearly in equations (21) and (21'). Each of these equations is the state equation for the variable W^n in the sense used in optimization. Yet we need to transform equation (19) into the second discrete equation of state.

The fluid's temperature changes are related to the second cumulative variable, which is the cumulative driving heat Q^n defined for the first n stages of the cascade: $Q^n = \Sigma q^k$, where $k = 1, 2..n$. This quantity satisfies the equality $Q^n - Q^{n-1} = q^n$, where q^n is given by equation (19). The sequence of heats q^n released to Carnot engines describe allocations of the cumulative heat Q^n between the stages, $i = 1, 2....n$. But, in terms of the fluid's temperature change ΔT^n at the stage n, each heat q^n equals $- Gc(T^n - T^{n-1})$; consequently

$$T^n - T^{n-1} = - \theta^n (T^n - \frac{T^e}{1 - \eta^n}) . \tag{22}$$

This is the second discrete equation of state of the cascade process. We have assumed here that the fluid specific heat, c, is constant. In terms of the driving temperature control, T', equation (22) has the form

$$T^n - T^{n-1} = \theta^n (T'^n - T^n), \tag{22'}$$

where $T'^n - T^n = - q^n/g^n$ is a measure of the driving heat received by the n-th Carnot engine. Here T^n is the second variable of state of the problem. To close the model it suffices to regard $\theta^n \equiv g^n/(Gc)$ as the increment of an independent variable, τ^n. The latter is the cumulative number of the heat transfer units over the stages; $\tau^n = \Sigma \theta^k$ for $k = 1, 2..n$. Thus, by definition, the variable θ^n is the controlled interval of the cumulative number of heat transfer units, and the third equation of state is

$$\tau^n - \tau^{n-1} = \theta^n . \tag{23}$$

In the present formulation τ^n is an independent variable which plays the role of a (nondimensional) process time in our discrete model. It is also a special cumulative variable which is a measure of the total residence time of the resource fluid in the cascade. Equation (23) is essential in all problems with a finite exchange area, which bear *constrained* sums of θ^n. In accordance with the therminology used in optimization each of equations (21) - (23) is a discrete equation of state for the cascade. These equations contain on their right-hand sides the state variables and controls. Since now, we shall exclusively apply the driving temperature representation in the optimization model, which means that we will use equations (21'), (22') and (23).

4. General difference model and Bellman's optimality principle

Now a general scheme of optimization of cascade processes by Bellman's dynamic programming method (DP) is considered. The developed theory will be applied to CAN cascades in the next sections. Our basic task is the analysis of *discrete* (multistage) processes with finite number of stages, although results at the continuous limit (for an infinite number of stages) will also be discussed. Taking into account the fact that the model of NCA cascade contains θ^n as a linear decision, we focus on particular discrete equations of state: those which are linear with respect to a particular decision, θ^n. In the most general situation, θ^n is a free interval of a state variable accepted as the independent variable.

Assume a general discrete process described by the variables $x_1^n,\ldots,\ x_i^n,\ldots x_s^n$ and the performance variable x_0^n, which has to be maximized at $n = N$. The set of the discrete state equations which admits the linearity of model with respect to θ^n has the form

$$\frac{x_i^n - x_i^{n-1}}{x_{s+1}^n - x_{s+1}^{n-1}} = f_i^n(\mathbf{x}^n, \mathbf{u}^n) \tag{24}$$

where $i = 0, 1\ldots s,\ s + 1$ and x_i^n are the coordinates of the state vector $\mathbf{x}^n = (x_0^n,\ x_1^n,\ldots,\ x_i^n,\ldots x_s^n,\ x_{s+1}^n)$. This vector includes the performance variable x_0^n and an independent variable $x_{s+1}^n \equiv \tau^n$. In our case, by definition of x_{s+1}^n, the rate f_{s+1} equals 1, hence the $s+1$-th coordinate of $\mathbf{x}$ satisfies $\Delta x_{s+1}^n/\Delta \tau^n \equiv 1$. The time intervals $\Delta x_{s+1}^n \equiv \Delta \tau^n$ appearing in this set are those satisfying equation (23). In a special case when the rates $f_1..f_s$ are identical with the controls $u_1..u_s$, the problem is a discrete version of the variational calculus. The variable x_{s+1}^n is only a dummy variable when all f_i are independent of τ (autonomous models) *and* the duration of the process, $\tau^N - \tau^0$, is unconstrained. However any duration-related constraint makes the variable x_{s+1}^n a genuine state variable even in the case of an autonomous system ($\mathbf{f}$ independent of τ). When usual description based on ordinary difference equations is applied, equation (24) is equivalent with the set of equations

$$x_i^n - x_i^{n-1} = f_i^n(\mathbf{x}^n, \mathbf{u}^n)\theta^n \tag{25}$$

The additional control, θ^n, which appears linearly in the above set, constitutes the unconstrained interval of the independent variable τ. The set (25) with arbitrary f_{s+1}^n may be considered as a proper generalization of the original model which still preserves the linearity of the model with respect to a free controls θ^n. In the case of our NCA cascade, where s=1, the state vector $\mathbf{x}^n$ is three-dimensional: $x_0^n \equiv W^n$, $x_1^n \equiv T^n$, and $x_2^n \equiv \tau^n$, for each stage n.

The standard difference structure of the set, equation (25), makes use the dynamic programming method possible for an optimization.

Whenever a discrete model has the structure linear with respect to the control θ^n, a remarkable similarity is observed between necessary optimality conditions in a discrete process and its limiting continuous counterpart. This means, in fact, that a single discrete model can handle both continuous and discrete cases. This also suggests that an analogy should exists between forms of discrete (difference) relations and limiting differential relations. In particular, as we shall see soon, optimal values of a performance coordinate satisfy in the discrete case Hamilton's and Hamilton-Jacobi formalisms in forms analogous to those known for continuous systems. In the case of NCA cascades, it is the optimal work function and a related a discrete exergy which satisfy these formalisms. Now we need to derive necessary optimality conditions for discrete processes, equations (23) - (25), which are much less known than their continuous counterparts. To find these conditions we apply the method of dynamic programming (DP) which is based on the Bellman's optimality principle [4].

Bellman's principle of optimality is crucial for both the existence of the finite time potentials and the derivation of the pertinent dynamic programming equation, which describe these potentials. When the optimal profit function is generated in terms of the initial states and initial time, this principle may be stated as follows: *In a continuous or discrete process which is described by an additive performance criterion, the optimal strategy and the optimal profit function are functions of the initial state, initial time and (in a discrete process) the total number of stages.* A basic consequence of this property is that each final segment of an optimal path (continuous or discrete) is optimal with respect to its initial state and initial time and (in a discrete process) the corresponding number of stages. The proof of this formulation by contradiction uses the additivity property of the performance criterion.

The formulation of the principle of optimality, stated above, refers to the so-called backward algorithm of the dynamic programming method. In this algorithm the recursive optimization procedure for applying of a govering functional equation begins at the final process state and terminates at its initial state. The process to which this may be applied may be arbitrary: it may be discrete by nature or may be obtained by a discretization of an original continuous process. The state transformations possess in this algorithm their most natural form, as they describe output states in terms of input states and controls at a stage. The optimization procedure at a stage and the optimal functions do recursively involve the information generated at earlier subprocesses. It is well-known [12] that in the continuous the so-called Hamilton-Jacobi-Bellman equation. However, as we shall see here, only for special discrete processes which are linear in θ^n, can a similar equation be derived. The optimality principle makes one possible to replace the simultaneous evaluation of all optimal controls by a sequence of local evaluations of optimal controls at stages, for evolving subprocesses. For the procedure using the backward algorithm of dynamic programming method, local optimizations take place in the direction opposite to the direction of the physical time or the direction of the flow of a resulting stream.

On the other hand, one may generate the optimal profit function in terms of the final states and final time. The optimality principle has than its dual form: *In a continuous or discrete process, which is described by an additive performance criterion, the optimal strategy and the optimal profit function are functions of the final state, final time and (in a discrete process) the total number of stages.* A basic consequence of this property is that each initial segment of the optimal path (continuous or discrete) is optimal with respect to its final state and final time and (in a discrete process) the corresponding number of stages.

The dual formulation refers to the so-called forward algorithm of the dynamic programming method, which we apply here to NCA cascade, Fig. 2. In this algorithm the recursive optimization procedure for solving the govering functional equation begins from the initial process state and terminates at its final state. The state transformations used have the form which describes input states in terms of output states and controls at a process stage. While this is not most typical representation of the state transformations, the transformations of this sort are directly obtainable for multistage processes with an ideal mixing at the stage, otherwise the inverse transformations (applicable to the backward algorithm) may be difficult to obtain in an explicit form. With the forward DP algorithm, one makes local optimizations in the direction of real time. Again, as in the case of the original form of the optimality principle, its dual form makes one possible to replace the simultaneous evaluation of all optimal controls by sequence of successive evaluations of the optimal controls for the evolving optimal subprocesses.

5. Definitions of optimal performance functions for work production

Let us now focus on the *N-th* stage process in the NCA cascade which starts at the thermodynamic state T_A and terminates at T_B. The *extremal* value of the work released may be characterized by a function $I_-^N(T_A, T_B, \mathcal{D})$ which refers to the problem formulation with prescribed initial temperature $T^0 = T_A$, final temperature $T^N = T_B$ and duration of the process $\tau^N - \tau^0 = \mathcal{D}^N$ (a positive quantity). The minus subscript stresses a particular minus sign. Under most typical conditions ($T > T^e$) this function describes the cooling of the resource fluid in an engine mode of the system ($T_B < T_A$). Since the process is autonomous, the form of I_-^N is such that it depends on the end times τ^N and τ^0 through the duration $\mathcal{D}^N = \tau^N - \tau^0$ only; that is, the initial time instant does not influence the value of I_-^N whenever the duration $\mathcal{D}^N$ remains fixed. This is associated with constancy of a dissipative Hamiltonian along an extremal path, which is a first integral of the discrete process. Since $max\Sigma = - min(-\Sigma)$ for an arbitrary quantity Σ at the same end states and times, the extremal function $I_-^N \equiv max\Sigma w^n$ may be described by two equivalent criteria, one of maximum type and one of minimum type, that is,

$$I_-^N[T_A, T_B, \mathcal{D}] \equiv max\sum_{n=1}^{N} f_0^n\theta^n = - min\sum_{n=1}^{N}{}' (-f_0^n\theta^n) \ , \tag{26}$$

where f_0^n is the work intensity function. Now consider an inverse process process which starts at T_B and terminates at T_A. Since the time always flows forward, the inversion of the time is not made, i.e. the duration $\mathcal{D}$ (of the same magnitude as previously) is still positive. At this point the irreversibility enters the issue. Should one be able to invert the sign of time along with the inversion of end temperatures, the value of the extremal function would remain invariant, i.e. $I_-^N(T_A, T_B, \mathcal{D}) = I_+^N(T_B, T_A, -\mathcal{D})$. However, since only *thermodynamic* end states are inverted, the invariance holds only for infinite durations, when the effects of finite rates and irreversibility are inessential. In general, however, the function $I_-^N(T_A, T_B, \mathcal{D})$ differs from the function $I_+^N(T_B, T_A, \mathcal{D})$. The formal reason for this is the fact that the sign of $\mathcal{D}$ has not been changed, so that the inversion process is not a cycle in the process state space, **X** (which includes the time). For the inverted thermodynamic process, from T_B to T_A, the extremal function $I_+^N \equiv min\Sigma(-w^n)$, is described by

$$I_+^N[T_B, T_A, \mathcal{D}] \equiv min\sum_{n=1}^{N} (-f_0^n\theta^n) = - max\sum_{n=1}^{N} f_0^n\theta^n \tag{26'}$$

Thus, it is the thermodynamic irreversibility which causes that I_-^N and I_+^N differ for the same finite durations. The above definitions are valid even in case of crossing the temperature T^e and change the process mode along an optimal path. Quite essentially, the function f_0^n characterizes the heating or cooling actions, depending whether $T' > T$ or $T' < T$. From equations (21') or (22') an expression for f_0^n can be found

$$f_0^n \equiv - c(1 - \frac{T^e}{\ })(T'^n - T^n) = - c(1 - \frac{T^e}{\ })(T'^n - T^n) - cT^e\frac{(T'^n - T^n)^2}{\ } \ . \tag{27}$$

With this we demonstrate below the identity of I_-^N and I_+^N for very small (quasistatic) rates in a limiting continuous process ($N \rightarrow \infty$). Based on equation (27), for a process from A to B ($T' < T$ and $T_B < T_A$, say, minus subscript at I), the functional of the extremal work is

$$
\begin{aligned}
I_-(T_A, T_B, \mathcal{D}) &= \max_u\{-\int_{\tau_A}^{\tau_B} c(1 - \frac{T^e}{T'})(T' - T)d\tau\} \\
&= c(T_A - T_B) - cT^e ln\frac{T_A}{T_B} - T^e \min_u \int_{\tau_A}^{\tau_B} c\frac{(T' - T)^2}{TT'}d\tau
\end{aligned}
\tag{28}
$$

and, for the inverse process from B to A ($T' > T$ and $T_B < T_A$; plus subscript at I)

$$
\begin{aligned}
I_+(T_B, T_A, \mathcal{D}) &= \min_u\{\int_{\tau_B}^{\tau_A} c(1 - \frac{T^e}{T'})(T' - T)d\tau\} \\
&= c(T_A - T_B) - cT^e ln\frac{T_A}{T_B} + T^e \min_u \int_{\tau_B}^{\tau_A} c\frac{(T' - T)^2}{TT'}d\tau .
\end{aligned}
\tag{29}
$$

When $T_D = T^e$ both above formulae define a nonclassical exergy of a finite-time process. They show that only in reversible quasistatic case ($T' = T$) both work functions are equal, i.e. $I_- = I_+$. The equality is caused by potential properties of the classical work in the thermodynamic state space.

The definitions of I_-^N and I_+^N are suitable when one wants to secure the same *positive* exergy function at the classical limit. Yet I_+^N incorporates the minimum entropy production added whereas I_-^N has it substracted. A single common definition of the optimal performance function is usually more suitable in optimization. We shall use the symbol $V^N(T^i, T^f, \mathcal{D})$ for the optimal performance function $V^N \equiv max\Sigma w^n$ in each case, i.e.we define

$$
V^N[T^i, T^f, \mathcal{D}] \equiv max\sum_{n=1}^{N} f_0^n\theta^n = - min\sum_{n=1}^{N} (-f_0^n\theta^n)
\tag{30}
$$

where f_0^n is the work generation function and indices i and f refer to initial and final states. The function V^N has always the minimum entropy production subtracted. This definition pays attention on distinction whether the end thermodynamic states are initial or final in a process, thus it is different than the previous one which stressed physicaly identical end temperatures. The particular extremal work which describes the generalized exergy should contain the environment temperature as one of the boundary states. This exergy is the maximal work $W_{max} = I_{-}^{N} = U(T^i, T^f, \tau^f - \tau^i, N)$ with $T^i \equiv T^0$ and $T^f \equiv T^N = T^e$ for the engine mode, and the negative minimal work $(-W)_{min} = I_{+}^{N} = - U(T^i, T^f, \tau^f - \tau^i, N)$ with $T^i \equiv T^0 = T^e$ and $T^f \equiv T^N = T$ for the heat-pump mode. Classical thermal exergy follows the vanishing process intensity.

The optimization analysis of the problem may use each of the above functions, still some others are suitable which describe extrema of boundary coordinates. They work in the whole space **x**, and appear in both original and dual formulations, which have to be introduced and compared in order to choose the most suitable criterion. First, we need to consider an original optimization problem in which a maximum of the final work coordinate $x_0^N \equiv W^N$ is sought for a fixed initial state $\mathbf{x}^0$. DP approach works then with *family* of the *N*-stage processes starting at various temperatures T^0 and terminating at a fixed temperature T_B. The maximum final work-coordinate x_0^N satisfies: $max(x_0^N) = x_0^0 + U^N$, where $x_0^0 \equiv W^0$ is an arbitrary initial value of work.

However an analysis shows that, in order to deal with explicit transformations of state, equations (25), working with a dual problem is more beneficial. In the dual problem a minimum of the initial performance coordinate, $x_0^0 \equiv W^0$, that satisfies $min(x_0^0) = x_0^N - U^N$, is sought for a fixed final state, $\mathbf{x}^N$, including a fixed x_0^N. The dual formulation exploits the structure of the state equations (25) in which transformations of stage inputs $\mathbf{x}^{n-1}$ are explicit in terms of stage outputs $\mathbf{x}^n$ and controls $\mathbf{u}^n$, whereas the inverse transformations would be implicit and hence difficult to handle. In the dual problem necessary optimality conditions have to be determined for minimum of the initial performance coordinate $x_0^0 \equiv W^0$. This is done in the next section.

6. Dynamic programming theory of necessary optimality conditions

In the space-time **x** the minimum condition of $x_0^0 \equiv W^0$ leads to the forward algorithm of the dynamic programming applied for the optimal performance function defined as

$$Q^n(x^n) \equiv min\ x_0^0 \equiv min\ W^0 . \tag{31}$$

Equation (31) states that the minimum of the initial work coordinate W^0 is generated in terms of the coordinates $\mathbf{x}^n$ which are the final coordinates of an *n-th* stage process. (This is valid for both cooling and heating processes.) Bellman's optimality principle applied to the optimal function (31) subject to the state equations (25) yields the following recurrence equation (forward DP algorithm)

$$\min_{u^n,\, \theta^n} \{ Q^{n-1}(\mathbf{x}^n - \mathbf{f}^n(\mathbf{x}^n, \mathbf{u}^n)\theta^n) - Q^n(x^n)\} = 0 \,. \tag{32}$$

As the zero-th component of the rate vector this equation contains the optimal profit intensity f_0 or the power release function. It is the negative of the or the power supply function $L = (-f_0)$. The function L is called the lagrangian of the problem. (In the case of NCA cascades, the signs of these functions are in general indefinite as each of them may comprize various ranges of work of thermal machines.) The DP method generates Q^n in terms of $\mathbf{x}^n$ via iterative minimizations performed at various constant $\mathbf{x}^n$. Example of analytical interations for our NCA cascade, in terms of a function $\mathbf{R}^n$ derived from Q^n, is given in Sec. 8. Here, however, we develop a general formulation of the necessary optimality conditions based on the dynamic programming equation (32). The virtue of the general formulation is its lucidity especially when optimality conditions are related to a Hamiltonian.

For a stationary minimum of $x_0^0 \equiv W^0$, the necessary optimality conditions follow from Eq. (32) by setting to zero the partial derivatives of the minimized expression with respect to the controls θ^n and $\mathbf{u}^n$. The stationarity conditions constitute the set of the following difference - differential equations for $Q^n = Q^{n-1}$

$$H^{n-1}(\mathbf{x}^n, \frac{\partial Q^{n-1}}{\partial \mathbf{x}^{n-1}}, \mathbf{u}^n) = 0 \tag{33}$$

and

$$\frac{\partial H^{n-1}(\mathbf{x}^n, \partial Q^{n-1}/\partial \mathbf{x}^{n-1}, \mathbf{u}^n)}{\partial \mathbf{u}^n} = 0 \,, \tag{34}$$

where H^{n-1} is the (x_0 independent) Hamiltonian function. It is sometimes called the enlarged Hamiltonian as it is defined in the state space $\mathbf{x}$ which includes τ; in this case H^{n-1} is the scalar product of all rates and partial derivatives of Q^{n-1} with respect of the state coordinates

$$H^{n-1}(\mathbf{x}^n, \frac{\partial Q^{n-1}}{\partial \mathbf{x}^{n-1}}, \mathbf{u}^n) \equiv \sum_{i=0}^{s+1} \frac{\partial Q^{n-1}}{\partial x_i^{n-1}} f_i^n = \frac{\partial Q^{n-1}}{\partial x_0^{n-1}} f_0^n + \sum_{i=1}^{s} \frac{\partial Q^{n-1}}{\partial x_i^{n-1}} f_i^n + \frac{\partial Q^{n-1}}{\partial \tau^{n-1}} \,. \tag{35}$$

Equation (33) is the stationary extremum condition with respect to the unconstrained time intervals θ^n whereas equation (34) defines the stationary control vector $\mathbf{u}^n$ in the optimal process. When this control is evaluated from equation (34) and substituted into equation (33) the latter becomes the discrete Hamilton-Jacobi equation

$$\frac{\partial \boldsymbol{Q}^{n-1}}{\partial \tau^{n-1}} + H^{n-1}(x_1^n \; ...x_s^n, \; \tau^n, \; \frac{\partial \boldsymbol{Q}^{n-1}}{\partial x_1^{n-1}} .... \frac{\partial \boldsymbol{Q}^{n-1}}{\partial x_s^{n-1}}) = 0 \tag{36}$$

which is nonlinear in terms of the derivatives $\partial \boldsymbol{Q}^{n-1}/\partial \mathbf{x}^{n-1}$. Equation (36) is written in terms of the *extremum* Hamiltonian function of energy type, H^{n-1}, rather than in terms of the enlarged hamiltonian $\boldsymbol{H}^{n-1}$. In the limiting case of infinitesimal θ^n, this equation goes over into the well-known Hamilton-Jacobi equation of corresponding continuous process.

However, in contrast with continuous processes, in the discrete case the solving approaches to the problem deal usually with the underlying equation (32), not with equation (36). Sometimes, the set of equations (33) and (34) is solved, see our example of optimizing of NCA cascade in the second part of this work [13]. As compared with continuous processes, which deal with a single function $\boldsymbol{Q}(x)$, in discrete processes a sequence of the optimal functions $\boldsymbol{Q}^n(x^n)$ forms the solution to equation (32) or equations (34) and (36). When a numerical procedure is organized, effective for problems with small dimensionality of the state (the case of NCA cascades), Bellman's recurrence equation (32) can be effectively solved, as shown by our example in Sec. 8. However, this recurrence equation is usually applied in its equivalent form which deals with the function $\boldsymbol{U}^N$, equation (30). For a general nonautonomous case of an arbitrary n-stage subprocess

$$\boldsymbol{V}^n(\mathbf{x}^n) \equiv max \sum_1^n f_0^n(\mathbf{x}^n, \mathbf{u}^n)\theta^n . \tag{37}$$

Here $\mathbf{x}^n$ is the final state of the n-stage subprocess (forward algorithm), and $\boldsymbol{U}^n$ is independent on x_0^n. The above function may be expressed in terms of the function $\boldsymbol{Q}^n(x^n)$ of equation (31) with the help of the formula

$$\boldsymbol{V}^n(\mathbf{x}^n) = max(x_0^n - x_0^0) = x_0^n - \boldsymbol{Q}^n(\mathbf{x}^n) . \tag{38}$$

This is consistent with our earlier statement that in the dual problem a minimum of the initial performance coordinate $x_0^0 \equiv W^0$ satisfies $\min(x_0^0) = x_0^N - \boldsymbol{U}^N$ for a fixed final state $\boldsymbol{x}^N$.

The function $\boldsymbol{U}^n(\mathbf{x}^n)$ does not contain the performance variable x_0^n, which means that it works in the state space of one dimension less than the function $\boldsymbol{Q}^n(\mathbf{x}^n)$. In terms of the function $\boldsymbol{U}^n(\mathbf{x}^n)$, Bellman's recurrence equation (32) becomes

$$\min_{\mathbf{u}^n,\theta^n}\{\, Q^{n-1}(\mathbf{x}^n - \mathbf{f}^n(\mathbf{x}^n, \mathbf{u}^n)\theta^n) - Q^n(\mathbf{x}^n)\}$$
$$= \min_{\mathbf{u}^n,\theta^n}\{\, x_0^{n-1} - V^{n-1}(\mathbf{x}^n - f^n(\mathbf{x}^n, \mathbf{u}^n)\theta^n) - x_0^n + V^n(\mathbf{x}^n)\} = 0\,, \tag{39}$$

where the variable x_0^{n-1} has to be expressed in terms of the state and the controls $\mathbf{u}^n$ at the stage, that is the function $f_0^n(\mathbf{x}^n, \mathbf{u}^n)$ has to be substituted in place of the difference x_0^n - x_0^{n-1}. Moreover, since now, one can redefine the state vector such that $\mathbf{x} = (x_1, x_2, ... x_{s+1} = \tau)$, thus excluding the performance coordinate x_0. The x_0-free recurrence equation is

$$V^n(\mathbf{x}^n) = \max_{\mathbf{u}^n,\theta^n}\{f_0^n(\mathbf{x}^n, \mathbf{u}^n)\theta^n + V^{n-1}(\mathbf{x}^n - \mathbf{f}^n(\mathbf{x}^n, \mathbf{u}^n)\theta^n)\}\,. \tag{40}$$

It represents the most popular form of recurrence equations for multistage optimization. Starting with $U^0=0$ the sequence $U^1, ... U^n ... U^N$ can be obtained by a well-established recurrence procedure [5] in which extremizations are with respect to controls at constant coordinates $\mathbf{x}^n$. Analytical solutions are seldom possible. However, an analytical solution of equation (40) can be obtained for the NCA cascade, as shown in Sec. 8.

Equation (40) lets one treat a general case when the controls $\mathbf{u}^n$ are constrained, i.e. they are restricted to reside in an allowable regime U. Writing equation (40) in the form

$$\max_{\mathbf{u}^n,\theta^n}\{f_0^n(\mathbf{x}^n, \mathbf{u}^n)\theta^n - (V^n(\mathbf{x}^n) - V^{n-1}(\mathbf{x}^n - \mathbf{f}^n(\mathbf{x}^n, \mathbf{u}^n)\theta^n)\} = 0 \tag{41}$$

one can determine the complete set of necessary optimality conditions including those with respect to state vector $\mathbf{x}^n$ [14, 15]. This realizes the passage to the algorithm of discrete maximum principle and related difference equations, see Part II of this work [13]. Here, however, we continue our work in the frame of dynamic programming, so it is the necessary optimality conditions with respect to controls $\mathbf{u}^n$ and θ^n which are relevant. For an unconstrained θ^n and constrained $\mathbf{u}^n \in U$ the set equivalent with equation (41) has the form of the three equations

$$f_0^n(\mathbf{x}^n, \mathbf{u}^n)\theta^n - (V^n(\mathbf{x}^n) - V^{n-1}(\mathbf{x}^n - \mathbf{f}^n(\mathbf{x}^n, \mathbf{u}^n)\theta^n) = 0\,, \tag{41a}$$

$$f_0^n(\mathbf{x}^n, \mathbf{u}^n) - \frac{\partial V^{n-1}}{\partial \mathbf{x}^{n-1}} .\mathbf{f}^n(\mathbf{x}^n, \mathbf{u}^n) = 0 \tag{41b}$$

and

$$\max_{\mathbf{u}^n}\{f_0^n(\mathbf{x}^n, \mathbf{u}^n)\theta^n - (V^n(\mathbf{x}^n) - V^{n-1}(\mathbf{x}^n - \mathbf{f}^n(\mathbf{x}^n, \mathbf{u}^n)\theta^n)\} = 0\,. \tag{41c}$$

Equation (41b) is the stationarity condition for the optimal intervals θ^n. Whenever θ^n is finite and positive, then, after using equations a and b in c, the last equation can be transformed to the form

$$\max_{\mathbf{u}^n}\{f_0^n(\mathbf{x}^n, \mathbf{u}^n) - \frac{\partial \mathbf{V}^{n-1}}{\partial \mathbf{x}^{n-1}}.\mathbf{f}^n(\mathbf{x}^n, \mathbf{u}^n)\} = 0 \tag{42}$$

which is the strong maximum principle for the enlarged Hamiltonian (35) with respect to the controls $\mathbf{u}^n$. Equation (42) contains the Hamiltonian (35) in the space without coordinate x_0, in which representation

$$\mathbf{H}^{n-1} \equiv f_0^n(\mathbf{x}^n, \mathbf{u}^n) - \frac{\partial \mathbf{V}^{n-1}}{\partial \mathbf{x}^{n-1}}.\mathbf{f}^n(\mathbf{x}^n, \mathbf{u}^n) = f_0^n - \sum_{i=1}^{s} \frac{\partial \mathbf{V}^{n-1}}{\partial x_i^{n-1}} f_i^n - \frac{\partial \mathbf{V}^{n-1}}{\partial \tau^{n-1}}, \tag{43}$$

where $\mathbf{U}^n \equiv max\Sigma(f_0^n\theta^n)$ is the *maximum* value of the cumulative profit (i.e. total work produced by the n-stage cascade). As f_{s+1} $(=1)$ is $\mathbf{u}$-indepedent, the optimality condition (42) can also be expressed in terms of the energy-like Hamiltonian H^{n-1} which has excluded the partial derivative of $\mathbf{Q}^{n-1}$ or $\mathbf{U}^{n-1}$ with respect to the time

$$H^{n-1} \equiv \frac{\partial \mathbf{Q}^{n-1}}{\partial x_0^{n-1}} f_0^n + \sum_{i=1}^{s} \frac{\partial \mathbf{Q}^{n-1}}{\partial x_i^{n-1}} f_i^n = f_0^n - \sum_{i=1}^{s} \frac{\partial \mathbf{V}^{n-1}}{\partial x_i^{n-1}} f_i^n. \tag{44}$$

Equation (38) was used to get the above expression for H^{n-1}. Equation (44) rejects the $\mathbf{u}$-independent term $\partial \mathbf{U}^{n-1}/\partial \tau^{n-1}$ from the Hamiltonian (43).

As shown by equation (41c), the maximum principle (42) can also be written in the form

$$\max_{u^n}\{f_0^n(\mathbf{x}^n, \mathbf{u}^n) - \frac{(\mathbf{V}^n(\mathbf{x}^n) - \mathbf{V}^{n-1}(\mathbf{x}^n - \mathbf{f}^n(\mathbf{x}^n, \mathbf{u}^n)\theta^n)}{\theta^n}\} = 0\,. \tag{45}$$

This form is suitable to discuss the discrete-continuous analogies in the next section.

7. Analogies between optimal continuous and discrete systems

Equation (45) may be compared with the limiting continuous relationship

$$\max_{\mathbf{u}}\{f_0(\mathbf{x}, \mathbf{u}) - \frac{d\mathbf{V}(\mathbf{x})}{d\tau}\} = 0 \tag{46}$$

which can be obtained from diferentiation of an equation defining the maximum performance function $\boldsymbol{V}$ of the continuous process

$$\max_{\{\mathbf{u}(\tau)\}}\left\{\int_{\tau^i}^{\tau^f} f_0(\mathbf{x},\mathbf{u})d\tau - V(\mathbf{x})\right\} = 0 \tag{47}$$

with respect to variable upper limit of integration τ^f and then replacement of the variable τ^f by τ. Expanding of the total derivative in equation (46) yields the Hamilton-Jacobi-Bellman equation (HJB equation) of the continuous problem

$$\max_{\mathbf{u}}\{f_0(\mathbf{x},\mathbf{u}) - \frac{\partial \boldsymbol{V}}{\partial \tau} - \sum_{i=1}^{s}\frac{\partial \boldsymbol{V}}{\partial x_i}f_i(\mathbf{x},\mathbf{u})\} = 0 \ . \tag{48}$$

For example, in a NCA cascade with an infinite sequence of infinitesimal NCA machines and the driving temperature control, T', the continuous HBJ equation is

$$\max_{T'}\{- c(1 - \frac{T^e}{T'})(T' - T) - \frac{\partial \boldsymbol{V}}{\partial \tau} - \frac{\partial \boldsymbol{V}}{\partial T}(T' - T)\} = 0 \ . \tag{49}$$

The power intensity function of this equation, given by the limiting form of Eq. (27), can be presented as sum of the potential and dissipative components

$$f_0 \equiv - c(1 - \frac{T^e}{T'})(T' - T) = - c(1 - \frac{T^e}{T})(T' - T) - cT^e\frac{(T' - T)^2}{TT'} . \tag{50}$$

The analogy between discrete and continuous systems can now be understood as follows. A discrete analogue of equation (47) is the sum

$$\max_{\{\mathbf{u}^k,\theta^k\}}\{\sum_{k=1}^{n} f_0^k(\mathbf{x}^k,\mathbf{u}^k)\theta^k - V^n(\mathbf{x})\} = 0 \tag{51}$$

(c.f. equation (37)). In the discrete case the differentiation with respect to the upper limit of integration, as in equation (47), is replaced by the rule for subtraction of two optimal performance functions, described by the DP equation (41). Since the extremum with respect to $\mathbf{u}^n$ is taken therein at a constant θ^n, one can extremize equaly well the ratio of this equation and θ^n, which is equation (45). Its continuous analogue is equation (46). Now the role of the free extremum with respect to θ^n is seen to be crucial. As equation (41b) is necessary for the optimality with respect to an unconstrained θ^n and equation (41a) must

$\mathbf{u}^n$). Thus the maximum principle follows for the Hamiltonian (43) with respect to controls $\mathbf{u}^n$, as expressed by equation (42) or its expanded form below

$$\max_{\mathbf{u}^n}\{f_0^n(\mathbf{x}^n, \mathbf{u}^n) - \frac{\partial \mathbf{V}^{n-1}}{\partial \tau^{n-1}} - \sum_{i=1}^{s} \frac{\partial \mathbf{V}^{n-1}}{\partial x_i^{n-1}} f_i(\mathbf{x}^n, \mathbf{u}^n)\} = 0 \ . \tag{42'}$$

Equation (42') is thus the discrete analogue of the HJB equation of Pontryagin's principle, equation (48). The HBJ equation for a multistage process with a finite number of NCA stages follows now immediately

$$\max_{T'^n}\{- c(1 - \frac{T^e}{T'^n})(T'^n - T^n) - \frac{\partial \mathbf{V}^{n-1}}{\partial \tau^{n-1}} - \frac{\partial \mathbf{V}^{n-1}}{\partial T^{n-1}}(T'^n - T^n)\} = 0 \ . \tag{52}$$

The whole theory of optimal NCA cacades is contained in this equation.

For processes consuming work it is convenient to replace the profit type function $\mathbf{U}^n$ by its negative, $\mathbf{R}^n = - \mathbf{U}^n$. The fuction $\mathbf{R}^n$ has meaning of a cost describing the minimum work supplied. It is related to the Lagrangian $L^n \equiv -f_0^n$ by the equation

$$\mathbf{R}^n(x^n) \equiv - \mathbf{V}^n(x^n) = - \max \sum_{1}^{n} f_0^n \theta^n = \min \sum_{1}^{n} L^n \theta^n \ . \tag{53}$$

As $\mathbf{U}^n$, $\mathbf{R}^n$ does not contain the performance variable x_0^n. In the special case when the process rates $f_1..f_s$ are the controls $u_1..u_s$, a proper Lagrangian $L^n = -f_0^n$ is convex with respect to rates. Substituting $f_0^n = -L^n$ into the recurrence equation (40) multiplied by the minus unity and using equation (53) one obtains an equivalent equation in terms of $\mathbf{R}^n$

$$\mathbf{R}^n(\mathbf{x}^n) = \min_{\mathbf{u}^n, \theta^n}\{ L^n(\mathbf{x}^n, \mathbf{u}^n)\theta^n + \mathbf{R}^{n-1}(\mathbf{x}^n - \mathbf{f}^n(\mathbf{x}^n, \mathbf{u}^n)\theta^n)\} \ . \tag{54}$$

Either of equations (40) or (54) may be used to define the so-called adjoint variables, the partial derivatives of $\mathbf{U}^n$ or $\mathbf{R}^n$ with respect to the state coordinates x_i^n. Since $\mathbf{U}^n(\mathbf{x}^n) = x_0^n - \mathbf{Q}^n(\mathbf{x}^n) = -\mathbf{R}^n(\mathbf{x}^n)$, the state adjoints can be defined as

$$\lambda_i^{n-1} \equiv \frac{\partial \mathbf{Q}^{n-1}}{\partial x_i^{n-1}} = \frac{\partial(x_0^{n-1} - \mathbf{V}^{n-1}(x^{n-1}))}{\partial x_i^{n-1}} = \frac{\partial(x_0^{n-1} + \mathbf{R}^{n-1}(x^{n-1}))}{\partial x_i^{n-1}} \tag{55}$$

($i = 0, 1...s, s+1$). In terms of the state and adjoint variables the Hamiltonian (43) is

$$H^{n-1}(\mathbf{x}^n, \lambda^{n-1}, \mathbf{u}^n) \equiv \sum_{v=0}^{s+1} \lambda_v^{n-1} f_v^n = \lambda_0^{n-1} f_0^n + \sum_{i=1}^{s} \lambda_i^{n-1} f_i^n + \lambda_\tau^{n-1}, \tag{56}$$

where $\lambda_0^{n-1} = 1$, $\lambda_i^{n-1} = -\partial U^{n-1}/\partial x_i^{n-1} = \partial R^{n-1}/\partial x_i^{n-1}$, $\lambda_{s+1}^{n-1} \equiv \lambda_\tau^{n-1} = -\partial U^{n-1}/\partial \tau^{n-1} = \partial R^{n-1}/\partial \tau^{n-1}$ and $f_{s+1} = 1$, for $i=1, 2..s$ and $n = 1, 2..N$. This Hamiltonian has to be a maximum with respect to the controls $\mathbf{u}^n$ which maximize a performance index of the profit type, such as the work production criterion, $W^n = \Sigma f_0^n \theta^n$, whose optimal function U^n is defined by equation (37). In an equivalent formulation, one minimizes a performance index of the cost type, such as the work consumption criterion, $(-W^n) = \Sigma(-f_0^n)\theta^n = \Sigma L^n \theta^n$, whose optimal function R^n is defined by equation (53). In that formulation, the unchanged Hamiltonian (56), which may operate with $-L^n$ in place of f_0^n, still has to attain the global maximum with respect to the controls $\mathbf{u}^n$, or a modified Hamiltonian $H'^{n-1} \equiv -H^{n-1}$ can be used that has to attain the global minimum with respect to $\mathbf{u}^n$. Thus the original maximization problem for $W^n = \Sigma f_0^n \theta^n$ requires maximum of the Hamiltonian H^{n-1} and the equivalent problem for minimization of $(-W^n) = \Sigma L^n \theta^n$ requires minimum of the modified Hamiltonian H'^{n-1} ($\equiv -H^{n-1}$), the one which contains L^n or $(-f_0^n)$ in place of f_0^n in equation (56). It may be shown [15,16] that the optimal difference set is canonical with two sort of equations, one defining the changes of state and one those of adjoint variables. The Hamiltonian (56) is the basis for the method using canonical equations, which is considered in Part II of this work [13]. Below we search for the analytical solution of the DP algorithms, equations (40) or (54) for the NCA cascade.

8. Work optimization in NCA cascades by dynamic programming

For the considered discrete NCA process, the optimal performance function R^N is

$$R^N \equiv -\max W^N = \min(-W^N) = \min_{\{T^n, \theta^n\}} \sum_{n=1}^{N} c\left(1 - \frac{T^e}{T'^n}\right)(T'^n - T^n)\theta^n \tag{57}$$

The sign of R^N defines the working mode of the cascade as a whole. In agreement with our conventions, R^N is positive in work-supply modes; this refers to the range of the fluid changes in which $T' > T$ whenever $T > T^e$; in which case $T' - T > 0$ assures the fuid's heating, whereas $T' - T < 0$ -- fluid's cooling. Under these restrictions $W < 0$ and $R^N > 0$. For stages working in engine mode the inequality $T' - T < 0$ assures fluid's cooling whereas the inequality $T' - T > 0$ -- fluid's heating. In the engine modes $W > 0$ and $R^N <$

process in the cascade is determined by the positivenes of the entropy production rather than by the sign of any work function.

We now apply the dynamic programming equation (54) to solve the problem of minimum work consumed $(-W^n) = \Sigma L^n \theta^n$ for a fixed time τ^N in the *N-th* stage cascade (assuming $\tau^0 = 0$). For equation (57), the "forward" recurrence equation (54) takes the form

$$R^n(T^n, \tau^n) = \min_{T'^n} [c(1 - \frac{T^e}{T'^n})(T'^n - T^n)\theta^n + R^{n-1}(T^n - (T'^n - T^n)\theta^n, \tau^n - \theta^n)] . \quad (58)$$

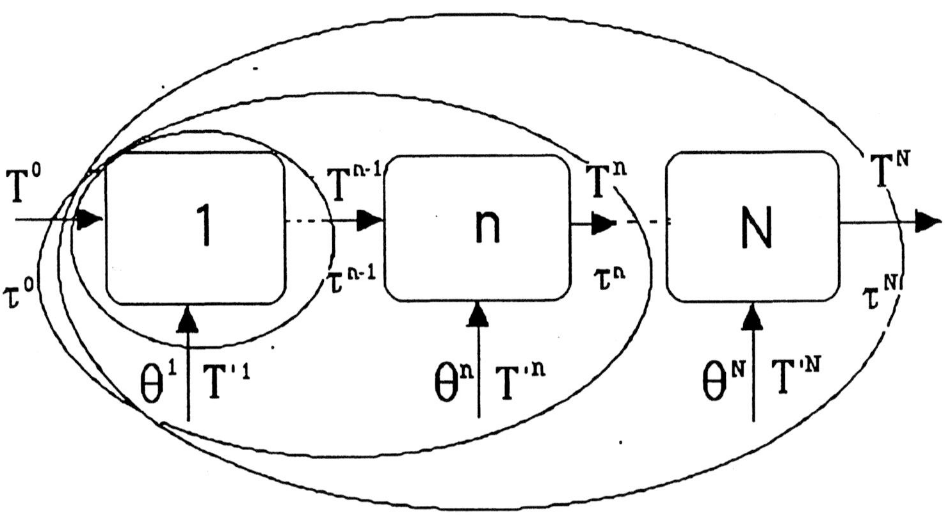

Fig. 2: Bellman's principle of optimality for multistage NCA process which is optimized by forward algorithm of dynamic programming. Elipse-shaped balance areas embrace successive subprocesses which evolve by inclusion of remaining stages.

Equation (58) may be applied to all modes, i.e. it can describe processes in which the work is supplied or released to the system. Of special attention are two processes: the one which starts at $T^0 = T^e$ and terminates at an arbitrary T^n and the one which starts at T^n and terminates at T^e. In these processes the functions R^n are generalizations of the classical thermal exergy in discrete processes with finite durations .

Now we successively apply equation (58) for subprocesses shown in Fig. 2. For $n = 1$ (the first stage) there is no choice of T'^1 other than $T'^1 = T^1 + (T^1 - T^0)/\tau^1$ because T^0 and τ^1 are fixed. Thus the work consumed in the one-stage process follows directly as

$$R^1 = \min_{T'^1} c(1 - \frac{T^e}{T'^1})(T'^1 - T^1)\tau^1 = c(1 - \frac{T^e}{(T^1 - T^0)/\tau^1 + T^1})(T^1 - T^0). \quad (59)$$

As an example, the finite-time exergy of single is obtained by appying the boundary conditions: $T^0 = T^e$ and $T^1 = T$ and $\tau^1 = \mathcal{D}$. The omission of the index means that a state variable or time is arbitrary at the end of the process. Thus the exergy of the one-stage process in the heat-pump mode is

$$E_x^1 = \boldsymbol{R}^1 = c\left(1 - \frac{T^e}{(T - T^e)/\mathcal{D} + T}\right)(T - T^e) = q^1\left(1 - \frac{T^e}{(T - T^e)/\mathcal{D} + T}\right). \tag{60}$$

In the limit of an infinite duration $\mathcal{D}$ one obtains the classical exergy of heat. When the procedure is applied to the engine mode with the conditions $T^1 = T^e$ (the exergy reference state), $T^0 = T$ and $\tau^1 = \mathcal{D}$, the result is

$$E_x^1 = -\boldsymbol{R}^1 = c\left(1 - \frac{T^e}{(T^e - T)/\mathcal{D} + T^e}\right)(T - T^e). \tag{60'}$$

Eqs. (60) and (60') show that the exergies of engine and heat pump modes do not coincide even at the limit of infinite holdup time of the fluid in the stage. This is because the mixing at stages associated with their discrete nature causes inevitable irreversibility. For an infinite holdup time of the fluid in a single perfectly mixed stage, the classical exergy vanishes in the engine-mode because the fluid's temperature in the stage is T^e. For $T > T^e$ the generalized exergy is negative which means that some work must be supplied to compensate irreversible losses necessary to accomplish the passage from T to T^e in a finite time.

The procedure is now applied for the two-stage subprocess, where the minimizing is with respect to u^2 and θ^2. In accordance with equation (58) one has to search for $\boldsymbol{R}^2$ such that

$$\begin{aligned} \boldsymbol{R}^2 = \min_{T^2,\,\theta^2} \Big\{ & c\left(1 - \frac{T^e}{T'^2}\right)(T'^2 - T^2)\theta^2 + \\ & \left(1 - \frac{T^e}{(T^2 - \theta^2(T'^2 - T^2) - T^0(\tau^2 - \theta^2)^{-1} + T^2 - \theta^2(T'^2 - T^2)}\right)(T^2 - \theta^2(T'^2 - T^2) - T^0)\Big\}. \end{aligned} \tag{61}$$

The second line of this equation contains the cost function (59) of the first stage expressed in terms of the state and controls at the second stage. The control variables, which have to be optimally chosen for the two-stage subprocess, are u^2 and θ^2.

The stationary optimal control T'^2 which minimizes the cost expression contained in the bracket $e \equiv \{...\}$ of the above equation must satisfy the equation

$$\frac{\partial e^2}{\partial T'^2} = -c\theta^2\left(\frac{T^e T^2}{(T'^2)^2} - \frac{T^0 T^e}{[(T^2 - \theta^2(T'^2 - T^2) - T^0)(\tau^2 - \theta^2)^{-1} + T^2 - \theta^2(T'^2 - T^2)]^2}\right) = 0 \quad (62)$$

To this we attach the extremum condition for the sum in equation (61) with respect to the stationary control θ^2, which is

$$\frac{\partial e^2}{\partial \theta^2} = cT^e\left(\frac{(T^2 - \theta^2(T'^2 - T^2) - T^0)^2 + (T'^2 - T^2)T^0(\tau^2 - \theta^2)^2}{[T^2 - \theta^2(T'^2 - T^2) - T^0 + (T^2 - \theta^2(T'^2 - T^2))(\tau^2 - \theta^2)]^2} - \frac{T'^2 - T^2}{T'^2}\right) = 0 \quad .(63)$$

The solution to equations (62) and (63) defines the optimal controls u^2 and θ^2 at the stage 2 in terms of the extended state T^2 and τ^2. From this information quantities at the stage 1 and properties of the two-stage system are determined including the optimal work function $\boldsymbol{R}^2$. The complete analytical solution for n=2 has the form

$$T^1 = \sqrt{T^2\, T^0} \quad (64)$$

$$\theta^2 = \theta^1 = \tau^2/2. \quad (65)$$

$$T'^2/T^2 = T'^1/T^1 = 1 + 2(\tau^2)^{-1}[1 - (\frac{T^0}{T^2})^{1/2}] \quad (66)$$

and

$$\boldsymbol{R}^2 = c(T^2 - T^0) - 2cT^e[1 - (\frac{T^0}{T^2})^{1/2}] + cT^e\frac{\{2[1 - (T^0/T^2)^{1/2}]\}^2}{\tau^2 + 2[1 - (T^0/T^2)^{1/2}]} \quad . \quad (67)$$

Equation (66) defines the driving forces in the optimal process and shows that their sign depends on the relation between the initial and final temperatures T^0 and T^2. For example, the condition $T^2 < T^0$ (cooling) is satisfied by the negative driving forces $T'^2 - T^2 < 0$ and $T'^1 - T^1 < 0$ as required by the second law. Appropriate end conditions should be applied in equation (67) if one wants to conform with the reference state of the exergy. These results imply that in optimal multistage cascades the absolute interstage temperatures satisfy the rule of the geometric mean: $T^n = (T^{n-1}T^{n+1})^{1/2}$, and the optimal time intervals are equal along an optimal path, $\theta^1 = \theta^2 ... \theta^n$. In fact, these properties are true for an arbitrary n.

The analysis of the three stage process uses transformed $\mathbf{R}^2$ to evaluate

$$\begin{aligned} \boldsymbol{R}^3 = \min_{T'^3,\, \theta^3} \{ & c\left(1 - \frac{T^e}{T'^3}\right)(T'^3 - T^3)\theta^3 + c(T^3 - (T'^3 - T^3)\theta^3 - T^0) \\ & - 2cT^e\left[1 - \left(\frac{T^0}{T^3 - (T'^3 - T^3)\theta^3}\right)^{1/2}\right] \\ & + cT^e \frac{\{2[1 - (T^0/(T^3 - (T'^3 - T^3)\theta^3))^{1/2}]\}^2}{\tau^3 - \theta^3 + 2[1 - (T^0/(T^3 - (T'^3 - T^3)\,\theta^3))^{1/2}]} \} \end{aligned} \tag{68}$$

which leads to the generalized work function for the three-stage cascade

$$\boldsymbol{R}^3 = c(T^3 - T^0) - 3cT^e\left[1 - \left(\frac{T^0}{T^3}\right)^{1/3}\right] + cT^e \frac{\{3[1 - (T^0/T^3)^{1/3}]\}^2}{\tau^3 + 2[1 - (T^0/T^3)^{1/3}]}. \tag{69}$$

The generalization of the above result for an arbitrary N is

$$\boldsymbol{R}^N = c(T^N - T^0) - cT^eN\left[1 - \left(\frac{T^0}{T^N}\right)^{1/N}\right] + cT^e \frac{\{N[1 - (T^0/T^N)^{1/N}]\}^2}{\tau^N + N[1 - (T^0/T^N)^{1/N}}. \tag{70}$$

This function is associated with the optimal trajectory satisfying the rule $T^n = (T^{n-1}T^{n+1})^{1/2}$ for arbitrary stages $n-1$, n and $n+1$, the result which leads to the temperatures along the path in terms of the boundary temperatures

$$T^1 = (T^N)^{1/N}(T^0)^{(N-1)/N},\ T^2 = (T^N)^{2/N}(T^0)^{2(N-1)/N - 1},\ldots T^{N-1} = (T^N)^{(N-1)/N}(T^0)^{1/N}. \tag{71}$$

For an infinite number of stages the limiting form of this sequence is an exponential line $T = T^0 exp(\xi\tau)$ known from the optimization of heat exchangers, simulated annealing and infinitesimal NCA sequences. We discuss this issue in Part II [13] along with a number of other properties of the solution. The limiting form of equation (70) for infinite N

$$\boldsymbol{R}^\infty = c(T - T^0) - cT^e ln\frac{T}{T^0} + cT^e \frac{[ln(T/T^0)]^2}{\tau^f + ln(T/T^0)} \tag{72}$$

is consistent with recent results for continuous systems [6, 7].

9. Final remarks

Frequently it is quite difficult or even impossible to find an analytic solution. In such costs through a direct extremizing procedure for a recurrence equation of the type of equation (58), or by iterative solving of stationarity conditions such as our equations (62) and (63), by the Newton-Raphson method. In the second case analytical properties of the problem are exploited. In each case the procedure generates the optimal work functions for the cascade process, the sequence $\mathbf{R}^1(T^1, \tau^1)$, $\mathbf{R}^2(T^2, \tau^2)$, ..$\mathbf{R}^N(T^N, \tau^N)$. The discussion of numerical aspects is available [4].

Based on equations (70) and (72) enhanced limits are found for real processes. They are quantified in terms of a finite-time exergy which excludes some processes allowed by classical exergy [6]. In thermostatics the two bounds on work (the upper bound on the work produced and the lower bound the work consumed) do coincide. But the static limits are often too far from reality to be really useful. The generalized exergy provides bounds stronger than classical. They do not coincide for work production and work consumption; they are 'thermokinetic' rather than 'thermostatic' bounds. Only for infinitely long durations or for processes with excellent transfer (infinite number of transfer units) thermokinetic bounds reduce to thermostatic bounds. The divergence of bounds means that a number of processes which are allowed by thermostatics is excluded by the second law. The range of processes excluded grows when durations becomes shorter, associated with higher mean process rates and higher entropy production. There is the remarkable practical need to extend this theory to separation and chemical systems.

Notation

$\mathcal{A}$ - stage criterion

a_v - specific area of heat exchange (per unit volume)

c- specific heat capacity

E_x - exergy

g_1, g_2 - partial conductances

g-overall conductance

G-fluid flow

H, h-hamiltonian function and its numerical value, respectively

H_{TU}-heat of transfer unit

q_1-driving heat in the engine mode of the process

Q-cumulative heat exchanged

$\mathbf{Q}$, $\mathbf{R}$- optimal performance functions

t-time

T' driving temperature defined in the text

T^e -constant temperature of infinite reservoir

$T_{1'}$, $T_{2'}$ -upper and lower temperature of the Carnot engine as part of the NCA engine

T^n - temperature of fluid leaving the stage n

T^{n-1} - temperature of fluid entering the stage n

u - control vector

$u = T' - T$ - overall driving force

w- power delivered

W-total specific work, total power per unit mass flux

$\mathcal{U}$- optimal performance function

x-state vector

x- transfer area coordinate

α - heat transfer coefficient

γ- cumulative conductance

λ– adjoint variable

$\eta = w/q_1$ -engine efficiency

τ -dimensionles time (x/H_{TU})

θ^n- time interval at stage n

Subscripts

C- Carnot point

i - coordinate i of the state vector

1, 2 first and second fluid, respectively

Superscripts

e-equilibrium, environment

f- final state

i- initial state

n - stage n

Acknowledgements

The author acknowledges a financial support from the European Commission in the framework of the Copernicus program "Carnet".

References

[1] M. J. Ondrechen, B. Andresen, M. Mozurkiewicz and R. S. Berry, Maximum work from a finite reservoir by sequential Carnot cycles, Am. J. Phys. **49**, 681-685 (1981).

[2] J.M. Gordon, Maximum work form isothermal chemical engines, J. Appl. Phys. **73**, 8-11 (1993).

[3] A. de Vos, J. Chen and B. Andresen, Analysis of combined systems of two endoreversible engines, Open Sys. Inform. Dynamics **4**, 3-14 (1997).

[3] L. S. Pontryagin, V. A. Boltyanski, R. V. Gamkrelidze, and E. F. Mischenko, *The Mathematical Theory of the Optimal Processes*, Wiley, New York (1962).

[4] R. E. Bellman, and S. E. Dreyfus, *Applied Dynamic Programming*, Panstwowe Wydawnictwa Ekonomiczne, Warsaw (1967).

[5] R. Aris, *Discrete Dynamic Programming*, Blaisdell, New York (1964).

[6] S. Sieniutycz, Hamilton-Jacobi-Bellman theory of dissipative thermal availability, *Physical Review* **56**, 5051-5064 (1997a).

[7] S. Sieniutycz, Irreversible Carnot problem of maximum work in finite time via Hamilton-Jacobi-Bellman theory, *J Non-Equilibrium Thermodyn.* **22**, 260-284 (1997b).

[8] A. de Vos, *Endoreversible Thermodynamics of Solar Energy Conversion*, Clarendon Press, Oxford (1992).

[9] I. I. Novikov, The efficiency of atomic power stations, Nucl. Energy II 7, 125-128 (1958). [English translation from At. Energ. **3**, 409-412 (1957)].

[10] F.L. Curzon, and B. Ahlborn, Efficiency of Carnot engine at maximum power output, *Amer. J. Phys.* **43**, 22-24 (1975).

[11] B. Andresen, P. Salamon, and R. S. Berry, Thermodynamics in finite time: extremals for imperfect heat engines, J. Chem. Phys. **66**, 1571-1577 (1977).

[12] W. Findeisen, J. Szymanowski, and A. Wierzbicki, *Theory and Computational Methods of Optimization*, Panstwowe Wydawnictwa Naukowe,Warsaw (1980).

[13] S. Sieniutycz, and Z. Szwast, Optimization of multistage thermal machines by a Pontryagin's-like discrete maximum principle, Part II of this work, Intern. J. of Energy, Environ. and Economics (1998), this volume.

[14] V.G. Boltyanski, *Maximum Pronciple and Optimal Control of Continuous Systems*, Nauka, Moscow (1971).

[15] S. Sieniutycz, The constant hamiltonian problem and an introduction to mechanics of optimal discrete systems, *Reports of Inst. of Chem. Engng. Warsaw Tech. Univ.* **3**, 27-53 (1974).

[16] Z. Szwast, *Discrete Algorithms of Maximum Principle with Constant Hamiltonian and their Selected Applications in Chemical Engineering,* PhD Thesis, Institute of Chemical Engineering at the Warsaw University of Technology, Warsaw (1979).

OPTIMIZATION OF MULTISTAGE THERMAL MACHINES BY A PONTRYAGIN'S-LIKE DISCRETE MAXIMUM PRINCIPLE

STANISLAW SIENIUTYCZ AND ZBIGNIEW SZWAST
Faculty of Chemical Engineering, Warsaw University of Technology,
00-645 Warsaw, 1 Warynskiego Street, Poland

Abstract.
Multistage optimization of endoreversible thermal machines is treated by a relatively unknown discrete maximum principle with a constant Hamiltonian, endowed in the Pontryagin's structure familiar from optimization of continuous systems. Total power output is maximized at constraints describing dynamics of heat transport and work production. A finite-rate model takes into account irreducible losses of the work potential caused by thermal resistances. The optimization procedure is a discrete counterpart of that applied for continuous systems: it involves discrete canonical equations of Hamilton's type and maximizes the Hamiltonian at each stage of a discrete path. The results show link between the process duration and the optimal value of the Hamiltonian. Extremum performance functions for optimal work, which incorporate a residual entropy production, are found in terms of number of stages and a holdup time describing the number of transfer units in the cascade. A *discrete* generalization of classical exergy is obtained for systems with finite number of stages and finite holdup time in the cascade. A hysteretic property is valid for this generalized exergy, that causes difference between the maximum work received from a multistage engine system and the minimum work added to a multistage heat-pump system. Due to unconstrained nature of the holdup time intervals, all discrete characteristics of the cascade achieve their continuous counterparts at the limit of infinite number of stages.

PACS number(s): 05.70.-a. KEY WORDS: Carnot cycle, finite-time thermodynamics, endoreversible engines, exergy (available energy), thermodynamic limits, second law.

1. Basic features of multistage endoreversible processes

Principles of modeling of steady-state endoreversible cascades composed of engines or heat pumps are described in a chapter in this volume [1] referred to as Part I. That text presents also a scheme of power production in a sequence of Novikov-Curzon-Ahlborn (NCA) stages and provides the list of basic designations. Here we only briefly summarize these basic issues. We consider a multistage and finite-time extension of the traditional problem of maximum work produced or consumed by a cascade system with two fluids: the first fluid being a resource fluid which flows through stages of the cascade and drives Carnot machines from which the work is taken out, and the second fluid being an infinite reservoir. The fluids are of finite thermal conductivity, hence there are finite thermal resistances in the system representing the dissipative effect of their thermal boundary layers. A finite holdup time is assumed at each process stage for of the cascade fluid. The multistage process is, in fact, a sequence of the Novikov-Curzon-Ahlborn processes (NCA processes) which are repeated in subsequent stages of the cascade. At each stage the fluid is well mixed. The fluxes of work are summed together, and the total power output per unit mass flux of the driving fluid, $W = \dot{W}/G$, is maximized. This quantity has the dimension of work per unit mass, so it describes a specific work in the steady process of power production. By definition, the work released in the engine mode is positive. For a cascade of engines, where the work is released, the work W has to be maximized. For a cascade of heat pumps, where the work is supplied to the system, the minimizing of the negative of work, $(-W)$, takes place. When the final state of the resource fluid is the state of equilibrium with the environment ($T^N = T^e$), the maximal specific work represents a finite-time exergy of the engine mode, for the fluid at flow. When the initial state of the fluid is at the state of equilibrium with the environment ($T^0 = T^e$), the minimal specific work to achieve T^N represents a finite-time exergy of the heat-pump mode.

2. A non-traditional description of discrete processes

As pointed out in Part I, NCA cascades are special discrete processes characterized by a mathematical model linear with respect to holdup time for each stage, θ^n for $n = 1,....N$, or some measure of this time. Assume that the variable t^n describes a cumulative holdup time after the first n stages of the cascade. The variable θ^n is the interval of the time t at the stage n; this must be positive and satisfy the following equation (for $n = 1,....N$):

$$\theta^n = t^n - t^{n-1} \tag{1}$$

We assume that the values of θ^n are free for each stage n, so the variable θ^n is an unconstrained control. However, in our formalism, the variable θ^n is not treated jointly with other controls. Those other are components of the control vector $\boldsymbol{u}^n$ whereas the decision variable θ^n is treated separately, because of its linearity in the model. The mathematical model of these processes is linear respect to the variable θ^n but may have any kind of dependence on the components of the control vector $\boldsymbol{u}^n$.

The discrete equations of state are represented by equation (2), and the performance index J, by equation (3). Both these equations are linear with respect to θ^n and arbitrary with respect to every component of the control vector $\boldsymbol{u}^n$

$$x_v^n - x_v^{n-1} = f_v^n(\mathbf{x}^n, \mathbf{u}^n, t^n)\theta^n \tag{2}$$

$$J = \sum_{n=1}^{N} f_0^n(\mathbf{x}^n, \mathbf{u}^n, t^n)\theta^n + F^N(\mathbf{x}^n, t^n) - F^0(\mathbf{x}^0, t^0)\,, \tag{3}$$

where $n = 1,...N$ and $v = 1,...s$. Note that the definition of the state vector $\boldsymbol{x}$ differs here from that used in Part I as it excludes the performance variable and time form the state coordinates. Thus, we will have equations related to the time coordinate explicit. This is to make the process description more lucid. As compared with Part I the performance index is taken here in a more general (Bolza) form, which involves a function F of end states. This generalization will be useful both for the present problem reformulations and for future formulations of the thermal machines theory with economic optimization criteria, which are generally more complicated than the work criteria analyzed here. But the basic difference with Part I is that we treat here the cascade problem by the maximum principle, not the dynamic programming as in Part I. Related Hamilton's canonical sets are developed and then solved.

Equations (2) and (3) constitute the mathematical model for the discrete processes linear with respect to the unconstrained decision θ^n. Optimization theory for this class of discrete processes, which, in contrast to traditional discrete systems, is similar to the Pontryagin's continuous theory, was developed by us in the context of multistage separation processes and chemical reactors [2-9]. Now we apply this theory to energy systems.

3. Necessary optimality conditions and strong maximum principle

We write the difference constraints (1) and (2) in the form of explicit transformations of state

$$t^{n-1} = t^n - \theta^n \tag{4}$$

$$x_V^{n-1} = x_V^n - f_V^n(\mathbf{x}^n, \mathbf{u}^n, t^n)\theta^n . \tag{5}$$

Defining the optimal performance function $Q^n(x^n, t^n)$ for the n-stage subprocess

$$Q^n(\mathbf{x}^n, t^n) = \max_{\mathbf{u}^k, \theta^k}\{\sum_{k=1}^{n} f_0^k(\mathbf{x}^k, \mathbf{u}^k, t^k)\theta^k + F^n(\mathbf{x}^n, t^n) - F^0(\mathbf{x}^0, t^0)\} . \tag{6}$$

and applying Bellman's optimality principle (see Part I) one obtains the recurrence equation

$$Q^n(\mathbf{x}^n, t^n) = \max_{u^n, \theta^n}\{f_0^n(\mathbf{x}^n, \mathbf{u}^n, t^n)\theta^n + F^n(\mathbf{x}^n, t^n) - F^{n-1}(\mathbf{x}^{n-1}, t^{n-1}) + Q^{n-1}(\mathbf{x}^{n-1}, t^{n-1})\} \tag{7}$$

which is valid subject to the transformation equations (4) and (5). Its alternative form

$$\begin{aligned} \max_{u^n, \theta^n}\{ & f_0^n(\mathbf{x}^n, \mathbf{u}^n, t^n)\theta^n + [Q^{n-1}(\mathbf{x}^{n-1}, t^{n-1}) - F^{n-1}(\mathbf{x}^{n-1}, t^{n-1})] \\ & - [Q^n(\mathbf{x}^n, t^n) - F^n(\mathbf{x}^n, t^n)]\} = 0 \end{aligned} \tag{8}$$

shows that after we define the function

$$\mathcal{U}^n(\mathbf{x}^n, t^n) \equiv Q^n(\mathbf{x}^n, t^n) - \mathcal{U}^n(\mathbf{x}^n, t^n) \tag{9}$$

equation (7) can be simplified to the form

$$\max_{u^n, \theta^n}\{ f_0^n(\mathbf{x}^n, \mathbf{u}^n, t^n)\theta^n + \mathcal{U}^{n-1}(\mathbf{x}^{n-1}, t^{n-1}) - \mathcal{U}^n(\mathbf{x}^n, t^n)\} = 0 . \tag{10}$$

(Compare equation (40) of Part I.) With the state transformations (4) and (5) we thus obtain

$$\max_{u^n, \theta^n}\{ f_0^n(\mathbf{x}^n, \mathbf{u}^n, t^n)\theta^n + \mathcal{U}^{n-1}(\mathbf{x}^n - f^n(\mathbf{x}^n, \mathbf{u}^n, t^n - \theta^n) - \mathcal{U}^n(\mathbf{x}^n, t^n)\} = 0 . \tag{11}$$

To take advantage of equation (11) for arbitrary x^n and t^n, a function is introduced to abbreviate the expression maximized in that equation:

$$\mathcal{A}^n(\mathbf{x}^n, \mathbf{u}^n, t^n, \theta^n) \equiv$$
$$f_0^n(\mathbf{x}^n, \mathbf{u}^n, t^n)\theta^n + U^{n-1}(\mathbf{x}^n - f^n(\mathbf{x}^n, \mathbf{u}^n, t^n - \theta^n) - U^n(\mathbf{x}^n, t^n)\}. \tag{12}$$

This is related to the forward algorithm of the dynamic programming, which we use here. Since equation (12) contains only the inputs and outputs at the (n-th) stage, we sometimes call the function $\mathcal{A}^n(x^n, u^n, t^n, \theta^n)$ the stage criterion. It has to satisfy the relationship

$$\mathcal{A}^n(\mathbf{x}^n, \mathbf{u}^n, t^n, \theta^n) \leq 0\,, \tag{13}$$

where the equality sign refers to optimal processes. Attaining by the function $\mathcal{A}^n(x^n, u^n, t^n, \theta^n)$ the maximum for every n=1,...N, provides the necessary condition for maximum of the performance index (3). Criterion $\mathcal{A}^n(x^n, u^n, t^n, \theta^n)$ provides frequently also locally-sufficient conditions of optimality, however, we will not discuss this issue here, since it is the necessary optimality conditions which are the task of the present analysis. It is assumed that $\mathcal{A}^n(x^n, u^n, t^n, \theta^n)$ can be differentiated with respect to every component of the state vector $\boldsymbol{x^n}$ and the control vector $\boldsymbol{u^n}$, as well as with respect to the time variable t^n and its increment θ^n. With this assumption, the necessary optimality conditions describing stationary extremum states maximizing Q^n, equation (7), are

$$\frac{\partial \mathcal{A}^n}{\partial x_i^n} = \{\frac{\partial f_0^n}{\partial x_i^n} - \sum_{j=1}^{s} \frac{\partial U^{n-1}}{\partial x_j^{n-1}}\frac{\partial f_j^n}{\partial x_j^n}\}\theta^n + \frac{\partial U^{n-1}}{\partial x_i^{n-1}} - \frac{\partial U^n}{\partial x_i^n} = 0 \tag{14}$$

and

$$\frac{\partial \mathcal{A}^n}{\partial t^n} = \{\frac{\partial f_0^n}{\partial t^n} - \sum_{j=1}^{s} \frac{\partial U^{n-1}}{\partial x_j^{n-1}}\frac{\partial f_j^n}{\partial t^n}\}\theta^n + \frac{\partial U^{n-1}}{\partial t^{n-1}} - \frac{\partial U^n}{\partial t^n} = 0\,. \tag{15}$$

They are complemented by the stationarity conditions for the controls θ^n

$$\frac{\partial \mathcal{A}^n}{\partial \theta^n} = f_0^n - \sum_{j=1}^{s} \frac{\partial U^{n-1}}{\partial x_j^{n-1}} f_j^n - \frac{\partial U^{n-1}}{\partial t^{n-1}} = 0 \tag{16}$$

and the strong-maximum condition for $\mathcal{A}^n$ with respect to u^n expressed in the form

$$-\frac{\partial U^{n-1}}{\partial t^{n-1}} + \max_{u^n}\{f_0^n - \sum_{j=1}^{s} \frac{\partial U^{n-1}}{\partial x_j^{n-1}} f_j^n\} = 0 \tag{17}$$

which is a discrete counterpart of the Hamilton-Jacobi-Bellman equation for continuous processes. As shown in Part I, Eq. (42) therein, equation (17) generalizes the stationary extremum condition $\partial \mathcal{A}^n/\partial u_l^n = 0$, or the expression

$$\frac{\partial \mathcal{A}^n}{\partial u_l^n}(\theta^n)^{-1} = \frac{\partial f_0^n}{\partial u_l^n} - \sum_{j=1}^{s} \frac{\partial \boldsymbol{U}^{n-1}}{\partial x_j^{n-1}} \frac{\partial f_j^n}{\partial u_l^n} = 0\,, \tag{18}$$

to the case of the global extremum of $\mathcal{A}^n$ in the allowable region $\boldsymbol{U}$ for the controls u_l^n. Equation (17) yields the condition (18) in the case of the stationary extremum of u^n.

With the adjoint variables defined by the equations

$$\lambda_i^{n-1} = -\frac{\partial \boldsymbol{U}^{n-1}}{\partial x_i^{n-1}} \tag{19}$$

and

$$\lambda_t^{n-1} = -\frac{\partial \boldsymbol{U}^{n-1}}{\partial t^{n-1}} \tag{20}$$

and the discrete Hamiltonian defined by the equation

$$H^{n-1}(\mathbf{x}^n, \lambda^{n-1}, \mathbf{u}^n, t^n) = f_0^n(\mathbf{x}^n, \mathbf{u}^n, t^n) + \sum_{v=1}^{s} \lambda_v^{n-1} f_v^n(\mathbf{x}^n, \mathbf{u}^n, t^n) \tag{21}$$

it is easy to transform equations (14) to (16) to the following canonical form

$$\frac{x_v^n - x_v^{n-1}}{\theta^n} = \frac{\partial H^{n-1}}{\partial \lambda_v^{n-1}}, \tag{22}$$

$$\frac{\lambda_v^n - \lambda_v^{n-1}}{\theta^n} = -\frac{\partial H^{n-1}}{\partial x_v^n}, \tag{23}$$

$$\frac{H^n - H^{n-1}}{\theta^n} = \frac{\partial H^{n-1}}{\partial t^n}, \tag{24}$$

and

$$\lambda_t^{n-1} + \max_{u^n} H^{n-1}(\mathbf{x}^n, \lambda^{n-1}, \mathbf{u}^n, t^n) = 0\,. \tag{25}$$

They hold for n=1,...N ; n=1,...s and l=1,...r: Equation (22) constitutes the Hamiltonian form of the state equation (2); equation (23) is its adjoint equation. Equation (24) describes the Hamiltonian increment at the stage n, whereas equation (25) states that the enlarged Hamiltonian $\boldsymbol{H}^{n-1} = H^{n-1} + \lambda_t^{n-1}$ of an extremal process is always constant and equal zero. Equation (25) includes the necessary condition for the stationary optimality of the decision vector $\boldsymbol{u}^n$, if its optimal value falls in the interior of the allowable range $\boldsymbol{U}$.

The boundary conditions are determined with respect to the end state coordinates and end times as the vanishing stationarity conditions for the extremum of J, equation (3), in terms of the original function Q, equation (6). Partial differentiations of the equation

$$\boldsymbol{U}^n(\mathbf{x}^n, t^n) \equiv Q^n(\mathbf{x}^n, t^n) - F^n(\mathbf{x}^n, t^n) \tag{26}$$

and setting to zero partial derivatives of Q^n along with use of definition of adjoint variables (19) and (20) yield

$$\lambda_v^N = \frac{\partial F(\mathbf{x}^N, t^N)}{\partial x_v^N}, \quad v \neq a, b \tag{27}$$

$$\lambda_v^0 = \frac{\partial F(\mathbf{x}^0, t^0)}{\partial x_v^0}, \quad v \neq c, d \tag{28}$$

$$H^N = -\frac{\partial F(\mathbf{x}^N, t^N)}{\partial t^N}, \quad \text{if } t^N \textit{ is free} \tag{29}$$

$$H^0 = -\frac{\partial F(\mathbf{x}^0, t^0)}{\partial t^0}, \quad \text{if } t^0 \textit{ is free}. \tag{30}$$

As pointed out above, these equations do not apply to fixed coordinates, such as x_a^N and x_b^N in equation (27). If any component of boundary state vector, x_v^N or x_v^0, is fixed, the respective component of adjoint vector, λ_v^N or λ_v^0, is undetermined. Analogously, if the boundary time, t^N or t^0, is fixed, the respective Hamiltonian, H^N or H^0, is undetermined. Equations (27) and (28), respectively, describe the boundary conditions for the adjoint vector $\boldsymbol{\lambda}^n$ at the end and at the beginning of the process, and equations (29) and (30), respectively, state boundary conditions for the Hamiltonian H^n at the end and at the beginning of the process.

When the number of stages tends to the infinity the discrete algorithm becomes the Pontryagin maximum principle consisting of the following equations

$$\frac{dx_v}{dt} = \frac{\partial H}{\partial \lambda_v}, \tag{31}$$

$$\frac{d\lambda_v}{dt} = -\frac{\partial H}{\partial x_v}, \tag{32}$$

$$\left(\frac{dH}{dt} = \frac{\partial H}{\partial t}\right), \tag{33}$$

and

$$\lambda_t + \max_u H(\mathbf{x}, \lambda, \mathbf{u}, t) = 0, \tag{34}$$

valid for $v = 1,...s$ and for $t^i \le t \le t^f$. Equation (33) is in brackets since it follows from the canonical equations, the property of continuous systems only. Regardless a formal difference in notation, the corresponding boundary conditions for the adjoint vector and the Hamiltonian are for continuous processes the same as those given above for the discrete processes, equations (27) to (30).

4. Multistage work optimization by discrete maximum principle

Now we shall use our discrete theory to optimize the N-stage sequence of the Novikov-Curzon-Ahlborn processes (NCA processes) with an ideal mixing of the participating fluids at the stage. The following equation, derived in Part I, describes the work supplied to the cascade system interacting with the environment having the temperature T^e

$$U^N \equiv \max W^N = \max_{\{T'^n, \theta^n\}} \sum_{n=1}^{N} \{-c(1 - \frac{T^e}{T'^n})(T'^n - T^n)\theta^n\} \tag{35}$$

The symbol c designates the specific heat, T'^n refers to the driving temperature control. In terms of T'^n the efficiency of a NCA stage is decribed by the Carnot formula (Part I). Here the symbol θ^n describes the interval of the *nondimensional* time τ^n at the stage n. The time τ^n is the quantity identical with the number of the heat transfer units. In the engine mode the inequality $T > T^e$ is associated with $T'^n - T^n < 0$ whereas the inequality $T < T^e$ with $T'^n - T^n > 0$; in this mode the work is released from the system, and, by definition $W \geq 0$. In the work pumping mode the inequality $T > T^e$ is associated with $T'^n - T^n > 0$, whereas the inequality $T < T^e$ with $T'^n - T^n < 0$; in this mode always $W < 0$. When the proper end conditions are satisfied, equation (35) describes a finite rate availability (exergy), which simplifies to the classical exergy when the process undergoes quasistatically in a cascade with infinite number of stages, i.e. when $u=0$ and $N \rightarrow \infty$.

The fluid temperature T and the cumulative number of transfer units τ (a measure of the total area of heat exchange) satisfy the discrete equations of state

$$\frac{T^n - T^{n-1}}{\theta^n} = T'^n - T^n \tag{36}$$

$$\frac{\tau^n - \tau^{n-1}}{\theta^n} = 1 \tag{37}$$

The solution to this discrete variational problem, associated with the optimal choice of the transfer areas at the stages for the prescribed total transfer area, leads to a discrete exergy function. This is a "finite size problem" in which intervals of the number of transfer units, θ^n, are measures of exchange areas at the stages. The problem represents the discrete counterpart of corresponding continuous problem [10, 11]. To solve this problem we exploit the algorithm of the discrete maximum principle, equations (22) - (25). They apply with the condition of constancy of the Hamiltonian function as we are dealing with an autonomous system.

The discrete Hamiltonian for the problem of the maximum work (35) is

$$H^{n-1} = \lambda^{n-1}[1 - c(1 - \frac{T^e}{T'^n})](T'^n - T^n) \tag{38}$$

With this fuction we investigate a subclass of all solutions for the same $H^{n-1} = h$. This subclass is described by the equations

$$\frac{T^n - T^{n-1}}{\theta^n} = \frac{\partial H^{n-1}}{\partial \lambda^{n-1}} = T'^n - T^n, \tag{39}$$

$$\frac{\lambda^n - \lambda^{n-1}}{\theta^n} = - \frac{\partial H^{n-1}}{\partial T^n} = \frac{cT^e}{(T'^n)^2}(T'^n - T^n), \tag{40}$$

$$\lambda^{n\text{-}1}[1 - c(1 - \frac{T^e}{T'^n})](T'^n - T^n) = h\,, \tag{41}$$

and

$$\frac{\partial H^{n\text{-}1}}{\partial T'^n} = \lambda^{n\text{-}1} - c(1 - \frac{T^eT^n}{(T'^n)^2}) = 0\,. \tag{42}$$

In a general case of an arbitrary optimization model, the condition $H^{n-1} = H^n = h$ means restriction to a subclass of the total family of the optimal trajectories. Such subclass is characterized by a common intensity h and various undetermined durations. In our simple case one deals with the single variable of thermodynamic state, T^n, and a constant number of stages N. When N is constant, the application of the condition $H^n = h$ restricts one to a definite trajectory.

After elimination of λ^{n-1} from the last two equations the integral of discrete motion is obtained

$$(1 - \frac{T^n}{T'^n})^2 = \frac{h}{cT^e} \tag{43}$$

Without superscript n this formula is consistent with the corresponding equation for continuous systems [10, 11]. The condition $T' > 0$ is sufficient for the stationary maximum of the Hamiltonian function, corresponding with the maximum work delivered from an engine system and minimum work supplied to a heat-pump system. With these constraints one obtains two possible realizations of control which correspond respectively to increasing and descreasing temperatures T^n in time τ^n, or with heating and cooling strategies for the resource fluid

$$T'^n = \frac{T^n}{1 - \pm\sqrt{\frac{h}{cT^e}}}\,, \tag{44}$$

where the upper sign refers to heating and lower for cooling processes. The first process can be accomplished in a heat-pump mode and the second in an engine mode of the cascade.

Defining an intensity constant ξ in terms of the numerical hamiltonian h as

$$\xi = \pm \frac{\sqrt{h(cT^e)^{-1}}}{1 - \pm\sqrt{h(cT^e)^{-1}}} \tag{45}$$

(again, the same rule holds for signs) and expressing equation (44) with respect to $u^n = T'^n - T^n$ one obtains $u^n = \xi T^n$, or, after using the state equation (36),

$$\frac{T^n - T^{n-1}}{\theta^n} = \xi T^n \tag{46}$$

This equation describes the discrete rate of an extremal process, which changes proportionally with the temperature, the result analogous to the equation

$$\frac{dT}{d\tau} = \xi T \tag{47}$$

which is valid for the continuous version. Equation (46) is the discrete generalization of the optimality condition (47) known for the optimal trajectories of heat exchangers, simulated annealing and infinitesimal NCA sequences [10-15].

The difference equation (46) should be solved simultaneously with the second canonical equation to which the adjoints λ^{n-1} and λ^n [obtained from the extremum condition of H^{n-1} with respect to T'^n, equation (42))] have been substituted

$$\frac{cT^eT^n}{(T'^n)^2} - \frac{cT^eT^{n+1}}{(T'^{n+1})^2} = \frac{cT^e}{(T'^n)^2}(T'^n - T^n)\theta^n . \tag{48}$$

With the state equation (36) the above result can be written in terms of the discrete change of the temperature exclusively, and then simplified. This leads to an equation describing the temperature relations for any two neighboring stages, n and n+1

$$\frac{T^{n-1}}{(T'^n)^2} = \frac{T^{n+1}}{(T'^{n+1})^2} \tag{49}$$

Using in this formula equation (44) for the stages n and $n+1$ one gets the basic result

$$(T^n)^2 = T^{n-1}\, T^{n+1} . \tag{50}$$

This result means that the interstage temperatures T^n between the stages n and $n+1$ are geometric means of the boundary temperatures for any two-stage subprocess. It can also be obtained by the dynamic programming (Part I) but the use of the maximum principle makes the derivation more transparent. The direct use of equation (50) for finding the optimal temperature sequence is described below.

For a prescribed T^0 and in terms of the temperatue T^1 assumed as a "first decision" the recursive use of the fomula $T^{n+1} = (T^n)^2/T^{n-1}$ yields the sequence

$$T^2 = (T^1)^2/T^0 \,, \tag{51}$$

$$T^3 = [T^2]^2/T^1 = [(T^1)^2/T^0]^2/T^1 = (T^1)^3/(T^0)^2 \,, \tag{52}$$

$$\begin{aligned} T^4 &= [T^3]^2/T^2 = [(T^1)^3/(T^0)^2]^2/[(T^1)^2/T^0] \\ &= [(T^1)^4/(T^0)^3] \,, \end{aligned} \tag{53}$$

and so on. Hence the final temperature T^N for the N-th stage process is

$$T^N = [(T^1)^N/(T^0)^{N-1}] \,. \tag{54}$$

This leads to the temperature T^1 as a function of the initial and final temperatures

$$T^1 = (T^N)^{1/N}(T^0)^{(N-1)/N} \,. \tag{55}$$

With this result one may return to the intermediate temperatures to obtain these in terms of the boundary temperatures

$$T^2 = (T^N)^{2/N}(T^0)^{2(N-1)/N\ -1} \,, \tag{56}$$

$$T^3 = (T^N)^{3/N}(T^0)^{3(N-1)/N-2} \,, \tag{57}$$

$$T^4 = (T^N)^{4/N}(T^0)^{4(N-1)/N-3} \,, \tag{58}$$

....

$$T^{N-1} = (T^N)^{(N-1)/N}(T^0)^{[(N-1)^2/N-(N-2)]} = (T^N)^{(N-1)/N}(T^0)^{1/N} \,. \tag{59}$$

Let us now write equation (46) in the form

$$\xi = (\theta^n)^{-1} (1 - T^{n-1}/T^n) = (\theta^{n+1})^{-1} (1 - T^n/T^{n+1}) \ ... \ . \tag{60}$$

Substituting into this result equation (50) we obtain

$$\xi = (\theta^n)^{-1} (1 - \sqrt{\frac{T^{n-1}}{T^{n+1}}}) = (\theta^{n+1})^{-1} (1 - \sqrt{\frac{T^{n-1}}{T^{n+1}}}) \ ... \ . \tag{61}$$

This means that the time intervals are equal along an extremal

$$\theta^n = \theta^{n+1} \ ....= \theta = \tau^N/N \ . \tag{62}$$

With equations (59) and (62) and the first equality of equation (60) applied the N-th stage cascade, the path independent value of the optimal intensity ξ can be expressed in terms of the boundary temperatures (T^0 and T^N), process duration τ^N and the total number of stages N

$$\xi = \frac{N}{\tau^N} \frac{T^N - T^{N-1}}{T^N} = \frac{N}{\tau^N} [1 - (\frac{T^0}{T^N})^{1/N}] \ . \tag{63}$$

For an infinite stage process with fixed end thermodynamic states one obtains from equation (63) the corresponding formula for the limiting continuous process

$$\xi = \frac{ln(T^N/T^0)}{\tau^f} \tag{64}$$

which holds for $T^N = T^f$ and $\tau^i \equiv 0$.) Use of the function $\xi(h)$, equation (45), in equation (63) links the total duration τ^N with the numerical value of the hamiltonian h for given boundary temperatures

$$\tau^N = N\frac{(1 - \pm\sqrt{h(cT^e)^{-1}})}{\pm\sqrt{h(cT^e)^{-1}}} (1 - (\frac{T^0}{T^N})^{1/N}) \ . \tag{65}$$

A formula inverse with respect to equation (65) is of importance

$$h = \frac{cT^e}{\left(\frac{\tau^N}{N[1 - (T^0/T^N)^{1/N}]} + 1\right)^2} . \tag{66}$$

Equation (66) defines a cost of the process time or the numerical value of the Lagrange multiplier associated with the duration constraint. This is also the numerical value of the hamiltonian constant h which has to be used when the equations set (38)-(42) is solved for the given duration τ^N.

The above results make one possible to obtain an the work potential $\boldsymbol{U}^n = max\,(W^n)$ and a related exergy function E_x for multistage process characterized by the finite number of stages, N, with an ideal mixing at each stage. This exergy function is described by an equation

$$E_x^N = c(T^N - T^e) - cT^eN[1 - (\frac{T^e}{T^N})^{1/N}] \pm cT^e \frac{\{N[1 - (T^e/T^N)^{1/N}]\}^2}{\tau^N \pm N[1 - (T^e/T^N)^{1/N}} \tag{67}$$

in which the upper sign refers to the pumping mode and the lower one to the engine mode. The above function is a discrete generalization of a continuous dissipative exergy that can otherwise be obtained from equation (67) at the limit of infinite number of stages (See also [10, 16 and 17] for finite-time exergy of continuous processes.) The results are the same as those obtained by the dynamic programming method (Part I). As in the case of continuous processes, the generalized exergy of processes departing from the equilibrium (upper sign) is larger than that of the processes approaching the equilibrium (lower sign). This is because one respectively adds or subtracts the the entropy production in the least term in equation (67).

5. Final remarks

The practical value of the solutions obtained here refers to energy limits for multistage thermal machines with small number of stages, where the results of continuous approaches, adduced above, are too approximate. The classical exergy yields an exact estimation of bounds on extremal work only for excellent transfer conditions, or for infinitely long times of the energy exchange. We have shown that for finite-time processes bounds on mechanical energy production can be obtained either by dynamic programming or by discrete maximum principle, the latter method being easier to use.

For short durations the heat-pump mode exergy $E_x = (-W)_{min}$, which defines the lower bound on the work consumption can be significantly higher than the minimal work of classical thermodynamics. This explains restrictive applicability of classical thermodynamic bounds when they are applied to real processes, and it shows that these bounds should be replaced by stronger bounds obtained from nonequilibrium thermodynamics [19]. The hysteretic properties of the generalized exergy as a finite-time work function are important. They are associated with different values of the extremal work obtained when a process which leaves the equilibrium is compared with its inverse which approaches the equilibrium. While in the reversible case these processes can be accomplished with exactly the same magnitude of work, in the dissipative case the works consumed and produced are no longer equal. An essential decrease of the maximal work received from an engine system and an increase of the minimal work added to a heat pump system takes place in the high-rate regimes and for short durations. Extensions towards separation processes and chemical reactions should be the topic of a future research.

Acknowledgements

The authors acknowledge the support of the European Commission in the framework of the Copernicus program "Carnet".

References

[1] S. Sieniutycz, Endoreversible modeling and optimization of multistage thermal machines by dynamic programming, Part I of this work, Intern. J. of Energy, Environ. and and Economics (1998), this volume.

[2] S. Sieniutycz, The constant hamiltonian problem and an introduction to the mechanics of optimal discrete systems, *Reports of Inst. of Chem. Engng. Warsaw Tech. Univ.* **3**, 27-53 (1974).

[3] S. Sieniutycz, *Optimization in Process Engineering*, 1-nd edn, Wydawnictwa Naukowo Techniczne, Warsaw (1978).

[4] S. Sieniutycz, A general theory of optimal discrete drying processes with a constant hamiltonian, *Drying* **84**: 62-75. Ed. A. Mujumdar, Hemisphere, New York (1984).

[5] Z. Szwast, *Discrete Algorithms of Maximum Principle with Constant Hamiltonian and their Selected Applications in Chemical Engineering,* PhD Thesis, Institute of Chemical Engineering at the Warsaw University of Technology, Warsaw (1979).

[6] Z. Szwast, Enhanced version of a discrete algorithm for optimization with a constant hamiltonian, *Inz. Chem. Proc.* **3**: 529-545 (1988).

[7] Z. Szwast, Exergy optimization in a class of drying systems with granular solids. Article in: Finite-Time Thermodynamics and Thermoeconomics, eds. Sieniutycz, S., and P. Salamon. *Adv. in Thermodyn.* **4:** 209-248. Taylor and Francis, New York, 1990.

[8] S. Sieniutycz and Z. Szwast, *Practice in Optimization: Process Problems*, Wydawnictwa Naukowo Techniczne, Warsaw (1982).

[9] S. Sieniutycz and Z. Szwast, A discrete algorithm for optimization with a constant hamiltonian and its application to chemical engineering, *Intern. J. Chem. Engng.* **23**: 155-166 (1983).

[10} S. Sieniutycz, Hamilton-Jacobi-Bellman theory of dissipative thermal availability, *Physical Review* **56**, 5051-5064 (1997 a).

[11] S. Sieniutycz, Irreversible Carnot problem of maximum work in finite time via Hamilton-Jacobi-Bellman theory, *J Non-Equilibrium Thermodyn.* **22**, 260-284 (1997 b).

[12] P. Salamon, P., A. Nitzan, B. Andresen and R. S. Berry, Minimum entropy production and the optimization of heat engines, *Phys. Rev. A* **21**, 2115-2129 (1980).

[13] B. Andresen and J. Gordon, Optimal paths for minimizing entropy production in a common class of finite-time heating and cooling processes, *Intern. J. Heat Fluid Flow* **13**, 294-299 (1992 a).

[14] B. Andresen and J. Gordon, Optimal heating and cooling strategies for heat exchanger design, *J. Appl. Phys.* **71**, 76-79 (1992 b).

[15] B. Andresen and J. Gordon, Constant thermodynamic speed for minimizing eqntropy production in thermodynamic processes and simulated annealing, *Phys Rev. E.* **50**, 4346-4351 (1994).

[16] B. Andresen, M. H. Rubin, and R. S. Berry, Availability for finite time processes. General theory and model, *J. Phys. Chem.* **87**, 2704-2713 (1983).

[17] R. S. Berry, Exergy and optimization of time-constrained processes, *Periodica Polytech. Ser. Phys. and Nucl. Sci.* **2**, 5-14 (1994).

[18] P. Salamon, P., A. Nitzan, B. Andresen and R. S. Berry, Minimum entropy production and the optimization of heat engines, *Phys. Rev. A* **21**, 2115-2129 (1980).

[19] K. -H. Hoffmann, Optima and bounds for irreversible thermodynamic processes. Article in: Finite-Time Thermodynamics and Thermoeconomics, eds. S. Sieniutycz and P. Salamon, *Advances in Thermodyn.* **4**: 22-65 (1990), Taylor and Francis, New York.

FINITE-TIME THERMODYNAMIC ANALYSIS OF A MHD POWER PLANT

MAMDOUH EL HAJ ASSAD
Helsinki University of Technology, Laboratory of Applied Thermodynamics, P O Box 4400, FIN-02015 HUT, Finland.

and

CHIH WU
Mechanical Engineering Department
U.S. Naval Academy, Annapolis, MD 21402

Abstract- Finite-time thermodynamic analysis has been applied to an MHD (magnetohydrodynamic) power plant. General expressions for power and efficiency of the power plant have been derived when MHD generator is operating at constant velocity and constant Mach number conditions. It is shown that when electrical efficiency of the MHD generator is 100 %, both cases are identical and the efficiency at maximum power is identical to Curzon and Ahlborn efficiency.

Keywords- MHD, velocity, Mach number, electrical efficiency

INTRODUCTION

Finite-time thermodynamic analysis has been first demonstrated by Curzon and Ahlborn [1] . They showed that when a heat engine is running infinitely slow, the engine produces work but not power . However when the engine is running at finite speed, the engine will produce power.

An analysis of ideal Brayton cycle [2] showed that the cycle efficiency is independent of thermal conductances and the maximum power is obtained when the cycle is placed between a heat source and a heat sink reservoirs.

The efficiency at maximum power of Brayton cycle with nonisentropic compression and nonisentropic expansion has been studied [3] and it has been found out that these nonisentropic processes lower the power and the cycle efficiency. It has been proved [4] that imposition of maximum power adds constrains on Brayton cycle temperatures.

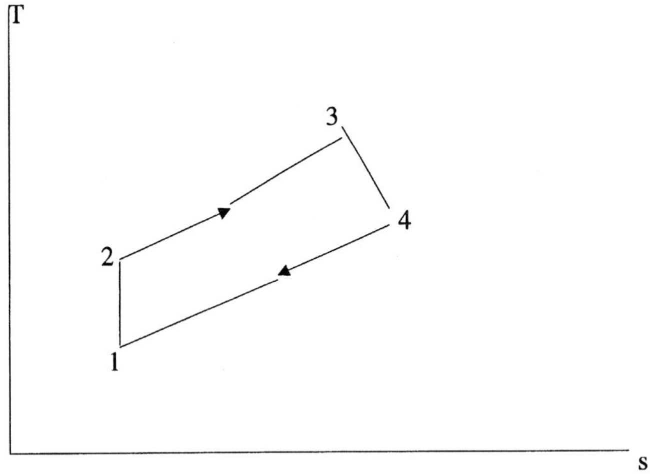

Fig. 1. T-s diagram of an MHD power cycle.

Recently, many investigators have applied FTT (finite -time thermodynamic) analysis to an internally reversible Brayton cycle [5] and an internally irreversible Brayton cycle [6].

The objective of this paper, is to study MHD power plant which is similar to Brayton cycle by using FTT analysis through which the power and efficiency are analyzed for constant velocity and constant Mach number as operating conditions of MHD generator.

MHD CYCLE ANALYSIS

A MHD power plant running between an infinite heat source at T_H and infinite heat sink at T_L is illustrated in Fig. 1.

All the processes in the cycle are internally reversible except for the MHD generator.

Process 1-2 is considered to be isentropic compression process, i.e. the compressor is ideal, process 2-3 is the addition process from the heat source, process 3-4 is the nonisentropic expansion process in MHD generator and process 4-1 is the heat rejection process to the heat sink.

The heat addition and heat rejection rates, considering counter flow heat exchangers, are, respectively,

$$Q_H = \frac{U_H A_H\left[(T_H - T_2) - (T_H - T_3)\right]}{\ln\left[\dfrac{T_H - T_2}{T_H - T_3}\right]} = C_W(T_3 - T_2) \tag{1}$$

$$Q_L = \frac{U_L A_L\left[(T_4 - T_L) - (T_1 - T_L)\right]}{\ln\left[\dfrac{T_4 - T_L}{T_1 - T_L}\right]} = C_W(T_4 - T_1) \tag{2}$$

where $U_H A_H$ is the heat conductance on the hot side heat exchanger, $U_L A_L$ is the heat conductance on the cold side heat exchanger (overall heat transfer coefficient -area product) and C_W is the heat capacity rate of the working fluid.

From Eqs. (1) and (2), we obtain

$$Q_H = C_W \varepsilon_H (T_H - T_2) \tag{3}$$

$$Q_L = C_W \varepsilon_L (T_4 - T_L) \tag{4}$$

where ε_H is the effectiveness of the hot side heat exchanger and ε_L is the effectiveness of the cold side heat exchanger, for counter flow heat exchangers, they are given by [7]

$$\varepsilon_H = 1 - \exp(-N_H) \tag{5}$$

$$\varepsilon_L = 1 - \exp(-N_L) \tag{6}$$

The number of heat transfer units N_H and N_L based on minimum heat capacity rates, which are defined as $N_H = \dfrac{U_H \varepsilon_H}{C_W}$ and $N_L = \dfrac{U_L \varepsilon_L}{C_W}$.

The power output and efficiency of MHD power plant are, respectively,

$$W = Q_H - Q_L = C_W \varepsilon_H (T_H - T_2) - C_W \varepsilon_L (T_4 - T_L) \tag{7}$$

and

$$\eta = \frac{W}{Q_H} \,. \tag{8}$$

By using Eqs. (1) and (3), and Eqs. (2) and (4), the following temperatures can be obtained as

$$T_1 = T_4(1-\varepsilon_L)+\varepsilon_L T_L \tag{9}$$

and

$$T_3 = T_2(1-\varepsilon_H)+\varepsilon_H T_H \tag{10}$$

For one dimensional steady state flow, the momentum and energy equations of MHD generator are written as

$$\rho u\frac{du}{dz}+\frac{dP}{dz}+JB=0 \tag{11}$$

and

$$\rho u\frac{d}{dz}(h+0.5u^2)+JE=0 \tag{12}$$

where ρ is the gas density, u is the gas velocity, h is the gas enthalpy, P is the gas pressure , J is the current density, B is the magnetic field and E is the electric field.

The electrical efficiency of MHD generator is given by [8]

$$\eta_e = \frac{JE}{uJB} \tag{13}$$

CONSTANT VELOCITY CASE

From Eqs. (11), (12) and (13), for constant gas in MHD generator, we obtain

$$\left(\frac{k-1}{k}\right)\eta_e\frac{dP}{P}=\frac{dT}{T} \tag{14}$$

where k is the specific heat ratio of the working gas.

Integrating Eq. (14) for the process 3-4 yields

$$\frac{T_3}{T_4}=X^{\eta_e} \tag{15}$$

For process 1-2, we obtain

$$X = \frac{T_2}{T_1} = \left(\frac{P_2}{P_1}\right)^{(k-1)/k} \tag{16}$$

By using Eqs. (7), (9), (10), (14), (15) and (16), the power output W can be written in terms of X only, and defining a dimensionless power output as $W_U{}^* = \frac{W}{C_W T_L}$, we obtain

$$W_U{}^* = \frac{aX^{\eta_e+1} + bX^{\eta_e} + cX + d}{X^{\eta_e} + fX} \tag{17}$$

The efficiency for constant gas velocity can be expressed as

$$\eta_U = \frac{aX^{\eta_e+1} + bX^{\eta_e} + cX + d}{aX^{\eta_e+1} + \gamma\varepsilon_H X^{\eta_e} - (d + \gamma\varepsilon_H)X} \tag{18}$$

where

$$a = -\varepsilon_H \varepsilon_L \tag{19a}$$

$$b = \gamma\varepsilon_H + \varepsilon_L \tag{19b}$$

$$c = -\gamma\varepsilon_H(1-\varepsilon_L) + \varepsilon_H\varepsilon_L - \varepsilon_L \tag{19c}$$

$$d = -\gamma\varepsilon_H\varepsilon_L \tag{19d}$$

$$f = (1-\varepsilon_L)(1-\varepsilon_H) \tag{19e}$$

where γ is defined as $\gamma = \frac{T_H}{T_L}$.

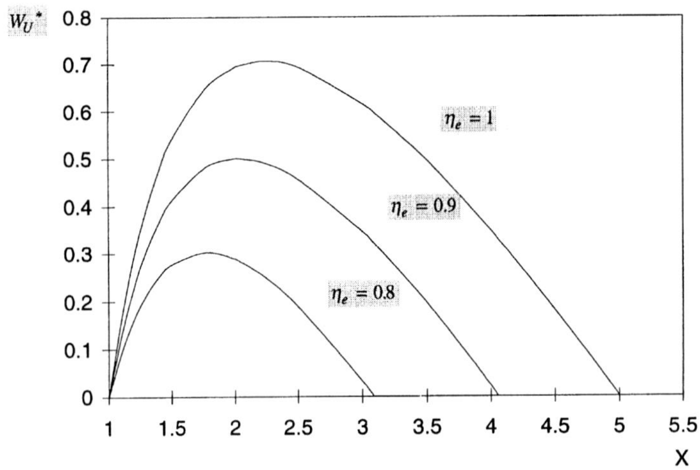

Fig. 2. Effect of X on W_U^* for $N_L = N_H = 1$.

Constant Mach Number Case

For constant Mach number (M) , Eqs. (11), (12) and (13) give

$$\frac{dP}{P} = \beta \frac{dT}{T} \tag{20}$$

where

$$\beta = \frac{k}{(k-1)\eta_e}\left[1 + 0.5(k-1)M^2(1-\eta_e)\right] \tag{21}$$

Integrating Eq. (20) for process 3-4, we obtain

$$\frac{T_3}{T_4} = X^{\alpha} \tag{22}$$

where

$$\alpha = \frac{\eta_e}{\left[1 + 0.5(k-1)M^2(1-\eta_e)\right]} \tag{23}$$

The dimensionless power output for constant Mach number can be obtained just by replacing η_e in Eq. (17) by α, this gives

$$W_M{}^* = \frac{W}{C_W T_L} = \frac{aX^{\alpha+1} + bX^{\alpha} + cX + d}{X^{\alpha} + fX} \tag{24}$$

The efficiency of MHD power plant for constant Mach number can be expressed as

$$\eta_M = \frac{aX^{\alpha+1} + bX^{\alpha} + cX + d}{aX^{\alpha+1} + \gamma\varepsilon_H X^{\alpha} - (d + \gamma\varepsilon_H)X} \tag{25}$$

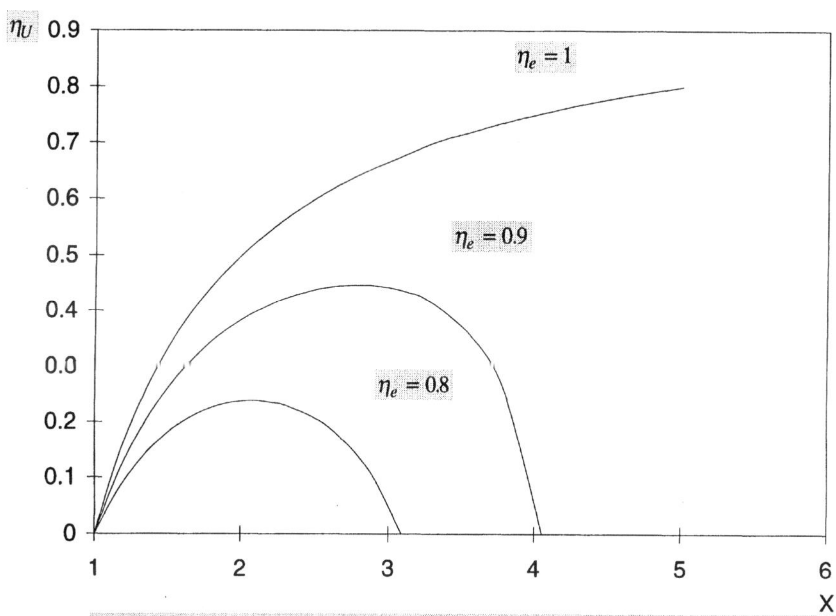

Fig. 3. Effect of X on η_U for $N_L = N_H = 1$.

MAXIMUM POWER OUTPUT

The maximum dimensionless power outputs for constant velocity and constant Mach number can be found by maximizing Eqs.(17) and (24) with respect to *X*, which give

$$af\eta_e X^{\eta_e+1} + aX^{2\eta_e} + (bf - c)(\eta_e - 1)X^{\eta_e} - d\eta_e X^{\eta_e-1} - fd = 0 \tag{26a}$$

$$af\alpha X^{\alpha+1} + aX^{2\alpha} + (bf - c)(\alpha - 1)X^{\alpha} - dX^{\alpha-1} - fd = 0 \tag{26a}$$

A special case when $\eta_e = 1$ and $\varepsilon_H = \varepsilon_L$, Eq. (26) gives the same result in both cases as $X_{opt} = \gamma^{1/2}$.

The efficiency of MHD power plant at maximum power output, is obtained as

$$\eta = 1 - \left(\frac{T_L}{T_H}\right)^{1/2} \tag{27}$$

which is known as Curzon and Ahlborn efficiency [1].

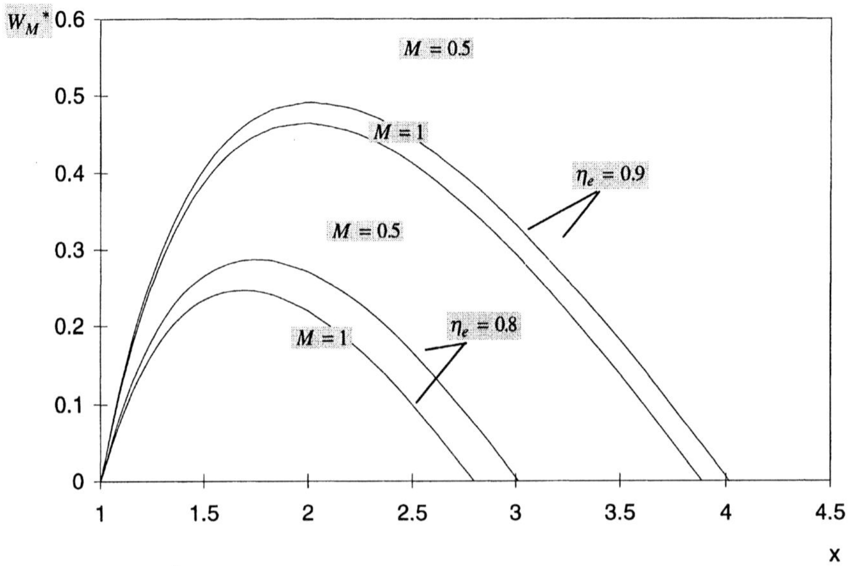

Fig. 4. Effect of X on $W_M{}^*$ for $N_L = N_H = 1$.

Discussion and Conclusions

In this section, we present the results for constant velocity and constant Mach number cases for $N_H = N_L = 1$ and $\gamma = \dfrac{T_H}{T_L} = 5$.

CASE A : CONSTANT VELOCITY

The effect of the temperature ratio X on the dimensionless power W_U^* is shown in Fig.2 for different values of MHD electrical efficiency η_e. The figure shows that there is an optimum value of the temperature ratio X for which the dimensionless power W_U^* is maximum. The figure also shows that as η_e decreases, the cycle produces less power. Figure 3. shows the variation of X with η_U. For $\eta_e = 1$, the efficiency of MHD power plant increases monotonically with X, however for $\eta_e < 1$, there exists a maximum.

As can be seen from Figs. 2 and 3, the maximum efficiency of MHD power plant occurs at higher X values than that of the maximum dimensionless power.

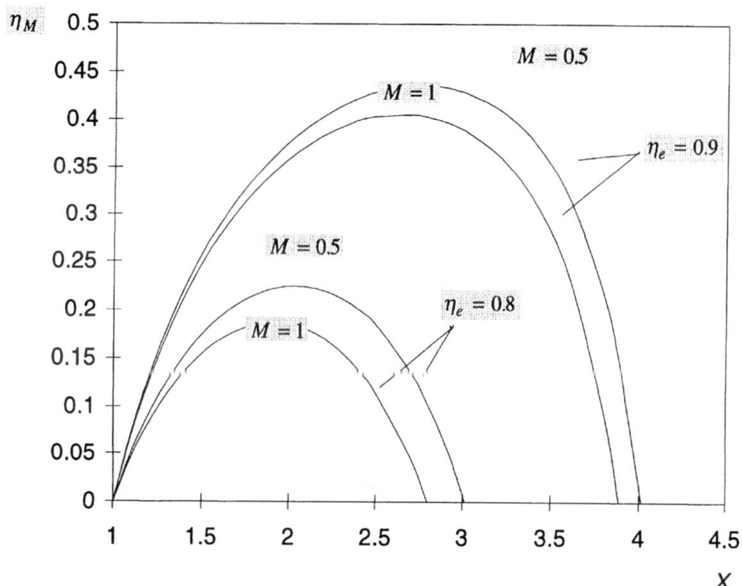

Fig. 5. Effect of X on η_M for $N_L = N_H = 1$.

CASE B : CONSTANT MACH NUMBER

The variation of the temperature ratio X with the dimensionless power W_M^* and thermal efficiency η_M are shown in Fig. 4 and 5, respectively. Both figures show that the power plant produces more power and has higher thermal efficiency for Mach number less than one.

One of the important points that worths mentioning is for $\eta_e = 1$, the working fluid has zero electrical resistivity and both cases are internally reversible by which they become identical to Joule - Brayton cycle. Hence both cycles will produce the same power $W_U^* = W_M^*$ and have the same thermal efficiency $\eta_U = \eta_M$ regardless of Mach number M.

REFERENCES

F. L. Curzon and B. Ahlborn, Am. J. Phys. **43**, 22 (1975).

A. Bejan, Int. Heat Mass Transfer **31**, 1211 (1988).

C. Wu and R. L. Kiang, Trans. ASME J. Engng Gas Turbine Power **113**, 501 (1991).

H. S. Leff, Am. J. Phys. **55**, 602 (1987).

C. H. Cheng and C. K. Chen, Energy Int. J. **21**, 241 (1996).

C. Wu, L. Chen and F. Sun, Energy Int. J. **21**, 71 (1996).

W. M. Kays and A. L. London, Compact Heat Exchangers, McGraw-Hill, New York, NY (1964).

K. H. Spring, Direct Generation of Electricity, Academic Press , 1965.

OPTIMAL THERMODYNAMIC PROCESSES IN FINITE-TIME

ALEXIS DEVOS AND BART DESOETE
Vakgroep voor elektronika en informatiesystemen
Universiteit Gent
Sint Pietersnieuwstraat 41
B-9000 Gent
Belgium

Abstract--Finite-time thermodynamics is illustrated by some examples of simple process optimalizations. The objective is minimum entropy production during thermal, chemical, and electrical processes. Non-linear electronic processes are investigated in particular, because of their importance in computers and displays.

1 INTRODUCTION

We will investigate various simple examples of the so-called *horse- and- carrot* problem [1]. The purpose is to move a 'carrot' in such a way that it attracts a 'horse' from one particular point to another, (a) in a well-defined time and (b) in an optimal way. We call a process 'optimal' if it generates a minimum amount of entropy from t_i to t_f:

$$S = \int (dS/dt)dt = \text{minimum}$$

Here t_i and t_f are the initial and final time, whereas dS/dt is the entropy creation rate.
According to various authors, e.g. [1] [2], such process obeys

$$dS/dt = \text{constant}$$

For a single carrot and a single horse, we have

$$dS/dt = \sum_i (dX_i/dt)\, Z_i,$$
$$= (dQ/dt)(-1/T_c+1/T_h) + (dN/dt)(\mu_c/T_c-\mu_h/T_h) + ...$$

where T_h, μ_h, ... are the temperature, chemical potential, ... of the horse and T_c, μ_c, ... are the same intensive quantities for the carrot. The flux (dQ/dt) is the heat flow from carrot to horse, the flux (dN/dt) is the particle flow from carrot to horse, ... Thus a constant (dS/dt) becomes

$$\sum_i (dX_i/dt)Z_i = \text{constant} \qquad (1)$$

According to Spirkl and Ries [3], this is only true provided the resistances to the driving forces as function of the fluxes are themselves independent of the fluxes. They propose the following (more general) rule:

$$\sum (dX_i/dt)[\partial Z_i/\partial(dX_j/dt)](dX_j/dt) = \text{constant} \qquad (2)$$

2 SIMPLE THERMAL EXAMPLE

In a purely thermal process, we have only one intensive variable, i.e. temperature T, and one extensive variable, i.e. heat Q. See Figure la. The horse-and-carrot problem is the following: how to change the 'horse temperature' T_h from one particular value to another, (a) in a well-defined time and (b) at the expense of a minimum entropy production. The entropy production rate obeys

$$(dS/dt) = (dQ/dt)(-1/T_c+1/T_h).$$

Here, T_c is the 'carrot temperature', i.e. the temperature we can control directly. The horse temperature, i.e. the temperature we want to control, we can only influence indirectly, i.e by means of the temperature T_c and the heat flow (dQ/dt). For example, T_h is the air temperature in a room whereas T_c is the water temperature in the central heating system.

First, we will assume a linear transport law (call it Fourier law or Newton law). Therefore we have the following set of equations:

$$Z = (-1/T_c+1/T_h)$$

$$(dQ/dt) = g(T_c-T_h)$$

$$Q = f(T_h).$$

Here g is a constant (called thermal conductance) and f(x) is some (not yet specified, but monotonously increasing) function, describing the heat capacitance of the horse.

Applying (1) immediately yields

$$(dQ/dt)(-1/T_c+1/T_h) = \text{constant} \qquad (3)$$

Considering that

$$Z(Q,dQ/dt) = -1/[f^{-1}(Q)+(1/g)(dQ/dt)]+1/[f^{-1}(Q)]$$

and therefore

$$\partial Z/\partial(dQ/dt) = (1/g)\{1/[f^{-1}(Q)+(1/g)(dQ/dt)]^2\},$$

application of (2) yields

$$(dQ/dt)[1/(gT_c^2)](dQ/dt) = \text{constant}$$

and thus

$$(1/T_c)(dQ/dt) = \text{constant}. \qquad (4)$$

The two results (3) and (4) are equivalent, as both give rise to

$$f'(T_h)(1/T_h)(dT_h/dt) = \text{constant} .$$

For convenience, let us consider the case of a linear function f(x). This leads to $(1/T_h)(dT_h/dt)$ = constant. For the initial and final conditions $T_h(t_i) = T_i$ and $T_h(t_f) = T_f$, this yields the following simple exponential function of time:

$$T_h(t) = T_i(T_f/T_i)^{(t-ti)(tf-ti)}.$$

This result has been presented by Sieniutycz [4] [5].

Because of $Q = CT_h$, also the stored heat Q is an exponential function of time, and hence also its derivative, i.e. the current (dQ/dt). Finally also the carrot temperature $T_c(t)$ is an exponential function. The reader will easily verify that $T_c(t) = (1 + \alpha)T_h(t)$, where α is a short notation for $(C/g)[\log(T_f/T_i)/(t_f-t_i)]$.

Let us use the above result in the framework of a practical example. Both carrot and horse are initially at ambient temperature T_a. First we like to rise T_h in a time lapse t from T_a to some elevated temperature T_e. Later, we like to lower T_h again to T_a in a same lapse of time. For both objectives we power the carrot with a Carnot engine. See Figure 1b. After Carnot's formula, the power dW/dt fueling the engine is $(dQ/dt)(1-T_a/T_c)$. Figure 2 shows the trajectories $T_h(t)$ and $T_c(t)$, as well as dQ/dt, dW/dt, and dS/dt, as a function of time. In order to avoid zero or negative carrot temperatures at the end of the cycle, we have to accept that $\tau > (C/g)\log(T_e/T_a)$.

Note that dS/dt is indeed constant, i.e. $g\,[\alpha_{up}^2/(1+\alpha_{up})]$, where α_{up} is $(C/g)[\log(T_e/T_a)/\tau]$. Note that dW/dt is positive for the heating experiment. The total work we have to supply, i.e. $W_{up} = \int(dW/dt)dt$, equals $C(T_e - T_a) - g\,\tau T_a\,[\alpha_{up}/(1+\alpha_{up})]$. During the cooling experiment, the Carnot engine first produces power, later consumes power. The fact that, at the end of the cooling, we have to supply work to the system, is a price we have to pay for the finite-time condition. The net work W_{down} we apply is still negative and equals $-C(T_e - T_a) - g\tau T_a\,[\alpha_{down}/(1+\alpha_{down})]$. Taking into account that $\alpha_{down} = -\alpha_{up}$, we find that the overall work dissipated during the heating+cooling cycle is $W = W_{up} + W_{down} = 2g\tau T_a\,[\alpha_{up}^2/(1-\alpha_{up}^2)]$.

Figure 3 shows the cyclic path in the (T_h,T_c)-plane. Notice the two straight lines with T_h/T_c = constant. Note that we here use the terms 'trajectory' and 'path' according to the definitions of Spirkl and Ries [3].

We also remark that the horse follows the carrot quite faithfully. Indeed, both during the rise and the fall experiment, we can write $T_h(t) = T_c(t - t_d)$, where t_d is a time delay. The reader will easily verify that $t_d = [\log(1+\alpha)/\log(T_f/T_i)]\,(t_f - t_i)$.

Finally, one can easily check that all results remain valid if we replace $Q = CT_h$ by

$$Q = CT_h + \text{constant},$$

a fact that gives the conclusions a far broader applicability.

3 MORE THERMAL EXAMPLES

If we replace the linear law of heat exchange by the non-linear Stefan-Boltzmann law of black-body radiation

$$(dQ/dt) = g(T_c^4 - T_h^4), \tag{5}$$

we get

$/\{[f^{-1}(Q)]^4+(1/g)(dQ/dt)\}+1/[f^{-1}(Q)]$

and thus

$$\partial Z/\partial(dQ/dt) = (1/4g)(1/\{[f^{-1}(Q)]^4+(1/g)(dQ/dt)\}^{5/4}$$

Substitution in (2) yields

$Q/dt)\{1/[4gT_c^5]\}(dQ/dt)$ = constant

and thus

$$(1/T_c^{5/2})(dQ/dt) = \text{constant} \tag{6}$$

In accordance with Eq. (20) of Spirkl and Ries [3]. We remark that this result cannot be found starting from rule (1)

Elimination of dQ/dt and T_c from Eq. (5) and (6), and (dQ/dt)= $f'(T_h)(dT_h/dt)$ gives a highly non-linear differential equation in T_h:

$$[(1/g)f'(T_h)(dT_h/dt)+(T_h^4)]^5 + \text{constant } [f'(T_h)(dT_h/dt)]^8 = 0.$$

For the simple capacitor law $Q = CT_h$, it gives rise to

$$[(C/g)(dT_h/dt)+T_h^4]^5 + \text{constant } (dT_h/dt)^8 = 0.$$

For $Q = CT_h^4$ (a model for a black-body cavity, which stores heat by storing photons), it leads to

$$[(4C/g)(dT_h/dt)+T_h]^5 + \text{constant } T_h^9(dT_h/dt)^8 = 0.$$

4 SIMPLE CHEMICAL EXAMPLE

In a purely chemical process, we have only one intensive variable, i.e. chemical potentil μ, and one extensive variable, i.e. particle number N. Now the 'horse potential' μ_h has to be changed from an initial value to a final value with a minimum of entropy production, the entropy production rate obeying

$$dS/dt = (dN/dt)(\mu_c/T - \mu_h/T).$$

Here, μ_c, is the 'carrot potential', i.e. the quantity we are able to control directly (instantaneously and without any bounds).

We will restrict ourselves here to the linear kinetic equation, where the driving force of the particle current is the chemical potential difference (in spite of the fact that this is discutable). Therefore we have the following set of equations:

$$Z = (1/T)(\mu_c - \mu_h)\,,$$
$$(dN/dt) = g(\mu_c - \mu_h)\,,$$
$$N = f(\mu_h).$$

Here g is a constant and f (x) is the (monotonously increasing) function describing the particle capacitance of the horse. Applying (1) yields

$$(dN/dt)(\mu_c - \mu_h) = \text{constant}\,,$$

whereas applying (2) yields

$$(dN/dt)[1/(gT)](dN/dt) = \text{constant}.$$

The two results are equivalent, as both result in

$$dN/dt = \text{constant}$$

and thus

$$\mu_c - \mu_h = \text{constant}$$

or

$$f'(\mu_h)(d\mu_h/dt) = \text{constant}. \tag{7}$$

For a linear capacitor [i.e. f(x) = Cx + constant], together with the initial and final conditions $\mu_h(t_i) = \mu_i$ and $\mu_f(t_f) = \mu_f$, this yields the following simple linear function of time:

$$\mu_h(t) = \mu_i + (\mu_f - \mu_i)(t-t_i)/(t_f - t_i) .$$

Because of $N = C\mu_h$, also the stored matter N is a linear function of time, and hence its derivative, i.e. the current dN/dt, is a constant. Finally also $\mu_c(t)$ is a linear function. The reader will indeed easily verify that $\mu_c(t) = \mu_h(t)-(c/g)[(\mu_f-\mu_i)/(t_f-t_i)]$.

Let us use the above result in the framework of a practical example. Both carrot and horse are initially at ambient potential μ_a. First we like to rise μ_h in a time lapse τ from μ_a to some elevated potential μ_e. Later, we like to lower μ_h again to μ_a ,in a same lapse of time. For both objectives we power the carrot with a reversible engine. The power dW/dt fueling the engine is $(dN/dt)(\mu_c-\mu_a)$. Figure 4 shows the trajectories $\mu_h(t)$ and $\mu_c(t)$, as well as the fluxes dN/dt, dW/dt, and dS/dt, as a function of time. Note that dS/dt is indeed constant, i.e. $[c^2/(gT)](\mu/\tau)^2$, where μ is a short notation for $\mu_e-\mu_a$. Note that dW/dt is positive for the 'rising' experiment. The total work we have to supply, i.e. $W_{up} = \int_\tau (dW/dt)(dt)$, equals $C\mu^2/2 + [c^2\mu^2/(g\tau)]$. During the 'lowering' experiment, the chemical engine first produces power, later consumes power. The net work invested during the potential decay is $W_{down} = -C\mu^2/2 + [c^2\mu^2/(g\tau)]$. The overall work consumed during the rising + lowering cycle is $W = W_{up} + W_{down} = 2[c^2\mu^2/(g\tau)]$, equal to T times the overall entropy creation S. This overall work investment W is inversely proportional to the finite time τ. Thus the higher the control speed, the higher price W we have to pay.

Figure 5 shows the cyclic path in the (μ_h, μ_c)-plane. Notice the two straight lines with $\mu_c - \mu_h$ = constant. Again we can say that $\mu_h(t) = \mu_c(t-t_d)$, now with $t_d = C/g$, for both the 'up' and the 'down' experiment.

The analogy between Figures 2 and 3, on the one side, and Figures 4 and 5, on the other side, can be fully understood by introducing the quantity 'thermodynamic length'. See Section 6.

5 MORE CHEMICAL EXAMPLES

If we replace the linear law of particle transport by the non-linear Fick law of diffusion

$$(dN/dt) = g[\exp(\mu_c/kT) - \exp(\mu_h/kT)], \tag{8}$$

with k Boltzmann's constant, then we get

$$Z(N,dN/dt) = (1/T)\{kT\log[\exp(f^{-1}(N)/kT) + (1/g)(dN/dt)] - f^{-1}(N)\},$$

such that

$$\partial Z/\partial(dN/dt) = (k/g)\{1/[\exp(f^{-1}(N)/kT) + (1/g)(dN/dt)]\} ,$$

and (2) becomes

$$(dN/dt)(k/g)[1/\exp(\mu_c/kT)](dN/dt) = \text{constant}$$

and thus

$$\exp[-\mu_c/(2kT)](dN/dt) = \text{constant}.$$

If $f(x) = Cx$ + constant, this leads to the differential equation

$$\exp(\mu_h/kT) + (C/g)(d\mu_h/dt) + \text{constant } (d\mu_h/dt)^2 = 0$$

If, on the contrary, $f(x) = C\exp(x/kT)$, i.e. if the particle number is governed by Nernst's law $N = C\exp(\mu_h/kT)$, then we find

$$kT + (C/g)(d\mu_h/dt) + \text{constant } \exp(\mu_h/kT)(d\mu_h/dt)^2 = 0.$$

6 THERMODYNAMIC LENGTHS

The notion of thermodynamic length enables us to unify the thermal and chemical models. The thermodynamic length L, followed by a thermodynamic reservoir, is a quantity taking into account the changes that occur to its intensive variables T, μ, ... It is expressed here in units of square root of energy divided by temperature, i.e. [(joule/kelvin)$^{1/2}$], although alternative choices are possible [3] [6] [7], such as [(joule)(second)/(kelvin)]$^{1/2}$, (joule)$^{1/2}$, and (meter)$^{-3/2}$. We have

$$dL^2 = (1/T)dT(dQ/T) + (1/T)d\mu\, dN + \dots .$$

See e.g. Eq. (3.30) in [7].

For the purely thermal case, the thermodynamic length is determined by

$$dL = (C^{1/2}/T)(dT),$$

such that

$$L = \int dL = \int (C^{1/2}/T)\, dT .$$

E.g. for a linear capacitor and a monotonously increasing T, we get

$$L = C^{1/2}\,[\log(T) - \log(T_a)].$$

For the purely chemical case, the thermodynamic length is determined by

$$dL = (C/T)^{1/2}\, d\mu ,$$

such that

$$L = \int dL = \int (C/T)^{1/2}\, d\mu .$$

E.g. for a linear capacitor and a monotonously increasing μ, we get

$$L = (C/T)^{1/2}\,(\mu - \mu_a).$$

Therefore, we can bring Figures 2 and 4 together into a single picture, i.e. Figure 6. Equally we can bring Figures 3 and 5 together into a single diagram, i.e. Figure 7. In Figure 6, we apply the tradition to also regauge the abscissa, by introducing the (dimensionless) 'thermodynamic time' T, which satisfies

$$dT = (g/C)dt.$$

We remark that the horse covers a distance $\Delta L = L_f - L_i$ in a time $\Delta T = T_f - T_i$, whereas the total distance covered by the carrot (in the same time) is longer by an amount $2(\Delta L/\Delta T)$ in the chemical case, and by an amount $2(\Delta L/\Delta T)[\log(1+\alpha)/\alpha]$ in the thermal case.

Surprising is the fact that the two non-linear transport laws (5) and (8), which have quite different origin and contain quite different mathematical functions, give rise to quite analog equations, when expressed in terms of thermodynamic length and time:

$$(dQ/dT) = C\, T_a^4\, [\exp(4L_c/C^{1/2}) - \exp(4L_h/(C^{1/2})]$$

$$(dN/dT) = C \exp(\mu_a/kT)\, [\exp(L_c/k(CT)^{1/2}) - \exp(L_h/k(CT)^{1/2})].$$

The entropy generation rate dS/dt is visualized by the distance $|L_c - L_h|$ between carrot and horse, called the thermodynamic lag [6]. For the linear chemical case we have

$$(dS/dT) = (L_c - L_h)^2,$$

or written in the style of eqn (8) of [6]:

$$(dS/dT) = (L_c - L_h)(dL/dT).$$

For an arbitrary case this is only valid for (infinitely) small distances, i.e. in the adiabatic limit. We finally remark that, in the linear chemical case, the thermodynamic delay T_d between horse and carrot is exactly unity. Also this property is only valid in the adiabatic limit, if we consider an arbitrary thermodynamic process.

7 ABOUT APPLICATIONS IN ELECTRONICS

Although thermal and chemical processes have a scientific, a technical and an economical interest of their own, we will now turn our attention to an application in electronics. The problem is stated as follows: how to charge a capacitor (with capacitance C) from an initial voltage V_i (at time t_i) to a final voltage V_f (at time t_f) and how to discharge it subsequently? For this purpose we have to our disposal a voltage source $V_c(t)$. Between this source (i.e. the carrot) and the capacitor (i.e. the horse) we assume an ohmic resistor (with conductance g).

For convenience we write $V_i = 0$, $V_f = V$, and $t_f - t_i = \tau$. The capacitor is assumed initially uncharged: $V_h(t_i) = 0$. The classical approach is as follows: the source consists of a constant voltage source V and an ideal switch. At $t = t_i$ the switch is closed, such that the capacitor is connected to a voltage source $V_c(t) = V$. After sufficient time ($t - t_i >> C/g$), the capacitor is charged: $V_h(t) \approx V$. See Figure 8a. At $t = t_f$, the capacitor is disconnected. The discharging happens analogously, the capacitor being connected during a time lapse τ to ground, i.e. to $V_c(t) = 0$. This classical strategy has two main disadvantages:

- First, in the finite time τ the capacitor is not charged (or discharged) completely.

- Second, the reader will easily verify that the whole procedure (i.e. the complete charge and- discharge cycle) costs an amount $\int g(V_c - V_h)^2\, dt = CV^2$ of work, delivered by the voltage source and dissipated in the resistor [$(1/2)CV^2$ being dissipated during charging and $(1/2)CV^2$ being dissipated during discharging]. This amount is independent of the value of g and thus applies also to the ideal case $g \rightarrow \infty$.

The work consumption and the heat production are of great concern in two cases:

- In a digital computer, C is the capacitive load of a logic gate. Such a capacitor is frequently charged and discharged, because $V_h = V$ is needed to represent logic 1, whereas $V_h = 0$ represents a logic zero [8].

- In a flat display, C is the capacitance of a pixel. Such a capacitor is frequently charged and discharged, because $V_h = V$ is needed to realise 'light on', whereas $V_h = 0$ is needed to switch the light off [9].

The thousands of logic gates in a computer and the thousands of pixels on a display are source of a lot of energy consumption and heat generation. We like to avoid this for several practical and economical reasons:

- Computers (and their displays) have become one of the main energy consumers in several environments, such as offices. Their heat dissipation is often a cause of trouble for the people around.

- Portable equipment (such as 'lap tops') should consume as little as possible, in order to avoid too often replacement of the battery.

- Cooling of computer hardware is necessary, because the reliability of the silicon circuits goes down dramatically if temperature rises. The more we can avoid heating, the less we have to cool.

The problem is particularly interesting to treat in finite-time thermodynamics, because we are faced with very stringent conditions along the time axis. We have only a very fixed time τ to our disposal to perform the (dis)charging of the tiny capacitors. Indeed.

- Computations are performed at a well defined and fixed pace, governed by the clock frequency.
- Pictures are written at a well defined pace, governed by the so-called frame rate.

We can conclude that both in a computer and a display, we have to charge and discharge a huge number of small capacitors in a well defined but very short time, with a minimum of entropy creation. This constitutes a particularly beautiful application of finite-time thermodynamics. The economical implications are huge, considering the ever increasing number of ever faster computers and displays in the world. The challenge is the following: how to change the carrot voltage $V_c(t)$ in order to do better (i.e. to dissipate less) than the classical $(1/2)CV^2$?

8 SIMPLE ELECTRONIC EXAMPLE

A purely electrical process, in principle, does not differ from a chemical process, except for names and notations. The intensive variable is the voltage V, the extensive variable is the electric charge Q. Entropy production obeys

$$(dS/dt) = (1/T)(dQ/dt)(V_c - V_h).$$

We will first look at the linear kinetic equation, i.e. the law of Ohm. Therefore we have the following set of equations:

$$Z = (V_c/T) - (V_h/T)$$

$$(dQ/dt) = g(V_c - V_h)$$

$$Q = f(V_h)$$

Here again g is a constant, whereas f (x) is a monotonously increasing function. We find

$$f'(V_h)(dV_h/dt) = \text{constant},$$

in accordance with (7). For f (x) = Cx (or f(x) = Cx + constant), the results are: linearly changing voltages V_h and V_c, with

$$V_h(t) = V(t/\tau)$$

$$V_c(t) = V_h(t) + (C/g)(V/\tau). \qquad (9)$$

For convenience of notation , we have not only chosen $V_i = 0$ and $V_f = V$, but also $t_i = 0$ and $t_f = \tau$. The result is not new: see [10] [11] [12].

The power dissipated is $dW/dt = V_c(dQ/dt)$. Therefore the work consumed during the voltage rise is $W_{up} = (1/2)CV^2 + [C^2V^2/(g\tau)]$; the work dissipated during the voltage decay $W_{down} = -(1/2)CV^2 + [C^2V^2/(g\tau)]$ is negative (if we assume $\tau > 2C/g$), meaning that part of the work is recovered. The work consumed during the whole cycle is $[2C^2V^2/(g\tau)]$. For $\tau >> C/g$, this energy, dissipated in the resistor, is much smaller than the energy $(1/2)CV^2$, which was temporarily stored in the capacitor. The limit $\tau \rightarrow \infty$ is called adiabatic charging - and- discharging.

We remark that, for a linear load capacitor, but for any (i.e. both for optimal and non-optimal) strategy, the total energy dissipated during one cycle is $W = \oint(dW/dt)dt = \oint V_c(dQ/dt)dt = \oint V_c C(dV_h/dt)dt$ is simply $C\oint V_c dV_h$, i.e. C times the surface area circumscribed by the path. Figure 8 shows the classical strategy (step), the quasi-optimal strategy (ramp), and finally the optimal strategy (step + ramp). Figure 9 shows the corresponding paths. Note that the line integral $\oint V_c dV_h$ is larger for the optimal strategy than for the quasi-optimal strategy. The extra amount of energy dissipated in the optimal process is the price we have to pay for finite time. Indeed, the quasi-optimal process is not a finite-time process.

In passing, we remark that, if the two conditions dQ/dt = constant and $Q = Cv_h$ are fulfilled, the path $V_c(V_h)$ has exactly the same shape as the $V_c(t)$ trajectory.

9 MORE ELECTRONIC EXAMPLES

Unfortunately, an ohmic resistor is not an accurate model for a transistor. Therefore, we will now investigate some simple non-linearities for the dissipative component.

9.1 A first class of current-voltage laws

In a first class of transport laws, we assume that the current dQ/dt is still a function of V_c and V_h through the difference $V_c - V_h$ alone:

$$(dQ/dt) = g\,\phi(V_c - V_h), \qquad (10)$$

where $\phi(x)$ is some (monotonously increasing) function [satisfying $\phi(0) = 0$]. The ohmic law is a special case of this class, i.e. $\phi(x) = x$. As we derive immediately that

$$Z(Q,dQ/dt) = (1/T)(\phi^{-1})[(1/g)(dQ/dt)],$$

we find that both strategies (1) and (2), i.e.

$$(dQ/dt)Z = \text{constant}$$

and

$$(dQ/dt)[\partial Z/\partial(dQ/dt)](dQ/dt) = \text{constant}$$

lead to the same simple result

$$dQ/dt = \text{constant}$$

and thus to

$$V_c - V_h = \text{constant}$$

Thus, like in the ohmic case, both V_c and V_h are linear functions of time, both during charging and discharging (if, at least, we assume that the non-linear resistor is combined with a linear capacitor). The path of the charge-discharge cycle is still a parallelogram, now with surface area $\oint V_c dV_h$ equal to

$$\{\phi^{-1}[CV/(gt)] - \phi^{-1}[-CV/(gt)]\}V.$$

As an example, we consider a linear conduction with threshold:

$$\phi(x) = \begin{cases} x + a & \text{for } x < -a \\ 0 & \text{for } -a < x < a \end{cases}$$

$$x - a \qquad \text{for } a < x \, .$$

See Figure 10b. It turns out that the threshold a costs us an additional lag between horse and carrot: $V_c - V_h = [CV/(g\tau)] + a$, an additional delay: $t_d = (C/g) + (a\tau/V)$, as well as an additional work: $W = 2[C^2V^2/(g\tau)] + 2CVa$. The dissipated work is non-zero, even for a very slow charging-and-discharging cycle: $\lim_{\tau \to \infty} W = 2CVa$.

A second example is a power law:

$$\phi(X) = X^n,$$

where n is a positive number, say e.g. n = 3, for convenience. Note that, in case n is not an odd integer, the above equation is, in fact, a short-hand notation for the more accurate expression:

$$\phi(x) = \text{sign}(x) \, [\, \text{abs}(x) \,]^n$$

See Figure 10c. It turns out that also the exponent n costs us an additional lag and an additional work: we find $V_c - V_h = [CV/(g\tau)]^{1/n}$ and $W == 2CV \, [CV/(g\tau)]^{1/n}$. For $n > 1$, this means that W goes to zero like $\tau^{-1/n}$, i.e. slower than $1/\tau$.

9.2 A second class of current-voltage laws

Let us now consider another non-linear law:

$$(dQ/dt) = g \, [\phi(V_c) - \phi(V_h)] \, . \tag{11}$$

Again ohmic conduction is recovered by $\phi(x) = x$. But, whereas, in the previous section, we had a function of the difference, we have here the difference of a function. As a consequence, in the (V_h , V_c)-plane, the lines of constant current are no more straight lines parallel to the bisector. As another consequence, the rules (1) and (2) do not give the same result anymore. Even more disturbing is the fact, that these two strategies cannot be derived easily anymore, as we cannot construct the function Z(Q , dQ/dt) anymore.

For sufficiently slow carrots, i.e. in the adiabatic limit, we can construct approximate solutions. Indeed:

$$(dQ/dt) \approx g \, \phi'(V_h) \, (V_c - V_h),$$

such that

$$Z = (1/T)(V_c - V_h) \approx (1/T)[(1/g)(dQ/dt)] \, / \, [\phi'(Q/c)]$$

As a result both strategies $(dQ/dt) \, Z = \text{constant}$ and $(dQ/dt)[\partial Z/\partial(dQ/dt)](dQ/dt) = \text{constant}$ lead to

$$(dQ/dt)^2/[\phi'(Q/c)] \approx \text{constant} \tag{12}$$

provided $\phi'(0)$ is finite and non-zero.

As an example, we consider the power law $\phi(x) = \text{sign}(x) \, |x|^n$, where n is again some positive number. Figure 11a illustrates that the lines of constant current do not coincide anymore with the lines $V_c - V_h = \text{constant}$. The larger is n, the more the lines show a sharp bend. We have

$$Z(Q,dQ/dt) = (1/T)\{[(1/C^n)Q^n + (1/g)(dQ/dt)]^{1/n} - Q/C\}$$

and thus rule (1) yields

$$(V_c^n - V_h^n)(V_c - V_h) = \text{constant} \tag{13}$$

and hence

$$(dV_h/dt)^n[V_h^n + (C/g)(dV_h/dt)] - [\text{constant} + V_h(dV_h/dt)]^n = 0,$$

whereas rule (2) yields

$$(dQ/dt)[1/(gT)](1/V_c^{n-1})(dQ/dt) = \text{constant}$$

or

$$(V_c^n - V_h^n)^2/V_c^{n-1} = \text{constant} \tag{14}$$

and hence:

$$[V_h^n + (C/g)(dV_h/dt)]^{n-1} + \text{constant } (dV_h/dt)^{2n} = 0.$$

We remark that the curves (13) are somewhat less curved than the curves (14), which themselves show somewhat less curvature than the lines of constant current, i.e. the lines obeying

$$V_c^n - V_h^n = \text{constant.} \tag{15}$$

See an example on Figure 12, for n = 3.

In the extreme case $n \to \infty$, each of the three curves consists of a horizontal part ($V_c \approx$ constant), a sharp kink and a part very close to the bisector ($V_c \approx V_h$). See Figure 13 for n = 19. The V_c versus V_h path now looks like Figure 11b.

Calculating $W = C \oint V_c \, dV_h$ yields approximately $2C[(c^2v^2/(g^2\tau^2)]^{1/n}$, for n large, indicating that the dissipated energy W goes to zero like $\tau^{-2/n}$.

As a second example, we combine the above non-linearity with a threshold:

$$dQ/dt = \begin{cases} g(V_c^n - V_h^n + a^n) & \text{for } V_c^n - V_h^n < -a^n \\ 0 & \text{for } -a^n < V_c^n - V_h^n < a^n \\ g(V_c^n - V_h^n - a^n) & \text{for } a^n < V_c^n - V_h^n \end{cases}$$

and where again the exponent n is large. See Figure 10d. Now W equals approximately $2C\{[CV/(g\tau)]+a^n)^2\}^{1/n}$, what can be approximated by $2Ca^2$, i.e. the surface area of two squares in Figure 11b.

10 CONCLUSION

We have seen how simple models for thermal, chemical, or electrical systems can lead to explicit finite-time strategies, guaranteeing minimum entropy creation. Optimal trajectories T(t), μ(t), or V(t) can be calculated. The thermodynamic length enables us to unify the models as well as the results.

Unfortunately, practical examples in electronics are not that easy to analyze, especially because a transistor current is a function of drain and source voltage, i.e. not function of the voltage difference alone.

ACKNOWLEDGEMENT

The present research is partly supported by the Commission of European Communities in the frame-work of the "Copernicus" Keep-In-Touch programme 'Carnet 2'.

REFERENCES

[1] L. Diosi, K. Kulacsy, B. Lukacs and A. Racz, Thermodynamic length, time, speed, and optimum path to minimize entropy production, *Journal of Chemical Physics* **105** (1996) pp. 11,220-11,225.

[2] B. Andresen, Finite-time thermodynamics and simulated annealing, in: J. Shiner: Entropy and entropy generation, Kluwer, Amsterdam (1996) pp. 111-127.

[3] W. Spirkl and H. Ries, Optimal finite-time endoreversible processes, *Physical Review E* **52** (1995) pp. 3485-3489.

[4] S. Sienuitycz: "Generalized thermodynamics of maximum work in finite time" *Meeting on Complex Systems in Natural and Economic Sciences,* Tata, 12-14 September 1996.

[5] S. Sienuitycz, Non-equilibrium thermodynamics as an extension of classical thermodynamics, *Open Systems and Information Dynamics* **4** (1997) pp. 185-207.

[6] P. Salamon and R. Berry, Thermodynamic length and dissipated availability, *Physical Review Letters* **51** (1983) pp. 1127-1130.

[7] G. Ruppeiner, Riemannian geometry in thermodynamic fluctuation theory, *Reviews of Modern Physics* **67** (1995) pp. 605-659.

[8] W. Athas, L. Svensson, J. Koller, N. Tzartzanis, and E. Chou, Low-power digital systems based on adiabatic-switching principles, *I.E.E.E. Transactions on V.L.S.I. Systems* 2 (1994) pp. M9.

[9] C. King, Electroluminescent displays, in: " S.I.D. seminar lectures notes" Society of Information Display, Santa Anna (1994) pp. M9.

[10] O. Elgerd, Control systems theory, McGraw-Hill, New York (1967) pp.445-494

[11] P. Patra and D. Fussell, On efficient adiabatic design of MOS circuits, *Proceedings of the 4th Workshop on Physics and Computation,* Boston, 22-24 November 1996, pp. 260-269.

[12] S. Paul, Optimal charging capacitors, *Proceedings of the European Conference on Circuit Theory and Design,* Budapest, 1-3 September 1997, pp. 918-922.

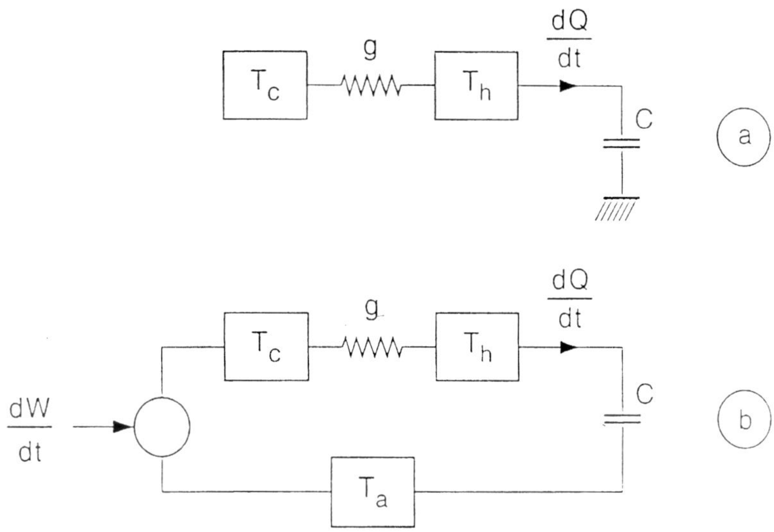

Figure 1: Thermal circuit: (a) partial circuit, (b) full circuit.

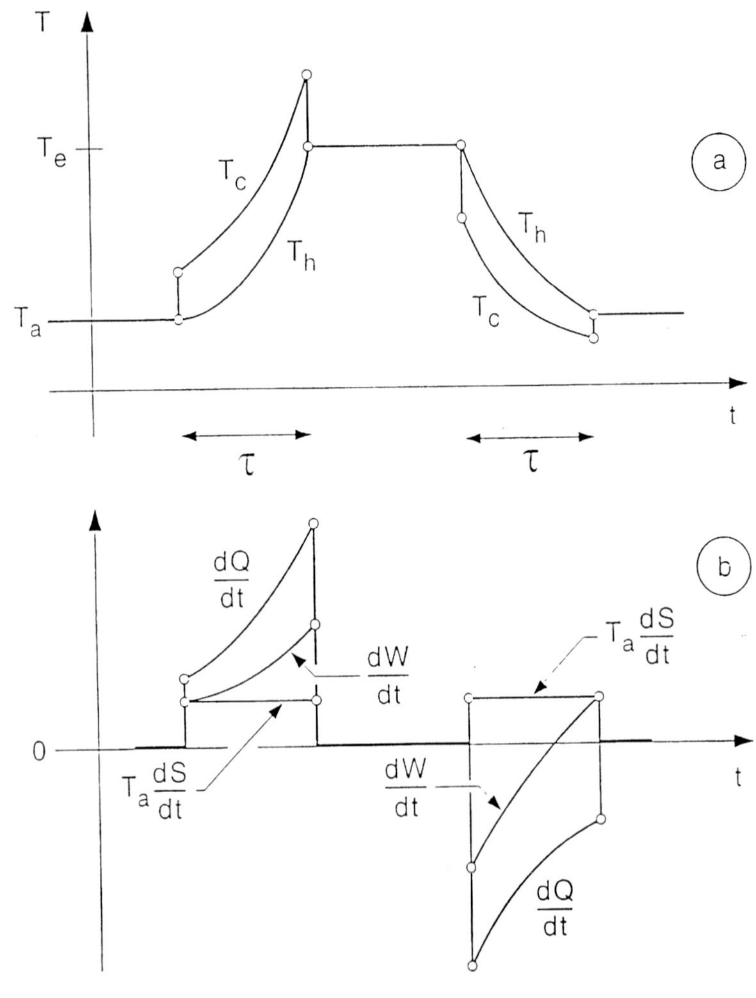

Figure 2: Optimal time evolution of (a) temperatures T_c and T_h and (b) heat flow $\frac{dQ}{dt}$, work supply $\frac{dW}{dt}$, and entropy creation $\frac{dS}{dt}$.

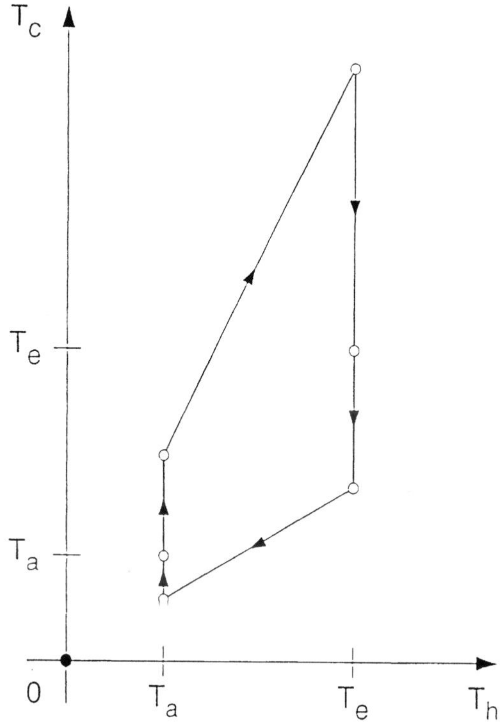

Figure 3: Optimal path in the (T_h, T_c)-plane.

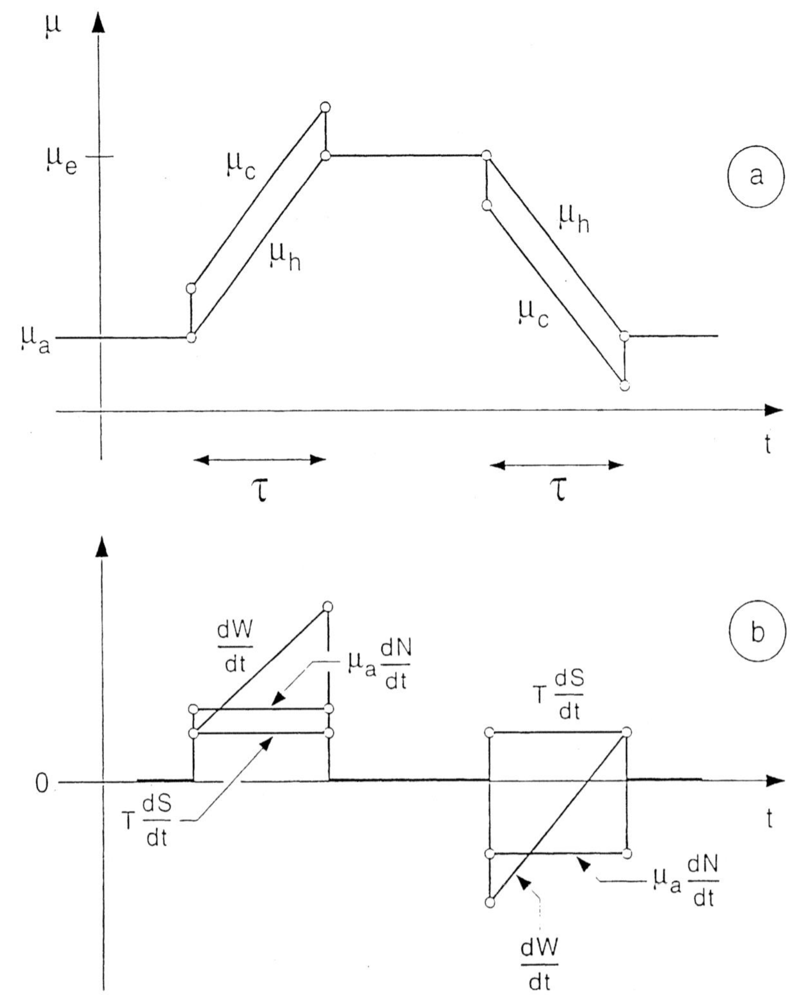

Figure 4: Optimal time evolution of (a) potentials μ_c and μ_h and (b) particle flow $\frac{dN}{dt}$, work supply $\frac{dW}{dt}$, and entropy creation $\frac{dS}{dt}$.

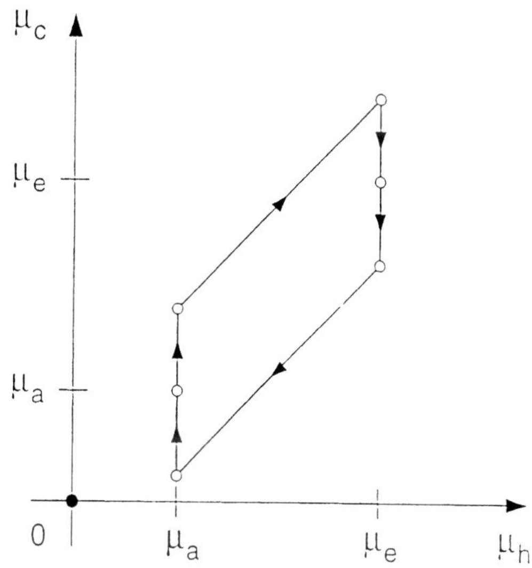

Figure 5: Optimal path in the $(\mu_h; \mu_c)$-plane.

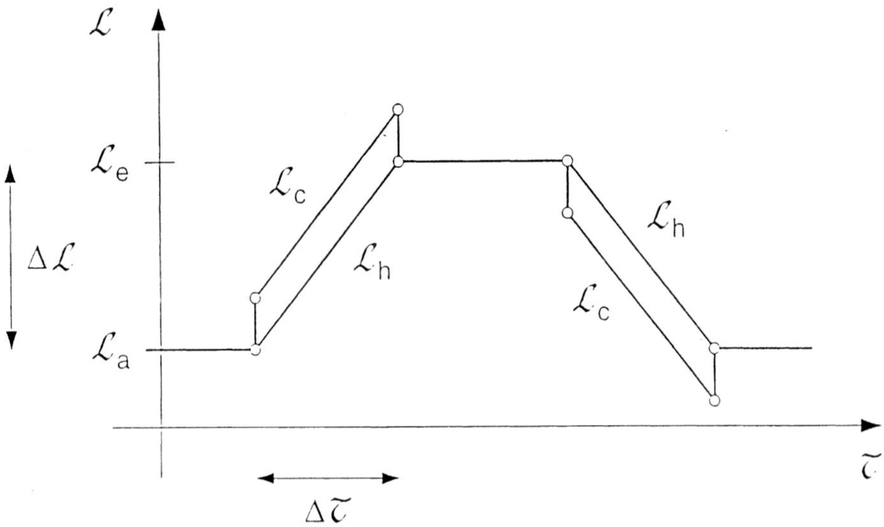

Figure 6: Optimal time evolution of distances $\mathcal{L}_c$ and $\mathcal{L}_h$.

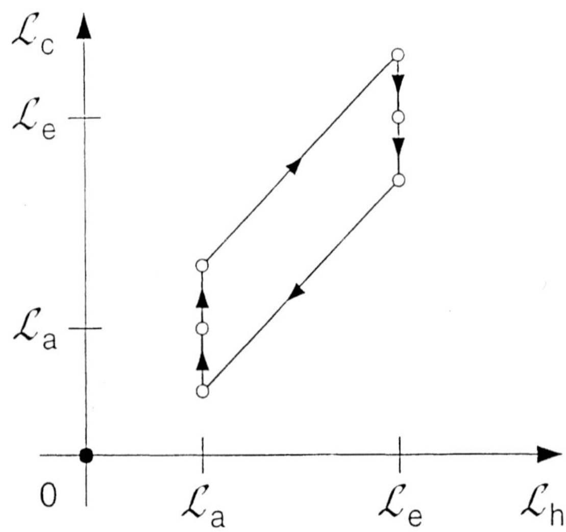

Figure 7: Optimal path in the $(\mathcal{L}_h, \mathcal{L}_c)$-plane.

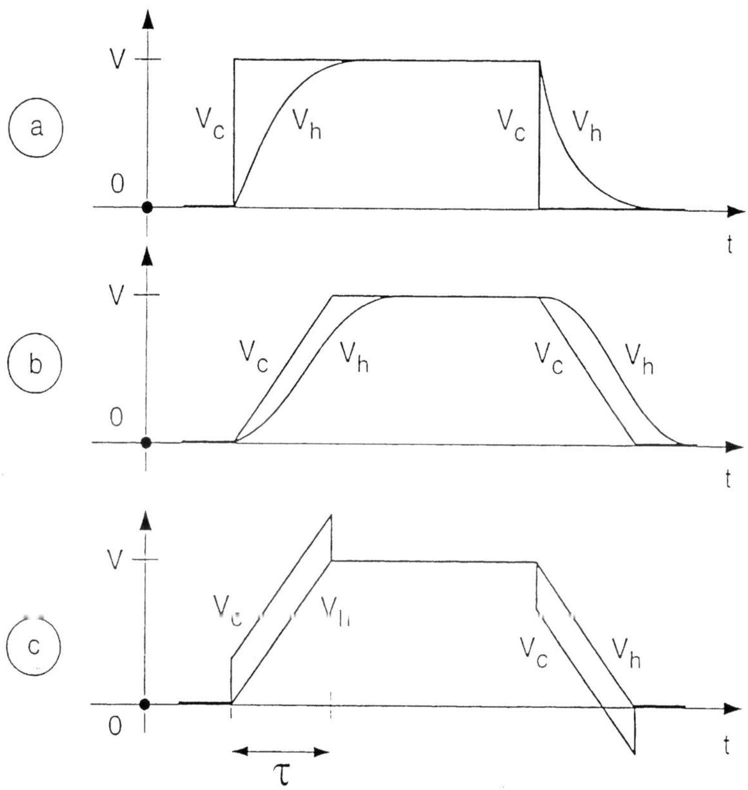

Figure 8: Time evolutions $V_c(t)$ and $V_h(t)$: (a) conventional strategy, (b) 'ramp' strategy, and (c) optimal strategy.

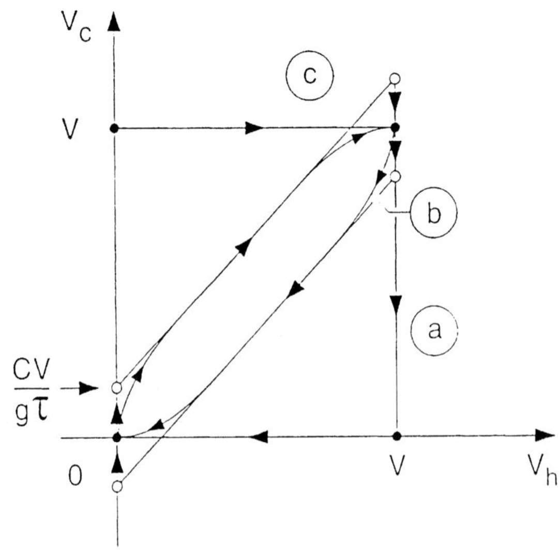

Figure 9: Paths in the (V_h, V_c)-plane: (a) conventional strategy, (b) 'ramp' strategy, and (c) optimal strategy.

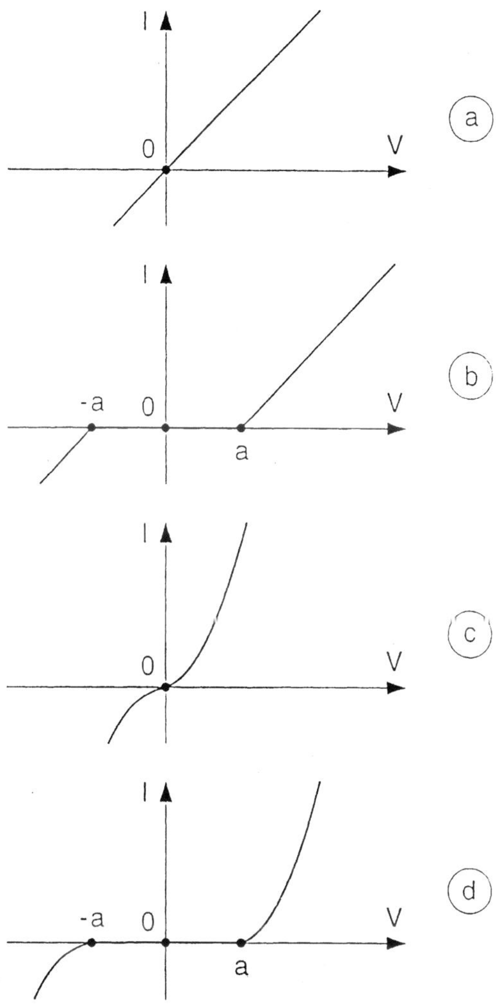

Figure 10: Current-voltage characteristics: (a) ohmic, i.e. linear, (b) linear with threshold, (c) non-linear, and (d) non-linear with threshold.

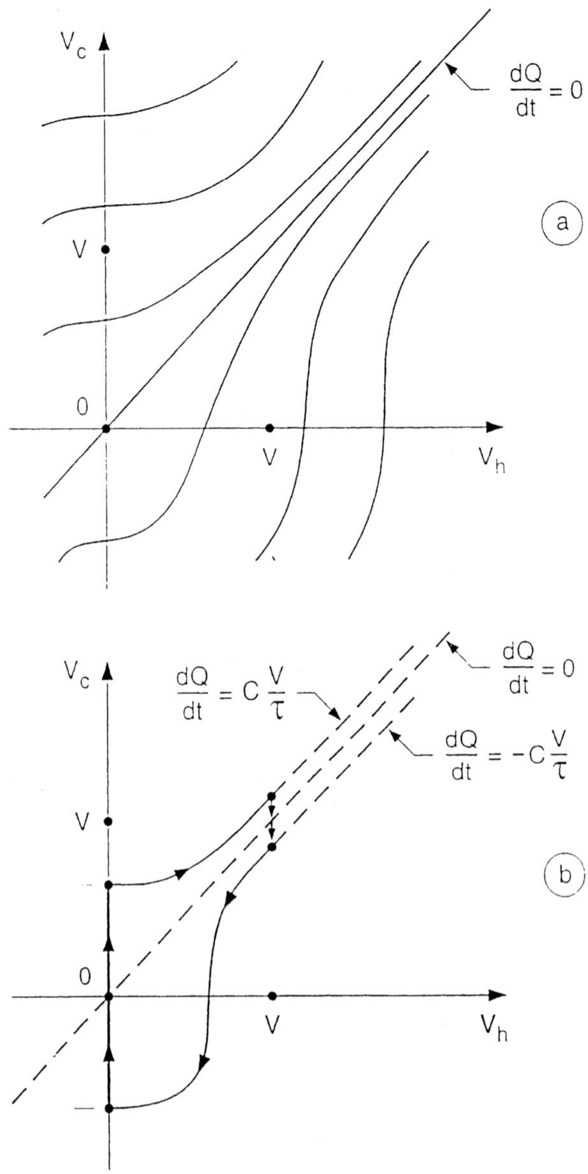

Figure 11: Lines in the (V_h, V_c)-plane: (a) the iso-current lines and (b) the cyclic path.

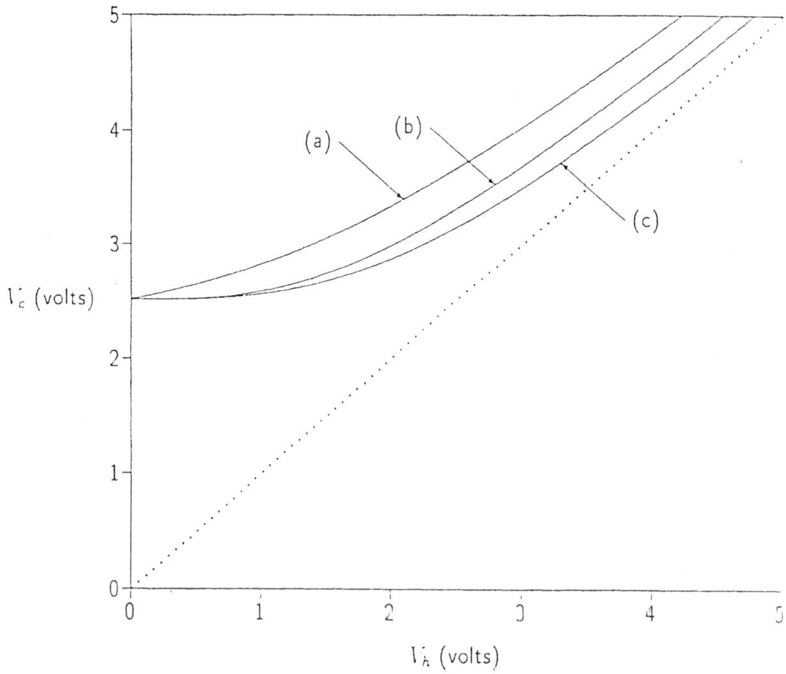

Figure 12: Lines (for $n = 3$) in the (V_h, V_c)-plane: (a) constant $\frac{dQ}{dt}$ Z, (b) constant $\frac{dQ}{dt}\frac{\partial Z}{\partial \frac{dQ}{dt}}\frac{dQ}{dt}$, and (c) constant $\frac{dQ}{dt}$.

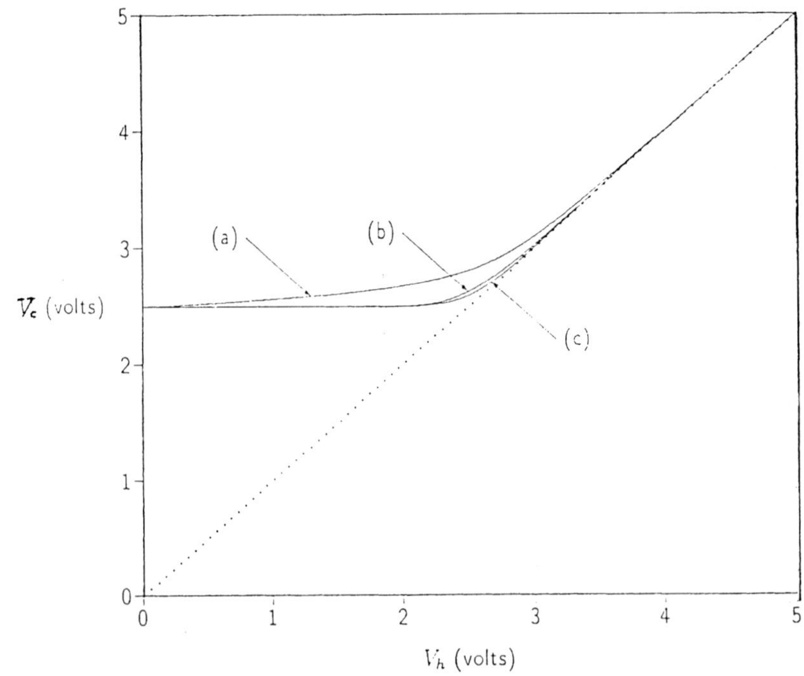

Figure 13: Lines (for $n = 19$) in the (V_h, V_c)-plane: (a) constant $\frac{dQ}{dt} Z$; (b) constant $\frac{dQ}{dt} \frac{\partial Z}{\partial \frac{dQ}{dt}} \frac{dQ}{dt}$, and (c) constant $\frac{dQ}{dt}$.

TRADE-OFFS IN WEIGHT-SENSITIVE DIESEL WASTE-HEAT RECOVERY SYSTEMS

JOHN B. WOODWARD
Professor, Naval Architecture & Marine Engineering
University of Michigan
Ann Arbor, MI 48109-2145

Abstract- A heat source of fixed flow rate and exergy supplies a Rankine bottoming cycle for which the nominal objective is to produce the greatest possible mechanical work output. However, with respect to an optimum exploitation of the source's exergy, an important factor is the investment in heat exchange surface as represented by the total NTU (number of thermal units) required. The problem is similar to the typical time-dependent theromodynamics problem of optimal temperature levels for maximum power from an isothermal heat source. In the case treated here, optimal exploitation is a function of economic balance between investment in heat transfer surface and income from mechanical output. As a consequence of a deliberate limitation of scope, this paper does not attempt to find such optimums; rather, it demonstrates the relationship between output and NTU investment. The results do illuminate what appear to be appropriate levels of NTU.

INTRODUCTION

We begin with the conventional notion that the principal objective in exploiting a thermal energy source is maximization of its conversion to work. Where the source is isothermal, and hence unlimited in extent, the objective may be better viewed as one of maximizing mechanical power with respect to the necessary investment in heat exchange surface. This concept was initially given prominence by Curzon and Ahlborn in 1975[1], and has been developed by a number of other authors ([2, 3, 4,5] may be cited from among many). Their efforts have given rise to the concept of time-dependent thermodynamics.

The heat source discussed in this paper is a hot gas, presumably the product of combustion. The gas is characterized by a certain exergy per unit mass. It is to be exploited by a diesel engine and possible bottoming cycles. This source is, of course, not isothermal because it flows at a finite rate determined by the size of the engine, but we can examine some aspects of its exploitation via notions based on time-dependent thermodynamics.

In a diesel-based case where maximum work per unit of heat source is the objective, the combustion gas passes in series through an engine cylinder, an exhaust turbine (turbocharger), and then a steam generator that is the source of working fluid for a Rankine-cycle engine. Maximization can be sought by adjusting the interfaces between the three elements, e.g. by varying engine stroke length to change the share of mechanical energy production between the primary cycle and its two followers. However, such unlikely practices are far beyond the scope of this paper, a paper that regards the thermal output of the turbocharger as being a fixed-rate source, non-isothermal, and treats only one aspect of maximizing the exploitation of that source by a Rankine bottoming cycle.

Fig. 1 and Fig. 2 illustrate the elements of the bottoming cycle. Fig. 1 indicates the four heat exchangers that are of interest: superheater, evaporator, preheater (collectively the "steam generator"), and condenser. Fig. 2 shows typical temperature patterns in these exchangers.

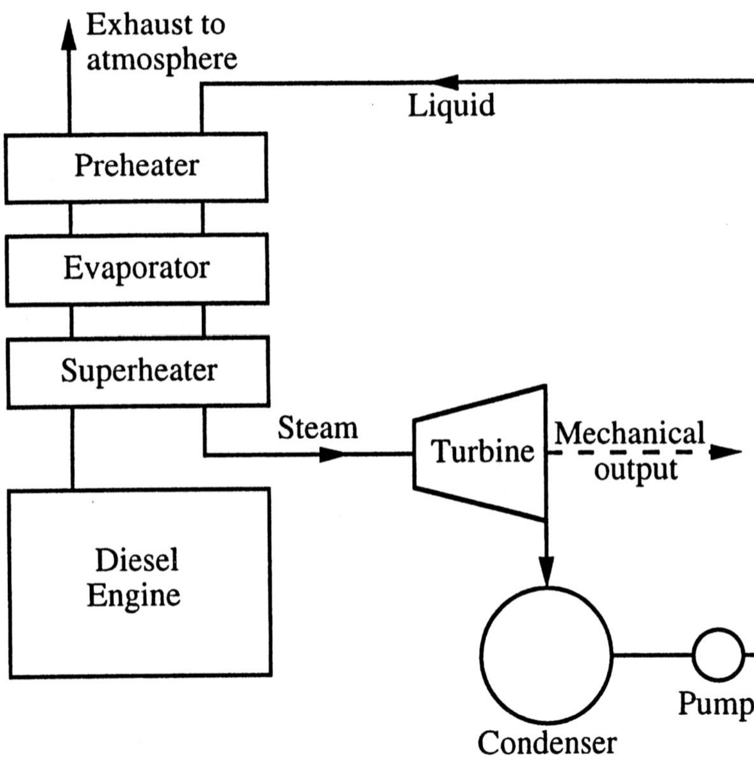

Fig 1. Elements of a diesel exhaust waste heat power system.

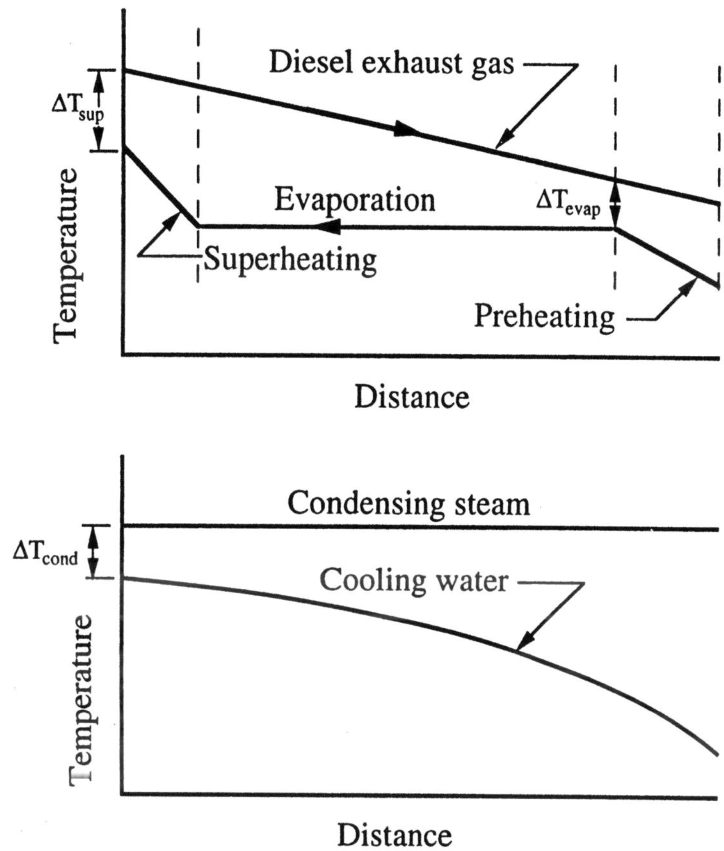

Fig 2. Typical temperature patterns in steam-producing and steam-condensing heat exchangers.

If an analyst's objective is to maximize the bottoming cycle work (or the power from a fixed input rate), then system parameters such as steam pressure, steam temperature, and exhaust pressure can be varied, with work or power to be calculated for each combination. A typical study is illustrated by Fig. 3, wherein exhaust (or condensing) pressure appears as the major determinant of work, with steam pressure and temperature being significant but comparatively minor. However, beyond providing Fig. 3 for introductory purposes, this paper does not treat the question of an optimal steam state.

This approach to Rankine bottoming cycle is incomplete in that it ignores the consequences of steam state on the investment in heat exchange surface necessary to attain favorable temperatures. If energy is dear and heat exchange hardware is cheap, such an approach may be satisfactory. More typically, cost of hardware is not to be ignored, and may have special significance in some applications. Examples of special significance occur in marine engineering, where waste-heat steam generators must be fitted into restricted space that is high in the ship. Provision of adequate machinery space, and compensations for degraded ship stability consequent on elevated weights, can amplify the consequences of direct investment in heat exchangers [6].

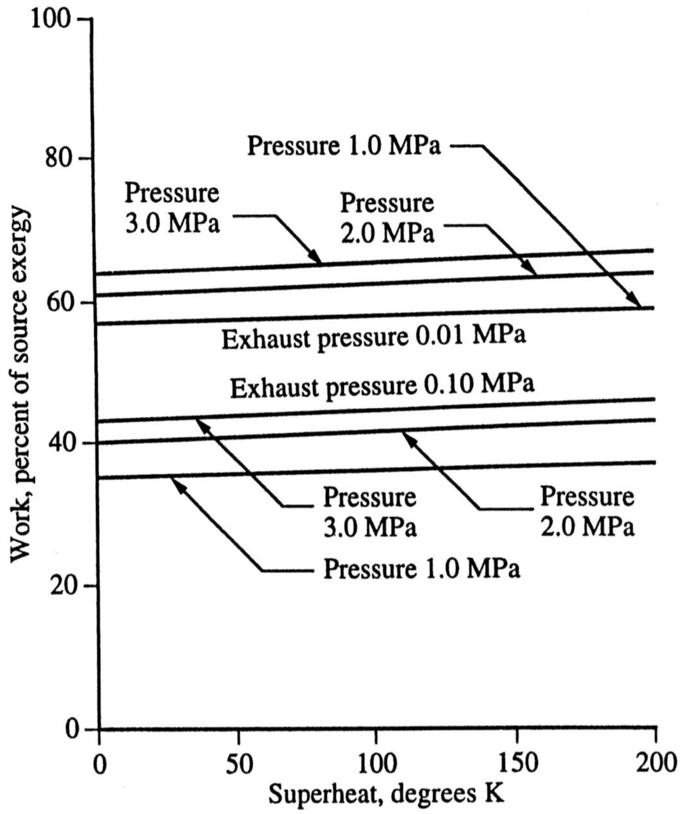

Fig 3. Conversion of waste-heat exergy to work: the fraction converted as a function of system pressure, superheat, and condensing pressures.

THE PROBLEM

A hot-gas heat source, fixed in mass flow rate and in entering temperature, is the heat source for an endoreversible Rankine bottoming cycle; nominally, the designers' objective is to maximize the fraction of source exergy that can be converted to shaft work. However, four heat exchangers are required, and the work maximization depends on the amount of heat transfer surface to be provided, as well as on working fluid pressures and temperatures. The extent of that surface is the essential element of the analysis here.

The *maximum* output is to be found by making the effectiveness of all heat exchangers 1.00 (or close thereto). The consequent extreme size, weight, and monetary investment would surely place the *optimum* at significantly lower effectivenesses, and hence at a less-than-maximum output. However, because the scope of this paper does not encompass the monetary value of work and of hardware, optimization is not attempted. The purpose of the paper is illustration only — to illustrate a concept of energy conversion while limiting conclusions to those that the illustration supports.

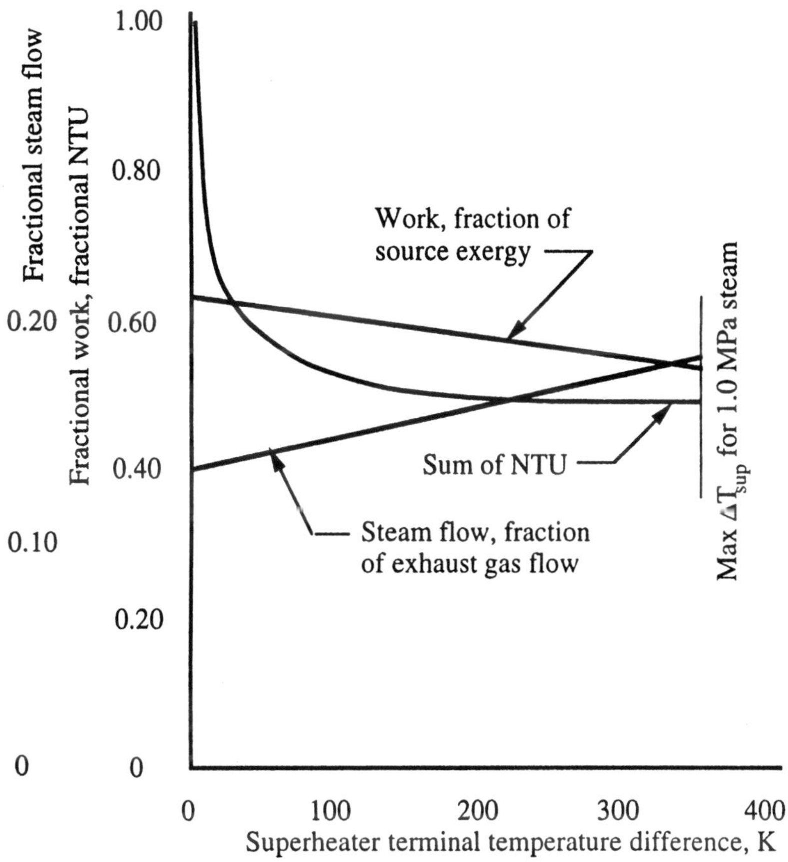

Fig 4. Steam flow, work fraction, heat exchanger NTU as functions of superheater terminal temperature difference.

The bottoming cycle working fluid is water. The possible benefits of alternative working fluids are not treated.

Because the heat source is exemplified by a fixed flow rate, the bottoming cycle output is also a rate, e.g. kJ/s or a power. However, since such a source is usually expressed as a *heat* flow rate, the output units are more appropriately expressed as the fraction of the heat source exergy per kilogram.

The independent variables affecting the magnitude of heat transfer surface are the following temperature differences:

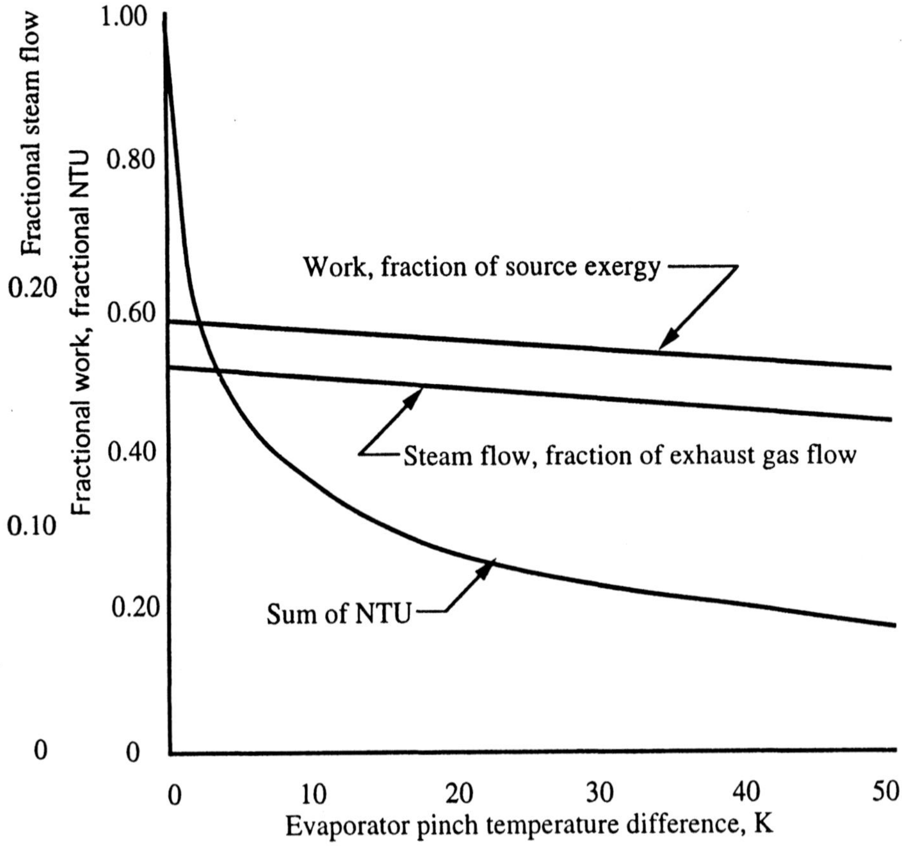

Fig 5. Steam flow, work fraction, heat exchanger NTU as functions of evaporator pinch temperature difference.

- Between source input temperature (taken to be 800 K) and superheated steam temperature (or alternatively, the degrees superheat of the steam).
- Between steam temperature of evaporation and exiting source temperature (often known as the "pinch temperature difference").
- Between steam condensing temperature and the exiting cooling water temperature (often known as a condenser's "terminal temperature difference"). Temperature of the entering cooling water is taken to be 300 K.

Temperature differences for the preheater are fixed by condenser and evaporator conditions.

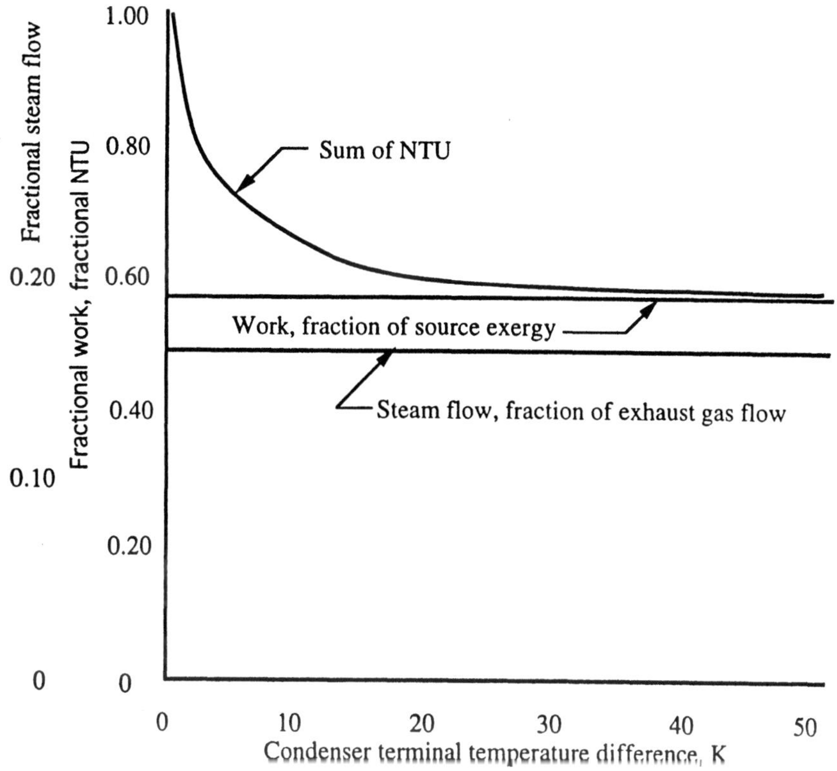

Fig 6. Steam flow, work fraction, heat exchanger NTU as functions of condenser terminal temperature difference.

METHOD OF ANALYSIS

Exergy of the heat source, heat flow rates, and working fluid flow rates and work quantities are found by standard thermodynamic equations (Eq 2–4).

The "heat exchanger resource" is represented by the NTU (number of thermal units), (source) a commonly-used parameter that is approximately proportional to the volume and weight of a heat exchanger [7]. The basic equation for this parameter is

$$NTU = U\,A / C_{min} \tag{1}$$

where U A is the exchanger's heat rate (e.g. kJ / h K) and C is a fluid enthalpy flow rate. There are two C for each exchanger; the subscript min indicates the smaller value. The *effectiveness* (fraction of possible heat transfer that is actually transferred) of a heat exchanger is a related parameter, with the relationship being shown by Eq. 5–10, combined with Eq. 1. Since NTU is more directly indicative of heat exchanger size, it rather than effectiveness is used in this paper.

The investigative approach is to vary the terminal ΔT of each of the four heat exchangers in order to find the consequences to NTU, steam flow, and fraction of source exergy converted to work.

For purposes of analysis, the diesel exhaust gas heat source is fixed in flow rate (10,000 kg/h) and state (atmospheric pressure, entering temperature 800 K) and is treated as a perfect gas (constant pressure specific heat 1.09 kJ/kg K). Steam is treated as a vapor whose state must be determined from steam tables [8]. Its pressure is 1.0 MPa, with superheat determined by choice of the minimum ΔT between steam and heat source, and with condensing pressure varied between 0.01 and 0.10 MPa. Independent variables are the minimum ΔT's (ΔT_{sup} and ΔT_{evap}) between heat source and steam, and the minimum ΔT (ΔT_{cond}) between condenser coolant and condensing steam. Surroundings temperature (which is also the condenser coolant inlet temperature) is 300 K.

EQUATIONS

Exergy of the heat source (exhaust gas entering the superheater), kJ/kg:

$$\psi = c_{pg}\,(T_{gin} - T_0) - T_0\,(s_{gin} - s_0) \tag{2}$$

Steam produced per kg of heat source, kJ/s

$$m_s = \frac{c_{pg}\,(T_{gin} - (T_f + {}^2T_{evap}))}{h_{sup} - h_f} \tag{3}$$

Work done by the steam, kJ/s

$$w = m_s(h_{sup} - h_{exh}) \tag{4}$$

Heat transferred by each of the four heat exchangers

$$Q_{sup} = UA_{sup}\left[\frac{(T_{gin} - T_{sup}) - (T_{x1} - T_f)}{\ln\dfrac{(T_{gin} - T_{sup})}{(T_{x1} - T_f)}}\right] = m_s(h_{sup} - h_g) \tag{5}$$

$$Q_{evap} = UA_{evap}\left[\frac{T_{x1} - T_f - {}^2T_{evap}}{\ln\dfrac{T_{x1} - T_f}{{}^2T_{evap}}}\right] = m_s(h_g - h_f) \tag{6}$$

$$Q_{pre} = UA_{pre}\left[\frac{{}^{2}T_{evap} - (T_{gout} - T_c)}{\ln\dfrac{{}^{2}T_{evap}}{T_{gout} - T_c}}\right] = m_s(h_f - h_c) \qquad (7)$$

$$Q_{cond} = UA_{cond}\left[\frac{T_c - T_0 + {}^{2}T_{cond}}{\ln\dfrac{T_c - T_0}{{}^{2}T_{cond}}}\right] = m_s(h_{exh} - h_c) \qquad (8)$$

Heat exchanger effectiveness

$$\varepsilon = \frac{1 - e^{-NTU\,(1-r)}}{1 - r\,e^{-NTU\,(1-r)}} \quad \text{where } r = C_{min}/C_{max} \qquad (9)$$

Special case : heat exchanger effectiveness for evaporator or condenser

$$\varepsilon = 1 - e^{-NTU} \qquad (10)$$

Results

A temperature difference of 10 degrees is a reasonable value for the evaporator and condenser, and it is taken as base values for those two units. The base value for the superheater is 347 degrees, i.e. the difference between the 800-degree source and the temperature of 1.0 MPa steam superheated 100 degrees. Eq. 3 produces the base-value steam flow, Eq. 4 produces the base-value work per unit of source exergy, and Eq. 1, supported by Eqs. 5 – 8, produces the base-value NTU for the four heat exchangers.

Figs. 4 – 6 show the consequences of varying each of these three ΔT while the other two remain at their base values. Reducing the superheater and evaporator temperature differences produces modest increases in steam flow and work; changing the condenser difference has no effect on flow and work since the change is assumed to occur on the water side with a fixed condensing temperature. NTU obviously must increase as temperature differences are reduced, and indeed, must approach infinity as temperature difference approaches zero. (Exchanger effectiveness similarly approaches 1.0.) The significance of these figures is in demonstrating, approximately at least, the temperature neighborhood in which an investment in NTU turns sharply upward, thereby illuminating the onset of "diminishing returns." The onset occurs in the 50-to-100 degree neighborhood for the superheater, and in the 10 degree neighborhood for the evaporator and condenser.

Fig. 7 shows the consequences for work of taking NTU to extreme high values and extreme low values. Low extremes are values given by effectiveness of 50 percent; high extremes are the NTU values occurring when all ΔT are 0.1 degree. The independent variable in this figure is condenser pressure, chosen to illustrate how the maximum work benefit of very low condenser pressure might be obtained also by a large investment in heat exchanger surface. For example, the figure shows that the work obtained at 0.01 MPa exhaust pressure can be obtained at 0.05 MPa through extreme (but still quite finite) investment in NTU. (However, discussion of Fig. 8 outlines the inadequacy of NTU as a key parameter for condensers.)

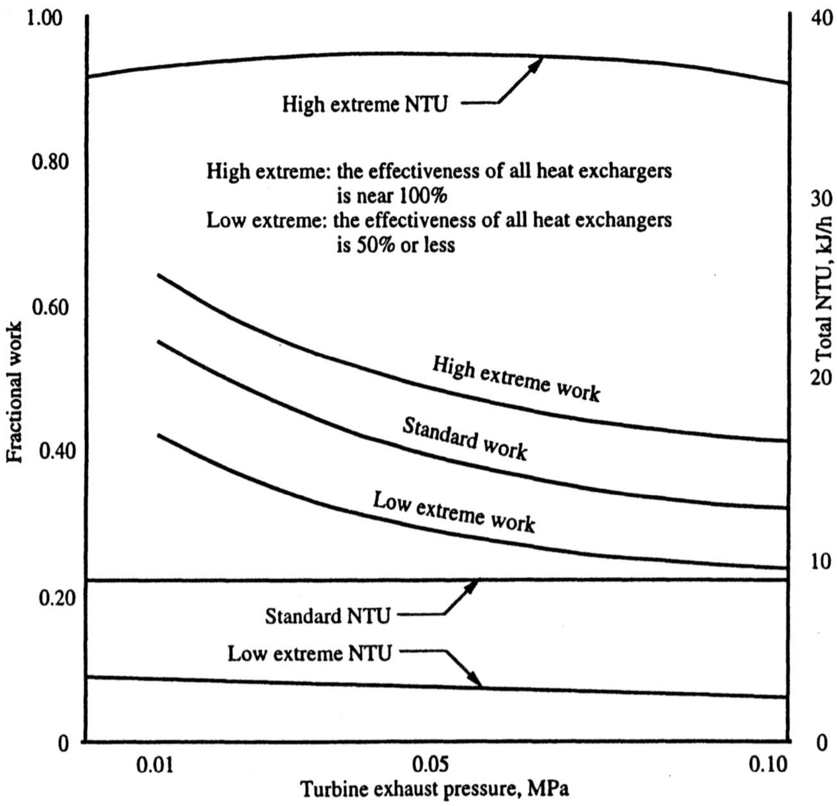

Fig 7. Total NTU as a function of turbine exhaust pressure. For extreme high and low values of heat exchanger effectiveness.

Fig. 3 has demonstrated the conventional wisdom that lowering condenser pressure maximizes Rankine cycle work. Fig. 8 shows the consequences to NTU as condenser pressure is lowered while the magnitude of the heat source remains constant, and the four heat exchanger end temperature differences are kept at their "standard" values. The rate of NTU change as a function of work change is also shown. Observe that condenser NTU decreases as condenser pressure decreases, a seeming anomaly because more surface is demanded as the temperature difference between condensing steam and the cooling water decreases. Accompanying this decrease is a major increase in the cooling water flow rate; it is this factor that drives NTU down while the heat transfer surface increases. A tentative conclusion is that the special conditions is this case, namely a fixed coolant temperature and a fixed minimum ΔT, negate the utilityof NTU.

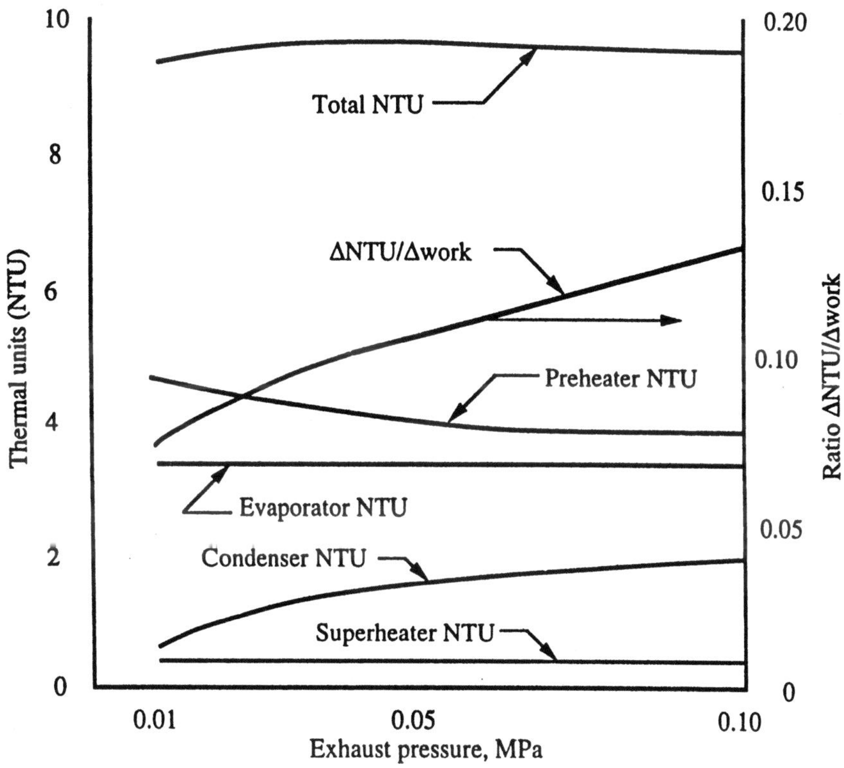

Fig 8. Individual and total NTU, also the rate of change of NTU with respect to changes in work, as functions of condensing pressure.

Figs. 9 and 10 show the ratio ΔNTU/Δwork (approximate partial derivitive) as a function of minimum temperature difference for the evaporator and for the superheater, respectively. The purpose is to demonstrate a guideline that may identify which heat exchanger to increment when the objective is to increase the exploitation of the fixed heat source. For example, at the temperature differences used as base values here (10° for the evaporator, 247° for the superheater), an attempt to increase output by reducing the evaporator pinch temperature difference requires approximately one unit increase in total NTU for every kJ/kg increase in work output. On the other hand, changing superheater minimum temperature difference shows a very small ΔNTU/Δwork relationship, and the difference is even slightly negative. (This anomaly occurs because it lies in the neighborhood where the effect on work of increasing superheat is offset by the consequent decrease in steam flow.) The upshot is that increased investment in evaporator surface is more productive of the two options.

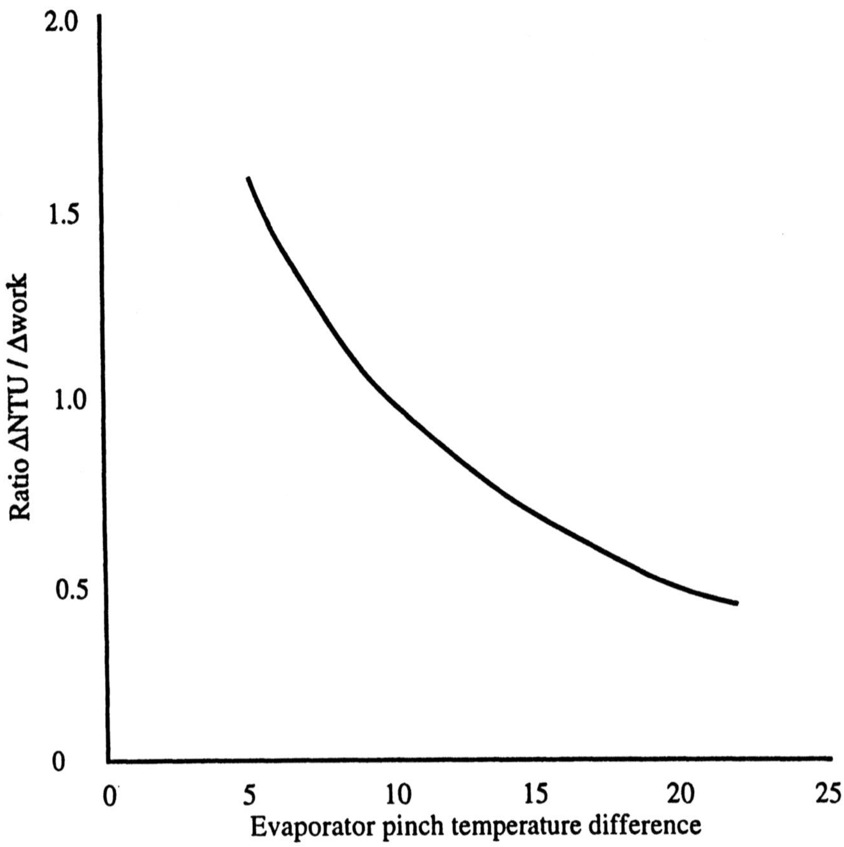

Fig 9. Rate of change of NTU with respect to changes in work, as a function of evaporator pinch temperature difference.

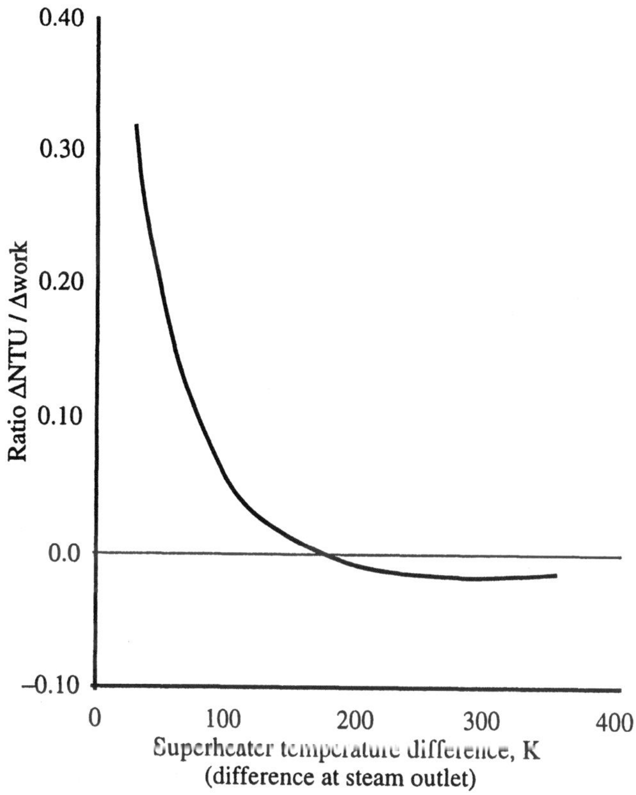

Fig 10. Rate of change of NTU with respect to changes in work, as a function of superheater temperature difference.

SUMMARY

For a heat source of fixed magnitude (temperature and flow rate), the fraction of its exergy converted to work is limited by the second law of thermodynamics. Given that the efficiencies of individual components are fixed by the current state of technology, the utilization of the fraction permitted is determined by the investment in heat exchange hardware, and that the investment is exemplified, approximately, by the NTU as defined by Eq. 1. Plotting NTU as a function of significant temperature differences (e.g. evaporator pinch temperature difference) is a direct way of showing the consequences of investment in heat exchange surface, and particularly in showing tendencies toward diminishing returns. It may also show which exchanger offers the highest benefit per unit of NTU. On the other hand, NTU may be an inappropriate in some instances, as here when particular limits place on the condenser caused the NTU to decrease in circumstances where heat transfer area to increase.

Symbols

c_{pg} constant pressure specific heat of the heat source
C_{min} product of specific heat and mass flow rate, the lower value
C_{max} product of specific heat and mass flow rate, the higher value
h_c enthalpy of liquid water at condensing pressure
h_{exh} enthalpy of turbine exhaust steam
h_g enthalpy of saturated vapor
h_f enthalpy of saturated liquid water
h_{sup} enthalpy of superheated steam
m_s steam mass flow rate per unit mass of heat source
NTU number of thermal units for a heat exchanger
Q_{cond} heat transferred by the condenser
Q_{evap} heat transferred by the evaporator
Q_{pre} heat transferred by the preheater
Q_{sup} heat transferred by the superheater
s_{gin} entropy of the entering heat source
s_0 entropy of the surroundings
T_c temperature of water entering the preheater
T_f temperature of saturated water
T_{gin} temperature of the entering heat source
T_{gout} temperature of heat source leaving preheater
T_{sup} temperature of superheated steam
T_{x1} temperature of heat source gas leaving the superheater
T_0 temperature of the surroundings
UA_{cond} product of heat transfer coefficient and heat transfer area, condenser
UA_{evap} product of heat transfer coefficient and heat transfer area, evaporator
UA_{pre} product of heat transfer coefficient and heat transfer area, preheater
UA_{sup} product of heat transfer coefficient and heat transfer area, superheater
w work per unit mass of heat source
ΔT_{cond} minimum temperature difference between fluids in the condenser
ΔT_{evap} minimum temperature difference between fluids in the evaporator
ΔT_{sup} $T_{gin} - T_{sup}$
ε heat exchanger effectiveness

REFERENCES

1. F L Curzon and B Ahlborn, Efficiency of Carnot Engine at Maximum Power Output, *American Journal of Physics*, 43, pages 22–24 (1975).
2. C Wu, Specific Power Optimization of Closed-Cycle OTEC Plants, *Ocean Engineering,* 17, pages 307–314 (1990).
3. C Wu, Optimization of the Endoreversible Otto Cycle, *Energy Conversion Management,* 34, pages 1255–1259 (1993).
4. O M Ibrahim, S A Klein, and J W Mitchell, Optimum Heat Power Cycles for Specified Boundary Conditions, American Society of Mechanical Engineers, *Journal of Engineering for Gas Turbines and Power,* 13, pages 514–521 (1991).
5. A Bejan, Theory of Heat Transfer — Irreversible Power Plants, *International Journal of Heat and Mass Transfer,* 31, pages 1211–1220 (1988).
6. E V Lewis et al, *Principles of Naval Architecture,* Society of Naval Architects and Marine Engineers (1988).
7. J P Holman, *Heat Transfer*, sixth edition, pages 545–558, McGraw-Hill Book Co, Singapore (1986).
8. J H Keenan, F G Keyes, P G Hill, and J C Moore, *Steam Tables* (SI units), John Wiley & Sons, New York (1978).

FINITE-TIME THERMODYNAMICS APPROACH FOR EXERGY OPTIMIZATION OF AN IRREVERSIBLE COGENERATION CYCLE

BAHRI SAHIN
Department of Naval Architecture, Yildiz Technical University, Besiktas, 80750, Istanbul, Turkey

and

ALI KODAL
Department of Aeronautical Engineering, Istanbul Technical University, Maslak, 80626, Istanbul, Turkey

Abstract- A performance analysis based on maximum exergy criterion has been carried out for an irreversible cogeneration cycle using finite-time thermodynamics. The optimum values of the design parameters of the cogeneration cycle at maximum exergy output were determined. In this context, the effects of design parameters and irreversibility on exergetic performance are investigated and the results discussed. The results obtained may provide a general theoretical tool for the optimal design and operation of real cogeneration plants.

Key Words: finite-time thermodynamics, exergy optimization, irreversible cogeneration

INTRODUCTION

Finite-time thermodynamics optimization studies of heat engines were started by Chambadal [1] and Novikov [2] in 1957 and were continued by Curzon and Ahlborn [3] in 1975. They extended the reversible Carnot cycle to endoreversible cycle by taking the irreversibility of finite-time heat transfer into account and as a result a new field: 'finite-time thermodynamics' was born and thus more realistic limits have been adapted for real engine performances. In recent years, there have been many optimization studies carried out for heat engines based on the endoreversible and irreversible models by considering finite-time and finite-size constraints. The reader interested in these optimization works could refer to a literature survey written by Bejan [4]. In the studies of finite-time thermodynamics for heat engines, most performance analyses have concerned about the maximum power output and corresponding efficiency. Wu [5,6], Chen and Wu [7] and Chen et al [8] have taken specific power output as the optimization criterion. Bejan [4] introduced entropy generation minimisation method. Angulo-Brown [9] proposed an ecological optimization criterion for performance analysis of heat engines. Sahin et al [10-12] carried out optimization studies for reversible and irreversible Joule-Brayton heat engines and for an endoreversible Carnot heat engine using the power density as a new criterion. Wu et al [13] examined the effect of the heat transfer law on the finite-time exergoeconomic performance of heat engines. More recently, Sahin et al [14] carried out an exergy optimization analysis for endoreversible cogeneration plants using finite-time thermodynamics. In this paper, the proposed exergy optimization method is extended by including internal irreversibility effects. In this context, the effects of design parameters and irreversibility on the exergetic performance are investigated.

THE THEORETICAL MODEL AND EXERGY OPTIMIZATION

The considered irreversible cogeneration cycle model and its T-S diagram are shown in Fig. 1. The cogeneration cycle operates between three heat reservoirs at temperatures T_H, T_L and T_K. The temperatures of the working fluids exchanging heat with the reservoirs at T_H, T_L and T_K are T_X, T_Y and T_Z, respectively. In the cycle, $\dot{Q}_H$ is the rate of heat transfer from the heat source at T_H to the warm working fluid at constant T_X in process 2-3, $\dot{Q}_L$ is the rate of heat transfer from the cold working fluid at constant T_Y to the heat sink at T_L in process 6-1, and $\dot{Q}_K$ is the rate of heat transfer from the working fluid at constant T_Z to the heat-consuming device at T_K in process 4-5.

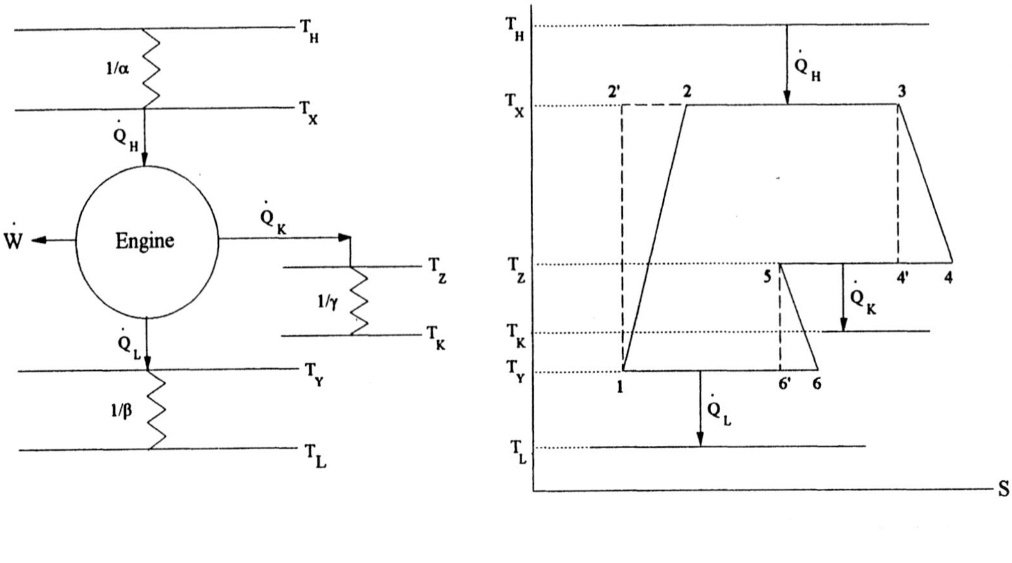

Fig.1. The irreversible cogeneration cycle and its T-S diagram.

The model of Fig. 1 represents cogeneration plants which have pass-out, condensing steam turbines. In these plants, some steam is extracted from the turbine at an intermediate pressure and transferred to a heat consumer for heating. The condensate is then returned to the plant. This operation corresponds to process 4-5 for a closed system, except that heat transfer takes place without mass transfer.

When heat transfer obeys Fourier's law,

$$\dot{Q}_H = \alpha(T_H - T_X), \tag{1}$$

$$\dot{Q}_L = \beta(T_Y - T_L), \tag{2}$$

$$\dot{Q}_K = \gamma(T_Z - T_K), \tag{3}$$

where α, β and γ are the thermal conductances of the heat source, heat sink and heat consumer, respectively. The power output of the cogeneration cycle, according to the first law, is

$$\dot{W} = \dot{Q}_H - \dot{Q}_L - \dot{Q}_K. \tag{4}$$

Substituting Eqs. (1), (2) and (3) into Eq. (4), the power output becomes

$$\dot{W} = \alpha(T_H - T_X) - \beta(T_Y - T_L) - \gamma(T_Z - T_K). \tag{5}$$

The exergy rate of power output is

$$\dot{E}_W = \dot{W}, \tag{6}$$

and the exergy rate for process heat is

$$\dot{E}_Q = \dot{Q}_K(1 - T_L/T_Z), \tag{7}$$

where T_L is the temperature of the environment. The total exergy rate delivered by the cogeneration system is

$$\dot{E}_T = \dot{E}_W + \dot{E}_Q. \tag{8}$$

Using Eqs. (3), (5), (6), and (7) in Eq. (8), it becomes

$$\dot{E}_T = \alpha(T_H - T_X) - \beta(T_Y - T_L) - \gamma(T_Z - T_K)T_L/T_Z. \tag{9}$$

Formally, the second law for irreversible cycle requires

$$\oint \frac{\partial \dot{Q}}{T} = \frac{\dot{Q}_H}{T_X} - \frac{\dot{Q}_L}{T_Y} - \frac{\dot{Q}_K}{T_Z} < 0 \tag{10}$$

One can rewrite the inequality in Eq. (10) as

$$I_{\Delta S}\dot{Q}_H/T_X - \dot{Q}_L/T_Y - \dot{Q}_K/T_Z = 0 \quad \text{with } I_{\Delta S} > 1, \tag{11}$$

where the internal irreversibility parameter $I_{\Delta S}$ defined as

$$I_{\Delta S} = \frac{(S_4 - S_5) + (S_6 - S_1)}{(S_3 - S_2)}. \tag{12}$$

For the cogeneration cycle, $I_{\Delta S}$ characterizes fully the degree of internal irreversibility. We also require that the ratio of power to heat demanded by the heat consumer is known and constant, i.e.

$$\dot{W}/\dot{Q}_K = [\alpha(T_H - T_X) - \beta(T_Y - T_L) - \gamma(T_Z - T_K)]/\gamma(T_Z - T_K) = R. \tag{13}$$

We maximise the total exergy rate in Eq. (9) with respect to T_X, T_Y and T_Z by using the constraints in Eqs. (11) and (13). The results are

$$T_Z^* = \left[T_K + B + C + \sqrt{(T_K + B + C)^2 - 4BT_K}\right]/2, \tag{14}$$

$$T_X^* = [\alpha T_H + \beta T_L - \gamma(1+R)(T_Z^* - T_K)]/[\alpha + \beta\sqrt{T_L/(I_{\Delta S}T_H)}], \tag{15}$$

$$T_Y^* = T_X^*\sqrt{T_L/(I_{\Delta S}T_H)}, \tag{16}$$

where the parameters

$$B = [\alpha T_H + \beta T_L + \gamma(1+R)T_K]/[(1+R)(\alpha I_{\Delta S} + \beta + \gamma)] \tag{17}$$

and

$$C = \alpha\beta\left(\sqrt{T_H} - \sqrt{I_{\Delta S}T_L}\right)^2/[\gamma(1+R)(\alpha I_{\Delta S} + \beta + \gamma)] \tag{18}$$

are defined for convenience. Substituting Eqs. (14), (15) and (16) into Eq. (9), the maximum total exergy rate becomes

$$\dot{E}_{T\max} = \gamma\left(T_Z^* - T_K\right)\left[(1+R)T_Z^* - T_L\right]/T_Z^* . \tag{19}$$

The exergy efficiency at maximum total exergy is

$$\varepsilon^* = \dot{E}_{T\max}/\dot{E}_H^* , \tag{20}$$

where $\dot{E}_H^*$ is the exergy input to the system at maximum exergy output conditions, i.e.

$$\dot{E}_H^* = \alpha\left(T_H - T_X^*\right)\left(1 - T_L/T_H\right) . \tag{21}$$

Discussion and Conclusions

The variation of the total exergy rate given in Eq. (9) with respect to exergy efficiency for various R (= ratio of power output to process heat) is shown in Fig. 2. From the figure, it may be seen that the maximum total exergy rate and exergy efficiency at maximum total exergy decrease as R increases. As R goes to infinity that is as the exergy rate for the process heat ($\dot{E}_Q$) approaches zero and thus the irreversible cogeneration cycle becomes simply an irreversible power cycle. In this case, the total exergy rate is equal to the power output (i.e. $\dot{E}_T = \dot{E}_W = \dot{W}$). The limit of the maximum total exergy rate in Eq. (19) as $R \rightarrow \infty$ is

$$\dot{E}_{T\max} = \dot{W}_{\max} = \left[\alpha\beta\left(\sqrt{T_H} - \sqrt{I_{\Delta S}T_L}\right)^2\right]/\left(\alpha I_{\Delta S} + \beta\right) \tag{22}$$

and that the exergy efficiency at maximum total exergy in Eq. (20) as $R \rightarrow \infty$ is

$$\varepsilon^* = \left(1 - \sqrt{I_{\Delta S}\tau}\right)/\left(1 - \tau\right), \tag{23}$$

where $\tau = T_L/T_H$. Since the relation between energy and exergy efficiencies is

$$\eta = \varepsilon\left(1 - \tau\right), \tag{24}$$

the energy efficiency at maximum total exergy for $R \rightarrow \infty$ becomes

$$\eta^* = 1 - \sqrt{I_{\Delta S}\tau}\,. \tag{25}$$

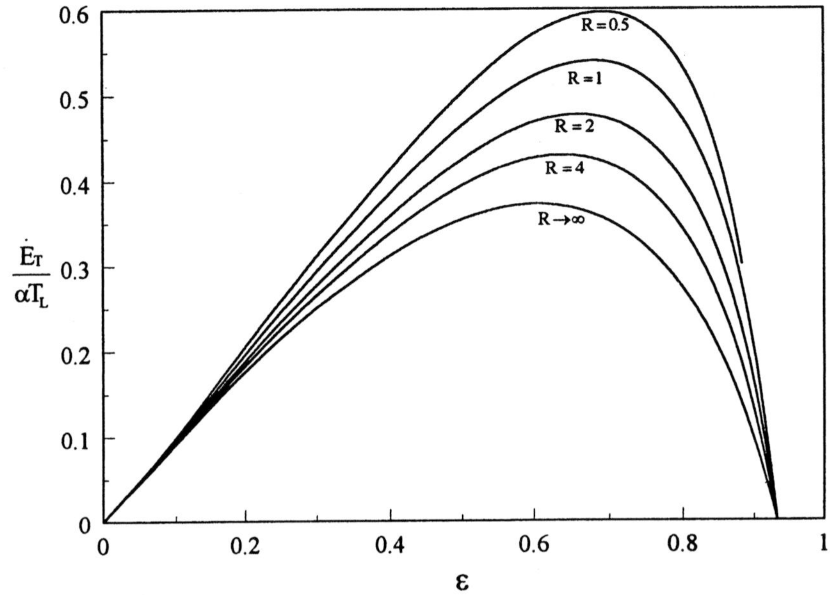

Fig. 2. Variation of the total exergy rate with respect to exergy efficiency for various R; $\psi = T_K/T_L = 1.67$, $\tau = T_L/T_H = 0.25$, $I_{\Delta S} = 1.2$ and $\alpha = \beta = \gamma$.

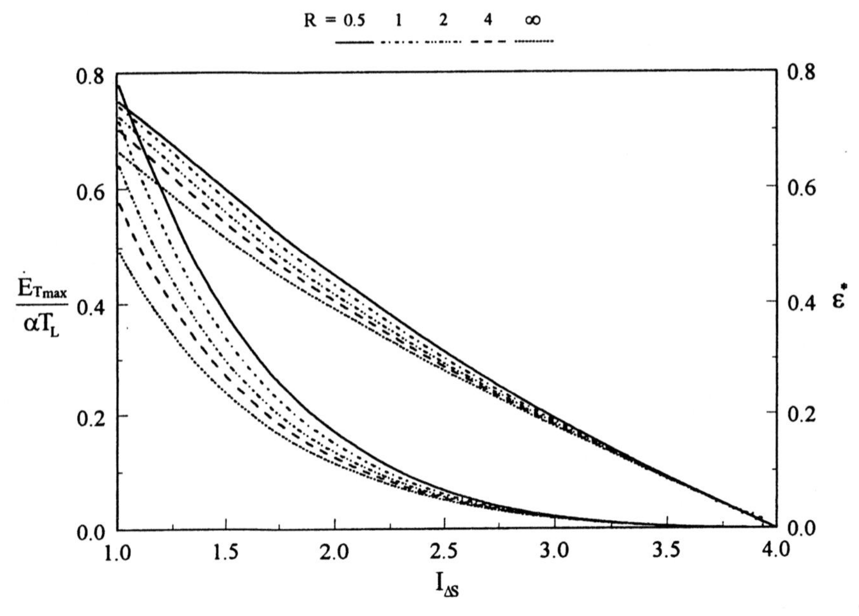

Fig. 3. Variations of maximum total exergy rate (lower family of curves) and exergy efficiency at maximum total exergy (upper family of curves) with respect to $I_{\Delta S}$ for various R; $\psi = 1.67$, $\tau = 0.25$ and $\alpha = \beta = \gamma$.

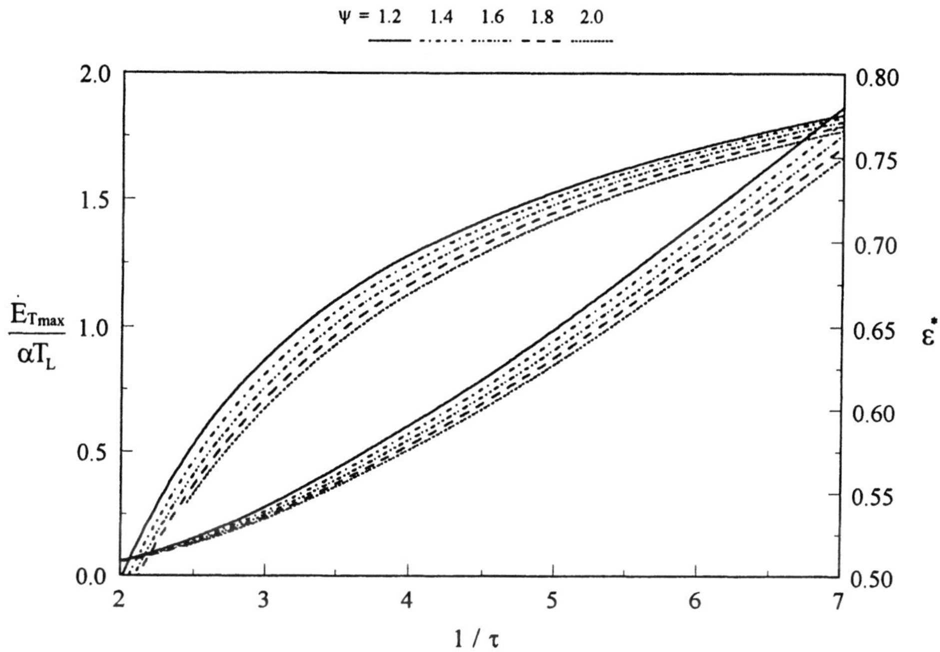

Fig. 4. Effects of τ and ψ on maximum total exergy rate (lower family of curves) and exergy efficiency at maximum total exergy (upper family of curves); $I_{\Delta S}$= 1.2, R= 1 and $\alpha = \beta = \gamma$.

This efficiency is the same as that given in Ref. 15 for an irreversible power cycle. Thus, the irreversible power cycle model is a special case of the irreversible cogeneration cycle for $R \rightarrow \infty$.

The effect of the internal irreversibility parameter ($I_{\Delta S}$) on the maximum total exergy rate ($\dot{E}_{T\max}$) and exergy efficiency at maximum total exergy (ε^*) can be seen in Fig. 3. For increasing $I_{\Delta S}$, both the maximum total exergy rate and the exergy efficiency at maximum total exergy decrease and become zero for $I_{\Delta S}=1/\tau$. It is also seen from the figure that the effect of R on the optimal exergetic performance is greater for lower $I_{\Delta S}$ values. In Fig. 4, the effect of the extreme temperature ratio (τ) on $\dot{E}_{T\max}$ and ε^* can be seen for various heat-consumer temperature parameter ($\psi = T_K/T_L$) values. Although both $\dot{E}_{T\max}$ and ε^* decrease for increasing τ and ψ, as expected. The reason of this decrease for increasing ψ is that the temperature T_Z^* at maximum total exergy given in Eq. (14) increases for increasing T_K, which is the heat-consumer temperature. The increase in T_Z^* causes the irreversible expansion work of the heat engine and thus the exergetic performance of the cogeneration system to decrease. Therefore, the heat-consumer temperature must be kept low to decrease T_Z^* in order to obtain high exergetic performance from the cogeneration system. This conclusion is consistent with the first law analysis carried out for cogeneration plants [16].

In order to determine optimal thermal conductances which are measures of heat exchanger sizes, we assume that total thermal conductance is constrained, i.e.

$$\alpha + \beta + \gamma = \lambda = \text{constant}, \tag{26}$$

and define the thermal conductance allocation ratios x, y and z for heat-source, heat-sink and heat-consumer sides, respectively, as

$$\alpha = x\lambda, \ \beta = y\lambda, \ \gamma = z\lambda, \tag{27}$$

where $z = 1 - x - y$ from Eq. (26). The maximum total exergy rate given in Eq. (19) can be optimised with respect to x and y by using Eq. (27) in Eqs. (14)-(19). This optimization has been carried out numerically and the results are submitted in Fig. 5 and Fig. 6. The variations of the optimal allocation ratios with respect to R, ψ, τ and $I_{\Delta S}$ are given in Fig. 6. From the figure, we see that the optimum conductance-allocation ratio on the heat-source side, x^* stays approximately constant for varying R, ψ and τ, however it decreases gradually for increasing $I_{\Delta S}$. On the other hand, the optimum conductance-allocation ratio on the heat-sink side, y^* increases and the one on the heat-consumer side, z^* decreases for increasing R, ψ, τ, and $I_{\Delta S}$.

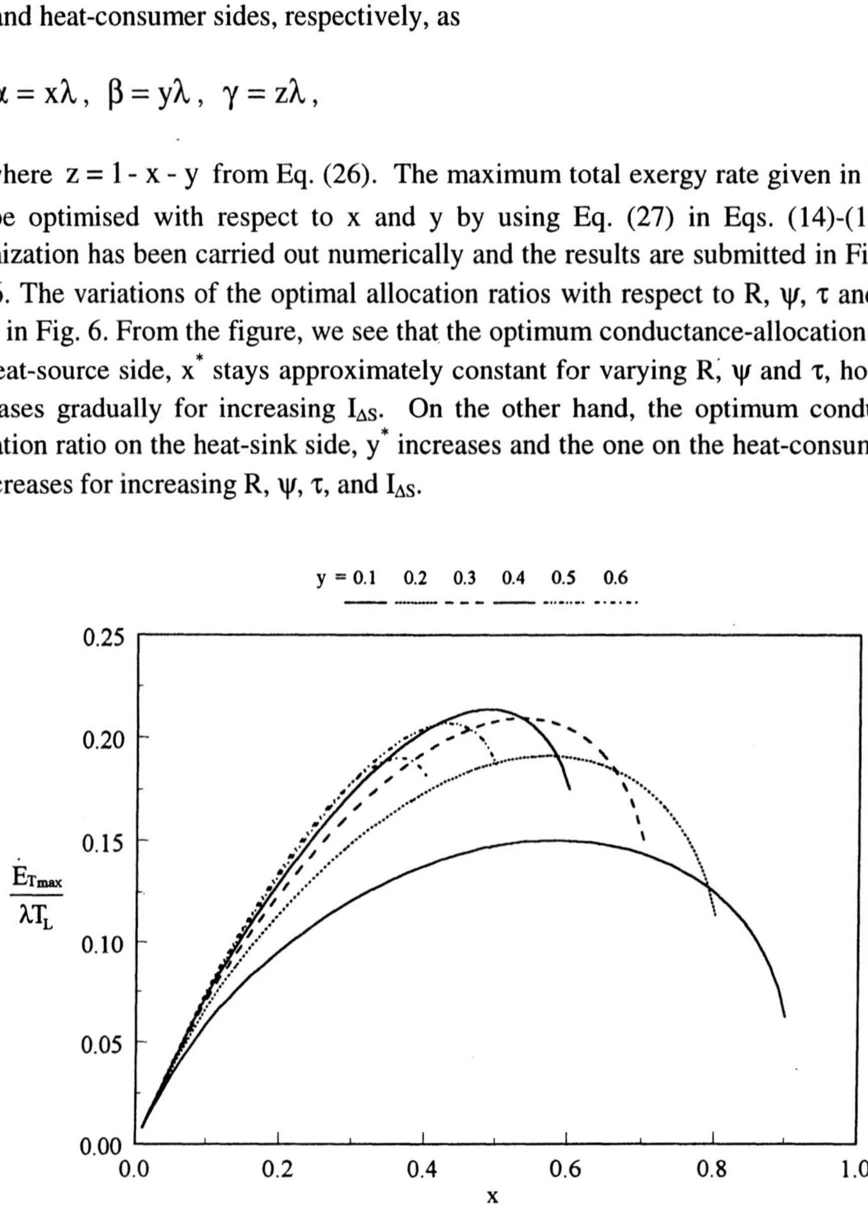

Fig. 5. Variation of maximum total exergy rate with respect to the allocation ratios x and y; $I_{\Delta S}$= 1.2, τ = 0.25, ψ = 1.67 and R = 1.

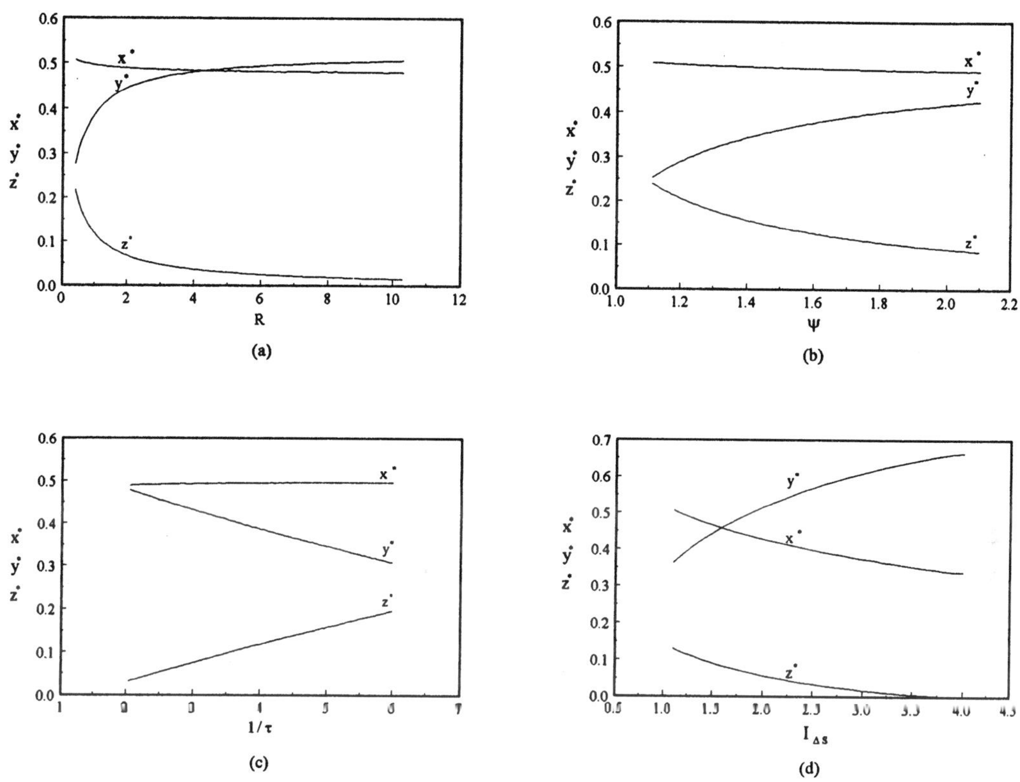

Fig. 6. Variations of the optimal allocation ratios with respect to R for $\tau = 0.25$, $I_{\Delta S}$=1.2 and ψ=1.67 (a), with respect to ψ for R=1, $\tau = 0.25$ and $I_{\Delta S}$= 1.2 (b), with respect to τ for R=1, ψ=1.67 and $I_{\Delta S}$= 1.2 (c), and with respect to $I_{\Delta S}$ for $\tau = 0.25$, ψ=1.67 and R=1.

NOMENCLATURE

$\dot{E}$	exergy rate
$I_{\Delta S}$	internal irreversibility parameter
$\dot{Q}$	rate of heat transfer
R	power to process heat ratio
S	entropy
T	temperature
$\dot{W}$	power generated from the cogeneration system
x, y, z	conductance allocation ratios for the heat-source, heat-sink and heat-consumer sides, respectively

α, β, γ thermal conductances of the heat-source, heat-sink and heat-consumer sides, respectively
λ overall thermal conductance
η energy efficiency
ε exergy efficiency
τ ratio of heat-sink to heat-source temperature
ψ ratio of heat-consumer to heat-sink temperature

Subscripts
H heat source
L heat sink
K heat consumer
max maximum
X warm working fluid
Y cold working fluid
Z process working fluid
T total

Superscripts
* maximum total exergy conditions

REFERENCES

1. P. Chambadal, *Les Centrales Nucleaires*, Armand Colin, Paris (1957).
2. I. I. Novikov, *J. Nucl. Energy II* **7**, 125 (1958).
3. F. L.Curzon and B. Ahlborn, *Am. J. Phys.* **43**, 22 (1975).
4. A. Bejan, *J. Appl. Phys.* **79**, 1191 (1996).
5. C. Wu, *Energy Convers. Mgmt.* **31**, 249 (1991)
6. C. Wu, *Energy Convers. Mgmt.* **34**, 1239 (1993)
7. J. Chen and C. Wu, *Energy Convers. Mgmt.* **37**, 1401 (1996)
8. J. Chen, C. Wu and R. L. Kiang, *Energy Convers. Mgmt.* **37**, 17 (1996)
9. F. Angulo-Brown, *J. Appl. Phys.* **69**, 7465 (1991).
10. B. Sahin, A. Kodal and H. Yavuz, *J. Phys. D: Appl. Phys.* **28**, 1309 (1995).
11. B. Sahin, A. Kodal, T. Yilmaz and H. Yavuz, *J. Phys. D: Appl. Phys.* **29**, 1162 (1996).
12. B. Sahin, A. Kodal and H. Yavuz, *Energy-The International Journal* **21**, 1219 (1996).
13. C. Wu, L. Chen and F. Sun, *Energy-The International Journal* **21**, 1127 (1996).
14. B. Sahin, A. Kodal, I. Ekmekci and T. Yilmaz, *Energy-The International Journal* **22**, 551 (1997).
15. C. Wu and R. L. Kiang, *Energy-The International Journal* **17**, 1173 (1992).
16. M. A. Habib, *Energy-The International Journal* **17**, 485 (1992).

HEATING RATE MAXIMIZATION FOR AN IRREVERSIBLE HEAT PUMP WITH A GENERAL HEAT TRANSFER LAW

ALI KODAL
Department of Aeronautical Engineering, Istanbul Technical University, Maslak, 80626, Istanbul, Turkey

Abstract-The performance of an irreversible heat pump cycle with a general heat transfer law is studied. The irreversibility effects due to finite-rate heat transfer and internal dissipations are considered in the analysis. The relation between the maximum rate of heating and the coefficient of performance, based on a general heat transfer law, is derived analytically. The performance at maximum heating rate is discussed and the effect of internal irreversibility is investigated.

Key Words: Finite-time thermodynamics, Irreversible heat pump, Heating rate maximization, General heat transfer law

INTRODUCTION

The endoreversible models have contributed a significant advance in the development of finite-time thermodynamics. By using these models, the influence of finite-rate heat transfer on the performance of heat engines, refrigerators and heat pumps were investigated [1-8]. In order to represent real thermodynamic systems in a model, other sources of irreversibility besides the irreversibility of finite-rate heat transfer were also included in some of these analyses. The performance studies about heat pumps have been increased in last decade due to recent developments in finite-time thermodynamics. Cheng and Chen [6] studied the performance of an irreversible heat pump by maximizing the coefficient of performance. Chen [7] investigated the optimal heat transfer areas for endoreversible heat pumps. Leff and Teeters [8] noted that there is no natural maximum for the coefficient of performance of an endoreversible heat pump. In this paper, the relation between the maximum rate of heating and the coefficient of performance of an irreversible heat pump cycle, based on a general heat transfer law, is derived analytically and the obtained results are evaluated. The reason for taking general heat transfer law in the analysis is because the irreversible cycles have different optimal configurations and characters under different heat transfer law [9-10].

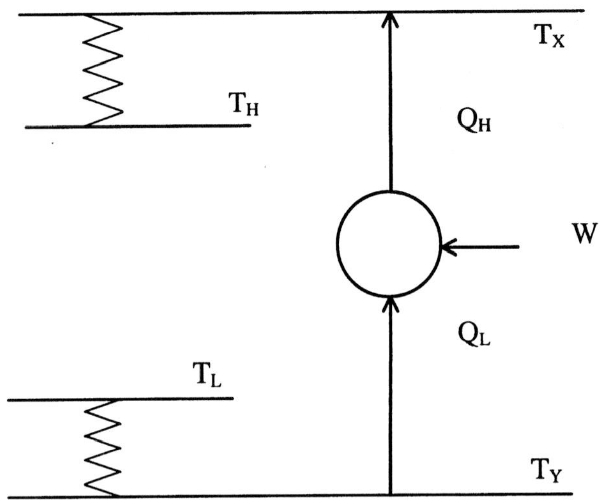

Fig. 1. Irreversible heat pump model.

THE MODEL

The model of the irreversible heat pump is shown in Fig. 1. The cycle operates between the high temperature T_H and the low temperature T_L. The temperatures of the working material exchanging heat with the reservoirs at T_H and T_L are T_X and T_Y, respectively. The heat Q_H released to the high temperature reservoir and the heat Q_L absorbed from the low temperature reservoir by the working material per cycle are assumed to satisfy the following relations:

$$Q_H = \alpha\left(T_X^n - T_H^n\right)t_H \tag{1}$$

and

$$Q_L = \beta\left(T_L^n - T_Y^n\right)t_L \tag{2}$$

where α and β are the heat conductances between the working material and the two heat reservoirs, and t_H and t_L are the times of releasing and absorbing heat processes, respectively. The cycle time is given by

$$t = t_H + t_L . \tag{3}$$

The non-zero integer n shows the type heat transfer law. When n<0, α and β are negative. Particularly, n=1 represents Newton's law, n=–1 stands for linear transfer law in irreversible thermodynamics and n=4 stands for the thermal radiation law.

An irreversibility parameter to account internal dissipation can be introduced as

$$I = \frac{Q_H/T_X}{Q_L/T_Y}. \tag{4}$$

Equation (4) shows clearly that I is equal to one when the cycle is internally reversible and greater than one when the cycle is internally irreversible. The coefficient of performance of the irreversible heat pump is defined as

$$\varepsilon = \frac{Q_H}{Q_H - Q_L} = \frac{1}{1 - T_Y/IT_X}. \tag{5}$$

Using equations (1)-(4), one can obtain the rate of heating of the irreversible heat pump cycle as

$$R = \frac{Q_H}{t} = \frac{\alpha(T_X^n - T_H^n)}{1 + t_L/t_H} = \frac{\alpha}{\dfrac{1}{T_X^n - T_H^n} + \dfrac{\alpha T_Y}{I\beta T_X(T_L^n - T_Y^n)}}. \tag{6}$$

HEATING RATE MAXIMIZATION

In order to find the relation between the optimal rate of heating and the coefficient of performance equation (5) is written as

$$\frac{T_Y}{IT_X} = 1 - \varepsilon^{-1}. \tag{7}$$

and substitution of equation (7) into equation (6) gives

$$\frac{R}{\alpha T_H^n} = \frac{1}{\dfrac{1}{(T_X/T_H)^n - 1} + \dfrac{c(1-\varepsilon^{-1})}{\tau^n - [(1-\varepsilon^{-1})I(T_X/T_H)]^n}} \tag{8}$$

where $\tau = T_L/T_H$ and $c = \alpha/\beta$. The variation of equation (8) for given τ, c and ε for various I values is shown in Fig. 2. From the figure, the maximum points can clearly be seen. Therefore, one can find these maximum points with respect to T_X for given ε.

By using the extremal condition

$$\left(\partial R/\partial T_X\right)_\varepsilon = 0, \tag{9}$$

one can find

$$\left(\frac{T_X}{T_H}\right)_{opt} = \left[\frac{\tau^n + \sqrt{c\left(1-\varepsilon^{-1}\right)^{n+1} I^n}}{\sqrt{c\left(1-\varepsilon^{-1}\right)^{n+1} I^n} + \left(1-\varepsilon^{-1}\right)^n I^n}\right]^{1/n} \tag{10}$$

and using equation (7)

$$\left(T_Y\right)_{opt} = \left(1-\varepsilon^{-1}\right) I \left(T_X\right)_{opt}. \tag{11}$$

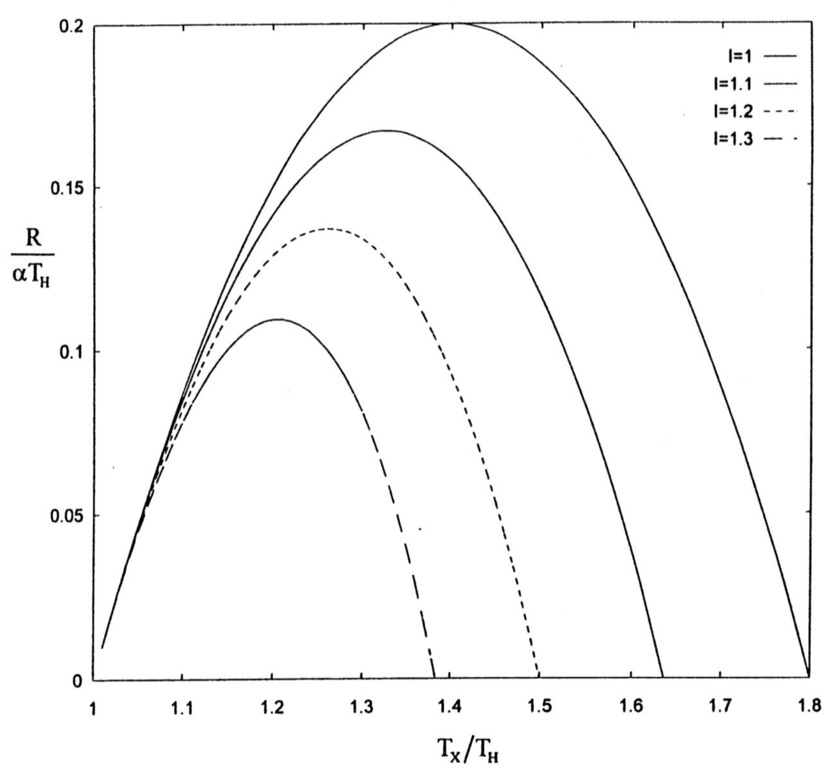

Fig. 2. Variation of R with T_X for n=1 with different I values, (τ=0.9, ε=2, c=1).

Substitution of equation (10) into equation (8) gives the relation between the optimal rate of heating, R and the coefficient of performance, ε i.e.

$$\frac{R_{opt}}{\alpha T_H^n} = \frac{\left\{\left[\tau/\left(1-\varepsilon^{-1}\right)I\right]^n - 1\right\}}{\left[1+\sqrt{c/\left(1-\varepsilon^{-1}\right)^{n-1} I^n}\right]^2}. \tag{12}$$

Since

$$\left(\partial\varepsilon/\partial T_X\right)_R = -\left(\partial R/\partial T_X\right)_\varepsilon / \left(\partial R/\partial\varepsilon\right)_{T_X}, \tag{13}$$

and the condition $\left(\partial\varepsilon/\partial T_X\right)_R = 0$ corresponds to equation (9) under the circumstances $\left(\partial R/\partial\varepsilon\right)_{T_X} \neq 0$. Thus equation (12) also determines the relation between the optimal coefficient of performance and the rate of heating. Particularly, for the heating transfer obeying Newton's law, that is when n=1 equation (12) becomes

$$\frac{R_{opt}}{\alpha T_H} = \frac{\beta}{\left[\sqrt{\alpha}+\sqrt{I\beta}\right]^2}\left[\frac{\varepsilon\tau}{\varepsilon-1} - I\right], \tag{14}$$

or

$$\varepsilon_{opt} = 1/\left\{1-\tau/\left[\left(R/\alpha T_H\right)\left(\sqrt{c}+\sqrt{I}\right)^2 + I\right]\right\}. \tag{15}$$

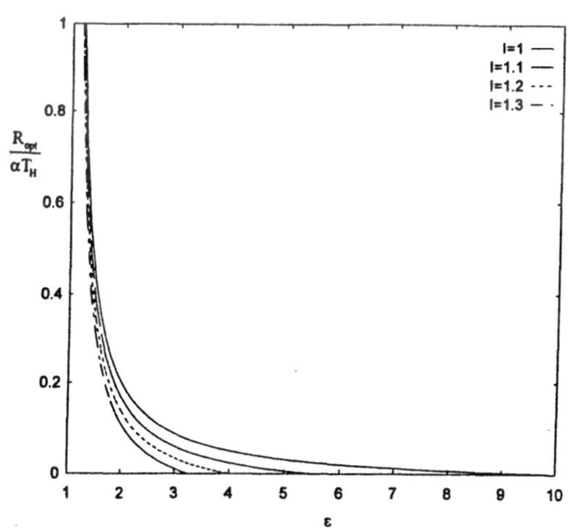

Fig. 3. Variation of R_{opt} with ε for n=1 with different I values, (τ=0.9, c=1).

The relation between R and ε determined by equation (14) or (15) is shown in Fig. 3. As it can be seen from the figure, R decreases monotonically as ε increases. When R=0 in equation (15), we can easily find $\varepsilon_{max}=1/(1-\tau/I)$ and this becomes the coefficient of performance of a reversible Carnot heat pump when I=1. Similarly if we let ε=1 in equation (14), we see that R becomes infinite as it is also be seen from Fig. 3. In such a case, heat pump cycle cannot be carried out. Under the circumstances of a given rate of heating, the bound of the coefficient of performance of an irreversible heat pump can be determined by using equation (15). In case n=−1, from equation (12) we have

$$\frac{R_{opt}}{\alpha T_H^{-1}} = \left[\left(1-\varepsilon^{-1}\right)I/\tau - 1\right] \Big/ \left[1+\left(1-\varepsilon^{-1}\right)\sqrt{Ic}\right]^2, \tag{16}$$

where in this case α and β are called kinetic coefficients and are negative. In this equation when R=0 we again have $\varepsilon_{max}=1/(1-\tau/I)$ but when we set ε=1 in equation (16), R this time is finite that is $R_{max}=-\alpha/T_H$. The relation between R and ε given by equation (16) is shown in Fig. 4. From the figure, the limits and the effect of internal irreversibility parameter can be seen clearly. The case n=4 is a model for heat transfer by thermal radiation. When n=4, from equation (12) we can obtain

$$\frac{R_{opt}}{\alpha T_H^4} = \frac{\left\{\left[\tau/\left(1-\varepsilon^{-1}\right)I\right]^4 - 1\right\}}{\left[1+\sqrt{c/\left(1-\varepsilon^{-1}\right)^3/I^2}\right]^2}. \tag{17}$$

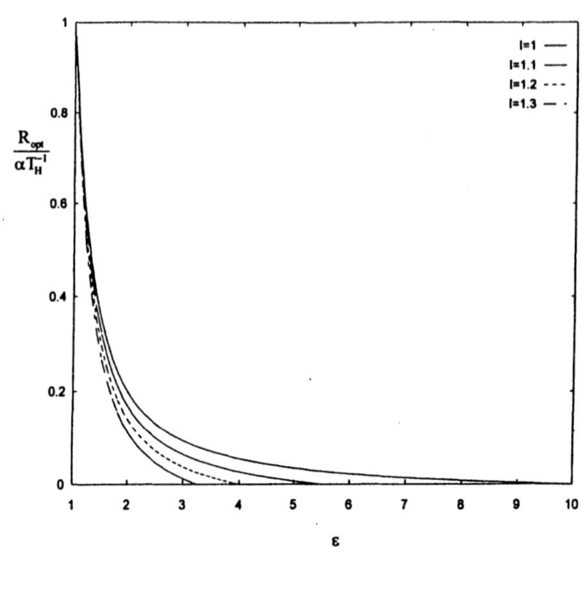

Fig. 4. Variation of R_{opt} with ε for n=−1 with different I values, (τ=0.9, c=1).

In equation (17), when R=0 we still have $\varepsilon_{max}=1/(1-\tau/I)$ and when we set $\varepsilon=1$ in equation (17), R becomes infinite. Equation (17) similar to the case when n=1, shows that the optimal heating rate of heat pump is also a monotonically decreasing function of the coefficient of performance ε.

CONCLUSION

The relation between the maximum rate of heating and the coefficient of performance is derived, based on a general heat transfer law for an irreversible heat pump cycle. Optimal performance is discussed and the effect of internal irreversibility is presented in terms of analytical equations and plots. It was seen that the optimal coefficient of performance always decreases as the rate of heating increases. In another words, the optimal heating rate decreases as the coefficient of performance increases and it attains its maximum value $\varepsilon_{max}=1/(1-\tau/I)$ when R=0 and this becomes the coefficient of performance of a reversible Carnot heat pump when I=1, i.e when the cycle is internally reversible. When the coefficient of performance is unity, the heating rate is infinite for n>0, but it has a finite value for n=–1. Although the heat pump cycles show different characteristics under different heat transfer law, the optimal coefficient of performance always decreases as the rate of heating increases. The above conclusions and mentioned limits using finite-time thermodynamics can provide some new analytical bases for the design of heat pumps with better performances.

REFERENCES

[1] Curzon F L and Ahlborn B 1975 *Am. J. Phys.* **43** 22

[2] Bejan A 1996 *J. Appl. Phys.* **79** 1191

[3] Sahin B, Kodal A and Yavuz H 1996 *Energy-The International Journal* **21** 1219

[4] Sahin B, Kodal A, Ekmekci I and Yilmaz T 1997 *Energy-The International Journal* **22** 551

[5] Yan Z and Chen J 1990 *J. Phys. D: Appl. Phys.* **23** 136

[6] Cheng C-Y and Chen C-K 1995 *J. Phys. D: Appl. Phys.* **28** 2451

[7] Chen J 1994 *Energy-The International Journal* **19** 1031

[8] Leff H S and Teeters W O 1978 Am. J. Phys. **46** 19

[9] Vos A D 1985 *Am. J. Phys.* **53** 570

[10] Chen L and Yan Z 1989 *J. Chem. Phys.* **90** 3740

OPTIMAL PERFORMANCE OF AN IRREVERSIBLE CARNOT ENGINE FOR ANOTHER LINEAR HEAT TRANSFER LAW

ZIJUN YAN AND LIXUAN CHEN
Department of Physics, Xiamen University,
Xiamen 361005, People's Republic of China

Abstract- Based on another linear heat transfer law instead of Newton's law, the optimal performance of a Carnot engine under the influence of thermal resistance, heat leak and other irreversibilities is studied, and the fundamental optimal relation of the engine is derived. From the relation, the bounds of power output and efficiency of the engine are determined and some new characteristics of the engine are expounded. The results obtained here may provide some new theoretical guidance for the optimal design and the selection of optimal operating condition of a real heat engine.

Key Words: Irreversible Carnot heat engine, fundamental optimal relation, performance bound, heat transfer law

NOMENCLATURE

C_i = coefficient of heat leak
I = degree of internal irreversibility
P = power output
Q = quantity of heat
q = heat flux
S = entropy
T = temperature
t = time for heat transfer
α = thermal conductance between working substance and hot reservoir
β = thermal conductance between working substance and cold reservoir
γ = ratio between cyclic time and heat transfer time
η = efficiency
τ = cyclic time

1. Introduction

In recent years, the fundamental optimal relation and some important performance bounds of an irreversible Carnot engine under the influence of thermal resistance, heat leak and other irreversibilities are deduced, which reflect the observed performance of some real heat engines quite well[1-6]. However, in these researches, the results derived are mainly based on Newton's heat transfer law. In fact, there are other heat transfer laws, such as the phenomenological law of irreversible thermodynamics

$$q \propto \Delta(1/T) \tag{1}$$

(where q is the heat flux and T is the absolute temperature), Stefan-Boltzmann thermal radiation law, and so on. It has been found that the performances of an endoreversible Carnot cycle are quite different for different heat transfer laws[7-11]. For example, Orlov's result pointed out that for the heat transfer law

$$q = \alpha\Delta(1/T) + \beta[\Delta(1/T)]^9 \tag{2}$$

(where $\alpha > 0$ and $\beta > 0$ are constants) the optimal cyclic configuration of a two-heat-source endoreversible cycle is no longer Carnot's form, but consists of three isothermal and three adiabatic processes[8]. Besides, some researchers show that it is better to use Dulong-Petit law

$$q = \alpha(\Delta T)^n \tag{3}$$

when the combined effect of conduction, convection and radiation is considered, where n is an exponent with value between 1.1 and 1.6[9]. Since there exist different heat transfer laws in different cases and the performances of a cycle are closely related to the heat transfer law, Thus, it is necessary to extend the study based on other heat transfer laws.

In this paper, based on heat transfer Equation (1), the model of an irreversible Carnot cycle under the influence of thermal resistance, heat leak and other irreversibilities is established, and the fundamental optimal relation of the power output and efficiency of the cycle is determined. At the same time, two new finite-time thermodynamic criteria for the selection of optimal operating parameters of a real heat engine are given. Besides, the effect of various irreversibilities and heat transfer law on the optimal performance of a heat engine is discussed, and some more satisfactory results are obtained.

2. Cycle Model

Consider a class of irreversible Carnot cycles shown in Figure 1, which satisfy the following four conditions:

(1) The cycle consists of two isothermal and two adiabatic processes. The temperatures of the working substance in the hot and cold isothermal processes are, respectively, T_1 and T_2, and the times of the two isothermal processes are, respectively, t_1 and t_2.

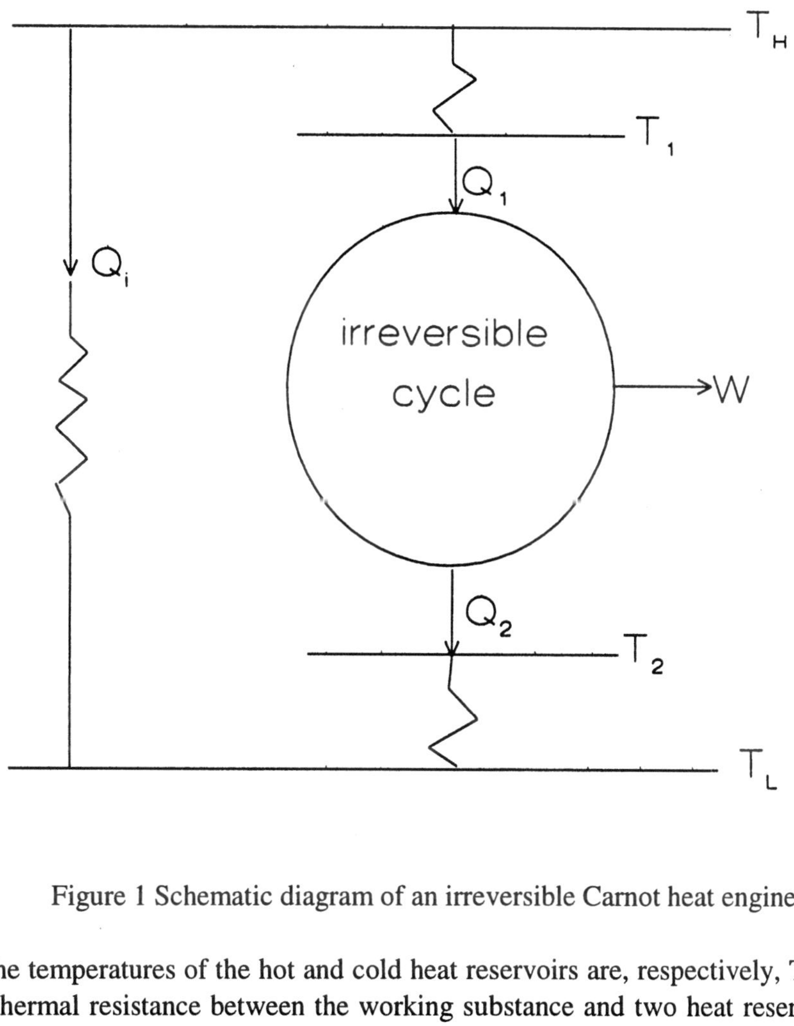

Figure 1 Schematic diagram of an irreversible Carnot heat engine

(2) The temperatures of the hot and cold heat reservoirs are, respectively, T_H and T_L. There is thermal resistance between the working substance and two heat reservoirs, and all heat transfer is assumed to be Equation (1). Therefore, the heat Q_1 absorbed from the hot reservoir and the heat Q_2 released to the cold reservoir by the working substance in the two isothermal processes can be written as

$$Q_1 = \alpha\left(\frac{1}{T_1} - \frac{1}{T_H}\right)t_1, \tag{4}$$

$$Q_2 = \beta\left(\frac{1}{T_L} - \frac{1}{T_2}\right)t_2, \tag{5}$$

where α and β are, respectively, the thermal conductance between the working substance and the hot and cold heat reservoirs.

(3) There exists heat leak between the two reservoirs. The heat Q_i leaked from the hot reservoir to the cold reservoir per cycle is giving by

$$Q_i = C_i(\frac{1}{T_L} - \frac{1}{T_H})\tau , \quad (6)$$

where C_i is the coefficient of heat leak, and τ is the cyclic time.

(4) Besides thermal resistance and heat leak, there are other irreversibilities, such as friction, eddy and other irreversible effects inside the cycle. Because the total effect of the other irreversibilities can be characterized by increasing the change in entropy of the cycle, we can introduce a parameter—the degree of internal irreversibility

$$I \equiv \Delta S_{out} / \Delta S_{in} = (Q_2 / T_2)/(Q_1 / T_1) \geq 1 \quad (7)$$

to describe the other irreversibilities, where ΔS_{out} and ΔS_{in} are, respectively, the entropy flows flowing out and into the working substance in the isothermal processes T_2 and T_1 per cycle. And, $\Delta S_{out} = \Delta S_{in} + \Delta S_i$, as shown in Figure 2, where ΔS_i is the entropy production of the cycle which is caused by various internal irreversibilities inside the working substance. According to the second law of thermodynamics, the entropy production in a cycle is always positive for a cycle with internal irreversibilitiy, so that $I > 1$, but for endoreversible cycles, $I = 1$.

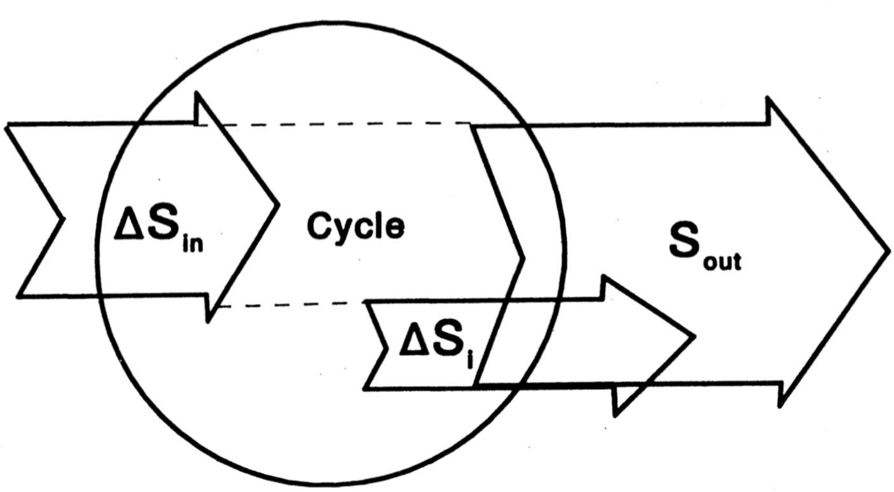

Figure 2 Schematic diagram of the entropy flows in an irreversible Carnot engine

In such a model, consideration is taken not only the thermal resistance but also the heat leak and other irreversibilities, and heat transfer law is based on Equation (1). Therefore, when heat transfer law is obeying Equation (1), some results which may reflect better various optimal performances of a real heat engine can be derived based on this model.

3. THE FUNDAMENTAL OPTIMAL RELATION

Using the cycle model described above, the total heat Q_H released from the hot reservoir and the total heat Q_L transferred to the cold reservoir are, respectively,

$$Q_H = Q_1 + Q_i \, , \tag{8}$$

$$Q_L = Q_2 + Q_i \, . \tag{9}$$

According to Equations (7)-(9), the efficiency of the cycle can be written as

$$\eta = \frac{Q_H - Q_L}{Q_H} = \frac{Q_1 - Q_2}{Q_H} = \frac{Q_1}{Q_H}(1 - \frac{Q_2}{Q_1}) = \frac{Q_1}{Q_1 + Q_i}\eta_0 = \frac{Q_1 / \tau}{Q_1 / \tau + Q_i / \tau}\eta_0 \, , \tag{10}$$

where

$$\eta_0 = 1 - Q_2 / Q_1 = 1 - IT_2 / T_1 \tag{11}$$

is the efficiency of the cycle without heat leak. From Equations (10) and (11), the power output of the cycle can be expressed as

$$P = (Q_1 / \tau)\eta_0 = [(Q_1 + Q_i) / \tau]\eta \tag{12}$$

Then, from Equations (4), (5), (7) and (11), one can get

$$\frac{Q_1}{\tau} = \frac{Q_1}{\gamma(t_1 + t_2)} = \frac{\alpha}{\gamma}[(\frac{1 - \eta_0}{IT_2} - \frac{1}{T_H})^{-1} + \frac{\alpha}{\beta}(1 - \eta_0)(\frac{1}{T_L} - \frac{1}{T_2})^{-1}]^{-1}, \tag{13}$$

where $\gamma = \tau / (t_1 + t_2)$. In general, the times of the two adiabatic processes are considered negligible, thus we assume $\gamma = 1$ [2,7].

From Equation (13) and the extreme condition $\partial\eta_0 / \partial T_2 = 0$, the optimal relation between η_0 and Q_1 / τ

$$\frac{Q_1}{\tau} = \frac{\alpha}{\gamma} \frac{[(1-\eta_0)(IT_L)^{-1} - T_H^{-1}]}{[1+\sqrt{\alpha/(I\beta)}(1-\eta_0)]^2} \tag{14}$$

can be got. Eliminating η_0 from Equations (10) and (14), and noticing $P = \eta Q_H / \tau$ and $\gamma = 1$, we can obtain the optimal relation between the efficiency η and power output P of the cycle, i.e., the fundamental optimal relation

$$\begin{aligned}&[(1+\delta-\delta\eta)P-(1+\delta)C_i(\frac{1}{T_L}-\frac{1}{T_H})\eta]^2\\&-\alpha[(\frac{1-\eta}{IT_L}-\frac{1}{T_H})P-C_i(\frac{1}{T_L}-\frac{1}{T_H})(\frac{1}{IT_L}-\frac{1}{T_H})\eta]\eta=0\end{aligned}, \tag{15}$$

where $\delta = \sqrt{\alpha/(I\beta)}$.

4. The Bounds of Power Output and Efficiency

Equation (15) shows that the characteristic curve of power P against efficiency η for an irreversible Carnot engine based on heat transfer Equation (1) is also a loop-line passing through the zero point as that based on Newton's heat transfer law shown as curve I in Figure 3. Therefore, there are also a maximum power point (power output is $P_{\max}$ and the correspondent efficiency is η_m) and a maximum efficiency point (efficiency is $\eta_{\max}$ and the correspondent power output is P_m)[3]. And, $\eta_{\max}$ is different from Carnot efficiency η_c ,and, in general, $\eta_{\max}$ is far below η_c but is in close proximity to η_m. However, the four parameters $\eta_{\max}$, η_m, $P_{\max}$ and P_m are all different from those derived based on Newton's law (see below),which mirror the observed performance of real heat engines with different heat transfer laws having different characteristics quite well.

Evidently, the efficiency η can not exceed the maximum efficiency $\eta_{\max}$ and the power output P can not exceed the maximum power output $P_{\max}$ for a real heat engine. But, it is not reasonable if $\eta < \eta_m$ or $P < P_m$, because in such a case, one can increase both power and efficiency. Therefore, the rational regions of the operating states should be set between the maximum power point and the maximum efficiency point, shown as the arc AB in Figure 3. Thus, the four parameters $\eta_{\max}$, η_m, $P_{\max}$ and P_m determine the upper and lower bounds of the efficiency and power output of a real heat engine. That is, the efficiency η and power output P of the engine should be satisfied the following two equations

$$\eta_m \le \eta \le \eta_{\max} , \tag{16}$$

$$P_m \le P \le P_{\max} , \tag{17}$$

respectively. Equation (16) and (17) provide two new criteria for the selection of operating conditions of an irreversible Carnot engine, and η_{max} , η_m, P_{max} and P_m are four important performance parameters. They have an important guiding significance for the optimal design and the selection of optimal operating states of a real heat engine.

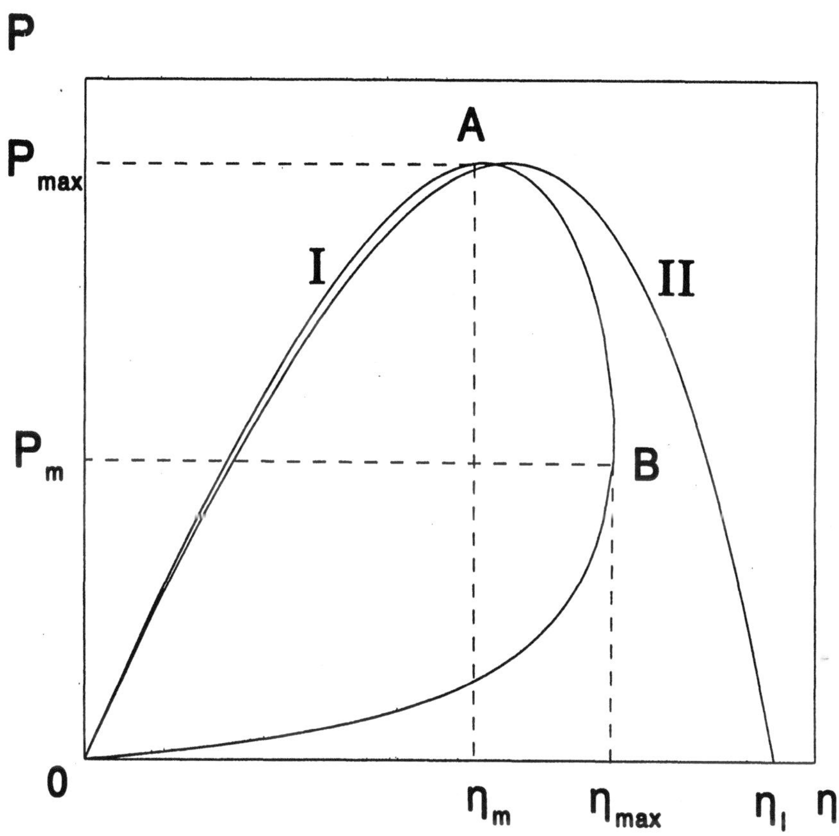

Figure 3 The power output versus efficiency curves of an irreversible Carnot heat engine.

Using the fundamental optimal relation Equation (15), the four parameters can be derived as follows:

$$P_{max} = \eta_I^2 / (4D), \tag{18}$$

$$P_m = \frac{C_i(T_L^{-1} - T_H^{-1})(1+\delta)\sqrt{1+\rho\eta_I / [DC_i(T_L^{-1} - T_H^{-1})]}\eta_{max}}{1+\delta-\delta\eta_{max}}, \tag{19}$$

$$\eta_{\max} = \frac{\eta_I - 2DC_i(T_L^{-1} - T_H^{-1})\{\sqrt{1 + \rho\eta_I / [DC_i(T_L^{-1} - T_H^{-1})]} - 1\}}{1 + 4DC_i(T_L^{-1} - T_H^{-1})(1 + \delta^{-1})}, \tag{20}$$

$$\eta_m = \eta_I / \{1 + \rho + 4DC_i(T_L^{-1} - T_H^{-1})[1 + (1 + \rho)^{-1}] / \eta_I\}, \tag{21}$$

where $\eta_I = 1 - IT_L / T_H$, $D = (1 + \delta)(1 + \delta IT_L / T_H)IT_L / \alpha$ and $\rho = (1 + \delta IT_L / T_H) / (1 + \delta)$.

5. DISCUSSION AND CONCLUSION

(1) When $C_i = 0$,i.e., the effect of heat leak is negligible, Equation (15) can be written as

$$P = \frac{\alpha\eta[(1 - \eta) / (IT_L) - 1 / T_H]}{[1 + \delta(1 - \eta)]^2}. \tag{22}$$

Equation (22) is the fundamental optimal relation of an irreversible Carnot engine[5] without heat leak. Its characteristic curve P against η is no longer loop-line, and $P_m = 0$ and $\eta_{\max} = \eta_I$,shown as curve II in Figure 3. Thus, it can be seen that the heat leak produces a distinctive influence on the performance of a heat engine, which changes the P vs η curve from a parabola into a loop. This is because the quantity of heat leak is proportional to the cyclic time τ , and $p = 0$ and $\eta = 0$ when $\tau \to \infty$. Therefore, the longer the period is, the larger the influence of heat leak . On the other hand, the longer the period is, the smaller the temperature differences for heat transfer are, such that the smaller the irreversible losses of heat transfer are. That is, the effect of thermal resistance on the performance of a cycle is more serious when the cyclic time τ is shorter.

(2) When $C_i = 0$ and $I = 1$, there only exists the irreversibility of heat transfer between the working substance and reservoirs, i.e., the cycle is endoreversible, Equation (15) changes into

$$P = \frac{\alpha\eta[(1 - \eta)T_L^{-1} - T_H^{-1}]}{[1 + \sqrt{\alpha / \beta}(1 - \eta)]^2}. \tag{23}$$

Equation (23) is just the fundamental optimal relation of an endoreversible Carnot cycle for the heat transfer law Equation (1), which has been discussed in Reference 7. This shows that under the heat transfer law Equation (1), Equation (15) describes not only the optimal performance of an irreversible Carnot cycle, but also an endoreversible one.

(3) If $\alpha \to \infty$ and $\beta \to \infty$ besides $C_i = 0$ and $I = 1$, Equation (15) can be written as

$$\eta = 1 - T_L / T_H. \tag{24}$$

This shows that the theory of irreversible Carnot cycle established in this paper includes also the theory of reversible Carnot cycle. In a word, it is a quite general theory.

(4) There is a distinctive difference between Equation (15) and the fundamental optimal relation of the irreversible Carnot cycle based on Newton's heat transfer law[3]. Both will be close to each other only when the temperature difference ΔT for heat transfer is sufficiently small. According to the equation

$$\frac{1}{T}-\frac{1}{T+\Delta T}=\frac{\Delta T}{T^2}+\frac{(\Delta T)^2}{T^3}+\ldots\ldots, \tag{25}$$

when $\Delta T << T$, we have

$$\alpha(\frac{1}{T}-\frac{1}{T+\Delta T})\cong\alpha'\Delta T, \tag{26}$$

(where $\alpha' = \alpha / T^2$), i.e., the two heat transfer laws tend to the same[11] . But , in general, one should consider the difference between the two linear heat transfer law.

(5) It must be emphasized that ΔS_{in} and ΔS_{out} in Equation (7) are the entropy flow into and out of the working substance respectively in T_1 and T_2 isothermal processes, but not the entropy changes of the working substance in the two isothermal processes. Some authors have introduced a parameter[12] $i = \Delta S_2 / \Delta S_1 - 1$ to describe the degree of internal irreversibility, where ΔS_1 and ΔS_2 are the entropy changes of the working substance in the two isothermal processes. Obviously, $1+i$ is different from I $(= \Delta S_{out} / \Delta S_{in})$ defined in this paper. So long as the two adiabatic processes of the cycle are reversible, there is $i = 0$ whether the two isothermal processes are reversible or not. Thus the parameter i can not describe the internal irreversibilities of the two isothermal processes of the cycle, but I can.

ACKNOWLEDGMENT

This work was supported by the Natural Science Foundation of Fujian , China.

REFERENCES

1. Z. Yan, *Journal of Engineering for Thermal Energy and Power*, 1994, **9**, 369.
2. J. Chen, *Journal of Physical D: Applied Physics, 1994,* **27,** 1144.
3. Z. Yan and L. Chen, *Journal of Physics A: Mathematics and General, 1995,* **28** 6167.
4. A. Bejan , *International Journal of Heat and Mass Transfer,* 1988, **31,** 1211.
5. C. Wu and R. L. Kiang, *EnergyThe International Journal,* 1992, **17,** 1173.
6. J. M. Gordon and M. Huleihil, *Journal of Applied Physics,* 1992, **72,** 829.
7. L. Chen, and Z. Yan, *Journal of Chemical Physics,* 1989, **90,** 3740.

8. V. N. Orlov, *Soviet Physics Dokl,* 1985, **30,** 506.
9. Angulo-Brown F and Paez-Hernandez R., *Journal of Applied Physics,* 1993, **74,** 2216.
10. A.De Vos, *American Journal of Physics,* 1985, **53,** 570.
11. J. Chen and Z. Yan, *Journal of Physics D: Applied Physics,* 1993, **26,** 1581.
12. O. M. Ibrabim, S. A. Klein and J. W. Mitchell, *Journal of Engineering for Gas Turbines and Power.* 1991, **113,** 514.

THERMODYNAMIC MODELING OF VAPOUR ABSORPTION COOLING SYSTEMS

N.E. WIJEYSUNDERA
Department of Mechanical and Production Engineering
National University of Singapore
10 Kent Ridge Crescent
Singapore 119260

Abstract- Thermodynamic modeling of vapour absorption cooling systems using the three-heat-reservoir concept is described in this paper. The external heat interaction between the heat reservoirs and the working fluid is represented by a linear heat transfer relationship. The analysis includes a general model that considers both external and internal irreversibilities of the cycle from which the results for the endoreversible and fully-reversible models are deduced. For the endoreversible model expressions are obtained for the maximum cooling capacity and the COP at maximum cooling capacity. The predictions of the models are compared with data from detailed simulations of single-stage and double-stage absorption machines. The endoreversible models predict the performance trends and limits well. The irreversible models that use semi-empirical internal entropy generation functions reproduce the performance data of real absorption machines well.

Key Words- Thermodynamic modeling, absorption cooling cycles, irreversibilities, entropy generation, performance limits.

INTRODUCTION

In recent years, there has been renewed interest in absorption refrigeration systems as possible alternatives to vapour compression systems due mainly to environmental and energy concerns. Advances in absorption refrigeration technology have increased their economic competitiveness in comparison to compression systems [1]. Design studies of absorption systems are usually performed by detailed simulation of the cycle. Such procedures require the solution of a large number of non-linear equations and detailed data on the properties of the working fluids. Computer models for the simulation of absorption systems have been presented by Grossman et al. [2].

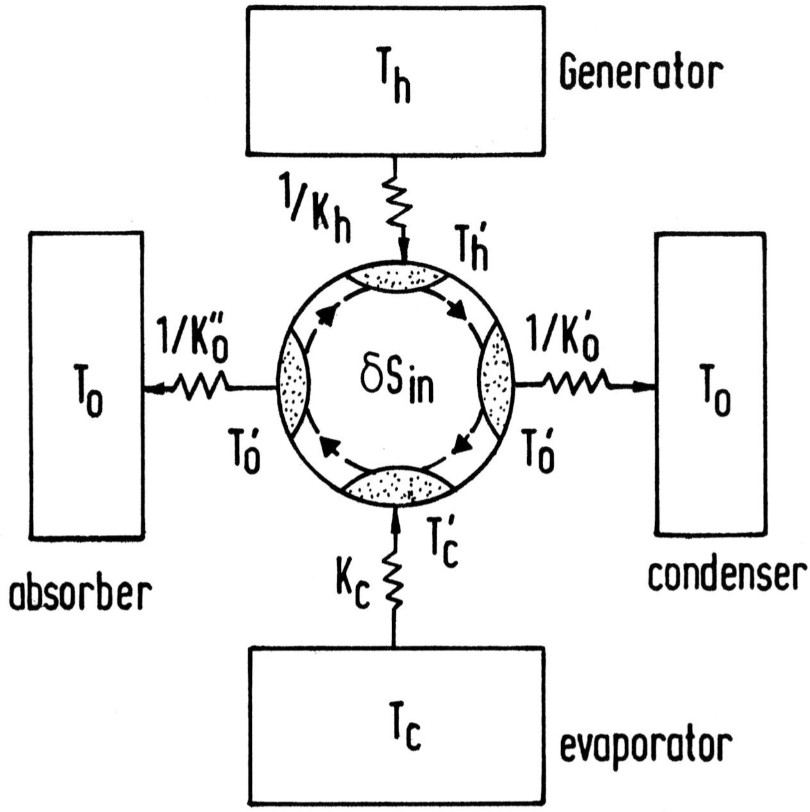

Figure 1. Schematic diagram of the three-heat-reservoir (THR) model of the absorption refrigeration system.

It is useful to develop simplified models for absorption systems that could be used for preliminary design studies to obtain the performance limits of the system. Also such models could be used as sub-component models in larger thermal system simulations where the absorption machine is only a component.

A simple model of the absorption cooling system can be developed by analyzing the three-heat-reservoir (THR) refrigeration cycle where all the processes are taken to be reversible [3]. However, the coefficient of performance (COP) of real absorption cycles are found to be much lower than the COP of the corresponding reversible cycle operating between the same heat reservoirs [4]. Chen and Yan [5] obtained the COP of the absorption cycle by decomposing it into a reversible power cycle and a reversible refrigeration cycle. Yan and Chen [6] presented an analysis of the THR-cycle including external irreversibilities due to heat transfer between the heat reservoirs and the working fluid. Wu [7] studied the optimization of waste-heat driven endoreversible absorption cooling cycles. A detailed comparison of the predictions of the endoreversible THR-cycle with computer simulation results [4] was made by Wijeysundera [8,9]. These studies showed that the endoreversible models predicted the performance limits and trends well.

Bejan , Vargas and Sokolvo [10] studies the optimal allocation of heat exchanger inventory using an endoreversible model of the absorption system.

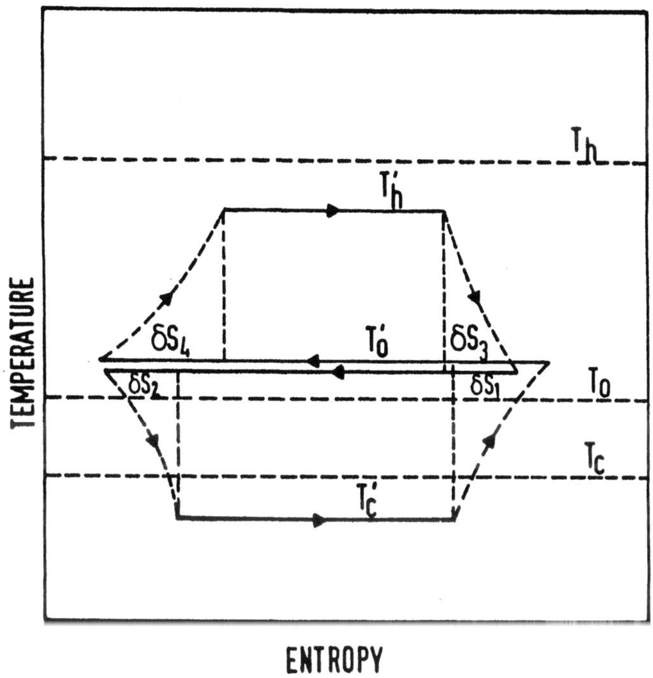

Figure 2. Ideal temperature-entropy diagram of the three-heat-reservoir model of the absorption refrigeration system.

The accurate prediction of the performance of real absorption systems using the THR-model is possible only if internal irrevesibilities of the cycle are taken into account. Gordon and Ng [11] developed a THR-model where the internal irreversibilities were accounted for by including additional heat transfers. Chua et al. [12] have used such a model to optimize the operation of the absorption cycle. Lin and Yan [13] obtained the optimal design conditions for an absorption cycle, which included internal irreversibilities. Wijeysundera [14] presented a semi-empirical model in which internal irreversibilities were characterized by an entropy generation function.

In the present paper the predictions of the various thermodynamic models for absorption cycles are compared with data obtained by detailed computer simulation, with a view to establishing their accuracy and limitations. The THR-models considered include, fully reversible, endoreversible and fully irreversible models of the absorption cooling system. The next section of the paper gives an outline of the different thermodynamic models.

ANALYSIS

In this section of the paper, a general analysis of the THR-cycle including both external and internal irreversibilities is developed. The results for the endoreversible cycle and the fully-reversible cycle are derived from the general model. A schematic diagram of the THR cycle is shown in Fig. 1. Four heat reservoirs are included to represent the generator, the condenser, the evaporator and the absorber of the absorption refrigeration system. In most practical systems, the temperatures of the absorber and the condenser differ only by a small amount. Therefore in the present model, the temperature of the absorber is assumed to be equal to the temperature of the condenser. This helps to make the governing equations considerably simpler. The ideal temperature-entropy diagram for the cycle undergone by the working fluid of the absorption machine is shown Fig. 2. The working fluid exchanges heat with four reservoirs isothermally and there is an external heat transfer irreversibility at each heat reservoir caused by the finite heat conductance between the working fluid and the heat reservoir. For the general model, the working fluid is assumed to undergo a series of irreversible adiabatic processes, which constitute the internal irreversibilities of the cycle. The work input to the cycle is neglected in the present analysis.

Application of the first law to the closed system consisting of the working fluid gives:

$$Q_h + Q_c - Q_o = 0 \tag{1}$$

where Q_o is the total heat transferred at the condenser and absorber reservoirs.

The heat-transfer rates between the heat reservoirs and the working fluid may be expressed in terms of the effective conductances and the temperature differences, namely;

$$Q_h = K_h(T_h - T_h') \tag{2}$$

$$Q_c = K_c(T_c - T_c') \tag{3}$$

$$Q_o = K_o(T_o' - T_o) \tag{4}$$

In equation (4) the working fluid is assumed to have the same temperature during the heat interactions with the absorber and the condenser and K_o is the sum of the effective conductances for these two heat interactions.

Although the heat reservoirs are used as idealizations in the THR model, in the real absorption machine the heat transfer to the working fluid takes place in a heat exchanger coil along which the temperature of the external fluid changes. Therefore, it is necessary to determine the equivalent reservoir temperature and the equivalent conductance that will give the same heat transfer to the working fluid as in the real machine. For this purpose, a single-stream, steady -flow heat exchanger analysis is performed for each component of the real machine. This gives the heat transfer in the generator of the real machine as;

$$Q_h = (mc)_g \varepsilon_g (T_{gi} - T_h') = (mc)_g (T_{gi} - T_{go}) \quad (5)$$

Comparing equations (2) and (5) it is seen that the equivalent reservoir temperature, T_h for the generator is the temperature of the heating fluid at inlet, T_{gi} , while the equivalent conductance is given by $K_h = (mc)_g \varepsilon_g$, where ε_g is the effectiveness of the generator heat exchanger. Similar expressions can be written for the equivalent temperatures and conductances of the other components of the absorption system.

Since the working fluid undergoes the cycle shown in Fig. 2 the sum of the changes of entropy for the different processes is zero. Therefore;

$$\delta s_1 + Q_h / T_h' + \delta s_2 + Q_c / T_c' + \delta s_3 - Q_o / T_o' + \delta s_4 = 0 \quad (6)$$

The total internal entropy generation is given by;

$$\delta s_{in} = \delta s_1 + \delta s_2 + \delta s_3 + \delta s_4 \quad (7)$$

The net entropy transfer to and from the working fluid due to heat transfer can be defined as;

$$\delta s_{tr} = Q_o / T_o' - Q_c / T_c' - Q_h / T_h' \quad (8)$$

For steady operation of the cycle the internal entropy generation must be equal to the net entropy transfer. This gives;

$$\delta s_{in} = \delta s_{tr} \quad (9)$$

The entropy generation due to heat transfer across finite temperature differences at the reservoirs can be written as;

$$\delta s_{ht} = Q_o (1/T_o - 1/T_o') + Q_c (1/T_c' - 1/T_c) + Q_h (1/T_h' - 1/T_h) \quad (10)$$

The above expression applies to the THR-model with heat reservoirs and not to the single-stream heat exchangers of the real machine. Therefore it is useful to compare the above expression with that for a real machine. Considering the generator of the real machine, the entropy generation in the single-stream heat exchanger is obtained as;

$$\delta s_{htg} = Q_h / T_h' - (mc)_g Ln(T_{gi} / T_{go}) \quad (11)$$

where T_{gi} and T_{go} are the inlet and outlet temperatures of the heating fluid in the generator.

Using equation (5), (11) can be written as;

$$\delta s_{htg} = (mc)_g \varepsilon_g (T_{gi} - T_h')/T_h' + (mc)_g Ln[1 - \varepsilon_g (T_{gi} - T_h')/T_{gi}] \quad (12)$$

For real absorption systems the term $\varepsilon_g(T_{gi}-T_h')/T_{gi}$ is much smaller than unity and therefore the second term in (12) may be approximated by a series expansion where only the first order term is retained. This gives the entropy generation in the form;

$$\delta s_{htg} = (mc)_g \varepsilon_g (T_{gi} - T_h')(1/T_h' - 1/T_{gi}) \quad (13)$$

It is interesting to note this expression has the form applicable for a reservoir (equation (10)) and the equivalent reservoir temperature and conductance are the same as those defined earlier based on heat transfer considerations alone.

The total entropy generation in the system is given by;

$$\delta s_{tot} = Q_o / T_o - Q_c / T_c - Q_h / T_h \quad (14)$$

For further analysis of the cycle it is convenient to express the above equations in terms of the following non-dimensional variables;

$$q_c = Q_c /(K_c T_h), \theta_h = T_h'/T_h, \theta_c = T_c'/T_h, \theta_o = T_o'/T_h, \alpha = T_o / T_h, \phi = T_c / T_h \quad (15)$$

$$s = K_o / K_c, r = K_h / K_c, \delta S_i = \delta s_i / K_c (i = in, tr, tot) \quad (16)$$

The working fluid temperatures at the evaporator and the condenser/absorber are obtained by using equations (1) to (5). These are expressed in the following non-dimensional form;

$$\theta_c = \phi - q_c \quad (17)$$

$$\theta_o = \alpha + q_c / s + r / s - r\theta_h / s \quad (18)$$

The analysis developed in this section can be used to derive the two special models. These are given in the next sections.

Endoreversible THR-Model

In the endoreversible THR-model, the working fluid is assumed to undergo reversible processes. The cycle has only heat transfer irreversibilities at the three heat reservoirs.

Therefore for the endoreversible THR-model;

$$\delta s_{in} = \delta s_{tr} = 0 \tag{19}$$

Equation (8) is first expressed in non-dimensional form and the working fluid temperatures at the evaporator and the condenser/absorber are eliminated using equations (17) and (18) to obtain the following expression for the entropy transfer function;

$$\delta S_{tr} = s^2\alpha/(r\theta_h - b_1) - r/\theta_h - a_1 \tag{20}$$

$$a_1 = q_c/(\phi - q_c) - r - s \tag{21}$$

$$b_1 = \alpha s + r + q_c \tag{22}$$

For the endoreversile THR-model, $\delta S_{tr}=0$. Applying this condition to equation (20) the temperature θ_h for a given cooling capacity is obtained as the solution of the equation;

$$a_1 r\theta_h^{\ 2} - (a_1 b_1 + \alpha s^2 - r^2)\theta_h - rb_1 = 0 \tag{23}$$

The COP for the given cooling capacity is computed by substituting in the expression;

$$COP = Q_c/Q_h = q_c/(1-\theta_h) \tag{24}$$

The COP attains the fully-reversible limit (Carnot) when the external heat transfer irreversibilities are zero. The cooing capacity also becomes zero at this limit. The maximum COP is obtained by using equations (1) and (14). This can be expressed as;

$$[COP]_c = \phi(1-\alpha)/(\alpha - \phi) \tag{25}$$

Maximum Cooling Capacity

The variation of the cooling capacity with the temperature ratio θ_h for the endoreversible cycle can be studied by considering equation (23). For a given set of parameters, the cooling capacity becomes zero when $\theta_h=0$ and $\theta_h=(r+s\alpha)/(r+s)$ (<1). When $\theta_h=0$, the cycle approaches the reversible limit, while for the second value there is direct heat transfer between the hot reservoir and the heat sink. Between these two values of θ_h, the cooling capacity reaches a maximum.

The condition for the maximum cooling capacity is obtained by differentiating equation (23) and substituting $dq_c/d\theta_h = 0$. This gives following condition for the maximum cooling capacity;

$$T_o^{'}/T_h^{'} = (T_o/T_h)^{0.5} = \sqrt{\alpha} \tag{26}$$

The maximum cooling capacity is obtained by substituting condition (26) in equation (23). The positive root of the following equation gives the maximum cooling capacity;

$$(c+1)q_{cm}^{\ 2} + (b+bc-c\phi-a^2)q_{cm} + (a^2-bc)\phi = 0 \tag{27}$$

$$c=(r+s), b=(r+s\alpha), a=(r+s\sqrt{\alpha}) \tag{28}$$

The COP of the cycle operating at its maximum cooling capacity can be written in the form:

$$(COP)_m = q_{cm}(1+s\sqrt{\alpha}/r)/[s(\sqrt{\alpha}-\alpha)-q_{cm}] \tag{29}$$

The preceding expressions will be used to study the performance limits predicted by the endoreversible THR-model.

THR-Model with External and Internal Irreversibilities

The temperature-entropy diagram for the THR-model with external and internal irreversibilities is shown in Fig.2. Compared to the external irreversibility, the internal irreversibility is more difficult characterize in terms of the temperatures of the cycle and therefore simplified procedures have to be found to compute the internal entropy generation. Alefeld and Radermacher [15] analyzed the absorption cycle using limited data and obtained expressions for the internal entropy generation which have the form,

$$\delta s_{in} = f_1(T_h^{'}-T_o^{'})^2/T_h^{'}T_o^{'} + f_2(T_o^{'}-T_c^{'})^2/T_o^{'}T_c^{'} \tag{30}$$

Here f_1 and f_2 are functions of the working fluid flow rates , the heat exchanger effectiveness and fluid properties. In reference [14], three semi-empirical entropy functions were used to represent the internal entropy generation.

The procedure proposed in the present work to include the internal irreversibilities, makes use of the relationship given equation (9). The entropy transfer given by equation (20) is equated to the entropy generation given by a semi-empirical function such as (30), to determine the constants in the semi-empirical function. A square-error-sum minimization method is used to obtain the best estimates of the constants.

The non-dimensional for of (30) may written as:

$$\delta S_{in} = f_1(\theta_h - \theta_o)^2 / \theta_h \theta_o + f_2(\theta_o - \theta_c)^2 / \theta_o \theta_c \tag{31}$$

The RHS of equation (31) may be expressed in terms θ_h by eliminating θ_c and θ_o using equations (17) and (18). When RHS of equation (20) is equated to the RHS of (31) according to condition (9), θ_h becomes the only unknown in the resulting equation. The direct solution of this equation is very tedious. However, the graphical procedure given below offers a convenient way to obtain the value of θ_h .

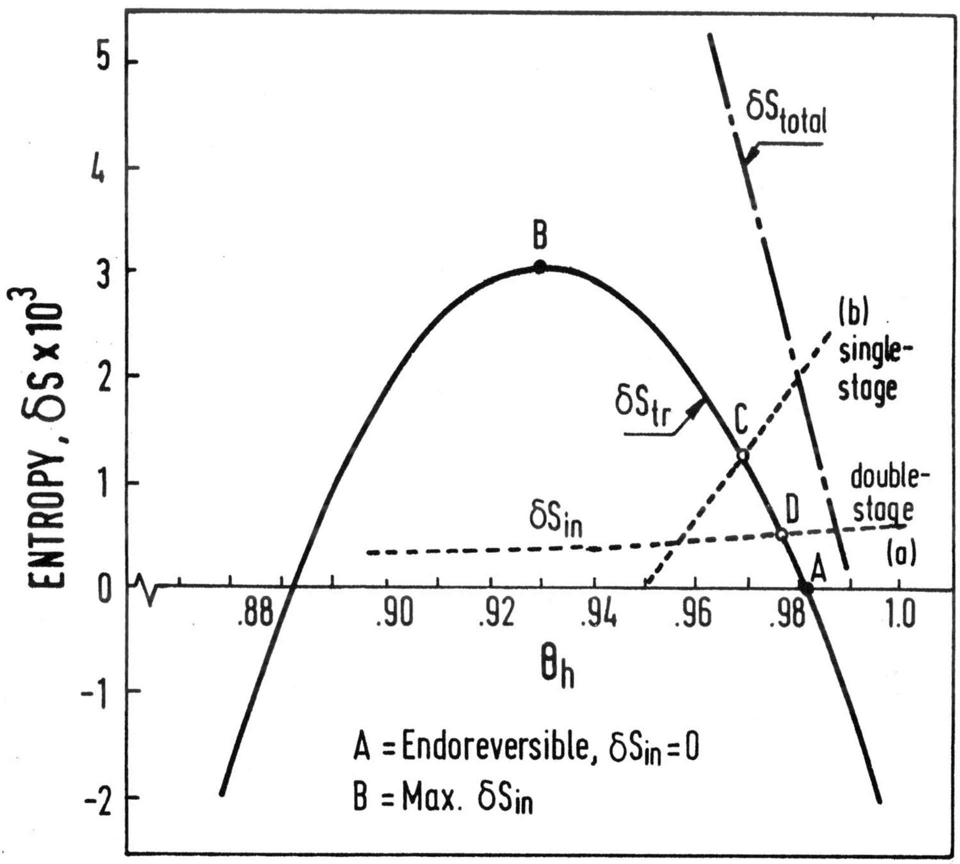

Figure 3. Variation of entropy transfer, internal entropy generation and total entropy generation for the three-heat -reservoir model.

Figure 3 shows the variation of δS_{tr} with θ_h (equation (20)) for a constant cooling capacity. The curve generally has three branches but only the physically relevant section is shown in the figure.

At point A, the entropy transfer is zero and therefore it corresponds to the endoreversible cycle. At point B, the entropy transfer is a maximum and therefore gives the cycle that has the largest internal entropy generation. The operating points for all physically meaningful THR-cycles fall between A and B.

Shown in Figure 3 are two typical variations of the function δS_{in} with θ_h for single-stage and double-stage absorption systems. The points of intersection C and D give the operating points for these two systems. It is seen that the value of θ_h at the point of intersection of the curves, lies between the value at A which is given by the solution of (23) and the value corresponding to the maximum at B which is given by

$$\theta_{hc} = b_1 /(r + s\sqrt{\alpha}) \tag{32}$$

A simple search method can used to locate the point of intersection because the limits of the value of θ_h have been established. Once the value of θ_h at the operating point is found the COP is computed by direct substitution in equation (24).

Another simple entropy function is obtained by assuming that the increase in entropy due to the irreversible processes shown in Fig.2 varies linearly with temperature. This may be expressed as:

$$\delta s_{in} = \beta_1 (T_h' - T_o') + \beta_2 (T_o' - T_c') \tag{33}$$

where β_1 and β_2 are to be estimated by fitting detailed simulation data to predictions by the model.

The non-dimensional form of equation (33) can be expressed in terms of one temperature θ_h using equations (17) and (18). This takes the form:

$$\delta S_{in} = (\beta_1 T_h / K_c)[\theta_h (1 - r/s) - (\alpha + q_c / s + r/s)] + (\beta_2 T_h / K_c) \\ [\alpha + q_c / s + r/s - \phi + q_c] \tag{34}$$

To determine the best estimates of the parameters β_1 and β_2 from simulated performance data or manufacturer-catalogue data the following least-square method is used. At each data point, the COP, the cooling capacity and the equivalent reservoir temperatures are known from the data source. Therefore by using equation (24) the non-dimensional temperature θ_h can be calculated. Substituting in equation (20), the value of δS_{tr} is computed directly. Similarly, by substituting in equation (34), an expression is obtained for δS_{in} in terms of the parameters β_1 and β_2. The best estimates of these parameters is determined by minimizing the square-error-sum, $\Sigma[\delta S_{tr} - \delta S_{in}]^2$ where the summation is over all the data points and therefore covers the entire range of temperatures and cooling capacities of the absorption machine. The predictions of the models are compared with simualtion data in the next section.

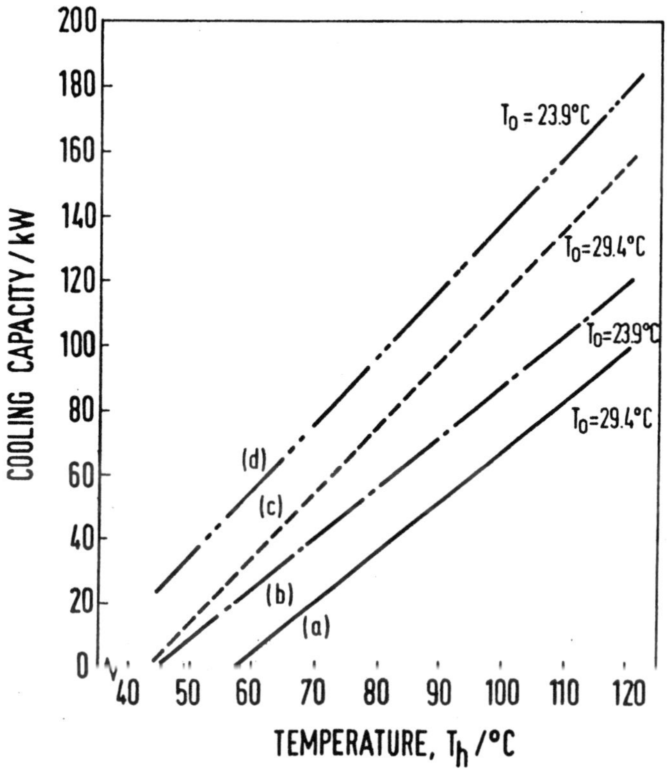

Figure 4. Comparison of maximum cooling capacity of endoreversible model with simulated data. K_h =8.48 kW/K, K_c=11.93 kW/K, K_o =23.98 kW/K, T_c =7.2 °C. *Curves*: (a) simulated capacity for T_o =29.4 °C, (b) simulated capacity for T_o =23.9 °C, (c) maximum capacity for T_o =29.4 °C, (d) maximum capacity for T_o =23.9 °C.

RESULTS AND DISCUSSION

The analytical models described in the preceding section were applied to a practical water-lithium bromide absorption cooling system to predict its coefficient of performance and cooling capacity over a range of operating conditions. The data for comparison was obtained from the work of Gommed and Grossman [4]. The effective thermal conductances and the reservoir temperatures assumed in the present computations are the same as those in [4]. The performance data extracted from the graphs in Ref.[4] were further confirmed using the computer simulation programme , ABSIM [16].

Figure 4 shows a comparison of the maximum cooling capacity predicted by the endoreversible model (equation (27)) with the simulation data. The nearly linear variation of the cooling capacity with hot reservoir temperature is well predicted by the model. As expected, the maximum cooling capacity is much larger than the real cooling capacity.

The endoreversibe model can be used to obtain a useful upper limit to the cooling capacity.

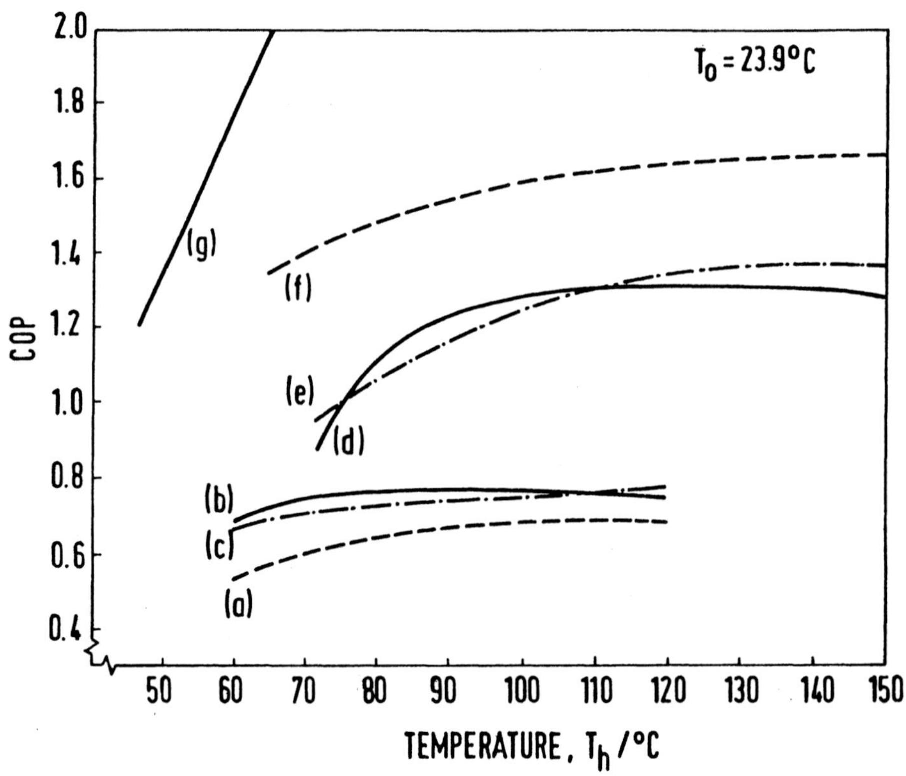

Figure 5. Comparison of COP predicted by different models. K_h =8.48 kW/K, K_c=11.93 kW/K, K_o =23.98 kW/K, T_c =7.2 °C, T_o =23.9 °C.

Curves: (a) COP at maximum cooling capacity for endoreversible model, (b) simulation results for single-stage system, (c) fully-irreversible model for single-stage system, (d) simulation results for double-stage system, (e) fully-irreversible model for double-stage system, (f) endoreversible model at the simulated cooling capacity (Fig. 4), (g) fully-reversible model.

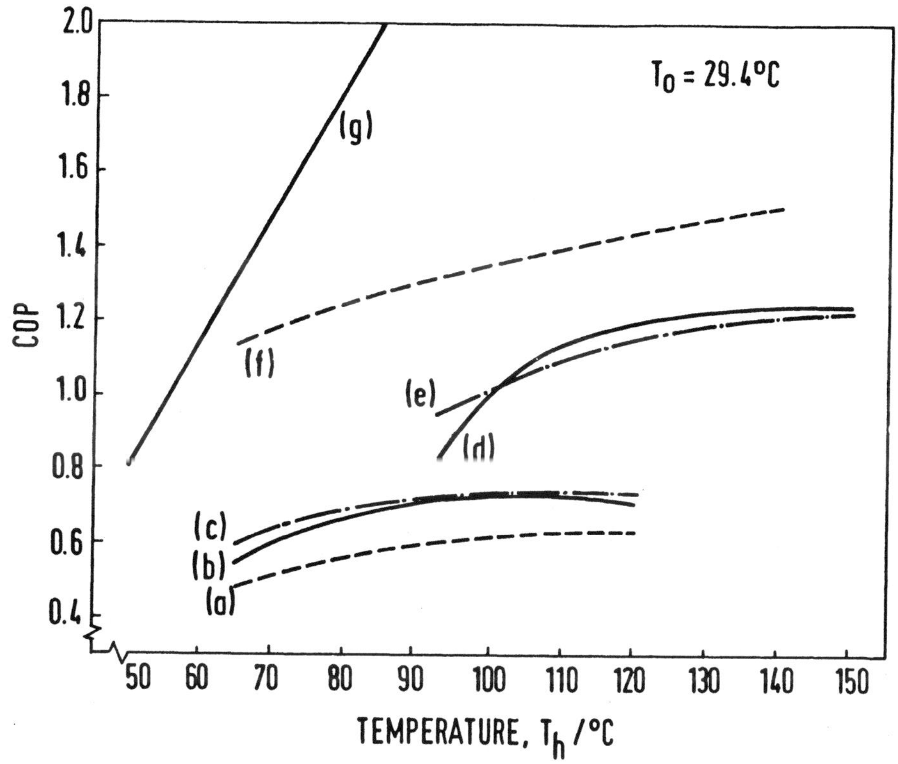

Figure 6. Comparison of COP predicted by different models. K_h =8.48 kW/K, K_c=11.93 kW/K, K_o =23.98 kW/K, T_c =7.2 °C, T_o =29.4 °C.
Curves: (a) COP at maximum cooling capacity for endoreversible model, (b) simulation results for single-stage system, (c) fully-irreversible model for single-stage system, (d) simulation results for double-stage system, (e) fully-irreversible model for double-stage system, (f) endoreversible model at the simulated cooling capacity (Fig. 4), (g) fully-reversible model.

The COP predicted by the various models is compared with the simulation results in Figures 5 and 6. The COP at of the endoreversible model at the maximum cooling capacity shown by curve (a) has the lowest value for the given operating conditions. It is interesting to note that the COP of the real single-stage machine is larger than that of the endoreversible cycle operating at its maximum cooling capacity. The COP of the endoreversible model at the given cooling capacity has a high value that is well above the COP of the real double-stage machine. The COP of the fully-reversible cycle (Carnot) is much higher than the upper limit predicted by the endoreversible cycle. The COP at a given cooling capacity and the COP at the maximum cooling capacity predicted by the endoreversible model can be taken as realistic upper and lower limits respectively.

The semi-empirical model, which includes both external and internal irreversibilities, predicts a COP variation that follows the simulation results closely. This is true for both single-stage and double-stage machines. The curves are for the linear entropy function (equation (34)), which gives a better prediction than the logarithmic function (equation (31)). The data considered cover hot reservoir temperatures from 65 to 150 ^{o}C , heat sink temperatures from 23.9 to 35 ^{o}C and cooling capacities from about 15 to 160 kW. The maximum deviation between the predicted data and the simulated date is about 10%, which occur for a few points at the lower end of the COP. Elsewhere the deviation is much smaller.

It is can be concluded that while the edoreversible model give realist upper and lower limits to the performance, the semi-empirical models can be used for more accurate performance prediction of absorption cooling systems.

CONCLUSION

The THR-model was used to predict the performance absorption cooling systems. The effect of using heat reservoirs to represent the single-stream heat exchangers of real machines was investigated. Subject to realistic assumptions, the equivalent temperatures of the reservoirs and the equivalent conductances may be chosen to give an accurate representation of both the heat transfer and the entropy production due to heat transfer. Expressions were derived for the COP of the endoreversible model , the fully-reversibel model and the semi-empirical model including both external and internal irreversibilities. The predictions of these models were compared with data from detailed computer simulation models. The endoreversible models gave realistic limits for the cooling capacity and the COP. The semi-empirical models using a linear internal entropy function gave good prediction of the COP of both single-stage and double-stage machines over a wide range of operating conditions.

NOMENCLATURE

a = constant defined in equation (28)
a_1 = constant defined in equation (21)
b = constant defined in equation (28)
b_1= constant defined in equation (22)
c = constant defined in equation (28)
COP = coefficient of performance

f_1 = constant in equation (30)
f_2 = constant in equation (30)
K = effective thermal conductance
$(mc)_g$ = mass flow rate times specific heat capacitance
q = cooling capacity
Q = heat transfer rate between reservoir and fluid
r = variable defined in equation (16)
s = variable defined in equation (16)
T= temperature
T' = temperature of working fluid

Greek symbols

α = variable defined in equation (15)
β_1 = constant in equation (33)
β_2 = constant in equation (33)
ε_g = effectiveness of the generator
ϕ = variable defined in equation (15)
θ = non-dimensional temperature
δs = entropy generation/transfer rate
δS = non-dimensional entropy generation/transfer rate

Subscripts

c = cold reservoir
cm = maximum cooling
h = hot reservoir
o = heat sink reservoir
gi = generator inlet
go = generator outlet
hc = at maximum entropy transfer
in = internal
tr = transfer
ht= heat transfer
htg = heat transfer at generator

REFERENCES

W.H.Wilkinson, "Absorption cooling: A viable option for today and tomorrow". ASHRAE Journal. 36(2), 41-44 (1994).

G.Grossman, K.Gommed and D.Gadoth, "A computer model for simulation of absorption systems in flexible and modular form". ASHRAE Transactions 93(2), 2389-2428 (1987).

Kenneth Wark, 'Advanced Thermodynamics for Engineers". McGraw-Hill, Inc. New York , (1995), p.359.

K. Gommed and G.Grossman, "Performance analysis of staged absorption heat pimps: water-lithium bromide systems". ASHRAE Transactions 96(1), 1590-1598 (1990).

J.Chen and Z.Yan , "Equivalent combined systems for three-heat-source heat pumps". J. Chem.Phys. 90(9), 4951-4955 (1989).

Z.Yan and J.Chen, "An optimal endoreversible three-heat-source refrigerator". J. Applied Physics 65(1), 1-4 (1989).

C.Wu , "Cooling capacity optimization of a waste heat absorption refrigeration cycle". Heat Recovery Systems & CHP 13(2), 161-166 (1993).

N.E.Wijeysundera, " Analysis of the ideal absorption cycle with external heat-transfer irreversibilities". Energy-The International Journal 20(2), 123-130 (1995).

N.E.Wijeysundera, "Performance limits of absorption cycles with external heat-transfer irreversibilities". Applied Thermal Engineering 16(2), 175-181 (1996).

A.Bejan,J.V.C.Vargas and M.Sokolov, "Optimal allocation of a heat-exchanger inventory in heat driven refrigerators". Int. J. Heat Mass Transfer, 38(16) , 2997-3004 (1995).

J.M.Gordon and K.C.Ng, "A general thermodynamic model for absorption chillers: theory and experiment". Hear Recovery & CHP 15(1), 73-83 (1995).

H.Tong Chua, Q.Han, K.Choon Ng, and J.M.Gordon, "Thermodynamic modeling and experimental evidence for the optimization and maximum-efficiency operation of absorption chillers". ECOS'96, Efficiency, Cost, Optimization, Simulation and Environmental Aspects of Energy Systems, June 25-27, Stockholm, pp. 157-166 (1996).

Guoxing Lin and Zijun Yan , "The optimal performance of an irreversible absorption refrigerator". J. Phys. D: Applied Phys: 30 , 2006-2011 (1997).

N.E.Wijeysundera, "Performance of three-heat-reservoir absorption cycles with external and internal irreversibilities". Applied Thermal Engineering, 17(12), 1151-1161 (1997).

G.Alefeld and R.Radermacher, "Heat Conversion Systems", CRC Press, Ann Arbor, 113-121(1995).

ABSIM-User's Guide and Reference (Version 1.2) - July 1995.

DESIGN PARAMETERS OF AN IRREVERSIBLE HEAT ENGINE WITH COMBINED HEAT TRANSFER

L. BERRIN ERBAY
Osmangazi University, School of Engineering
26030 Bademlik,Eskisehir, Turkey

Abstract- The design parameters of an irreversible heat engine have been researched and power optimization has been performed with the consideration of heat exchange by combined radiation and convection which is a technologically important case. Further derivations including the relations between optimal temperature ratio of cold to hot working fluids, reservoir temperatures, the ratio of radiative to Newtonian heat transfer coefficients, the ratio of thermal conductances, the emissivity ratio of heat reservoirs and optimal efficiency have been given under the maximum power conditions. The analysis has shown that the maximum power output of the heat engine has been sensitive to the temperature, small values of the convective heat transfer coefficient of hot - end, small values of the emittance of heat source and highly depends on irreversibilities. By considering the maximum power output of an irreversible heat engine with combined heat transfer, the thermal design parameters τ, β_C and ε_R must be held around 0.2, 0.02, and 0.01, respectively.

Keywords: Carnot, irreversibility, combined heat transfer, convection, radiation

1. INTRODUCTION

Following the fundamental studies in finite - time thermodynamics,[1-11] the attention has been paid to the effect of heat transfer laws on the performance of the endoreversible cycles.[12-17] Since there are main sources of irreversibility such as heat leaks, friction and turbulance of the working fluid in real heat engines, internal irreversibilities have been studied by using either an irreversibility parameter[18-21] or thermal efficiencies of the basic components of heat engines[22] other than Carnot. These elaborated studies consider mainly Newtonian and radiation heat transfers, separately. The fact considered in the present study is that heat exchange by combined convection and radiation is a technologically important example to analyze the systems in nature. [23]

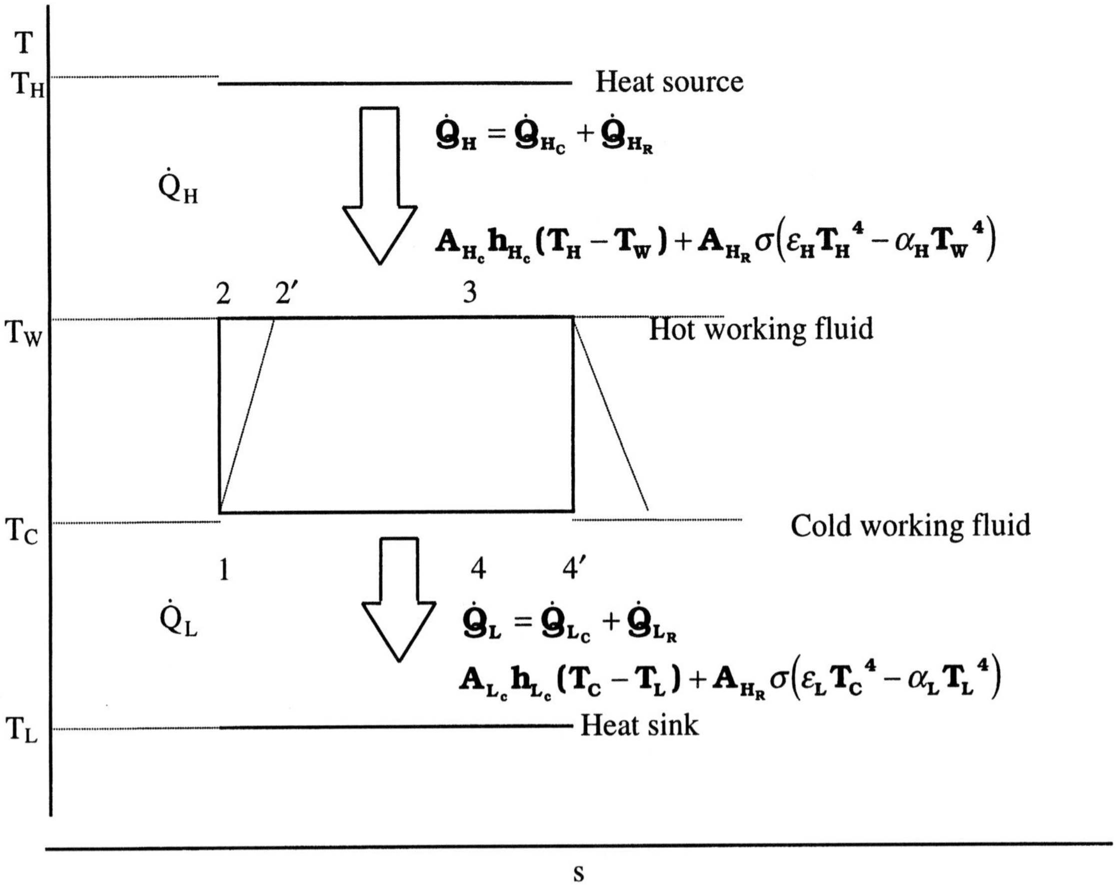

Figure 1. T - s diagram of the irreversible heat engine with combined heat transfer.

In this paper the optimal design parameters of an irreversible Carnot engine with combined heat transfer representing the real systems have been researched under the maximum power conditions. The irreversible heat engine model using a single irreversibility parameter [18-21] has been substituted for the endoreversible model with combined heat transfer proposed by Erbay and Yavuz[24]. Besides the effect of combined heat transfer on the performance of endoreversible cycle, the mechanisms have been evaluated according to their relative importance by introducing the ratios of radiative heat transfer coefficient to the thermal conductance for hot - end (β_H) and cold - end (β_L), the ratio of hot to cold end thermal conductances (β_C) and the emissivity ratio of the heat reservoirs (ε_R). After the optimal temperature ratio of cold working fluid to hot one (ψ_{opt}) has been obtained, further relations including the β_H, β_L, β_C, ε_R, thermal efficiency (η) and the maximum power (P_{max}) have been derived. The results are summarized by figures by considering different β_C and ε_R values. comprehensively.

2. THERMODYNAMIC ANALYSIS

The cyclic system of the Carnot heat engine under consideration is modeled in such a way that it has minimum construction detail being sufficient to identify the external and internal irreversibilities and shown in Fig. 1. For easy understanding, subscripts H, L, C and R are used for the parameters related to the sides of high and low temperature heat reservoirs, for the processes of convection and radiation, respectively.

Heat transfer analysis for the hot - end

The heat ($\dot{Q}_H$) is transferred from the high temperature heat reservoir at T_H to the hot working fluid at T_w by the combination of two different heat transfer mechanisms, which are Newtonian and radiative heat transfers and given as

$$\dot{Q}_H = A_{H_C} h_{H_C} (T_H - T_W) + A_{H_R} \sigma (\varepsilon_H T_H^4 - \alpha_H T_W^4) \tag{1}$$

where A is heat transfer area, h_{Hc} is the thermal conductance , σ is Stefan - Boltzmann coefficient, ε and α values are the emittance and absorption coefficients of heat source, respectively. The emittance will be balanced uniformly by absorption ($\varepsilon \cong \alpha$) for a thin slice of gas.[25]

In most engineering applications a radiation heat transfer coefficient is preferably used. The radiation heat transfer coefficient is defined as

$$h_{H_R} = \sigma \varepsilon_H \left(T_H^3 + T_H^2 T_W + T_H T_W^2 + T_W^3 \right) \tag{2}$$

The heat transfer areas for both of convection and radiation are physically the same, i.e. $A_{Hc} = A_{HR} = A_H$. With $\beta_H = h_{H_R} / h_{H_C}$, $\varphi = T_W / T_H$ and using equations (2) and (1), $\dot{Q}_H$ becomes

$$\dot{Q}_H = A_H T_H h_{H_c} (1 + \beta_H)(1 - \varphi) \tag{3}$$

Heat transfer analysis for the cold - end

The heat $\dot{Q}_L$ transferred from the cold working fluid at T_C to the heat sink at T_L by the combined Newtonian and radiative heat transfers is

$$\dot{Q}_L = A_{L_C} h_{L_C} (T_C - T_L) + A_{L_R} \sigma (\varepsilon_L T_C^4 - \alpha_L T_L^4) \tag{4}$$

where A is heat transfer area, h_{Lc} is the thermal conductance. The radiation heat transfer coefficient of cold side of the system is written as

$$h_{L_R} = \sigma \varepsilon_L \left(T_C^3 + T_C^2 T_L + T_C T_L^2 + T_L^3 \right) \tag{5}$$

With $\tau = T_L / T_H$, $\psi = T_C / T_W$, $\beta_L = h_{L_R} / h_{L_C}$, $A_{Lc} = A_{LR} = A_L$ and using Eqs. (5) and (4), $\dot{Q}_L$ is rearranged as

$$\dot{Q}_L = A_L T_H h_{L_C} (1 + \beta_L)(\psi\varphi - \tau) \tag{6}$$

III) THERMODYNAMIC ANALYSIS OF THE INTERNALLY IRREVERSIBLE HEAT ENGINE

The engine absorbs heat ($\dot{Q}_H$) in conjuction with an entropy flow $\dot{Q}_H / T_W$ and rejects heat ($\dot{Q}_L$) in conjunction with an entropy flow $\dot{Q}_L / T_C$. Since the engine is internally irreversible due to the internal dissipations of the working fluid, the second law requires

$$\oint \frac{\partial \dot{Q}_H}{T} = \frac{\dot{Q}_H}{T_W} - \frac{\dot{Q}_L}{T_C} < 0 \tag{7}$$

One can write the inequality in Eq.(5) as

$$\dot{Q}_H / T_W = R(\dot{Q}_L / T_C) \quad \text{with } R<1 \tag{8}$$

where R representing the ratio of entropy flows characterizes fully the degree of internal irreversibility and is named as the cycle - irreversibility parameter.[18] R= 1.0 corresponds to the reversible cycle. R is $0<R<1.0$ for an irreversible engine. Inserting Eqs. (3) and (6) into Eq. (8) yields

$$A_R \beta_C (1 + \beta_H)(1 - \varphi)\psi = R(\psi\varphi - \tau)(1 + \beta_L) \tag{9}$$

where $A_R = A_H / A_L$

$$\beta_C = h_{H_C} / h_{L_C}$$

$$\beta_H = h_{H_R} / h_{H_C} = \gamma_H (1 + \varphi + \varphi^2 + \varphi^3)$$

$$\beta_L = h_{L_R} / h_{L_C} = \gamma_L (\psi^3\varphi^3 + \psi^2\tau\varphi^2 + \psi\tau^2\varphi + \tau^3)$$

$$\gamma_H = \sigma\varepsilon_H T_H^3 / h_{H_C}$$

$$\gamma_L = \sigma \varepsilon_L T_H^3 / h_{L_C}$$

Equation (9) presents a non - linear constraint for the interaction between irreversibilities, temperature ratios, heat transfer areas and relative importance of the heat transfer processes. For further derivations, φ, the temperature ratio of the hot reservoir to the hot working substance, is required to be solved from the Eq. (9), which is

$$\varphi = \varphi\left(A_R, \beta_C, \beta_H, \beta_L, \psi, \tau, R\right) \tag{10}$$

Rearranging Eq. (9) for φ gives

$$\left(A_R \beta_C \gamma_H + R\gamma_L \psi^3\right)\psi\varphi^4 + \left(A_R \beta_C + R\right)\psi\varphi \text{ - } A_R \beta_C \left(1+\gamma_H\right)\psi - \left(\gamma_L \tau^3 + 1\right)R\tau = 0 \tag{11}$$

After Eq. (11) is linearized by expanding higher order terms in Taylor series and neglecting the second and higher order terms[26], the constraint takes the following form

$$\begin{aligned} &\left[A_R \beta_C \left(4\gamma_H + 1\right) + R\left(4\gamma_L \psi^3 + 1\right)\right]\psi\varphi - \\ &\qquad \left[A_R \beta_C \left(4\gamma_H + 1\right) + 3R\gamma_L \psi^3\right]\psi - R\left(\gamma_L \tau^3 + 1\right)\tau = 0 \end{aligned} \tag{12}$$

Thermal efficiency found easily by using Eq. (8) in the first law of the thermodynamics becomes

$$\eta^* = 1 - \frac{\psi}{R} \tag{13}$$

The power output of the heat engine is

$$P^* = \dot{Q}_H \left(1 - \frac{\psi}{R}\right) \tag{14}$$

Introducing the convenient dimensionless heat input as

$$\phi = \dot{Q}_H / A_H T_H \left(h_{H_C} + h_{H_R}\right) = \left[1 - \varphi\left(A_R, \beta_C, \beta_H, \beta_L, \psi, \tau, R\right)\right] \tag{15}$$

dimensionless power becomes

$$P = \left(1 - \frac{\psi}{R}\right)\left[1 - \varphi\left(A_R, \beta_C, \beta_H, \beta_L, \psi, \tau, R\right)\right] \tag{16}$$

By taking φ from equation (12) and inserting into equation (16) gives the explicit form of dimensionless power as

$$P = \frac{R\left(1-\frac{\psi}{R}\right)(\psi-\tau)\left(1+\gamma_L\psi^3+\gamma_L\psi^2\tau+\gamma_L\psi\tau^2+\gamma_L\tau^3\right)}{\left(R+A_R\beta_C+4A_R\beta_C\gamma_H\right)\psi+4R\gamma_L\psi^4} \tag{17}$$

Power is maximized by solving $\partial P/\partial\psi = 0$. Taking derivative of equation (17) gives

$$\psi^8-\alpha\psi^5+\chi\psi^4-\kappa\psi^3+\pi\psi^2-\omega=0 \tag{18}$$

where

$$\alpha=\frac{1}{\gamma_L R}\left(1-A_R\beta_C-4A_R\beta_C\gamma_H\right)$$

$$\chi=\frac{3}{4}\left(\alpha+\kappa+\frac{2}{\gamma_L}\right)$$

$$\kappa=\frac{4\tau}{\gamma_L}\left(1+\gamma_L\tau^3\right)$$

$$\pi=\frac{1}{4\gamma_L^2 R}\left(A_R\beta_C+4A_R\beta_C\gamma_H+R\right)$$

$$\omega=\frac{1}{4}\gamma_L\pi\kappa$$

Solving ψ from equation (18) will give the optimum temperature ratio ψ_{opt} for given parameters R, τ, A_R, β_C, , β_H, β_L, ε_H and ε_L. Substituting $\psi = \psi_{opt}$ in equation (17) and equation (13) yields the maximized power output (P_{max}) and thermal efficiency at the maximum power conditions, η , respectively.

Figure 2a

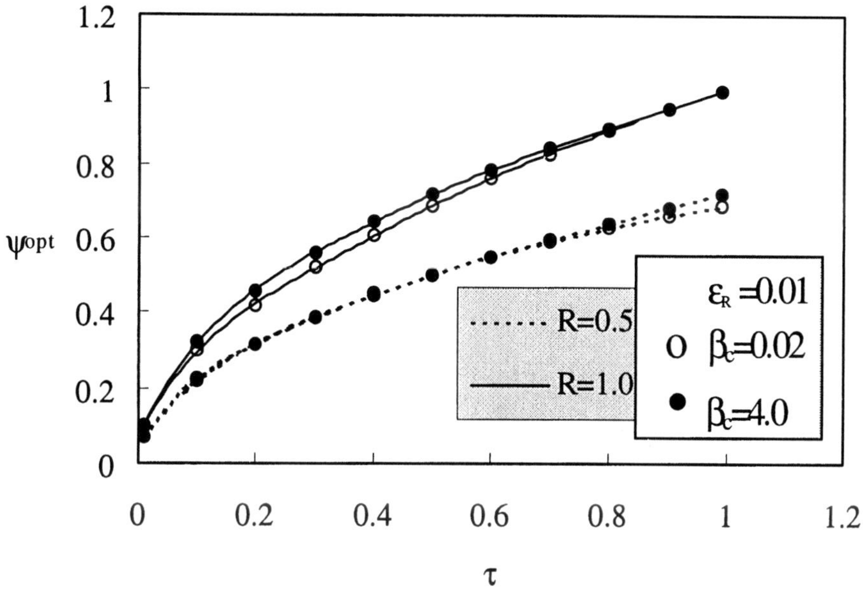

Figure 2b

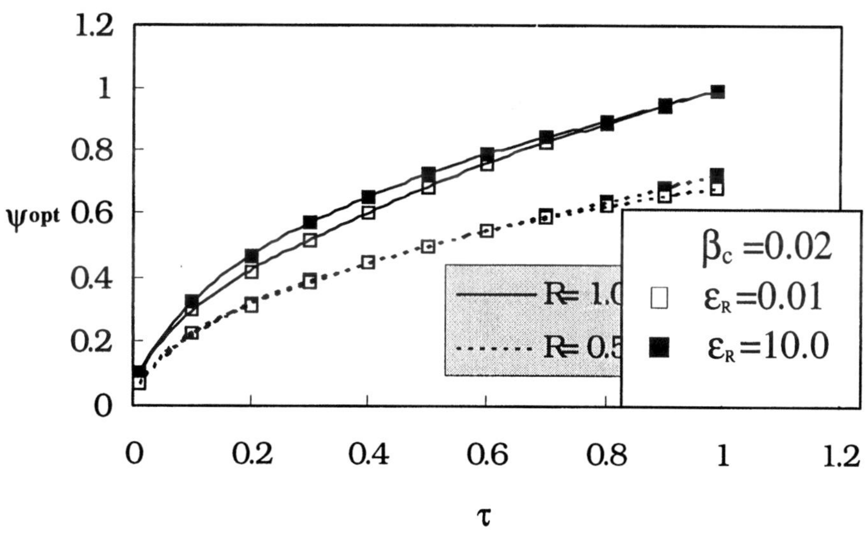

Figure 2. The variation of ψ_{opt} with respect to τ by considering (a) β_C and (b) ε_R.

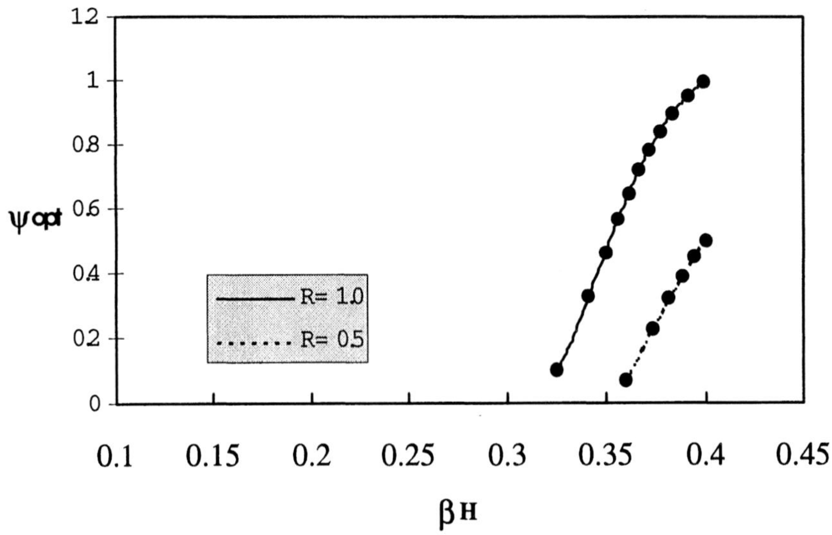
1.2
1
0.8
0.6
0.4
0.2
0
ψopt
R= 1.0
R= 0.5
0.1
0.15
0.2
0.25
0.3
0.35
0.4
0.45
βH

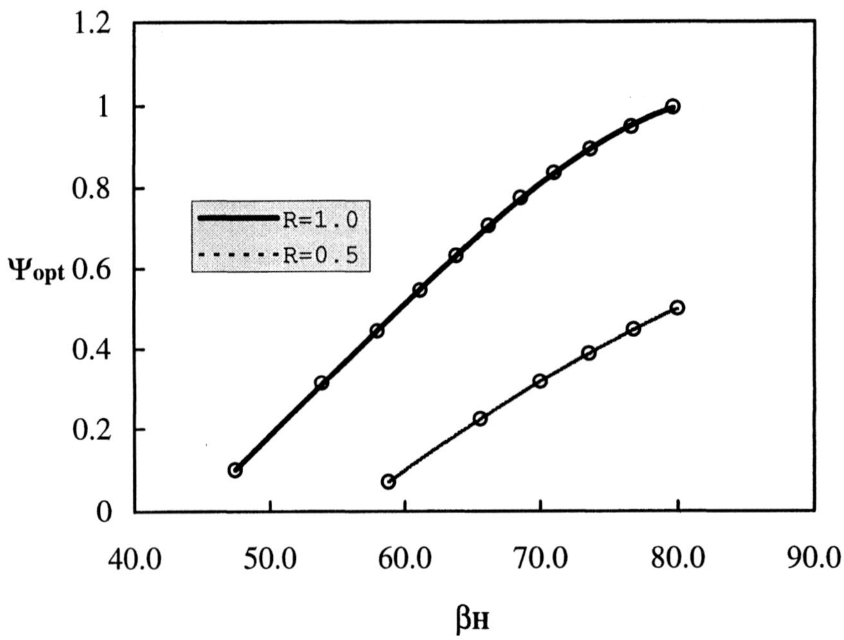
1.2
1
0.8
0.6
0.4
0.2
0
Ψopt
R=1.0
R=0.5
40.0
50.0
60.0
70.0
80.0
90.0
βH

Figure 3a

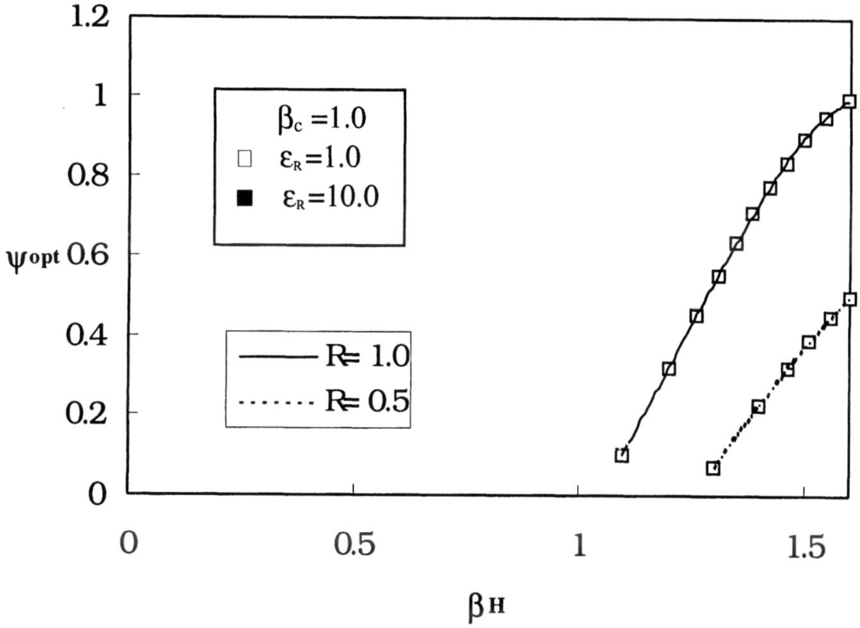

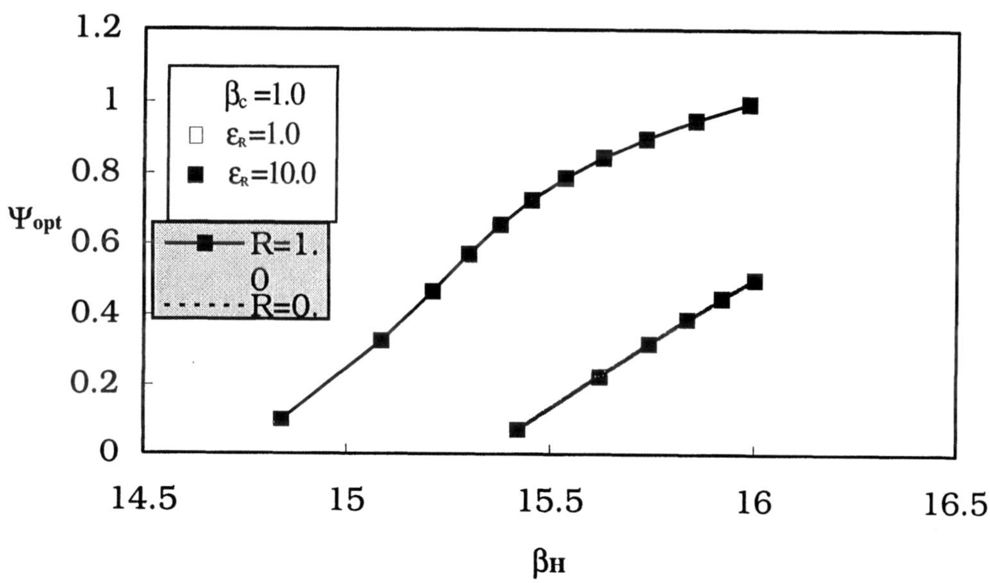

Figure 3b

Figure 3. The relation between ψ_{opt} and β_H at different (a) β_C and (b) ε_R.

3. Results and Discussion

Differentiating the dimensionless power with respect to the temperature ratio of the heat engine yields the optimum working temperature ratio (ψ_{opt}). The results yield the relations between the thermal design parameters and are summarized by figures by considering different β_C and $\varepsilon_R=\varepsilon_H / \varepsilon_L$ values. The values of ψ_{opt}, the optimum temperature ratio of the cold working fluid to the hot working fluid, are strongly dependent on τ, the ratio of the heat reservoir temperatures. As τ increases, in other words the temperature difference between reservoirs decreases, the value of ψ_{opt} increases almost linearly. When Fig. 2a and 2b are observed, it is clearly seen that not the heat transfer laws but the irreversibilities have dominant effect on the optimum working conditions. Increasing irreversibilities cause decrease definitely in ψ_{opt} values. All curves coincide at lower limit of τ.

Figures 3a and 3b show the variation of ψ_{opt} with respect to β_H. As given in the two parts of Fig. 3a, when the emissivities of heat reservoirs are equal, $\varepsilon_R=1.0$, and the hot - end thermal conductance is greater than that of the cold - end ($\beta_C>1.0$), a certain ψ_{opt} can be obtained at small β_H values. For $\beta_C<1.0$ radiative heat transfer at the hot - end is high at the certain ψ_{opt} value. The two parts of Fig. 3b explain that when the thermal conductances of hot and cold - ends are taken equal, $\beta_C=1.0$, small β_H values are more sensitive to very little changes and effects ψ_{opt}. At all cases, the irreversibilities cause certain decrease in ψ_{opt}.

ψ_{opt} increases by a slight increase in the radiative heat transfer of the cold - end at the small values and is effected by irreversibilities rather than β_C and ε_R as seen in Fig. 4a and 4b. ψ_{opt} behaves asmptotically near zero values of β_L. When R=0.5 is taken, ψ_{opt} does not increase more than 0.5 at the values $\beta_L<0.4$. That is, high thermal conductance of cold - end of the irreversible heat engine is important for the case of large temperature difference between the temperatures of hot and cold working fluids. If the radiative heat transfer coefficient is greater than that of the thermal conductance, ψ_{opt} values increases and the temperatures of working fluids come closer.

ψ_{opt} increases by a slight increase in the radiative heat transfer of the cold - end at the small values and is effected by irreversibilities rather than β_C and ε_R as seen in Fig. 4a and 4b. ψ_{opt} behaves asmptotically near zero values of β_L. When R=0.5 is taken, ψ_{opt} does not increase more than 0.5 at the values $\beta_L<0.4$. That is, high thermal conductance of cold - end of the irreversible heat engine is important for the case of large temperature difference between the temperatures of hot and cold working fluids. If the radiative heat transfer coefficient is greater than that of the thermal conductance, ψ_{opt} values increases and the temperatures of working fluids come closer.

Figure 4a

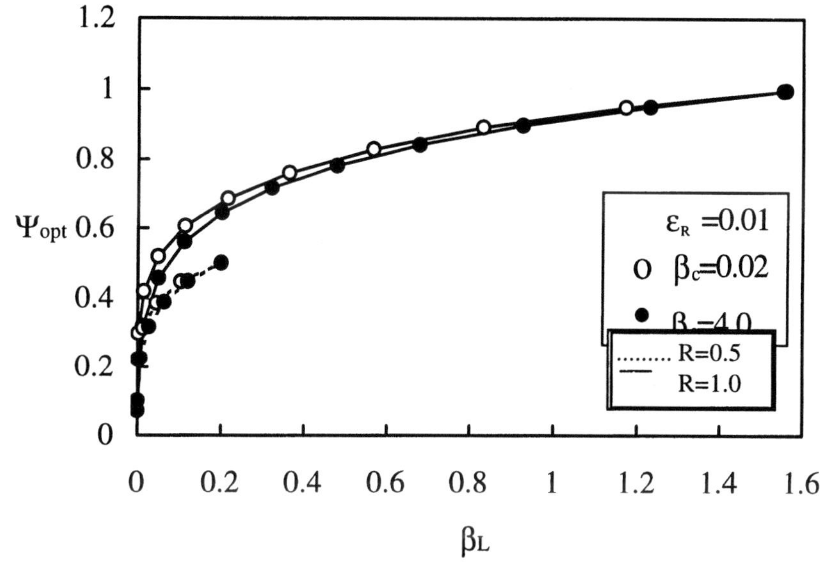

Figure 4b

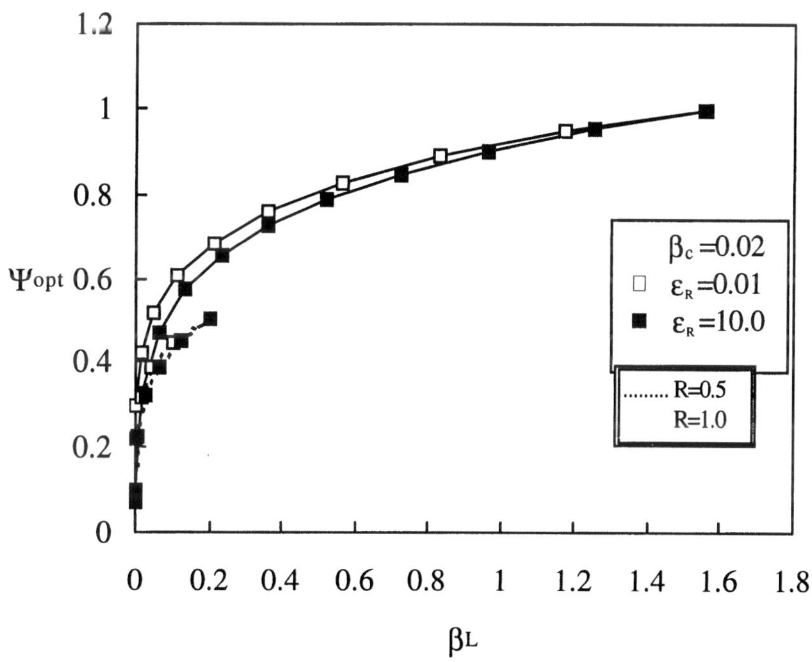

Figure 4. The effect of β_L on the variations of ψ_{opt} at various (a) β_C and (b) ε_R values.

Figure 5a

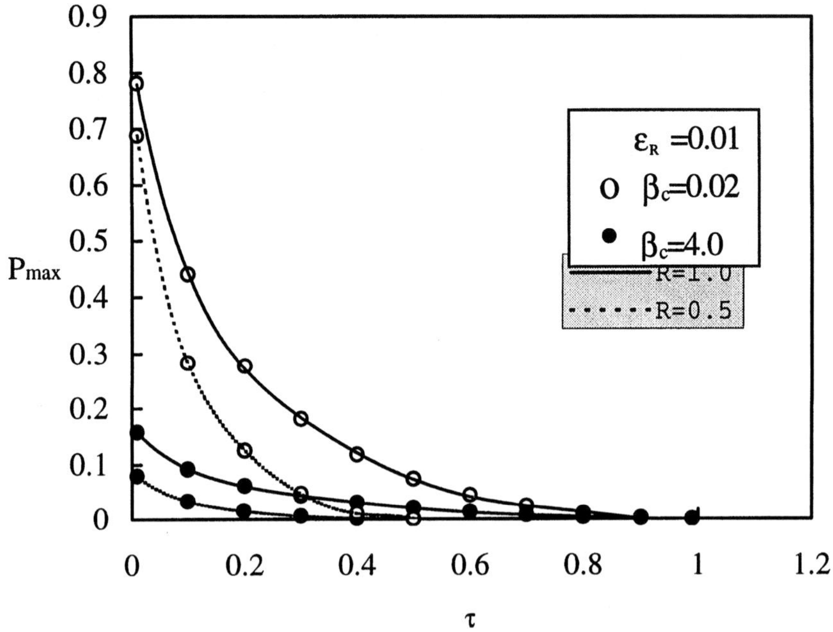

Figure 5b

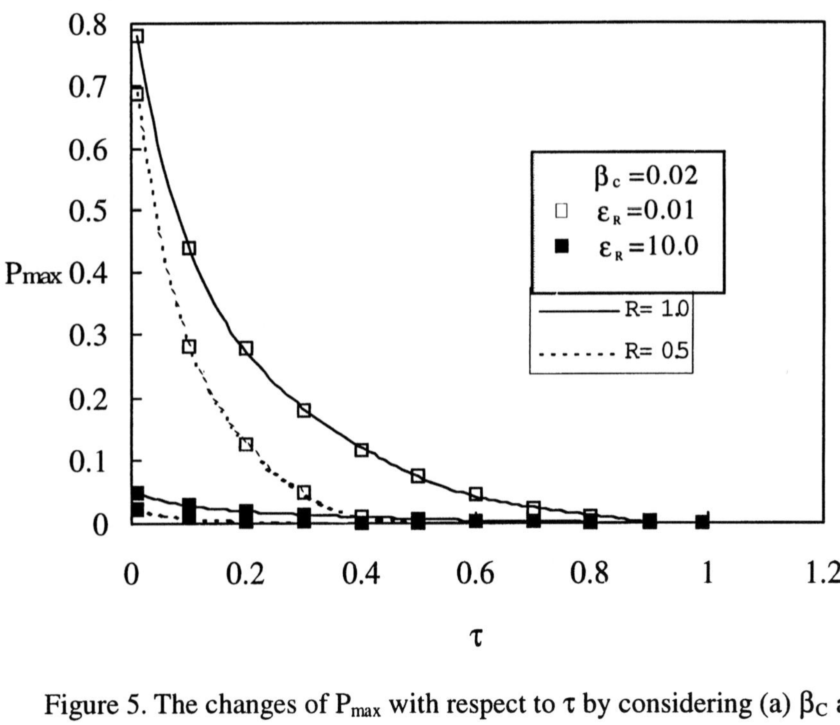

Figure 5. The changes of P_{max} with respect to τ by considering (a) β_C and (b) ε_R .

Figure 6a

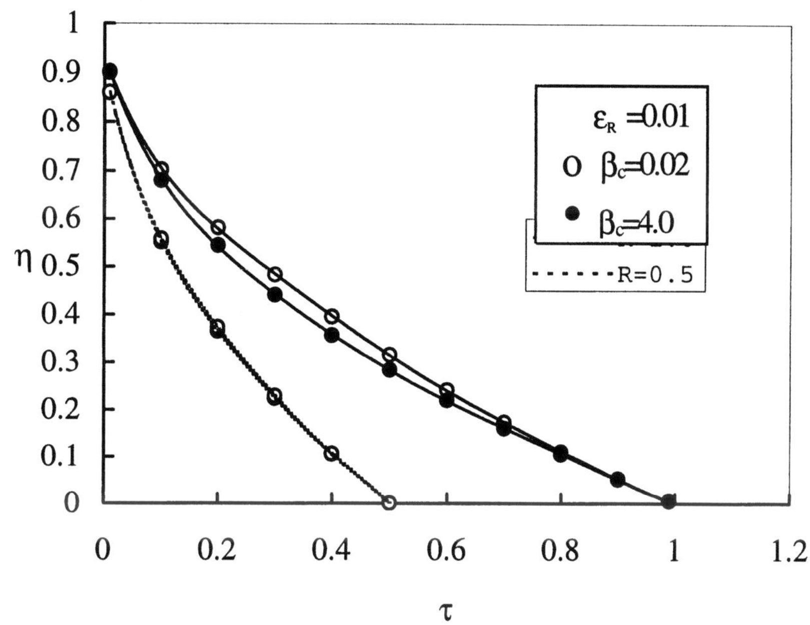

Figure 6b

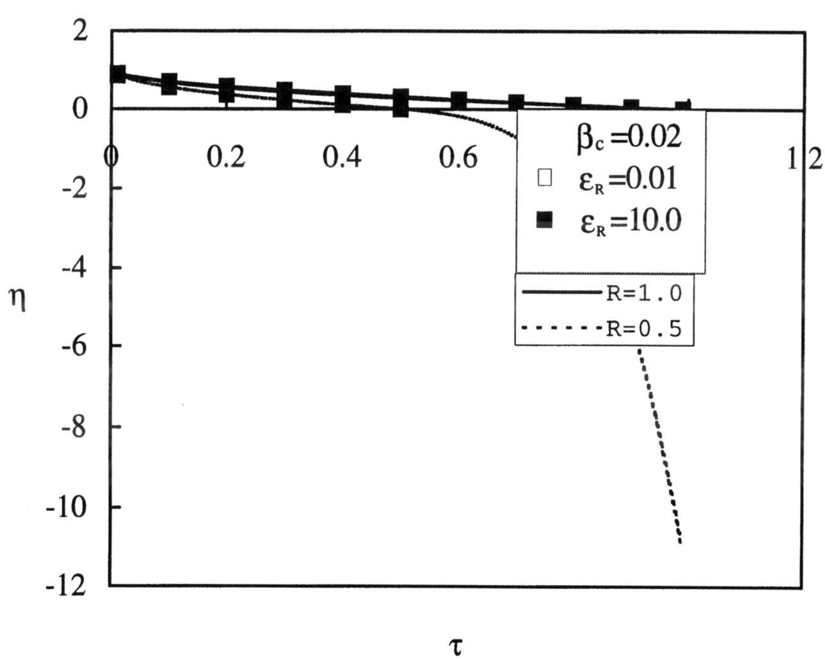

Figure 6. The variation of the efficiency at maximum power condition, η, according to the different (a) β_C and (b) ε_R values .

Figure 7a

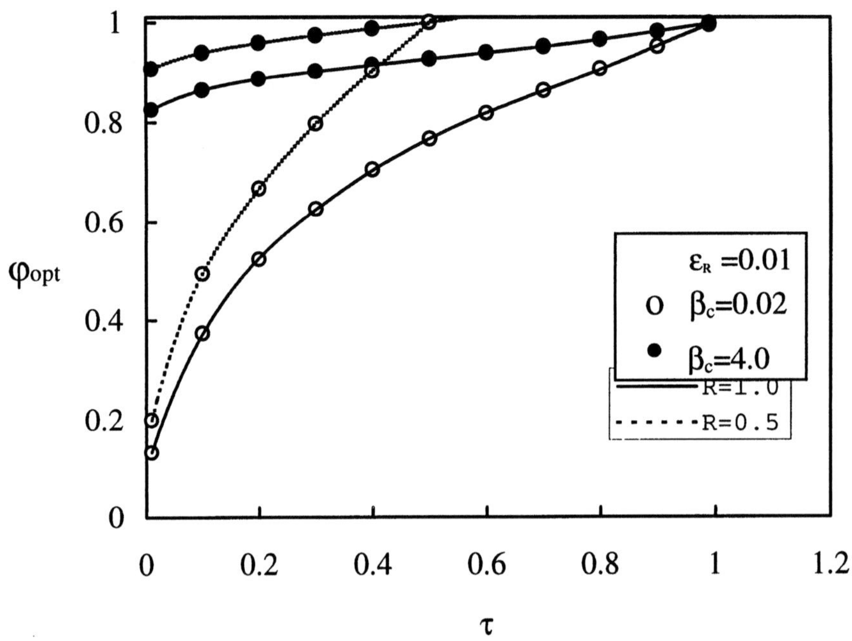

Figure 7b

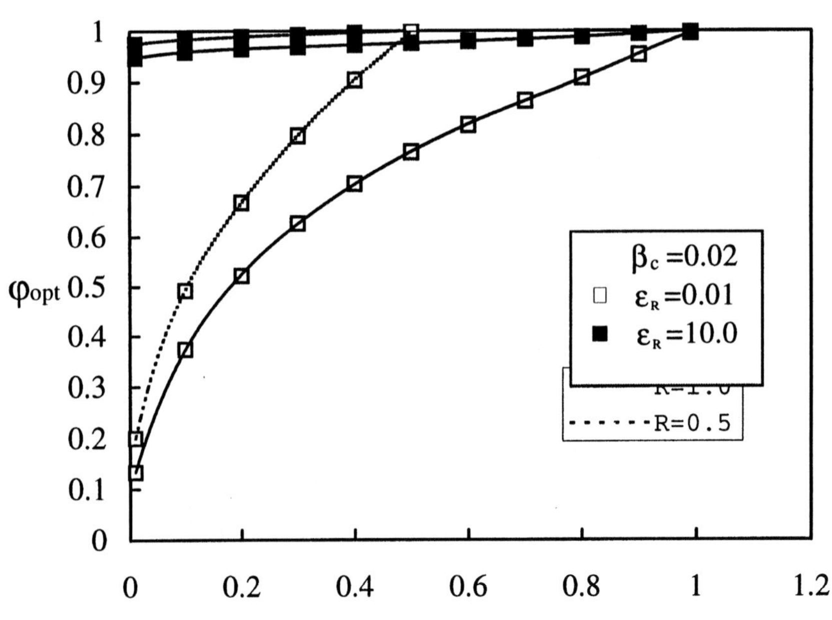

Figure 7. The variation of φ_{opt} with respect to τ by considering (a) β_C and (b) ε_R .

Figure 6 explains the changes of thermal efficiency at maximum power conditions with respect to τ. The efficiency at maximum power conditions, η, depends inevitably on the irreversibilities. This dependence is great especially at the low temperature differences between the reservoirs, i.e at high τ values. When Curzon - Ahlborn[2] efficiency ($\eta_{C\text{-}A}=1-\tau^{0.5}$) is obtained, it is found that the curves for endoreversible heat engine is around the $\eta_{C\text{-}A}$ values[24] but the irreversible ones are greatly far from these values.

In Figs. 7a and 7b how φ_{opt}, the optimum location of heat engine on the absolute temperature scale under the maximum power conditions, changes is found. The variation of φ_{opt} is drawn by inserting ψ_{opt} values into Eqn.(12). ψ_{opt} values of the irreversible heat engine ($R<1.0$) with combined heat transfer has a value of 0.5 for all β_C and ε_R values.

4. CONCLUSIONS

In the present study, the impact of combined heat transfer on the performance of irreversible Carnot engine has been investigated and the thermal dizayn parameters have been researched. The results have been summarized in figures which are generated by considering the variations of β_C and ε_R. It has been obtained that the maximum power output of an irreversible Carnot engine with combined heat transfer has been strongly dependent to the reservoir temperatures and emissivities beside the irreversibilities. It has been shown that the emissivity of heat sink, ε_L and thermal conductance of cold - end h_{Lc} must be higher than that of the heat source and hot - end. The temperature ratio of the heat reservoirs must be small, in other words the temperature difference between heat reservoirs must be high. Typically, the values for τ, β_C and ε_R must be held around 0.2, 0.02 and 0.01 under the maximum power conditions. The irreversibilities must be eliminated as high as possible.

NOMECLATURE

A heat transfer area
h_c thermal conductance
h_R radiation heat transfer coefficient
P dimensionless power output
P_{max} maximized power output
$\dot{Q}_H$ heat transferred from the heat source at T_H to the hot working fluid at T_w
$\dot{Q}_L$ heat transferred from the cold working fluid at T_C to the heat sink at T_L
R the cycle - irreversibility parameter, $0<R<1.0$ for an irreversible engine
T_w temperature of the hot working fluid
T_C temperature of the cold working fluid
T_H heat source temperature
T_L heat sink temperature

α absorption coefficient

β_C h_{H_C}/h_{L_C}

β_L h_{L_R}/h_{L_C}

β_H h_{H_R}/h_{H_C}

ε emissivity

ε_R $\varepsilon_H/\varepsilon_L$

σ Stefan - Boltzmann coefficient

τ T_L/T_H

ψ T_C/T_W

φ T_W/T_H

η^* thermal efficiency

η thermal efficiency at maximum power conditions

$\eta_{C\text{-}A}$ Curzon - Ahlborn efficiency $=1-\tau^{0.5}$

subscripts

H heat source side

L heat sink side

C convection

R radiation

REFERENCES

[1] Novikov I.I 1958 *J. Nuclear Energy* 7 125
[2] Curzon F L and Ahlborn B 1975 *Am. J. Phys.* 43 22
[3] Andresen B Berry R S Nitzan A and Salamon P 1977 *Physical Review A* 15 2086
[4] Krusin D G Procaccia I and Ross J 1978 *J. Chem. Phys.* 69 3898
[5] Ondrechen M J Rubin M H and Band Y B 1983 *J. Chem. Phys.* 78 4721
[6] Andresen B Salamon P and Berry R S 1984 *Physics today* 62
[7] Leff H S 1987 *Am. J. Phys.* 55 602
[8] Bejan A 1988 *Int. J. Heat Mass Transfer* 31 1211
[9] Wu C 1988 *Energy* 13 681
[10] Landsberg P T and Leff H S 1989 *J. Phys. A: Math. Gen.* 22 4019
[11] Bejan A 1995 *Int. J. Heat Mass Transfer* 38 433
[12] De Vos A 1985 *Am. J. Phys.* 53 570
[13] Rubin M H 1979 *Physical Review A* 19 1272
[14] Göktun S Özkaynak S and Yavuz H 1993 *Energy - The Int. J.* 18 651
[15] Salamon P and Nitzan A 1981 *J. Chem. Phys.* 74 3546
[16] Chen L and Yan Z 1989 *J. Chem. Phys.* 90 3740
[17] Agnew B Anderson A and Frost T H 1997 *Applied Thermal Engineering* 17 3
[18] Özkaynak S Göktun S and Yavuz H 1994 *J. Phys. D: Appl. Phys.* 27 1139

[19] Wu C and Kiang L R 1992 *Energy* 17 1173
[20] Gordon J M and Huleihil M 1992 *J. Appl. Phys* 72 829
[21] Chen J 1994 *J. Phys. D: Appl. Phys.* 27 1144
[22] Chen J Sun F and Wu C 1997 *Applied Thermal Engineering* 17 277
[23] Bisio G 1996 *Proceedings of ECOS'96* (KTH - Royal Ins.of Tech.) 127
[24] Erbay L B and Yavuz H 1997 *J. Phys. D: Appl. Phys.* 30 2841
[25] Lienhard J H 1981 *A Heat Transfer Textbook* (Prentice - Hall, Inc.)
[26] Öz•••k M N 1973 *Radiative Heat Transfer* (Wiley - Interscience Publ.)

Finite Time Thermodynamics: Limiting Possibilities of the Regenerative Systems

A. M. Tsirlin and V. Kazakov

Abstract

In this paper a thermodynamic system with regeneration of the working body is defined. We show that systems where energy transformation occurs as well as economic systems where commodities are exchanged obey this definition. The conditions under which such systems operate optimally are derived. Both systems with stationary and non-stationary reservoirs are considered.

1 Introduction and formulation of the problem

In this paper we consider *systems with regeneration.* The structure of such a system is shown in Figure 1. It consists of a finite number of reservoirs or sources (it could be just one reservoir) and the working body. When the working body is in contact with reservoir an exchange of mass, energy or some other types of resources between the working body and reservoir takes place. The reservoirs are not in contact with each other directly. The working body operates in such a fashion that on average over large enough period of time its state does not change. That is, after its parameters have been changed it restores (regenerates) their values.

Let us give two examples of such systems which will be considered in more detailed form later:

1. Heat engine.

In this system sources are heat reservoirs. If the working body is in contact with the reservoir then there is a heat flow which depend on the temperatures of the working body and the reservoirs. The goal of this system is to produce work. If the power of the heat engine is finite then the establishment of the heat flows in the system leads to the entropy production in it (dissipation of energy). The parameters of the working body are constant on average over the cycle of operation.

2. Trading company.

Consider a finite number of markets (reservoirs of resource) and a wholesaler or trading firm that buys commodity (resource) in some markets and resells it in another. The stock of commodity in its warehouse is constant on average. The goal of this system is to generate profit. The establishment of sufficiently intensive flows of goods requires some expenses (trade expenses). In this system these expenses are the dissipative flux.

In order to emphasize the variety of regenerative systems we give some more examples of such systems:

- Heat and refrigerate apparatus and heat transformers;
- Absorption-desorption cycles of gas cleansing, where liquid absorbent consequently passes through absorber and desorber regenerating its parameters. Such cycle occurs in animals where lungs are used as absorber, body tissues as desorber and blood which circulates between them as the working body;
- Bank which gets funds in the lenders market and re-lends them in the borrowers market where interest rate is higher;
- Broker which buys shares at one price and sells them at another price in one or a number of sharemarkets;
- Closed gas and water supply systems that do not pollute the environment.

The working body can be uniform or nonuniform spatially at any instant of time. In the former case it is called the working body with concentrated (lumped) parameters and in the latter case the working body with distributed parameters. An internal combustion engine and systems with circulating working body are close to the first class of systems. Turbines, rectification columns, etc. belong to the second class.

In this paper we will consider only the systems with lumped parameters of the working body. Nevertheless, the proposed optimization methods are also applicable to the systems with distributed parameters of the working body.

Let us define an important subclass of regenerative systems - the *ther-*

modynamic regenerative systems. They have the following characteristic features:

1. The variables which describe both reservoirs and the working body can be divided into two groups: extensive and intensive. The variables from the first group (volume, mass, amount of resource, etc.) are proportional to the size of the subsystem. The values of the variables from the second group (temperature, concentrations, prices, etc.) do not change in we combine two subsystems with the same values of these parameters.

We denote extensive variables of the i-th reservoir as x_i and the intensive ones as v_i. For the working body they are denoted as y and u correspondingly.

2. Contact between two thermodynamic systems yields exchange fluxes g_i between them. The values of these fluxes depend on the values of the intensive variables of contacting subsystems only and not on the extensive ones. The dimension m of the vector-function g_i is equal to the dimension of the vector of extensive variables x_i, and

$$g_i(u, v_i) = 0, \quad \text{if} \quad u = v_i; \tag{1}$$

$$\frac{dx_i}{dt} = k_i g_i(u, v_i), \quad i = 1, \ldots, n, \tag{2}$$

where $k_i(t)$ is switching function which is equal zero if there is no contact between i-th reservoir and the working body and equal one otherwise;

$$\frac{dy}{dt} = \Sigma_{i=1}^{n} k_i g_i(u, v_i). \tag{3}$$

3. The objective (key) flux r and dissipative flux σ in regenerative thermodynamic systems depend on the exchange fluxes $\tilde{g}_i = k_i g_i$, $i = 1, \ldots, n$, in such a fashion that

$$\left.\begin{aligned} \sigma(\tilde{g}, u, v) = r(\tilde{g}, u, v) = 0 \ \text{ if } \ \tilde{g} = 0 \\ \sigma(\tilde{g}, u, v) > 0 \ \text{ if } \ \tilde{g} \neq 0. \end{aligned}\right\} \tag{4}$$

We will demonstrate that all the above-described examples of regenerative systems are thermodynamic ones which obey conditions (1)-(4).

The problem of optimization of thermodynamic regenerative system consists of such choice of the contact functions $k_i(t)$ and intensive variables of the working body $u(t)$ that the dissipative flux is minimal subject to given value of the objective flux $r = r_0$ (the problem of minimal dissipation).

Before this problem is stated the problem of limiting values of the objective flux which can be achieved in the system $r^{\max}$ and $r^{\min}$ has to be solved.

We will state these problems formally, discuss their mathematical features and write down their optimality conditions. Then we will give the results of optimization of some of the particular systems listed above.

A wast number of publications is devoted to the investigation of the thermodynamic regenerative systems of physical nature with fixed time or rate of fluxes [1]-[7]. In this paper we develop a more general approach which makes possible to consider economic and other systems which satisfy the above-described conditions with arbitrary number of stationary and non-stationary reservoirs.

2 Optimality conditions for regenerative thermodynamic systems

Let us write down the formal statement of the problem of limiting value of the objective flux

$$r_0 = \frac{1}{\tau}\int_0^\tau r(k,u,v)dt \to \max_{k,u} \tag{5}$$

subject to the conditions of regeneration $y_j(0) = y_j(\tau)$

$$\overline{\frac{dy_j}{dt}} = \frac{1}{\tau}\int_0^\tau \Sigma_{i=1}^n k_i g_{ij}(u,v_i)dt = 0 \quad j = 1,\dots,m. \tag{6}$$

$$k_i = \{0;1\}, \quad i = 1,\dots,n. \tag{7}$$

Conditions of regeneration (6) are the conditions of zero average rates of change of all m extensive variables of the working body. Overline denotes time averaging over the regeneration period τ.

The problem of minimal dissipation has similar form

$$\overline{\sigma} = \frac{1}{\tau}\int_0^\tau \sigma(k,u,v)dt \to \min_{k,u} \tag{8}$$

subject to constraints (6), (7) and additional condition

$$\frac{1}{\tau}\int_0^\tau \Big[r(k,u,v) - r_0\Big]dt = 0. \tag{9}$$

If $r^{\min} \leq r_0 \leq r^{\max}$ then a solution of the problem (6)–(9) exists.

Substitution of the optimal solution $k^*(t), u^*(t)$ into equations (2), (3) gives the optimal pathways $x^*(t)$ and $y^*(t)$.

If the key flux is not removed from an open system ($r_0 = 0$) then the fluctuations of the parameters of the system correspond to the minimum of the dissipative flux and can be found as the solution of the problem of minimization of the averaged dissipation where regenerative period $\tau \to \infty$. This statement corresponds to the Prigogin principle of minimal entropy production for traditional thermodynamic systems [8] where entropy production σ is substituted with its average value $\overline{\sigma}$.

The optimal solution of both stated problems (since mathematically they are equivalent) essentially depends on the assumption of how the intensive variables of the sources change over time. Let us consider two cases:

1. Vector v is constant and does not depend on x and t (stationary sources with infinite capacity).

2. Vector $v(t)$ is a function of time (random or deterministic) which does not depend on x (non-stationary source with infinite capacity).

In both cases exchange fluxes g_i, which change values of x_i, does not affect v_i (actually this is the definition of reservoir with infinite capacity). If source has finite capacity then the right-hand side of equation (2) depends on x. Therefore these equations have to be taken into account when considering both minimal dissipation problem and limiting objective flux problem. Let us note that the limiting values of the efficiency criteria for the systems with infinite capacity sources are always "better" than the values for the systems with finite capacity sources and that the difference between them is a monotonically increasing function of τ. Therefore if the limiting possibilities of regenerative thermodynamic systems are being analysed then it is sufficient to consider the problems with infinite capacities of the sources only.

A. Stationary sources

In this case the stated problems are the averaged problems of nonlinear programming. The optimality conditions for such type of problems were obtained in [9], [10]. Let us write their optimality conditions without derivations. We denote here the vector-function of the control variables as $\tilde{u}(t) = (k(t), u(t))$:

1. The optimal solution $\tilde{u}^*(t)$ of the formulated problems is a piece-wise constant function which takes not more than $(m+1)$ values $\tilde{u}_\nu$ (called the *basic* values) in problem (5) – (7) and not more than $(m+2)$ basic values in problem (6) – (9). From equations (3) it follows that the state variables of the working body are piece-wise constant functions of time in an optimal regime.

2. The sequence in which function $\tilde{u}^*(t)$ takes basic values does not matter. What matters are the fractions γ_ν of the interval τ during which $u^*(t) = u_\nu$. Hence if the number of the basic values is more than one then there are infinite number of optimal solutions. All of them have the same value of the optimality criterion.

3. In order to find $\tilde{u_\nu}$ in the problem (5) – (7) it is necessary to construct its Lagrange function L_1 and then to find the minimum with respect to Lagrange multipliers λ of the maximum of L_1 with respect to $\tilde{u}$. We get

$$\min_\lambda \left[\max_{\tilde{u}} L_1(\lambda,\tilde{u},v)\right] = \min_\lambda \left[\max_{\tilde{u}}(r(\tilde{u},v) + \Sigma_{j=1}^m \lambda_j \Sigma_{i=1}^n k_i g_{ij}(u,v_i))\right]. \quad (10)$$

Similarly, for the problem (6) – (9) we get (after taking into account that the minimum of dissipation is sought here)

$$\max_\lambda \left[\min_{\tilde{u}} L_2(\lambda,\tilde{u},v)\right] =$$

$$= \max_\lambda \left[\min_u(\sigma(\tilde{u},v) + \Sigma_{j=1}^m \lambda_j \Sigma_{i=1}^n k_j, g_{ij}(u,v_i) + \lambda_{m+1}(r(\tilde{u},v) - r_0))\right]. \quad (11)$$

It is assumed that the solutions of these problems are not singular and multiplier λ_0 in front of r and σ is set to one.

Let us note that the conditions (10), (11) do not require functions which determine these problems to be differentiable on $\tilde{u}$. These conditions are also applicable if all or some of the components of control take discrete values (such are contact functions $k_i(t)$).

4. After the basic values of the control variables $\tilde{u}_\nu$ are found from the conditions (10), (11), the corresponding fractions γ_ν of the regeneration period τ are determined by the equations

$$\Sigma_{\nu=0}^m \gamma_\nu \Sigma_{i=1}^n k_i^\nu g_{ij}(u^\nu, v_i) = 0, \quad j = 1,\ldots,m, \quad (12)$$

$$\Sigma_{\nu=0}^m \gamma_\nu = 1, \quad \gamma_\nu \geq 0. \quad (13)$$

Since the maximal number of the basic values in the problem (6) – (9) is $(m+1)$ the upper limits in summing on ν in (12), (13) should be set to $(m+1)$, and one more equation

$$\Sigma_{\nu=0}^{m+1}\gamma_\nu r(\tilde{u}^\nu, v) = r_0. \tag{14}$$

should be added to the set of problem's constraints.

Hence the most complex problem here is the calculation of the basic values from conditions (10) or (11). When this is done the weight multipliers (time fractions γ_ν) can be found easily from the set of linear equations (12) – (14).

B. Non-stationary sources

The optimal solution of the problems with the stationary sources γ^*, $\tilde{u}^*$ depends on the value of the vector v and does not depend explicitly on time. In order to derive optimality conditions for the problem with non-stationary sources we introduce the distribution $P(v)$ of the function $v(t)$ over the set V of all its feasible values such that the fraction $\delta\tau(v_0)$ of the interval τ during which $v_0 \le v < v_0 + \Delta$ is

$$\delta\tau(v_0) = \int_{v_0}^{v_0+\Delta} P(v)dv.$$

For the stationary random process $v(t)$ and $\tau \to \infty$ the introduced distribution $P(v)$ is the density of the distribution of this random process. Constancy of the vector $v(t) = v_0$ corresponds to the distribution $P(v) = \delta(v - v_0)$.

For the system with non-stationary sources the optimality conditions of the problem (5) – (7) have the following form [10]:

$$\min_\lambda \int_V P(v)\Big[\max_{\tilde{u}} L_1(\lambda, u, v)\Big]dv =$$

$$= \min_\lambda \int_V P(v)\Big[\max_{\tilde{u}}\Big(r(\tilde{u}, v) + \Sigma_{j=1}^m \lambda_j \Sigma_{i=1}^n k_i g_{ij}(u, v_i)\Big)\Big]dv. \tag{15}$$

And for the problem (6) – (9) with non-stationary sources the optimality conditions have the form [10]:

$$\max_\lambda \int_V P(v)\Big[\min_{\tilde{u}} L_2(\lambda, u, v)\Big]dv =$$

$$= \min_{\lambda} \int_V P(v)\Big[\min_{\tilde{u}} \Big(\sigma(\tilde{u}, v) + \Sigma_{j=1}^m \lambda_j \Sigma_{i=1}^n k_i g_{ij}(u, v_i) + \lambda_{m+1}(r(\tilde{u}, v) - r_0)\Big)\Big] dv. \tag{16}$$

In a system with one source only the introduction of the switching function is meaningless because the exchange flux g always can be made equal zero by choosings $u(t) = v(t)$. If, as it is most often the case, the dependence $r(g)$ for fixed $v = v_0$ is a convex function then the maximum of the average value of r subject to condition of zero average value of g is also equal zero. Thus, it is meaningless to use such regenerative systems. The situation is quite different if the source is non-stationary. As we will show later, the conditions (15) make possible to produce positive objective flux in a system with one source only.

3 Using optimality condition for the analysis of thermal-mechanical systems

Let us consider how these optimality conditions can be applied to particular thermal-mechanical regenerative systems. Since such systems have been studied for a long time part of these results are known. Nevertheless it is convenient to use them as examples that demonstrate the benefits of the proposed approach based on the methods of averaged optimization.

3.1 Heat engine with one heat source

We denote the temperature of the source as T_0 and the temperature of the working body as T. The objective flux in this problem is the flux of energy $q(T_0, T)$ because from the energy balance it follows that the average value of the heat flux is equal to the average power $\bar{r}$ of the engine. Extensive variables of the source and the working body are internal energies E_0 and E and entropies S_0 and S.

We assume that the distribution $P(T_0)$ of the temperature of the working body that corresponds to the function $T_0(t)$ is given.

The problem of limiting value of the objective flux here takes the following form:

$$r_0 = \int_V P(T_0) q(T_0, T) dT_0 \to \max \tag{17}$$

subject to constraint

$$\int_V P(T_0)\frac{q(T_0,T)}{T}dT_0 = 0. \tag{18}$$

Here V is the set of all feasible values of the reservoir's temperature.

Condition (18) is the condition of the regeneration of the working body (average rate of entropy change is equal zero). Similar condition with respect to its internal energy holds because the average power produced in the system is equal to the average flux of heat. This is taken into account in (17).

The optimality condition (15) takes the form

$$\min_\lambda \int_V P(T_0)\Big\{\max_T \Big[q(T_0,T) - \lambda\frac{q(T_0,T)}{T}\Big]\Big\}dT_0. \tag{19}$$

Let us emphasize that the conditions (19) are also applicable if the dependence $q(T_0,T)$ is not differentiable on T, if the set of feasible values of T is constrained or if T can take only discrete set of values.

If we assume that the law of heat transfer is a smooth function of T then the conditions of maximum on T take the form

$$T^2\frac{\partial q}{\partial T} - \lambda\Big(T\frac{\partial q}{\partial T} - q(T_0,T)\Big) = 0, \tag{20}$$

$$T^3\frac{\partial q}{\partial T^2} - \lambda(\frac{\partial^2 q}{\partial T^2}T^2 - 2\frac{\partial q}{\partial T}T + 2q) < 0. \tag{21}$$

These expressions determine the dependence of the optimal temperature T^* of the working body on T_0 and λ. After substitution of $T^*(T_0,\lambda)$ into (19) λ^* is to be found from the condition of the minimum of the average value of the expression which is inside curly braces in (19).

For the law of heat transfer which is linear on T

$$q = \alpha(T_0)(T_0 - T) \tag{22}$$

from (20) it follows that for any α

$$\frac{T_0}{T^2} = \frac{1}{\lambda},$$

or

$$T^*(\lambda, T_0) = \sqrt{\lambda T_0}. \tag{23}$$

The expression inside the curly braces in (19) takes the form

$$\alpha(T_0)(T_0 - \sqrt{\lambda T_0}) - \lambda\frac{\alpha(T_0)(T_0 - \sqrt{\lambda T_0})}{\sqrt{\lambda T_0}} = \alpha(T_0)(\sqrt{T_0} - \sqrt{\lambda})^2, \qquad (24)$$

and the integral (19) is

$$\overline{(\alpha(T_0)T_0)} - 2\sqrt{\lambda}\overline{(\alpha(T_0)\sqrt{T_0})} + \lambda\overline{(\alpha(T_0))}. \qquad (25)$$

The overline here denotes the time averaging or the ensemble averaging with the distribution $P(T_0)$.

Function (25) is concave on λ. From condition of its minimum it follows

$$\sqrt{\lambda^*} = \frac{1}{\overline{\alpha(T_0)}}\overline{\alpha(T_0)\sqrt{T_0}}.$$

The optimal temperature of the working body depends on the temperature of the source as

$$T^*(t) = \frac{\overline{(\alpha(T_0)\sqrt{T_0})}}{\overline{\alpha(T_0)}}\sqrt{T_0(t)} \qquad (26)$$

The limiting value of the power of a heat engine with one source is determined by the following expression

$$r_0^{\max} = \overline{\alpha(T_0)T_0} - \frac{1}{\overline{\alpha(T_0)}}\overline{(\alpha(T_0)\sqrt{T_0})}^2. \qquad (27)$$

If α is a constant then the limiting power is proportional to the standard deviation D of the function $(T_0(t))^{1/2}$

$$r_0^{\max} = \alpha D[\sqrt{T_0}].$$

This result was obtained in [6].

If the temperature of the source is constant then this standard deviation is equal zero and no power can be produced by a heat engine.

Let us consider heat engine where the temperature of the source can take two values T_+ and T_- with probabilities P_+ and P_- correspondingly $(P_+ + P_- = 1)$. Here

$$r^{\max} = P_+\alpha_+T_+ + P_-\alpha_1T_- - \frac{1}{P_+\alpha_+ + P_-\alpha_-}(P_+\alpha_+\sqrt{T_+} + P_-\alpha_-\sqrt{T_-})^2. \qquad (28)$$

Or, denoting $\beta_+ = P_+\alpha_+, \beta_- = P_-\alpha_-$, after simple transformations (see [7]) we get

$$r^{\max} = \frac{\beta_+\beta_-}{\beta_+ + \beta_-}\left(\sqrt{T_+} - \sqrt{T_-}\right)^2. \tag{29}$$

The problem of minimal dissipation subject to the given power $r_0 < r^{\max}$ can be solved similarly. Since the entropy of the working body does not change on average the dissipative flux is equal to the rate of change of the entropy of the source

$$\sigma = \int_V P(T_0)\frac{q(T_0,T)}{T_0}dt \to \min . \tag{30}$$

The minimum of this functional has to be found subject to constraints (18) and

$$\int_V P(T_0)q(T_0,T)dt = r_0. \tag{31}$$

Let us write down the optimality condition (16) for this problem

$$\max_{\lambda_1,\lambda_2} \int_V P(T_0)\Big\{ \min_T \Big[q(T_0,T)\Big(\frac{1}{T_0} + \lambda_1\frac{1}{T} - \lambda_2(1-r_0)\Big)\Big]\Big\}dT_0.$$

The condition of minimum on T of the expression inside the square brackets yields the equation which determines $T^*(T_0,\lambda)$,

$$T^2\frac{\partial q}{\partial T}(\lambda_2(1-r_0) - \frac{1}{T_0}) - \lambda_1(T\frac{\partial q}{\partial T} - q(T_0,T)) = 0.$$

3.2 Heat engine with two reservoirs

Assume the system has two or more sources. Here it is required to choose not only the temperature of the working body but the switching functions also. For two sources with the temperatures T_+ and T_- the problem of limiting power of the heat engine takes the form

$$r_0 = \overline{(k_+q_+(T_+,T) + k_-q_-(T_-,T))} \to \max_{T,k_+,k_-} \tag{32}$$

subject to the condition of regeneration of the working body

$$\Big(\overline{\frac{k_+q_+(T_+,T)}{T}}\Big) + \Big(\overline{\frac{k_-q_-(T_-,T)}{T}}\Big) = 0. \tag{33}$$

Here again the overline denotes time averaging over the period of the cycle; $k_+(t)$ and $k_-(t)$ are switching functions with control contacts with the hot and cold source correspondingly. The control variables and the Lagrange multiplier λ in this problem are determined by the condition (15) which takes the form

$$\min_{\lambda}\Big\{\max_{k_+,k_-,T}\Big[k_+q_+(T_+,T)(1-\lambda/T)+k_-q_-(T_-,T)(1-\frac{\lambda}{T})\Big]\Big\} \tag{34}$$

The conditions of maximum on k_+ and k_- yield the requirement

$$k_+=\begin{cases}1 & \text{if} \quad (T_+-T)(T-\lambda)>0\\ 0 & \text{if} \quad (T_+-T)(T-\lambda)<0\end{cases}$$

$$k_-=\begin{cases}1 & \text{if} \quad (T_--T)(T-\lambda)>0\\ 0 & \text{if} \quad (T_--T)(T-\lambda)<0\end{cases} \tag{35}$$

Here we took into account that the sign of the heat flux coincides with the sign of the temperature difference between the source and the working body.

Since $T_+>T>T_-$ a contact between the working body and the hot source is established when its temperature $T>\lambda$. When $T<\lambda$ a contact between the working body and the cold reservoir is established. If $k_+=1, k_-=0$ then the conditions of maximum on T have the form (20), where $T_0=T_+$. Similarly during a contact with the cold source $T_0=T_-$ in these conditions, which determines $T_1^*(T_+,\lambda)$ and $T_2^*(T_-,\lambda)$. Both terms inside the square brackets in condition (34)

$$L_+(\lambda)=k_+q_+(T_+,T_1^*(T_+,\lambda))\Big(1-\frac{\lambda}{T_1^*(T_+,\lambda)}\Big)$$

and

$$L_-(\lambda)=k_-q_-(T_-,T_2^*(T_-,\lambda))\Big(1-\frac{\lambda}{T_2^*(T_-,\lambda)}\Big)$$

are positive, but the former one rises if λ decreases (because $T_1^*>\lambda$ and $q_+>0$), and the latter one rises (since $T_2^*<\lambda$ and $q_-<0$). Hence (34) is reduced to

$$\min_{\lambda}\Big\{L_+(\lambda),L_-(\lambda)\Big\}.$$

This minimum is achieved if $L_+(\lambda)=L_-(\lambda)$. This condition determines the optimal λ^* and hence $T_1^*(T_+,\lambda^*)$ and $T_2^*(T_-,\lambda^*)$.

For the linear law of heat transfer (22) the limiting value of $r_0^{\max}$ depends on the fractions γ_+ and γ_- of the regeneration period when $k_+ = 1$ and $k_- = 1$ correspondingly. Here the maximal power $r_0^{\max}$ can be found using the formula (28) where P_+ and P_- are substituted with γ_+ and γ_- correspondingly. Let us find the optimal values of γ_+ and γ_- without writing this expression explicitly. Since $\gamma_+ + \gamma_- = 1$ they are determined by the condition

$$\frac{\partial r_0^{\max}}{\partial \gamma_+} = \frac{\partial r_0^{\max}}{\partial \gamma_-}.$$

From this equation it follows [6] that the ratio of these durations of the contacts does not depend on the temperatures of the sources and is equal to

$$\frac{\gamma_+}{\gamma_-} = \sqrt{\frac{\alpha_-}{\alpha_+}}. \tag{36}$$

Here the limiting power is

$$r^{\max} = \frac{\alpha_- \alpha_+}{(\sqrt{\alpha_-} + \sqrt{\alpha_+})^2}(\sqrt{T_+} - \sqrt{T_-})^2. \tag{37}$$

Since coefficients α_+ and α_- represent the contact surfaces of the heat engine they depend implicitly on its cost. Let us introduce the additional constraint on the cost of the engine which has the following form

$$C_+ \alpha_+ + C_- \alpha_- = A, \tag{38}$$

where C_+ and C_- are specific cost of creation a unit of contact surface with hot and cold source. After finding the optimal α_- and α_+ from the condition

$$\frac{\partial r^{\max}}{\partial \alpha_+} : \frac{\partial r^{\max}}{\partial \alpha_-} = \frac{C_+}{C_-},$$

we get

$$\frac{\alpha_+^*}{\alpha_-^*} = \left(\frac{C_-}{C_+}\right)^{2/3}, \tag{39}$$

and the corresponding value of the limiting power is

$$r_0^{\max} = \frac{A}{(\sqrt{C_- + C_+^{1/3} C_-^{2/3}} + \sqrt{C_+ + C_-^{1/3} C_+^{2/3}})^2}(\sqrt{T_+} - \sqrt{T_-})^2. \tag{40}$$

It is easy to show that the efficiency of such engine does not depend on heat transfer coefficients. Therefore it does not depend on A, C_+ and C_- and is determined by the following expression which was obtained in [1] and [2]

$$\eta_{NCA} = 1 - \frac{\sqrt{T_-}}{\sqrt{T_+}}.$$

The problem of the optimal distribution of the contact surfaces was formulated in [5].

Similar approach makes possible to solve the problem of minimal dissipation subject to fixed power r_0 for the heat engine with two sources. In [4] it was shown that this solution corresponds to the maximum of its efficiency. Let us give this solution for the heat transfer law (22) without derivation. The limiting efficiency r_0 depends on the power of the engine, the temperatures of the sources and its limiting power $r_0^{\max}$ as

$$\eta_{\max} = \frac{2\delta k}{\delta k + 1 - \sqrt{(1-k)(1-k\delta^2)}}, \tag{41}$$

where

$$k = r_0 \big/ r^{\max}; \quad \delta = \frac{\sqrt{T_+} - \sqrt{T_-}}{\sqrt{T_+} + \sqrt{T_-}}.$$

Unlike the previous problem cost coefficients C_+ and C_- affect the efficiency via $r^{\max}$. Because η_{max} depends on r^{max} monotonically the optimal distribution of surfaces is determined by the condition (39).

3.3 Heat engine with a number of non-stationary sources

Let us consider the most general form of thermal-mechanical regenerative system with n sources. The temperatures of these sources vary over time. Here it is required to choose $k_i(t)$ and the temperature of the working body $T(t)$. Heat apparatus which utilize the heat of waste gases in metallurgy are close to this type of heat engine.

Note that the definition of efficiency as the ratio of r_0 to the average flux of heat removed from the hot source becomes meaningless for the system with $n > 2$ because the heat can be transferred to the working body from a number of reservoirs with different temperatures $T_{0i} > T$ and the corresponding heat fluxes are not equivalent.

We introduce the vector-function of the temperatures of the reservoirs $T_0(t)$ with components $T_{0i}(t)$, the set V of the feasible values of T_0 and the distribution $P(T_0)$ of these values $P(T_0)$ on V. Then the problem of limiting power takes the form

$$r_0 = \int_V P(T_0)\Sigma_i k_i q_i(T_{0i}, T)dT_0 \to \max_{k,T} \tag{42}$$

subject to constraint

$$\int_V P(T_0)\Sigma_i \frac{k_i q_i(T_{0i}, T)}{T} dT_0 = 0. \tag{43}$$

Optimality conditions (15) for this problem take the following form

$$\min_\lambda \int_V P(T_0) \max_{k,T} \left\{\Sigma_i k_i q_i(T_{0i}, T)(1 - \lambda/T)\right\} dT_0 \tag{44}$$

The maximum conditions on k yield the equality

$$k_i^* = \begin{cases} 1, & \text{if} \quad q_i(T_{0i}, T)(1 - \frac{\lambda}{T}) > 0, \\ 0, & \text{if} \quad q_i(T_{0i}, T)(1 - \frac{\lambda}{T}) < 0, \quad i = 1, \ldots, n \end{cases} \tag{45}$$

Assuming that the functions q_i obey the standard requirements of monotonicity on T_{0i} and T and are zero if $T = T_{0i}$ from (45) it follows that if $T > \lambda$ then the working body has to be in contact with all "hot" reservoirs whose temperatures $T_{0i} > T$ receiving heat from all of them. If $T > \lambda$ then it has to be in contact with all the "cold" reservoirs whose temperatures $T_{0i} < T$ rejecting heat into them. If at some instance of time the working body is in contact with the i-th reservoir and $T_{0i} > T$ then it is in contact with any of j-th sources whose temperatures $T_{j0} > T_{i0}$. The same is true for heat removal. Let us assume now that $T_{j0} > T_{i0}$ if $j > i$. Condition (45) divide the set of all reservoirs into two subsets of "hot" and "cold" ones. λ is the boundary between their temperatures.

Let us introduce for each source the heat supply function $q_i^+(T)$ and the heat removal function $q_i^-(T)$

$$q_i^+ = \frac{1}{2}(q_i + |q_i|), \quad q_i^- = \frac{1}{2}(q_i - |q_i|),$$

and let us denote the total heat fluxes of supply and removal as $q^+(T)$ and $q^-(T)$ correspondingly

$$q^+(T) = \Sigma_i q_i^+(T_{0i}, T), \quad q^-(T) = \Sigma_i q_i^-(T_{0i}, T).$$

These functions are shown in Figure 2.

The optimal temperatures of the working body during heat supply and heat removal (we denote them as T_+ and T_-) maximize the expression which is enclosed in curly brackets in (44). If it is convex then in all points where $q^+(T)$ and $q^-(T)$ are smooth, i.e. for almost all T we get

$$\frac{\partial L^+}{\partial T_-} \Rightarrow \frac{dq^+}{dT_+} - \frac{\lambda}{T_+}(\frac{dq^+}{dT_+} - \frac{q^+}{T_+}) = 0, \tag{46}$$

$$\frac{\partial L^-}{\partial T_-} \Rightarrow \frac{dq^-}{dT_-} - \frac{\lambda}{T_-}(\frac{dq^-}{dT_-} - \frac{q^-}{T_-}) = 0,$$

The optimal value of λ is determined by the requirement that the values of $L^+_{\max}(\lambda)$ and $L^-_{\max}(\lambda)$ averaged over T_0 are equal

$$\int_V P(T_0)q^+(T_+)(1 - \frac{\lambda}{T_+}) - q^-(T_-)(1 - \frac{\lambda}{T_-})\}dT_0 = 0. \tag{47}$$

Here the averaging over T_0 is meaningful because T_+ and T_- depend on vector T_0 and λ. From (46) it follows that for an arbitrary law of heat transfer the optimal ratio of the temperatures obeys the equation

$$\frac{T_-}{T_+} = \frac{\frac{dq^+}{dT_+}(\frac{dq^-}{dT_-} - \frac{q^-}{T_-})}{\frac{dq^-}{dT_-}(\frac{dq^+}{dT_+} + \frac{q^+}{T_+})}. \tag{48}$$

Using overline to denote averaging over T_0 we get from (47) the expression for λ

$$\lambda = \frac{\overline{q^+(T_+)} - \overline{q^-(T_-)}}{\overline{(\frac{q^+(T_+)}{T_+})} - \overline{(\frac{q^-(T_-)}{T_-})}}. \tag{49}$$

The joint solution of the equations (45), (46) and (49) determines the optimal switching functions $k_i^*(T, T_0)$, the optimal temperatures of the working body $T_+^*(T_0), T_-^*(T_0)$ and the optimal value of the Lagrange multiplier λ.

Similar expressions determine the optimal solution of the problem of minimal dissipation with fixed power. The only difference is that in this problem in addition to the constraint (43) there is one more averaged constraint on the power of heat engine. Unlike the problem (42)–(43) where the working body is in contact sequentially with all the "hot" reservoirs ($T_{0i} > T_+$) and then with all the "cold" reservoirs ($T_{0i} < T_-$) and the reservoirs with the "middle" temperatures ($T_- \leq T_{0i} \leq T_+$) are not used, in the problem of minimal dissipation the working body is in contact only with the reservoirs whose temperatures fall into the intervals $f_1(r_0) > T_{0i} > T_+$ and $T_- > T_{0i} > f_2(r_0)$. That is, not only the reservoirs with the temperatures between T_+ and T_- are not used but the sources with very high and very low temperature are not used either. The algorithmic expressions for the boundaries $f_1(r_0)$ and $f_2(r_0)$ are given in [11].

Different types of thermal-mechanical systems (heat pumps, refrigerators, heat transformers) are very close to the heat engines from the point of view of their optimization. Their limiting possibilities can be estimated using the approach of section 2 (see. for example [4]).

4 Using optimality conditions for the analysis of economic systems

The goal of this section is not only to obtain limiting estimates of the operational indexes of a commodity exchange system but also to demonstrate deep analogy between economic exchange systems and physical and chemical thermodynamic systems. Because of that we will consider the economic analogies of the problems which were considered in previous section.

4.1 Buying and selling in a market with varying prices

We assume that a scalar resource (commodity) is being traded and denote its equilibrium price in the market as C_0. The trading firm buys and re-sells this commodity at price C. The flow of buying (selling) is denoted as $g(C_0, C)$. The sign of this flow is determined by the sign of the difference $(C - C_0)$

$$g(C_0, C) \begin{cases} > 0 & \text{if} \quad C > C_0 \\ = 0 & \text{if} \quad C = C_0 \\ < 0 & \text{if} \quad C < C_0. \end{cases}$$

The flow of capital $q(C_0, C) = -Cg(C_0, C)$ is connected to flow of commodity g. The state of the firm is characterized by the stock of commodity G and its current capital K, (extensive variables). Here

$$\frac{dG}{dt} = g(C_0, C), \quad \frac{dK}{dt} = -g(C_0, C)C. \tag{50}$$

Note that the right-hand sides of equations (50) do not depend on G and K.

Assume that $C_0(t)$ is a known deterministic or random function of time which takes values from the set V with the distribution $P(C_0)$. The problem of maximizing the average profit (average rate of capital increase) subject to the condition of not accumulating stocks ($\bar{g} = 0$) takes the form which is analogical to the problem (17), (18)

$$I = -\int_V P(C_0)Cg(C_0, C)dC_0 \to \max_C \tag{51}$$

subject to the condition of regeneration

$$\int_V P(C_0)g(C_0, C)dC_0 = 0. \tag{52}$$

The optimality condition (15) takes the form

$$\min_\lambda \int_V P(C_0)\left\{ \max_C \left[g(C_0, C)(\lambda - C) \right] \right\} dC_0, \tag{53}$$

and for the smooth function g

$$\frac{\partial g}{\partial C}(C + \lambda) + g(C_0, C) = 0. \tag{54}$$

If the flow of commodity depends linearly on the price difference

$$g(C_0, C) = \alpha(C_0)(C - C_0),$$

(where $\alpha(C_0)$ implicitly characterizes the number of points of trade and therefore the cost of trade equipment) we get from the condition (53)

$$C^*(t) = \frac{C_0(t) + \lambda}{2}, \quad \lambda = \frac{\overline{\alpha(C_0)C_0}}{\overline{\alpha(C_0)}}, \tag{55}$$

and the limiting profit is

$$I_{\max} = \frac{1}{4}\left[\overline{\alpha(C_0)C_0^2} - \frac{1}{\overline{\alpha(C_0)}}(\overline{\alpha(C_0)C_0})^2\right]. \tag{56}$$

The overline here denotes the ensemble averaging over the set V. If α is constant then from (56) it follows that

$$I_{\max} = \frac{\alpha}{4}D[C_0], \tag{57}$$

where D is the standard deviation of the function C_0.

In reality not one but a number of commodities (vector of commodities) are usually being traded and their prices are the components of the vector-function of prices $C_j(t)$, $(j = 1, \ldots, m)$. If the flow of the j-th commodity depends on C_j and C_{j0} only and the distribution $P(C_0) = \Pi_j P_j(C_{0j})$ then the problem with vector of commodities is decomposable into m scalar problems. Condition (53) does not contain differentiation operator and can be used when prices C_j are bounded and depend on each other $(C \in V_c)$. The maximum should be sought for each C_0 on the set V_c. The value of the vector λ depends on the averaged characteristics of the equilibrium prices C_0 which can be approximated using historic data of the previous market behaviour and then corrected in real-time.

In many cases only the dependence $g(C)$ can be found from the actual trading data. Here C_0 is determined as a root of the equation $g(C) = 0$. Since $g(C)$ is a monotonic function this is a unique root.

The considered problem is very close to the problem of buying and selling bonds whose yield vary. We did not take into account the restriction on the capital K including the possibility of bankruptcy. Taking these restrictions into account can only reduce the estimate of the limiting profit (56), (57).

4.2 Commodity exchange via intermediary

We consider now the problem with two markets with equilibrium prices C_{01} and $C_{02} > C_{01}$. The exchange of commodity occurs via the intermediary (wholesaler or trading firm) who set the price at which it buys in the first market C_1 and C_2 at which it re-sells in the second one. It is clear that if the intermediary tries to maximize the increase of its capital ΔK then it buys and sells at the equilibrium prices and for any dependencies $g_1(C_1, C_{01})$ and

$g_2(C_2, C_{02})$ the ratio of the intermediary's net profit to its initial capital K_0 is

$$\eta_0 = \frac{\Delta K}{K_0} = \frac{C_{20}}{C_{10}} - 1 \tag{58}$$

This expression is an analogy of Carnot efficiency of the heat engine. Unlike the Carnot efficiency it can be greater than one.

But when the trading is done at equilibrium prices g_1 and g_2 and hence the flow of the intermediary's profit are infinitely close to zero. The problem of the limiting value of the (objective) flow of commodity is

$$r = \overline{q(C, C_0)} = -\Sigma_{i=1}^{2} \overline{g_i(C_i, C_{0i}) C_i} \to \max_{C_i} \tag{59}$$

subject to the condition of regeneration (not accumulation of the commodity by the intermediary)

$$\Sigma_{i=1}^{2} \overline{g_i(C_i, C_{0i})} = 0 \tag{60}$$

Here overline denotes the time averaging operation. The switching functions are not used because it is assumed that intermediary can simultaneously sell and buy the commodity. If this is not the case then the switching function k_i should be introduced into the problem.

The optimality conditions for the problem (59)–(60) are

$$\min_{\lambda} \left[\max_{C_i} \Sigma_{i=1}^{2} g_i(C_i, C_{0i})(\lambda - C_i) \right]. \tag{61}$$

And for the smooth functions g_i they lead to the equations

$$g_1(C_1, C_{01}) : \frac{\delta g_1}{\delta C_1} + C_1 = g_2(C_2, C_{02}) : \frac{\delta g_2}{\delta C_2} + C_2 \tag{62}$$

which jointly with (60) determine the optimal prices of buying and selling and the corresponding limiting flow of commodity $r^{\max}$.

In particular if

$$g_1 = \alpha_1 (C_1 - C_{10}), \quad g_2 = \alpha_2 (C_2 - C_{20}) \tag{63}$$

then from conditions (62), (63) it follows that

$$C_1^* = \frac{C_{10}(2\alpha_1 + \alpha_2) + C_{20}\alpha_2}{2(\alpha_1 + \alpha_2)}, \quad C_2^* = \frac{C_{20}(2\alpha_2 + \alpha_1) + C_{10}\alpha_1}{2(\alpha_1 + \alpha_2)},$$

$$r^{\max} = \frac{\alpha_1\alpha_2}{4(\alpha_1+\alpha_2)}(C_{20}-C_{10})^2.$$

Here the flux of net profit per cost of purchasing (the return on the invested capital) is

$$\eta_r = \frac{r_0^{\max}}{C_1^* g_1(C_1^*, C_{10})} = \frac{(C_{20}-C_{10})(\alpha_1+\alpha_2)}{C_{10}(2\alpha_1+\alpha_2)+C_{20}\alpha_2} = \eta_0 \frac{(\alpha_1+\alpha_2)}{2\alpha_1+\alpha_2(1+C_{20}/C_{10})}. \tag{64}$$

Since $C_{20} > C_{10}$

$$\eta_r < \frac{1}{2}\eta_0. \tag{65}$$

If a number of commodities is traded then the flows g_{ij} of purchasing and selling of j-th commodity depend on the choice of prices on this commodity. If the intermediary tries to maximize r_0 subject to constraints imposed onto its trade possibilities (on α_{ij}) then it solves the problem of the optimal distributing these possibilities

$$\Sigma_j r^{\max}(\alpha_{1j}\alpha_{2j}, C_{20}^j, C_{10}^j) \to \max_{\alpha_{ij}} \tag{66}$$

subject to constraint

$$\Sigma_j(s_{1j}\alpha_{1j} + s_{2j}\alpha_{2j}) \leq A, \tag{67}$$

where s_{ij} is the total cost of creating a trading places for buying and selling of j-th commodity, and $r_j^{\max}$ corresponds to the solution of the problem (59), (60).

Note that trade expenses of buying and selling play the role of the dissipative flux σ in the economic exchange problems which are determined by the product of the corresponding exchange flow g and the difference between the actual and equilibrium price [12]:

$$\sigma = (C - C_0)g(C, C_0). \tag{68}$$

It is easy to see that $\sigma > 0$ for both $C > C_0$ (the buying of resource) and $C < C_0$ (the selling of resource) and that minimization of σ for the fixed flux of net profit r_0 corresponds to the maximal return on the invested capital η.

4.3 A number of markets with non-stationary prices

let us consider a generalization of the considered problems where the system consists of intermediary and m markets. Equilibrium prices C_{0j} $j = 1, \ldots, m$ in these markets can change in some regular or random fashion. In both cases the vector distribution of prices $P(C_0)$ over the set of feasible prices V can be introduced. Assume that the exchange flow of commodity between the j-th market and the intermediary has the following form

$$\tilde{g}(C_j, C_{0j}) = k_j g_j(C_j, C_{0j}),$$

where $k_j = \{0; 1\}$ is the switching function. It is required to choose such k_j and C_j that the net flow of profit is maximal

$$r = -\int_V P(C_0)\Sigma_j k_j g_j(C_j, C_{0j})C_j dC_0 \to \max \tag{69}$$

subject to the condition of regeneration (condition of not accumulating the stock)

$$\bar{\dot{G}} = \int_V P(C_0)\Sigma_j k_j g_j(C_j, C_{0j}) dC_0 = 0 \tag{70}$$

and the constraint on the number of market used for trading

$$\Sigma_j k_j \leq N < m. \tag{71}$$

We assume that the switching functions k_j are determined by the averaged characteristics of the market and do not change over time. We denote the set of markets being currently used for trading (for which $k_j = 1$) as J^*.

For this formulation of the problem the optimality conditions (15) take the following form

$$\min_{\lambda,\mu} \max_{k} \left\{ \int_V P(C_0) \max_{C} \Sigma_j k_j \left[g_j(C_j, C_{j0})(\lambda - C_j) - \mu \right] dC_0 + \mu N \right\}, \qquad \mu \geq 0. \tag{72}$$

The maximum conditions on C yields the equations analogical to (69)

$$g_j(C_j, C_{0j}) : \frac{\partial g_j}{\partial C_j} = \lambda - C_j, \quad j = 1, \ldots, m. \tag{73}$$

The choice of the switching functions is determined by the following requirement

$$k_j = 1 \quad \text{if} \quad \overline{g_j(C_j^*(C_0, \lambda), C_{j0})(\lambda - C_j)} \geq \mu.$$

The overline here corresponds to the ensemble averaging over the set of prices V with the distribution $P(C_0)$.

After taking (73) into account the inequality which determines the set J^* of market that are used for trading takes the form

$$\overline{g_j^2(C_j^*(C_0,\lambda),C_{0j}) : \frac{\partial g_j}{\partial C_j}} \geq \mu. \tag{74}$$

Let us specify these conditions for the linear dependence of flows on prices

$$g_j = \alpha_j(C_j - C_{j0}).$$

Assuming that the changes of C_{j0} are independent of each other we get from (73)

$$C_j^*(C_{j0},\lambda) = \frac{C_{j0}+\lambda}{2}, \tag{75}$$

and the condition (74) takes the form

$$\overline{(C_j - C_{j0})^2} \leq \frac{\mu}{\alpha_j}. \tag{76}$$

The substitution of (72) into the equation (75) yields the condition

$$\min_{\lambda} \Sigma_{j\in J^*}\left[\overline{\frac{(\lambda - C_{j0})^2}{4}}\alpha_j\right] \tag{77}$$

which determines the optimal value of λ. The independence of C_{j0} makes possible to average each term independently. It is easy to see that the condition (77) yields equation

$$\lambda = \Sigma_{J\in J^*}\alpha_j\overline{C_{j0}}\Big/\Sigma_{j\in J^*}\alpha_j \tag{78}$$

The summing here is done on the set J^* only. After elimination of C_{j0} from (76) J^* is determined by the condition

$$\lambda^2 + D_j + M_j^2 - 2\lambda M_j \geq \frac{\mu}{\alpha_j} \tag{79}$$

where D_j and M_j are the standard deviation and the expected value of the price C_{j0}. If μ increases then set J^* is shrinking and for some μ^* the number of market which are used becomes less or equal to N. Hence the conditions (78), (79) and (71) allow to find the optimal solution.

5 Conclusion

The conditions of optimal functioning have been obtained for a wide class of regenerative thermodynamic systems with reservoirs of infinity capacity. The parameters of the reservoirs can be stationary or non-stationary. These conditions can be used to derive some known results as well as optimal solution of new problems from this class. It is shown that the commodity exchange systems in financial and economical domains obey general definition of thermodynamic system and their optimal regimes are determined by similar conditions.

References

[1] I. I. Novikov, J. Nucl. Energy II 7, 125 (1958).

[2] F. L. Curzon and B. Ahlborn, Am. J. Phys., 43, 22 (1975).

[3] K. H. Hoffman, S. J. Watowich, R. S. Berry, J. Appl. Phys. 58, 2125 (1985).

[4] L. I. Rozonoer and A. M. Tsirlin, Avtomatika and Telemekanika 44, 50 (1983)

[5] A. Bejan. Am. J. Phys. 64, (8), 1996.

[6] P. Salamon, A. Nitzan, B. Andresen, R. S. Berry. Physical Rev. A, vol. 21, N6. 1980.

[7] B. Andresen, M. H. Rubin, R. S. Berry, Journ. of Phys. Chem. vol. 87, N15 (1983).

[8] G. Nicolis, I. Prigogine. self–organization in nonequilibrium systems, J. Wiley, 1977.

[9] A. M. Tsirlin, Dokladi AN SSSR, 1992, **323**, 1.

[10] Tsirlin A. M., Methods of the averaged optimization and their applications, Moscow, Fizmatlit, 1997.

[11] S. A. Amelkin, B. Andresen, A. M. Tsirlin, Izv. Russian AN, Energetics, (to appear).

[12] A. M. Tsirlin, Automation and Remote Control, 1995, 3.

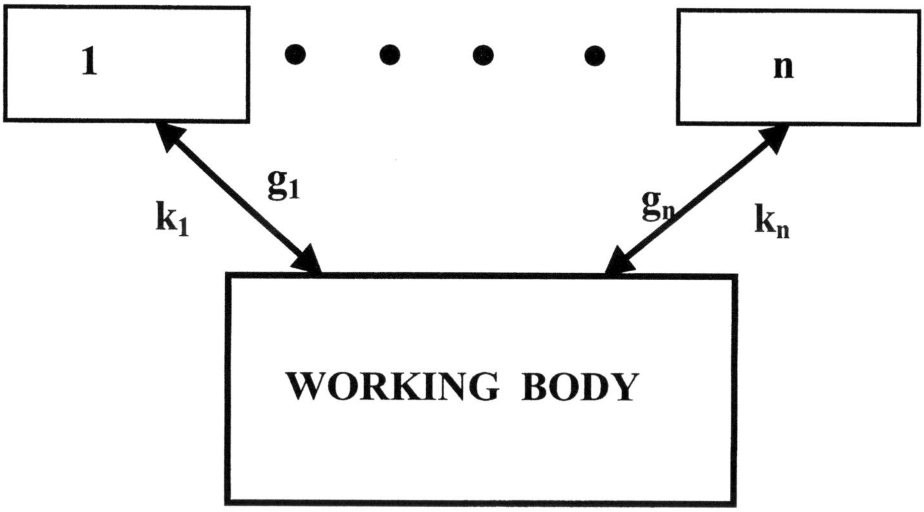

Figure 1: Two structure of the regenerative system.

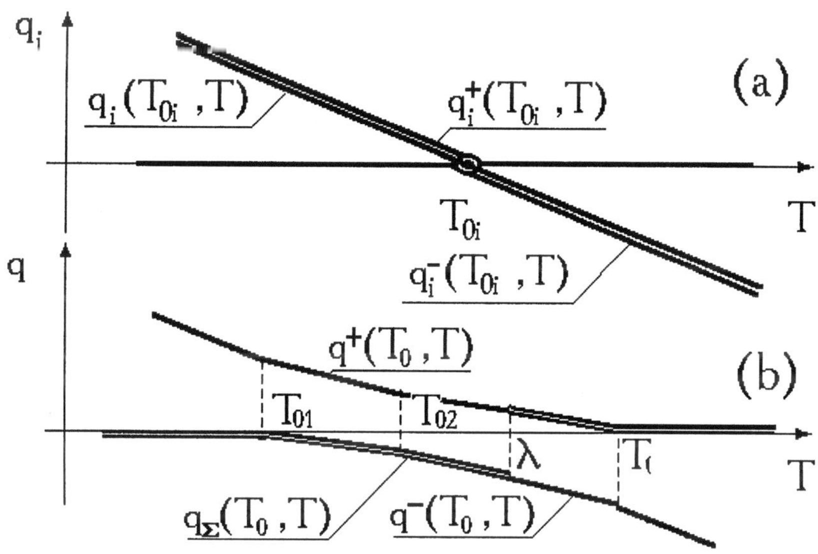

Figure 2: Two total heat supply and heat removal functions $q^+(t)$ and $q^-(T)$.

FINITE-TIME THERMODYNAMIC ANALYSIS OF A SOLAR HEAT PUMP WITH THREE HEAT SOURCES

SELAHATTIN GÖKTUN
I.T.Ü. Maritime Faculty, Department of Marine Engineering
81716 Tuzla-Istanbul, Turkey

and

HASBI YAVUZ
I.T.Ü. Institute for Nuclear Energy
80626 Maslak-Istanbul, Turkey

Abstract-- By employing finite-time thermodynamic approach and the energetic optimization technique, the optimal performance of a focusing solar collector-driven, a reversible and an irreversible heat pump with three heat sources is investigated. The overall coefficient of performance (COP) of the systems are adopted as the objective function of the performance analysis. The optimum operating temperature for the collectors of the solar-driven reversible and irreversible systems under consideration is carried out. The results provide a theoretical base for designing a real solar heat pump with three heat source heating system.

Keywords - COP, irreversible, three-heat-source, heat pump.

INTRODUCTION

A solar three-heat-source heating system consists of a focusing solar collector (FSC) and a heat pump with three heat sources (HPWTHS). A typical example of the solar HPWTHS system is a solar chemical heat pump. When a solar collector as a heat source is integrated with a heat pump one objective is to determine the operating conditions that will provide optimum system performance. In general, the solar-collector efficiency is decreased by the use of high temperatures for the working fluid. However, for heat pumps, the COP increases with the supply temperature. This incompatibility in the temperature-COP relation indicates that any solar-collector heat pump combination will have an optimum operating temperature. This problem has been extensively studied by many authors [1-13]. None of the previous investigations concerned solar-driven

HPWTHS system with internal irreversibilities, for determining the optimum system performance. The present study is a theoretical investigation for determine the optimum performance of a FSC-driven HPWTHS system. For this purpose, we will establish two cyclic models; one is a reversible model, other is an irreversible model for determining the optimum performance.

THE PERFORMANCE OF A FSC-DRIVEN REVERSIBLE HPWTHS

An reversible heat pump with three heat sources (RHPWTHS) may be treated as a combined cycle of a Carnot heat engine and a Carnot heat pump; the heat engine provides the power for heat pump and the heat pump produces the heating load with the power input as shown in Fig. 1. Carnot heat engine operates between reservoirs at temperatures T_H (the FSC temperature) and T_p. Carnot heat pump operates between reservoirs at temperature T_A (ambient temperature) and T_p (heated place temperature).

According to Fig.1, we obtain the COP of an RHPWTHS as

$$\varepsilon_r = (\dot{Q}_2 + \dot{Q}_3) / \dot{Q}_1 = (1 - \eta_r) + \eta_r \beta_r \tag{1}$$

where η_r and β_r are the efficiency and the coefficient of performance of Carnot heat engine and heat pump under consideration in the combined cycle, respectively. $\dot{Q}_1$ is the heat absorbed from the reservoir (FSC) at temperature T_H, $\dot{Q}_2$ is the heat rejected from the heat engine to the heated place and $\dot{Q}_3$ is the heating load of the heat pump.

The first and second laws of thermodynamics are used to analyze this system. The governing equations are

$$\dot{W} = \dot{Q}_1 - \dot{Q}_2 \tag{2}$$

$$\dot{W} = \dot{Q}_3 - \dot{Q}_4 \tag{3}$$

$$(\dot{Q}_1 / T_H) = \dot{Q}_2 / T_P \tag{4}$$

$$(\dot{Q}_4 / T_A) = \dot{Q}_3 / T_P \tag{5}$$

where $\dot{W}$ is the power produced by the Carnot heat engine which is used to drive the heat pump. $\dot{Q}_4$ is the heat absorbed by the Carnot heat pump from the reservoir at temperature T_A.

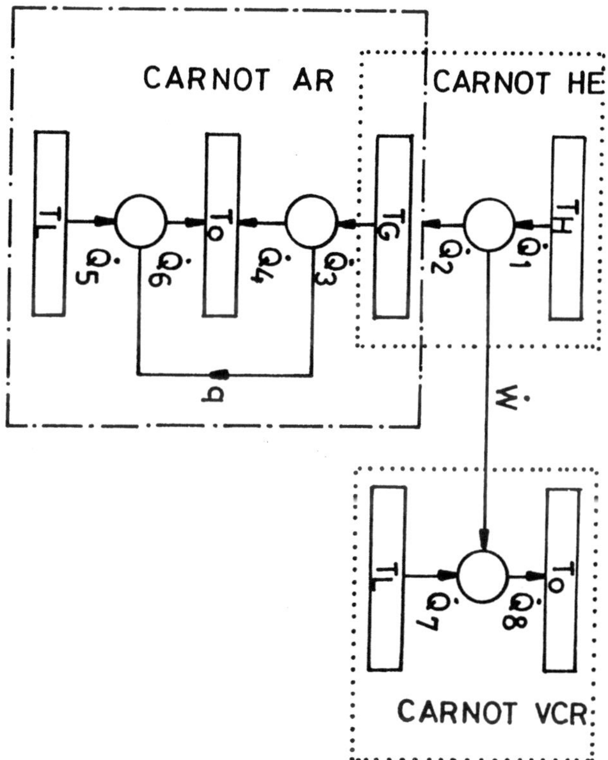

Fig.1. A schematic diagram of a reversible heat pump with three heat sources

Using Eqs. (1)-(5) the COP of the RHPWTHS is seen to be

$$\varepsilon_r = (T_P / T_H)\left[(T_H - T_A)/(T_P - T_A)\right] \tag{6}$$

For steady-state conditions, the thermal efficiency for a FSC is given by [13]

$$\eta_{coll} = (\tau\alpha) - \left[U(T_H - T_A)/\rho I\right] \tag{7}$$

where $(\tau\alpha)$ is the transmittance-absorptance product, T_H and T_A are collector and ambient temperatures, respectively. I is the average value of the total solar insolation and

ρ is the concentration ratio of the FSC. The overall system COP ψ for FSC-driven RHPWTHS is

$$\psi = \eta_{coll}\varepsilon_r \tag{8}$$

Using Eqs. (6) and (7)

$$\psi = \{(\tau\alpha) - [U(T_H - T_A)/\rho I]\}\{[\ (T_P/T_H)(T_H - T_A)/(T_P - T_A)\ ]\}. \tag{9}$$

For the extremal condition

$$(\partial\psi/\partial T_H) = 0 \tag{10}$$

the optimum FSC temperature $T_{H(opt)}$ for RHPWTHS becomes

$$T_{H(opt)} = T_A(1+\rho\gamma)^{0.5} \tag{11}$$

where

$$\gamma = (\tau\alpha)I/UT_A \tag{12}$$

Equation (12) is the operating parameter of a FSC [12].

Substituting Eq. (11) into Eq. (9) the maximum overall COP of the FSC-driven RHPWTHS is seen to be

$$\psi_{\max} = (\tau\alpha)T_P\left[(1+\rho\gamma)^{0.5} - 1\right]^2/(\rho\gamma)(T_P - T_A) \tag{13}$$

The Performance of a FSC-Driven Irreversible HPWTHS

An irreversible heat pump with three heat sources (IHPWTHS) can be thought as a combination of a finite-size, irreversible heat engine and an irreversible heat pump; the heat engine provides the maximum power for the heat pump and the heat pump produces the heating effect with the power input, as shown in Fig.2. Heat engine operates between reservoirs at temperature T_H and T_P and the temperatures of the working fluid in this engine with two isothermal processes are T_1 and T_2 . Heat pump operates between reservoirs at temperatures T_P and T_A and the temperatures of the working fluid in this heat pump with two isothermal processes are T_2 and T_3 .

According to Fig.2, we obtain the COP of an IHPWTHS as

$$\varepsilon = (\dot{Q}_2 + \dot{Q}_3) / \dot{Q}_1 = (1-\eta) + \eta\beta \tag{14}$$

where η and β are the efficiency and the COP of the heat engine and heat pump under consideration in the combined cycle, respectively.

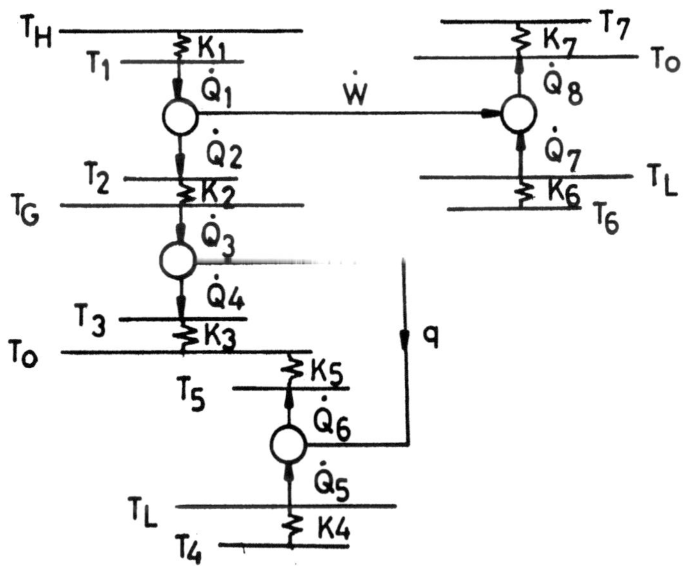

Fig.2. A schematic diagram of an irreversible heat pump with three heat sources.

Fourier's law, the first and the second laws of thermodynamics are used to analyze this system. The governing equations are

$$\dot{Q}_1 = K_1(T_H - T_1) \tag{15}$$

$$\dot{Q}_2 = K_3(T_2 - T_P) \tag{16}$$

$$\dot{Q}_3 = \dot{Q}_2 \tag{17}$$

$$\dot{Q}_4 = K_2(T_A - T_3) \tag{18}$$

$$\dot{W} = \dot{Q}_1 - \dot{Q}_2 \tag{19}$$

$$\dot{W} = \dot{Q}_3 - \dot{Q}_4 \tag{20}$$

$$(\dot{Q}_1 / T_1) = R(\dot{Q}_2 / T_2) \tag{21}$$

$$(\dot{Q}_4 / T_3) = R(\dot{Q}_3 / T_2) \tag{22}$$

$$K = K_1 + K_3 \tag{23}$$

Where K_1, K_2, K_3 are the heat conductances between the working fluid and reservoirs at temperatures T_H, T_A and T_P, respectively. K represents the total size of the heat-exchanger equipment in the heat engine [15]. R is the cycle-irreversibility parameter which can characterize fully the degree of internal irreversibility resulting from the working fluid [15].

Using Eqs. (15), (16), (19) and (21) the power $\dot{W}$, produced by the heat engine is

$$\dot{W} = \dot{Q}_1[1 - (\theta / R)] \tag{24}$$

where

$$\theta = T_2 / T_1 \tag{25}$$

On substituting Eqs.(15) and (16) into Eq.(21), and dividing T_H, after some manipulation Eq.(21) becomes

$$\alpha\theta[1 - (T_1 / T_H)] = R[(T_2 / T_H) - \tau] \tag{26}$$

where $\alpha = K_1 / K_3$ is the dimensionless hot-end /cold-end heat-exchanger size ratio of the heat engine and $\tau = T_P / T_H$. Substituting Eq.(25) into Eq.(26) and then solving for T_1 / T_H yields

$$(T_1 / T_H) = (\alpha\theta + R\tau) / [\theta(\alpha + R)] \tag{27}$$

Using Eqs.(23) and (27) the dimensionless heat-transfer rate to the heat engine $\varnothing$ and the dimensionless power output P of the heat engine can be obtained from Eqs. (15) and (24), respectively, as

$$\varnothing = (\dot{Q}_1 / KT_H) = \alpha R(\theta - \tau) / (\alpha + 1)(\alpha\theta + R\theta) \quad (28)$$

$$P = \dot{W} / KT_H = \varnothing(1 - \theta R) \quad (29)$$

Using the extremal conditions

$$(\partial P / \partial\theta) = 0 \text{ and } (\partial P / \partial\alpha) = 0 \quad (30)$$

we find for

$$\theta_{opt} = (R\tau)^{0.5} \quad (31)$$

that

$$\alpha_{opt} = R^{0.5} \quad (32)$$

Substituting Eqs. (31) and (32) into Eqs. (28) and (29) yields

$$\varnothing_{opt} = R^{0.5}(R^{0.5} - \tau^{0.5}) / (1 + R^{0.5})^2 \quad (33)$$

$$P_{\max} = (R^{0.5} - \tau^{0.5})^2 / (1 + R^{0.5})^2 \quad (34)$$

Where $P_{\max}$ is the maximum power generated by the heat engine.

The thermal efficiency of the finite-size, irreversible heat engine at its maximum power output is

$$\eta_m = (P_{\max} / \varnothing_{opt}) = 1 - (\tau / R)^{0.5} \text{ . for } \tau < R \quad (35)$$

Solving Eqs. (17), (18), (20) and (22) under the condition that irreversible heat pump which is driven by the finite-size irreversible heat engine at its maximum power yields

$$T_2 = T_P(1 + \frac{P_{\max} KT_H}{(1 - R\omega)K_3 T_P}) \quad (36)$$

$$T_3 = T_A(1 - \frac{P_{\max} K T_H R\omega}{(1 - R\omega) K_2 T_A}) \tag{37}$$

$$\dot{Q}_{3m} = K_2(T_2 - T_P) \tag{38}$$

$$\beta_m = \dot{Q}_{3m} / (P_{\max} K T_H) = 1/(1 - R\omega) \tag{39}$$

where $\omega = T_3 / T_2$ is the temperature ratio of the heat pump, $\dot{Q}_{3m}$ is the heating load delivered by the heat pump which is driven by the heat engine at its maximum power condition and β_m is the corresponding COP. The temperature ratio ω of the heat pump is a function of the parameters $R, K_1, K_2, K_3, T_H, T_A$ and T_P. When these parameters are given, ω can be determined from the Eqs. (34), (36) and (37).

On substituting Eqs. (35) and (36) into Eq. (14), the overall COP of the IHPWTHS can be obtained as

$$\varepsilon = [1/(1 - R\omega)][1 - (\tau R)^{0.5}\omega] \tag{40}$$

Using Eqs. (7) and (40), the overall system COP δ for FSC-driven IHPWTHS can be written as

$$\delta = \eta_{coll}\varepsilon \tag{41}$$

or

$$\delta = \{(\tau\alpha) - [U(T_H - T_A)/\rho I]\}\{[1/(1 - R\omega)][1 - (\tau R)^{0.5}\omega]\} \tag{42}$$

For the extremal condition

$$(\partial\delta / \partial T_H) = 0, \tag{43}$$

the optimum FSC temperature $T^*_{H(opt)}$ for IHPWTHS becomes

$$T^*_{H(opt)} = 0.25\omega^2 R T_P \left[1 + (T_A / T^*_{H(opt)})(1 + \rho\gamma)\right]^2 \tag{44}$$

Substituting Eq.(44) into Eq.(42) yields the maximum overall COP $\delta_{\max}$ for the FSC-driven IHPWTHS.

DISCUSSION

It is technically feasible to utilize a solar heat pump with three heat sources for heating applications. Its use is particularly advantageous in the regions of the world where a shortage of installed electric power exists. The determination of the COP of a solar HPWTHS for heating applications is of considerable theoretical and practical interest. A completely reversible solar HPWTHS system has the highest possible overall COP, as calculated by Eq.(13); however it does not produce any heating load. A real solar HPWTHS system is usually designed to function at its optimal COP. The optimum operating solar collector temperature and corresponding overall COP may now be calculated for the totally irreversible solar HPWTHS system. This study presents a reasonable COP bound for a real solar HPWTHS. It provides a basis for designing a real solar HPWTHS.

REFERENCES

1. Howell, J.R. and Bannerot, R.B., Solar Energy, 1977, 19, 149.
2. Gordon, J.M., Solar Energy, 1988, 40, 457.
3. Lee, W.Y. and Kim, S.S., International Journal of Energy Research, 1991, 15, 257.
4. Badescu, V., Energy - The International Journal, 1992, 17, 601.
5. Chen, J., Journal of Applied Physics, 1992, 72, 3778.
6. Eldighidy, S.M., Solar Energy, 1993, 51, 175.
7. Yan, Z. and Chen, J., Journal of Applied Physics, 1994, 76, 8129.
8. Chen, L., Sun, F. and Wu, C., International Journal of Ambient Energy, 1996, 17, 73.
9. Wu, C., Chen, L. and Sun, F., Energy Conversion and Management, 1996, 37, 1671.
10. Göktun, S., Renewable Energy, 1996, 7, 67.
11. Wu, C., Chen, L. and Sun, F., Applied Thermal Engineering, 1997, 17, 203.
12. Göktun, S. and Özkaynak, S., Energy - The International Journal, 1997, 22, 481.
13. Göktun, S. and Yavuz, H., Journal of Physics D:Applied Physics, 1998, 30, 1.
14. Bejan, A., Advanced Engineering Thermodynamics (New York:Wiley), 1988.
15. Wu, C. and Kiang, R.L., Energy - The International Journal, 1992, 17, 1173.

A FINITE TIME ANALYSIS OF AN IDEAL ABSORPTION REFRIGERATION CYCLE

B. AGNEW, A. ANDERSON, T.H. FROST
Department of Mechanical, Materials and Manufacturing Engineering,
University of Newcastle upon Tyne, UK.

and

R.M. TOZER AND WATERMANN GORE
Mechanical and Electrical Consulting Engineers, London.

Abstract--The characteristics of an irreversible steady-flow absorption refrigeration cycle are examined when approximated to a combination of ideal driving and cooling cycles. The classical technique of Curzon and Ahlborn has been applied to the Carnot Cycles making up the ideal combination. The initial driving cycle consisted of a Carnot Cycle driven from a high temperature heat source which was coupled to a reverse Carnot Cycle that acted as the cooling cycle. Double stage cycles were also examined in which the energy rejected from the initial driving cycle was utilised to drive a second stage driving cycle which also contributed power to the cooling cycle. Comments are made regarding the utilisation and design of absorption refrigeration cycles.

NOTATION

COP	Coefficient of performance $= Q_e/Q_g$ as a refrigerator or Q_a/Q_g as a heat pump
l	Heat of solution
P	Power
Q	Rate of heat flow
r	Heat of evaporation
S	Entropy
T	Temperature
T_{ci}	Temperature of cycle i low temperature isotherm
T_{hi}	Temperature of cycle i high temperature isotherm
ΔT_{ci}	Cold side temperature difference for cycle i
ΔT_{hi}	Hot side temperature difference for cycle i
θ_{ci}	Cold side heat transfer factor
θ_{hi}	Hot side heat transfer factor

η_i	Efficiency of cycle i
η_o	Optimised single cycle efficiency
$\eta_{(o2)}$	Efficiency of two coupled optimised cycles

Subscripts

a	absorber
C	low temperature reservoir
c	condenser
cc	cooling cycle
dc	driving cycle
e	evaporator
g	generator
H	high temperature reservoir
HP	heat pump
i	intermediate temperature reservoir
refr	refrigerator

1. Introduction

The optimisation of thermodynamic cycles is recognised as playing an important role in the teaching of thermodynamics. Since the introduction of Finite Time thermodynamics in 1975 [1] a great deal of work has been expended in examining the operational bounds of thermodynamic cycles [2-8]. The interaction of energy transfer mechanisms with the cycle configuration is now well established such that it is possible to explain why practical engines operate in a region between the two ideal limits of reversible, maximum efficiency and irreversible, maximum power. In most cases the actual operating point is determined by non-thermodynamic considerations such as initial costs, longevity, fuel costs, etc.

Other studies have introduced other forms of irreversibility such as piston friction [9,10] and have established the form of the optimum thermodynamic cycle for maximum power output. Real cycles are of course subject to significantly different constraints than model cycles. These have some influence on the cycle performance and result in the cycle process paths being far from ideal, but it has been illustrated [11] that the efficiency of several ideal practical cycles (Otto, Joule-Brayton, Diesel and Atkinson) exhibit a maximum work efficiency that is the same as or very close to the C-A result.

To date the majority of this work has been applied to power producing cycles. It is the objective of this work to extend the analysis to the field of cooling cycles by applying the methods of Finite Time Thermodynamics to an ideal representation of an absorption refrigeration cycle.

2. Recap of Previous Work

This paper draws on the outcome of several different but complementary pieces of work. For completeness a brief description of the essential features of this work is reproduced in the following section.

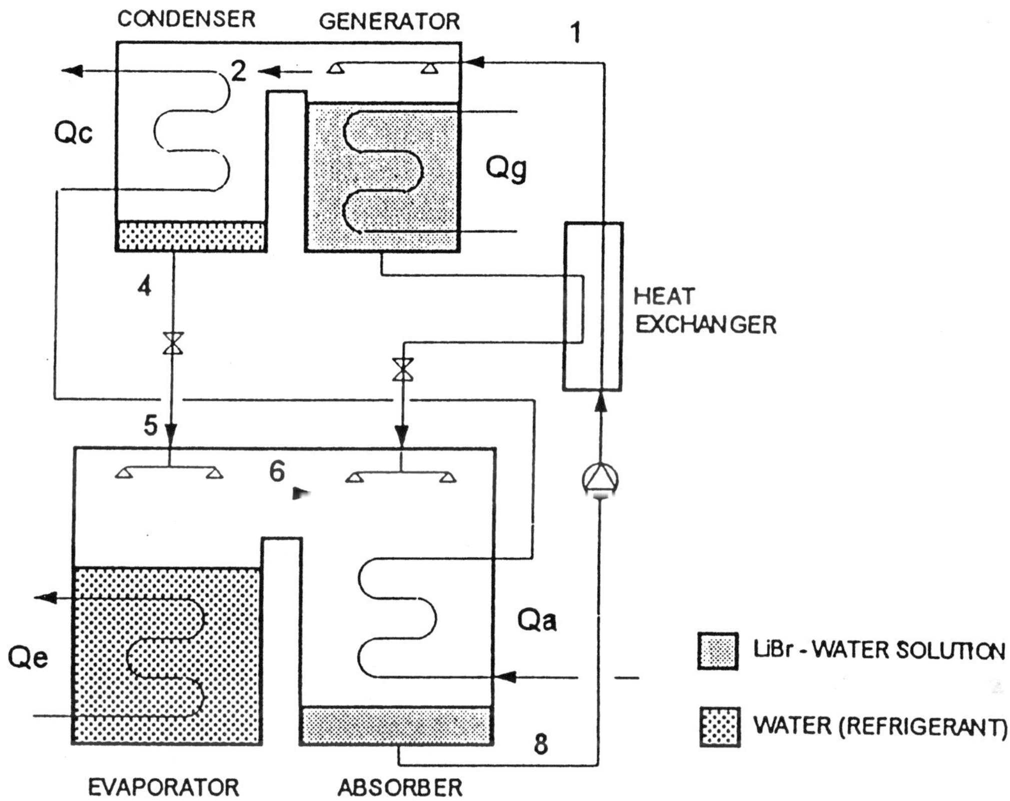

Figure 1- Schematic Diagram of a single stage LiBr-Water Absorption Refrigeration Unit

2.1 The Refrigeration Cycle.

The basic design of an absorption refrigeration unit is well established following recent interest for solar energy applications. Many studies have been performed to establish the optimisation configuration of the system, see for example Dalichaouch [16]. Such a system is shown in figure 1 in which an aqueous solution of refrigerant and absorbant fluids are circulated through the cycle. The process of a continuously operating absorption system is characterised by the points 1 to 8 described below and illustrated in figure 1.

1-8: A strong solution leaves the generator and enters the absorber via the heat exchanger. The strength of the solution is weakened by the absorbtion of the water vapour coming from the evaporator. The absorber temperature defines the point 8 and determines the concentration of the weak solution. The weak solution is transported from the absorber to the generator via the heat exchanger, (8-1). It is then boiled in the generator (1-2) which it then leaves to be condensed in the condenser (2-4). The liquid water then flows from the condenser to the evaporator (4-5) where it is evaporated and transferred to the absorber to produce the weak solution. The concentrations of the weak and strong solutions are determined by the absorber and generator temperatures respectively and the state of the weak solution entering the generator is dependent upon the condenser pressure.

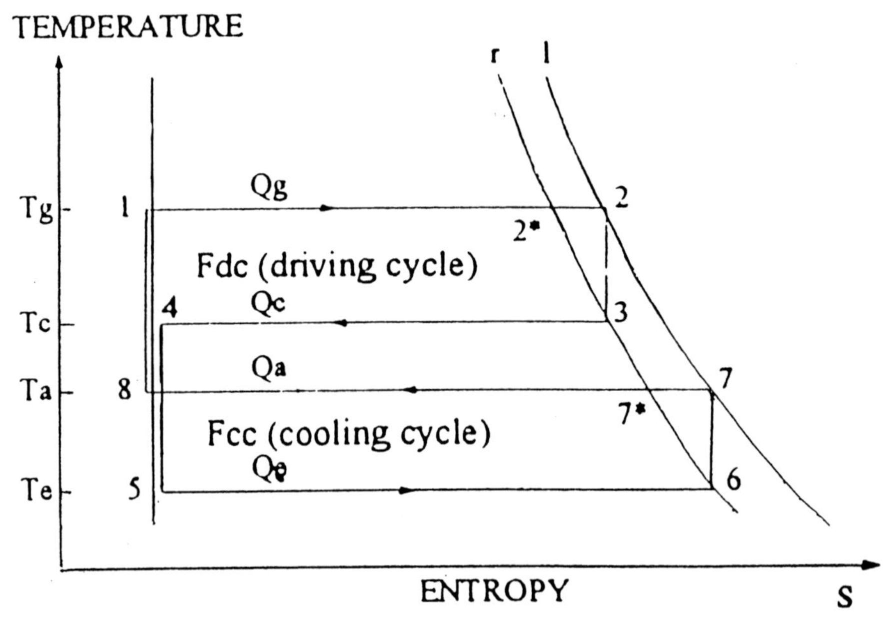

Figure 2- Single Stage Absorption Refrigeration Unit on a T-S Diagram

2.2 Driving and Cooling Cycles.

Drawing on the work of Tozer and James [12] the absorption refrigeration cycle can be considered to be a combination of a Carnot driving cycle and a reverse Carnot cooling cycle. The ideal single stage cooling cycle can be represented as shown in figure 2 on a T-S diagram. The processes within this ideal cycle are as follows:

8-1 Isentropic heating of the refrigerant in the solution
1-2 Heat introduced into the generator per unit mass flow rate of the working fluid, heat of evaporation $r = (S_2 - S_1)T_g$, plus the heat of solution $l = (S_2 - S_2^*)T_g$
2-3 Isentropic Cooling to the saturation temperature T_c of the superheated vapour
3-4 Condensation
4-5 Isentropic cooling of the refrigerant
5-6 Evaporation
6-7 Isentropic heating of the saturated vapour
7-8 Absorption by the solution of the superheated vapour. Both the heat of solution l and the heat of evaporation r are rejected to the environment.

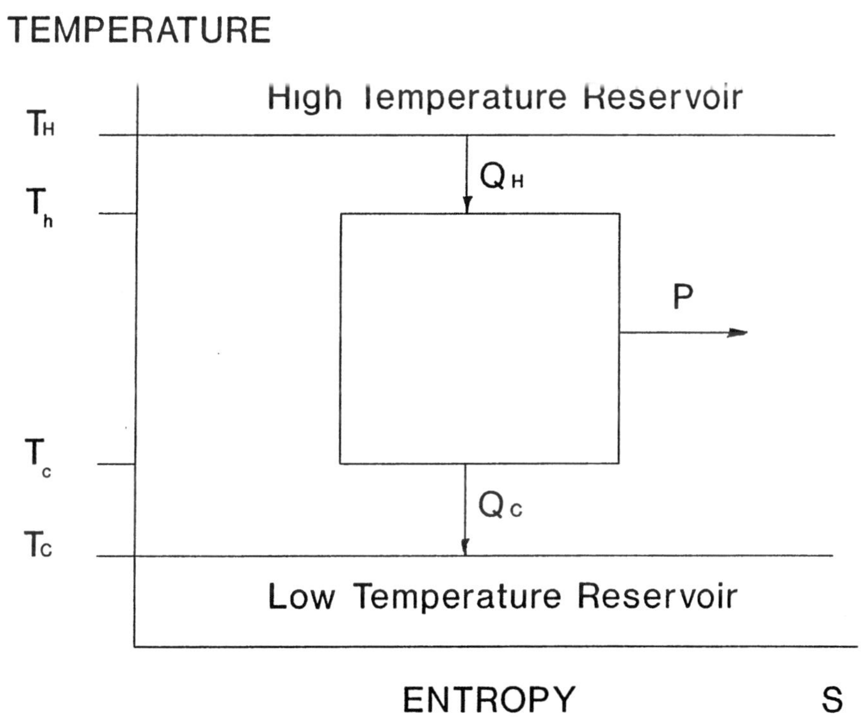

Figure 3- Irreversible Carnot Driving Cycle

The driving and cooling cycles are coupled thermodynamically through the heat and power balance equations as shown below.

$$Q_g + Q_e = | Q_c + Q_a | \quad (1)$$

$$P_{dc} = | P_{cc} | \quad (2)$$

2.2.1 THE CARNOT DRIVING CYCLE

A steady flow Carnot driving cycle, optimised to give maximum work for given heat transfer characteristics, shown schematically in figure 3., is fully described in [13]. It is sufficient for the present purpose to describe the main features of this cycle that are of importance in this work. In this cycle it is assumed that a finite heat exchange rate characterised by heat transfer factors θ_h and θ_c (defined as the heat transfer per unit temperature difference) exists between the cycle isotherms at T_{h1} and T_{c1} and the high and low temperature reservoirs at T_{H1} and T_{C1}. A temperature difference $\Delta T_{h1} = (T_H - T_h)$ then exists between the high temperature reservoir and the high temperature isotherm and similarly $\Delta T_{c1} = (T_C - T_c)$ between the low temperature isotherm and the low temperature reservoir. In the optimised case the cycle efficiency is found to be

$$\eta_o = 1 - \sqrt{\frac{T_C}{T_H}} = 1 - \frac{T_c}{T_h} \quad (3)$$

and the following relationships for the temperature differences also apply

$$\Delta T_{h1} = \frac{T_H\left(1 - \sqrt{\frac{T_C}{T_H}}\right)}{1 + \sqrt{\frac{\theta_{h1}}{\theta_{c1}}}} \quad (4)$$

$$\Delta T_{c1} = \frac{T_C\left(\sqrt{\frac{T_H}{T_C}} - 1\right)}{1 + \sqrt{\frac{\theta_{c1}}{\theta_{h1}}}} \tag{5}$$

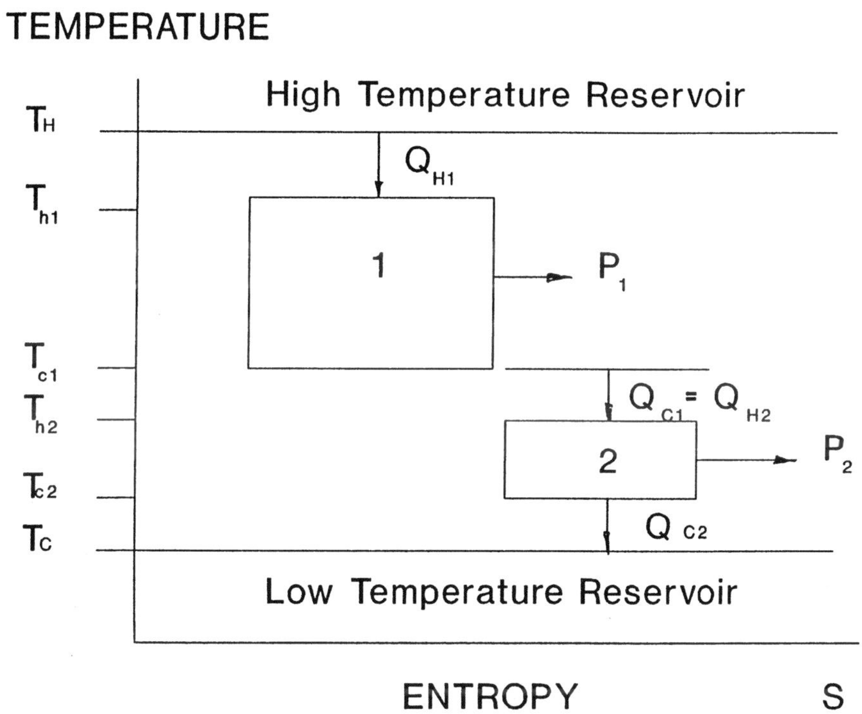

Figure 4- Coupled Carnot Driving Cycles

2.2.2 Coupling of the Driving Cycles.

Coupling of two or more cycles have been examined in a previous work by Rubin et al [14] and also by Agnew [15]. It was demonstrated in this that coupling the cycles in series with a common intermediate temperature reservoir suffers from severe restrictions with regard to the heat transfer factors and the necessary interaction with the environment. There is no gain to be had in overall efficiency or power output compared with a single optimised cycle operating with the same heat transfer factors. A more useful arrangement is as shown in figure 4 in which the low temperature isotherm of the first optimised cycle becomes the high temperature reservoir of the second cycle. The efficiencies of the two separate cycles will then be

$$\eta_1 = 1 - \left(\frac{T_C}{T_H} \right)^{\frac{1}{2}} , \quad \eta_2 = 1 - \left(\frac{T_C}{T_H} \right)^{\frac{1}{4}} \tag{6}$$

The overall efficiency of the two cycle system is now given by

$$\eta_{(o2)} = \eta_1 + (1 - \eta_1)\eta_2 = 1 - \left(\frac{T_C}{T_H} \right)^{\frac{3}{4}} \tag{7}$$

3. Finite-time Optimisation of The Cycle

The following analysis of the absorption cycle follows that performed on the Carnot (CA) cycle in that temperature differences are introduced between the ideal cycle isotherms and adjacent thermal reservoirs. The cycle can then be represented as shown in figure 5 in which the ambient temperature is represented by T_C and the high temperature reservoir by T_H. The driving cycle operating over an internal temperature range of (T_{h1} - T_{c1}) is shown on the left of the diagram and it is seen to be the conventional CA cycle operating between thermal reservoirs at T_H and T_C. The power produced by this cycle drives the cooling cycle shown to the right in the diagram. The cooling cycle is drawing energy from the ambient reservoir and rejecting it to a high temperature reservoir at an intermediate temperature T_i. The cycle isotherm T_{c2} is then at a lower temperature than T_C and T_i is at a lower temperature than the cycle high temperature isotherm at T_{h2}. In common with the CA analysis it is necessary to specify two of the temperatures mentioned above. In this analysis the evaporator temperature T_{c2} and the ambient reservoir temperature T_c were specified and the other temperatures in the cycle were determined by the optimisation procedure. In practical Lithium-Bromide cycles the evaporator temperature is usually of the order of 4°C and the ambient temperature was chosen within the range 17°C to 51 °C that is typical of ambient temperatures in Middle Eastern climates. The analysis will then provide the relationships between the ambient temperature and the generator temperature T_{h1}, the optimum temperature of the absorber coolant, T_i, and the resulting COP when the cycle is optimised for maximum heat transfer rate from the cooling cycle to the intermediate reservoir

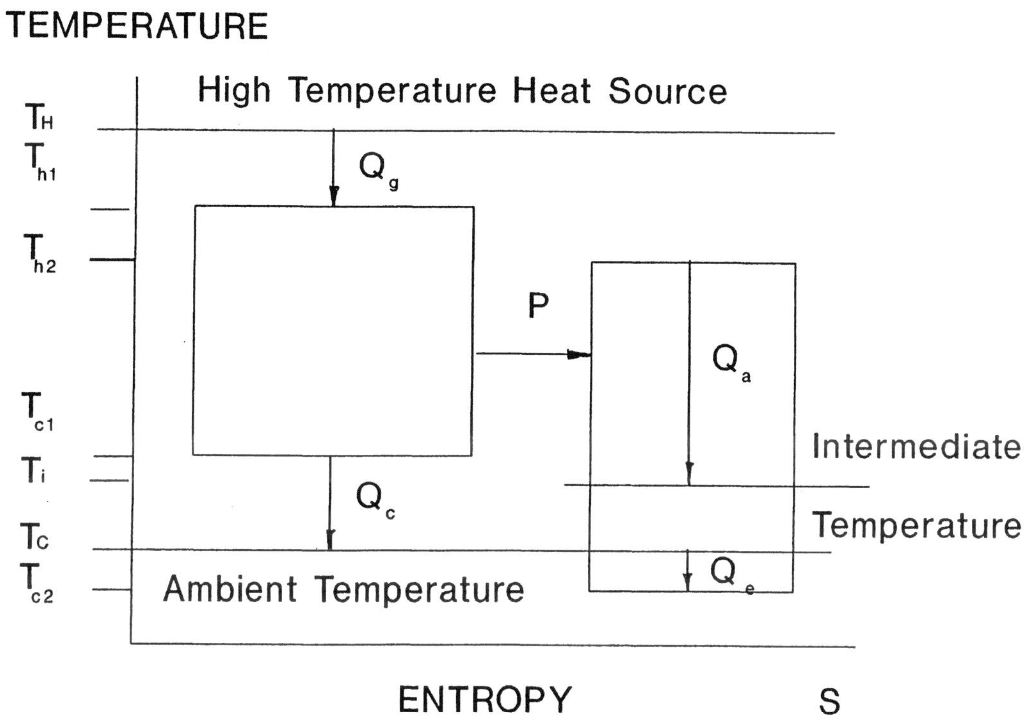

Figure 5- Single Stage Irreversible Absorption Refrigeration Unit on a T-S Diagram

3.1 Optimisation Procedure.

The objective of the optimisation procedure is to maximise the rate at which thermal energy is rejected from the cooling cycle to the intermediate temperature reservoir. This consists of the driving cycle power plus the thermal power absorbed by the cooling cycle from the ambient reservoir. The driving cycle is optimised for maximum power, it therefore satisfies the CA criteria and operates with temperature differences as specified in equations 2 and 3. Referring to figure five the power output of the cycle is:

$$P_{dc} = \theta_{h1} \Delta T_{h1} \left(1 - \sqrt{\frac{T_C}{T_H}} \right) = Q_g \qquad \textbf{(8)}$$

$$Q_{c2} = \theta_{c2}\left(T_C - T_{c2} \right) = Q_e \tag{9}$$

The energy transfer rate between the ambient reservoir and the cooling cycle is then and that between the cooling cycle and the intermediate reservoir

$$Q_{h2} = \theta_{h2}\left(T_{h2} - T_i \right) = Q_a \tag{10}$$

The objective function to be maximised was then Q_a.

$$f(P) = \theta_{h1}\,\Delta T_{h1}\left[1 - \sqrt{\frac{T_C}{T_H}} \right] + \theta_{c2}\,(T_C - T_{c2}) = \theta_{h2}\,(T_{h2} - T_i) \tag{11}$$

The constraints on the optimisation were obtained from the characteristics of the Carnot cycles. The driving cycle must satisfy the CA relationship of the cycle efficiency as shown in equation 3

$$\frac{T_{c1}}{T_{h1}} = \sqrt{\frac{T_C}{T_H}} \tag{12}$$

and the cooling cycle must satisfy the conditions specified in [12] ie:

$$\frac{T_{h1}}{T_{c1}} = \frac{T_{h2}}{T_{c2}} = COP \tag{13}$$

The COP in the above equation is the ratio of $Q_{c2}/Q_{h1} = Q_e/Q_g$ that is the COP of the cycle acting as a refrigerator.

3.2 DOUBLE STAGE CYCLE.

Multistage absorption cycles are becoming increasingly common as they can raise the COP of the overall cycle beyond 1. In practical cycles residual heat from the condenser is used to drive lower temperature cycles. A double stage absorption cycle consists of two single stage cycles, the cooling cycles of both these are identical and share the same evaporator and absorber. The driving cycles are however different in that one cycle is the conventional higher temperature cycle that is rejecting heat to drive the generator of the lower temperature cycle. In the case of ideal cycles the temperature difference between the high temperature condenser and the low temperature generator is negligible. A schematic of a double stage unit can be seen in figure 6 and the corresponding T-S diagram in figure 7. The ideal Carnot cycle representation of this cycle arrangement is shown in figure 8. The driving cycles are coupled as discussed previously in section 2.3 and the power output of the power cycles drives the single cooling cycle.

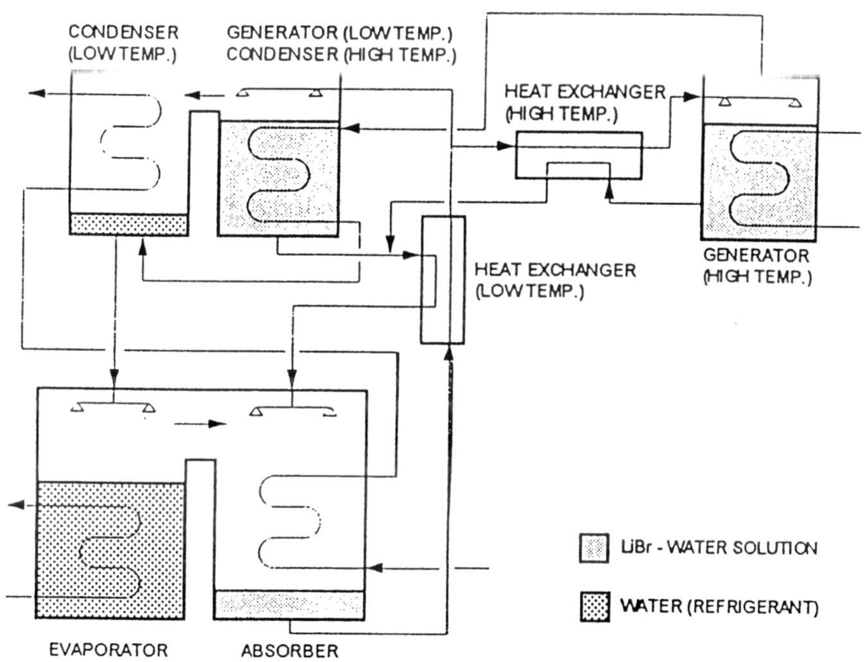

Figure 6- Schematic Diagram of a Double Stage Absorption Refrigeration Unit

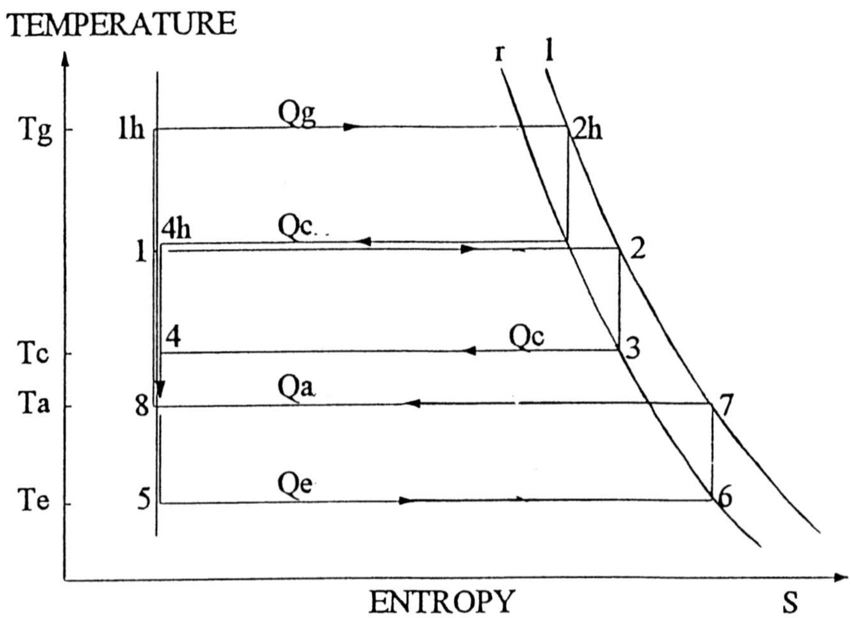

Figure 7- Double Stage Absorption Refrigeration Unit on a T-S Diagram

4. Results

The results obtained from the optimisation process are displayed in figures 9 to 12. The relationships between the ambient temperature T_C abd the temperature of the high temperature heat source T_H is shown in figure 9. It can be seen from this figure that the absorption refrigeration cycle is essentially a device that operates in the maximum power mode over a small extreme temperature difference. This being determined by the small temperature difference of the heat addition process in the evaporator and the relationship between the temperature differences and the reservoir temperatures, equations 4 and 5. For the range of ambient temperatures considered the temperature of the high temperature source did not exceed 650 K. The implications of this result are that absorption refrigeration units are best suited for the utilisation of low grade energy or waste heat from a primary power producing cycle such as exhaust gas waste heat. It would also indicate that systems that operate with higher maximum cycle temperatures such as gas fired units are optimised away from the maximum power condition. The relationship between the ambient temperature T_C and the temperature T_i of the intermediate thermal reservoir that is acting as a sink for the absorber, shown in figure 10, indicates that in this optimised condition T_i and T_C have similar values. It would seem that the atmosphere is a suitable thermal reservoir for the absorber. From a practical standpoint this would indicate that there are good thermodynamic reasons for the development of air cooled as opposed to the more common liquid cooled absorption systems. The increase in component size necessary to accommodate air cooling may be offset by the larger specific output.

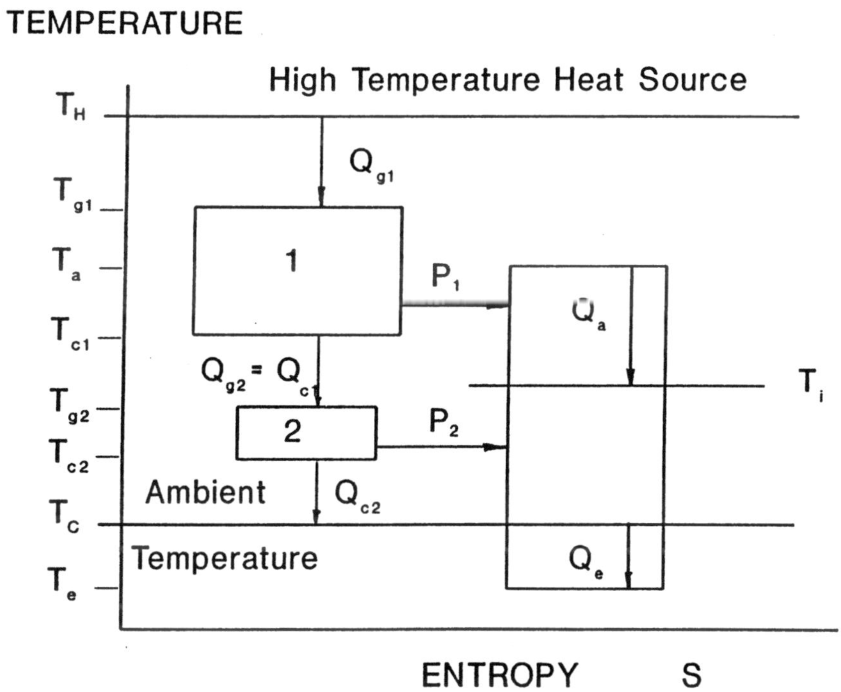

Figure 8- Double Stage Irreversible Absorption Refrigeration Unit on a T-S Diagram

Figures 11 and 12 show the trends that are to be expected from this type of analysis showing the limiting conditions that are expected from Carnot type cycles. At high COP the specific output is relatively small and this increases as the COP decreases. The COP of a single stage machine is seen to be always less than one and will reach this limiting value at the point of zero output, ie when the temperature of the evaporator T_e is the same as T_C. The double stage machine, figure 12, shows a similar trend with the limiting value of COP at zero output being in this case two.

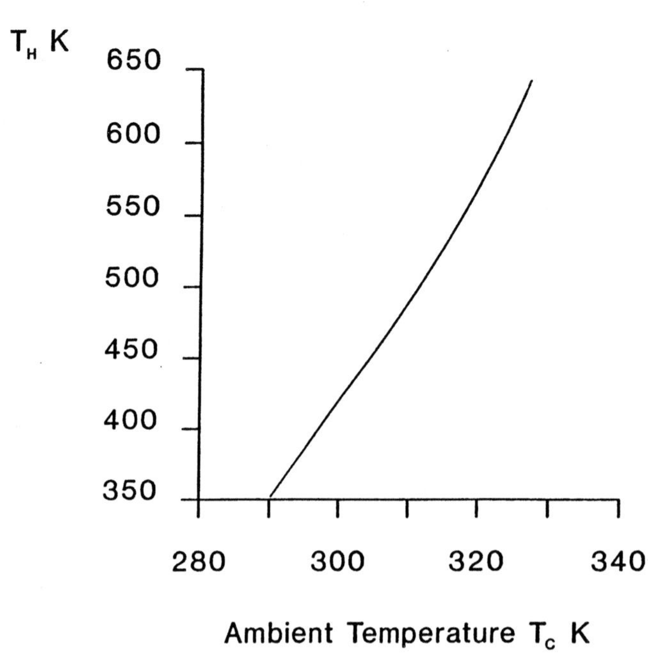

Figure 9- The Relationship between T_H and T_C at Maximum Output

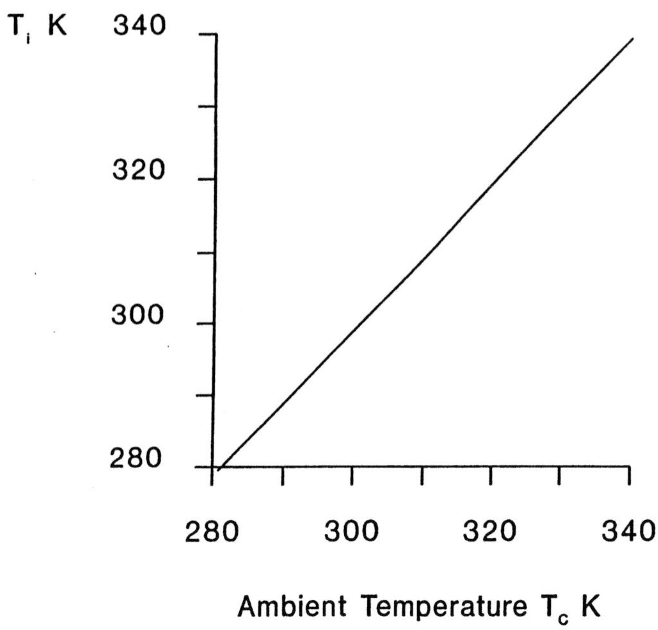

Figure 10- The Relationship between T_i and T_C at Maximum Output

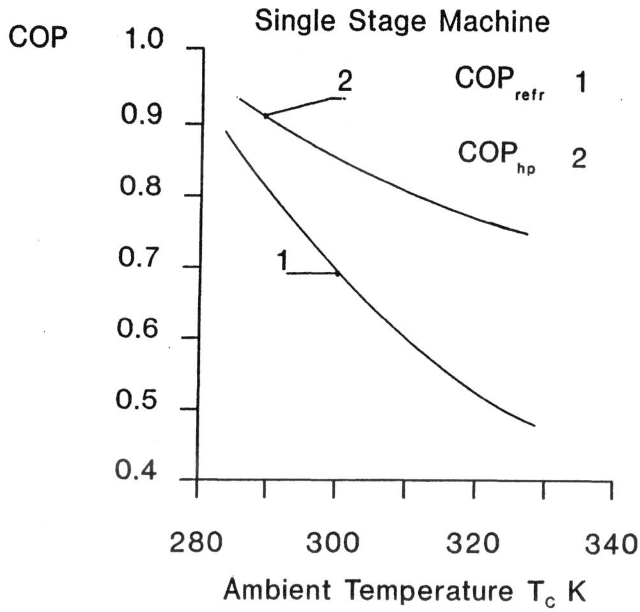

Figure 11- COP of a Single Stage Machine

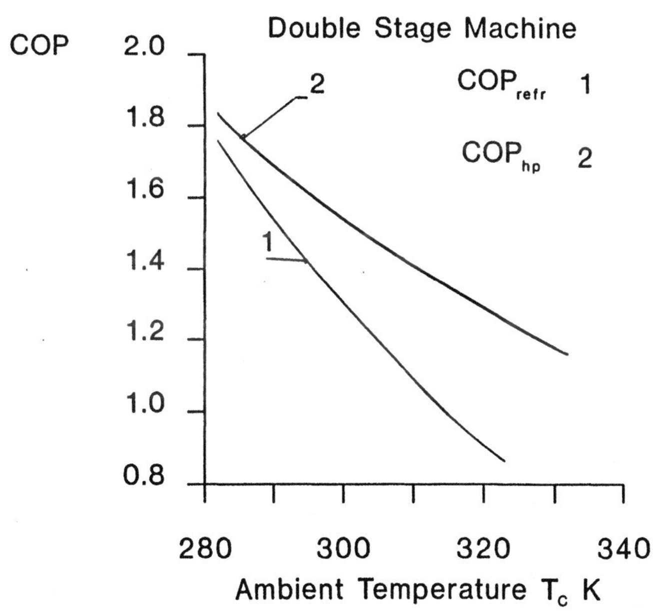

Figure 12- COP of a Double Stage Machine

5. Conclusions

A finite time analysis of an absorption refrigeration system has been presented based on approximating the system to a combination of Carnot driving and cooling cycles. The analysis indicates that the COP of a single stage absorption unit intrinsically cannot exceed the value 1 and this will only be obtained in the limiting condition of zero output. At this condition a double stage machine would have a COP of 2. The results of the analysis also indicate that absorption machines are essentially low temperature devices suitable for recovery of low grade energy or waste heat recovery and that there is scope for developing air cooled systems.

6. References

1. Curzon F. L., Ahlborn B., Efficiency of a Carnot Engine at Maximum Power Output, American Journal of Physics, 43, 1975.
2. Leff H.S., Heat Engines and the Performance of External Work. American Journal of Physics, 46 (3), 1978.
3. Band Y. B., Kafri O., Salamon P., Optimisation of a Model External Combustion Engine, Journal of Applied Physics, 53(1) 1982.
4. Salamon P., Band Y. B., Kafri O., Maximum Power from a Cycling Working Fluid. Journal of Applied Physics, 53(1) 1982.
5. Gordon J. M., Maximum Power Point Characteristics of Heat Engines as a General Thermodynamic Problem, American Journal of Physics, 57 (12) 1989.
6. Gordon J. M., Observations of Heat Engines Operating at Maximum Power, American Journal of Physics, 58 (4), 1990.
7. Swanson L. W., Thermodynamic Optimization of Irreversible Power Cycles with Constant External Reservoir Temperatures, ASME Journal of Engineering for Gas Turbines and Power, 113, 1991.
8. Ibrahim O. M., Klein S. A., Mitchell J. W., Optimum Heat Power Cycles for Specified Boundary Conditions, ASME Journal of Engineering for Gas Turbines and Power, 113, 1991.
9. Andersen B., Salamon P., Berry R. S., Thermodynamics in Finite Time: Extremals for Imperfect Heat Engines. Chem. Phys. 66, 1977.
10. De Voss A., Reflections on the Power Delivered by Endoreversible Engines. J. Phys. D: Appl. Phys. 20, 1987.
11. Agnew B., Anderson A., Frost T. H.; Cycle Optimisation and Comparison of Ideal Thermal Cycles at Maximum Specific Output. The Int. J of Eng. Education, 3, 1994.
12. Tozer R. M., James R. W., Thermodynamics of Absorption Refrigeration (Ideal Cycles). ASME Int. Absorption Heat Pump Conference 1994.
13. Agnew B., Anderson A., Frost T. H.; Optimisation of a Steady Flow Carnot Cycle with External Irreversibilities for Maximum Specific Output. Journal of Applied Thermal Engineering, 17, 1997.

14. Rubin M.H., Andersen B.; Optimal Staging of Endothermal Heat Engines. Journal of Applied Physics, 53, 1982.
15. Agnew B., An Examination of the Characteristics of a Non-flow Carnot Cycle with External Irreversibilities when used in a Combined Cycle Mode. Internal Report Tb87 University of Newcastle upon Tyne, 1996.
16. Dalichaouch M., A Theoretical and Experimental Examination of Absorption Refrigeration Systems for Application with Solar Energy Units, Ph.D Thesis, The University of Newcastle upon Tyne. 1987

ON THE POWER AND EFFICIENCY OF THERMOELECTRIC DEVICES

R.Y. NUWAYHID AND F. MOUKALLED
Mechanical Engineering Department
American University of Beirut
Beirut, Lebanon

Abstract- In this paper, the model for Carnot-like engines developed in [1] is modified to account for internal irreversibilities. The modified model is used to analyze, as a typical example of direct conversion heat engines, the thermoelectric generator. The effects of heat leak and finite rate heat transfer on the performance of the generator are studied. In specific, the power versus efficiency behaviour of the heat engine is thoroughly investigated. The efficiency at maximum power and the maximum efficiency of the generator are shown to be distinct. Finally, a comparison between the newly developed model and the one presented in [2] is performed.

Keywords: Thermoelectric generator, Heat engines, Finite-time thermodynamics.

NOMENCLATURE

I	Electric current (Amps).
K, K_C, K_H	Heat transfer coefficients (W/K).
m	Load ratio.
P	Work output (Joules).
Q_1	Heat entering the engine (Joules).
Q_2	Heat leaving the engine (Joules).
Q_C	Heat entering the heat sink (Joules).
Q_H	Heat leaving the heat source (Joules).
R	Electrical resistance (Ohms).
S	Seebeck coefficient (Volts/K).
T_1	Temperature of the hot junction (K).
T_2	Temperature of the cold junction (K).
T_C	Temperature of the heat sink (K).
T_H	Temperature of the heat source (K).

Z	Figure of merit (K^{-1}).
η	Efficiency of the heat engine.
η_C	The Carnot efficiency.
η_{max}	The maximum efficiency.
η_{mp}	The efficiency at maximum power.

INTRODUCTION

Heat engine performance based on the first and second laws of thermodynamics has been the subject of intense research work that resulted in the introduction of what has been named finite-time or endoreversible thermodynamics. The original idea of an endoreversible heat engine, was to consider the irreversibility associated with heat transfer to be between the working material and the heat reservoirs while processes within the engine were assumed to take place ideally. Further refinements to the theory allowed irreversibilities within the engine to be accomodated. With this concept, very complicated thermal systems may be simplified and made easier to understand. The approach, however, does not allow detailed predictions of the problem considered and, rather, a simplified view of the system performance can only be obtained. Nevertheless, this simplified view may be of great value for a designer trying to choose among several suggested systems before going into detailed analysis.

Several methods, of varying degrees of sophistication, for studying the performance of endoreversible heat engines have been suggested [1-6]. Seemingly the different approaches have been shown to lead to similar results and conclusions if certain common ground laws are followed correctly. Bejan recently put all this together under what may be considered a new method called the "Entropy Generation Minimization" (EGM) [7]. The method, a combination of thermodynamics, heat transfer, and fluid flow essentially unifies what is known in engineering as "thermodynamic optimization", and in physics as "finite-time thermodynamics".

Many workers have used the concept of endoreversible heat engine for predicting a variety of interesting phenomena. Gordon [2], applied finite-time thermodynamics to analyze a thermoelectric generator. Gordon and Zarmi [8] and DeVos and Flater [9], modeled the earth and its envelop using a Carnot-like engine with its heat input being solar radiation and its work output representing the wind generated. Nuwayhid and Moukalled [10], added a heat leak term into the model of DeVos and Flater [9] and studied the effect of a planet thermal conductance on conversion efficiency of solar energy into wind energy. The theoretical upper bound on conversion efficiency reported in [9] was shown to be well above the actual values predicted by the modified model. The realistic influence of heat leak on engine performance has motivated Moukalled et. al. [1] to develop a model in which a heat leak term is added to account for external thermal losses from the engine to the surroundings. In addition, the model allows the exchange of heat between the engine, heat reservoirs, and the surroundings to occur via combined modes of heat transfer.

Gordon [2] was one of the first to study the power versus efficiency behavior of heat engines and used the thermoelectric generator, which is considered to represent a typical example of a direct-conversion heat engine and entails interesting electrical as well as thermal phenomena, as a particular case. In his work, Gordon clearly demonstarted that real heat engines with sources of irreversibilities that include friction and heat leak exhibit fundamentally different power versus efficiency curves than those predicted neglecting friction and heat leak. Even though Gordon's model has produced satisfactory results, the way heat leak is accounted for is not physically convincing. The objectives of this paper are twofold. The first one, is to shed some light into the heat leak term of Gordon's model while the second objective is to modify the heat engine's model developed by Moukalled et. al. [1] and to use it in reanalyzing the thermoelectric generator considered by Gordon [2].

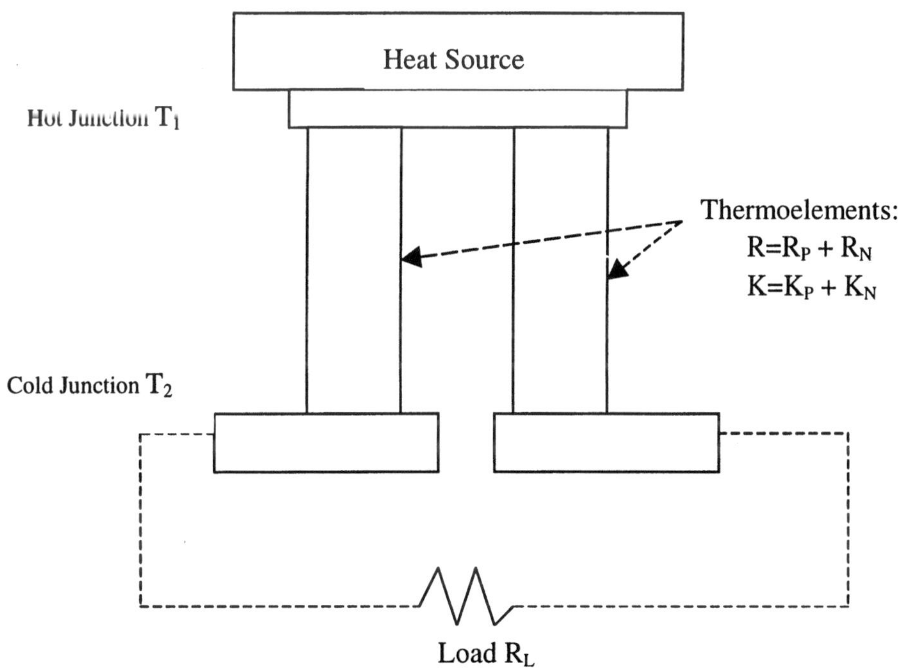

Figure 1 A schematic of a basic thermoelectric heat engine.

In the remainder of this article, the thermoelectric generator and Gordon's model are introduced and the latter is thoroughly discussed. Then, the model presented in [1] is modified to account for internal irreversibilities and used to study the thermoelectric generator.

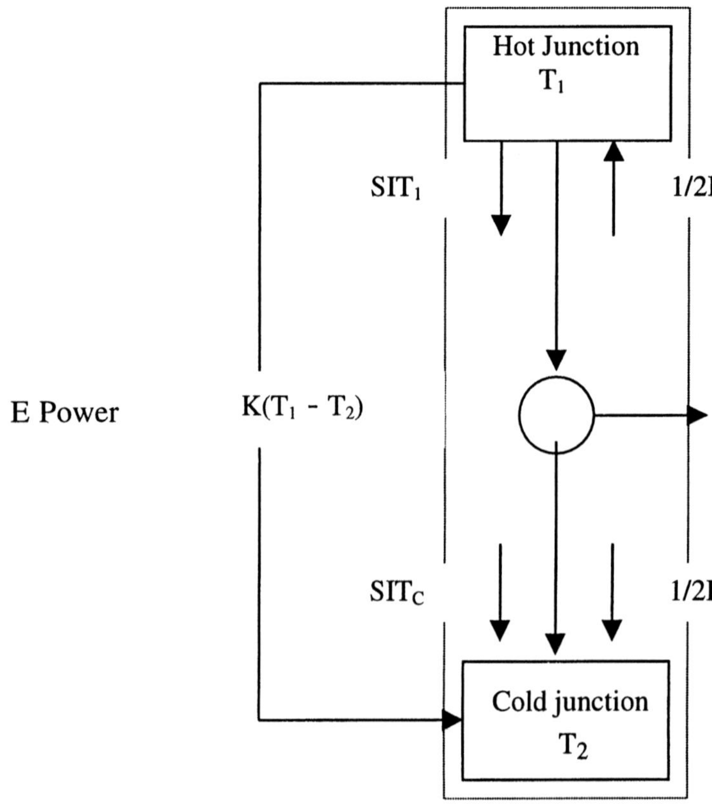

Figure 2 A schematic of an internally irreversible/externally reversible thermoelectric heat engine.

The Thermoelectric Generator

The thermoelectric effect was discovered by Seebeck in 1860 when he noticed that, upon the joining of two dissimilar metal wires and the heating of the junction, an electromotive force develops. Producing very low thermal-to-electric conversion efficiency, the thermoelectric generator was of little use. Only in temperature measurement has there been a significant application of the phenomenon. Advances in the field have been very slow and limited by material considerations. It was only with the introduction of semiconductors and the resultant rise in efficiency that renewed interest was stirred. Today the practical applications are few. One is in remote low level power generation using long-lived heat sources such as radioisotopes, and the second is in small portable refrigeration systems running on low power inputs. Figure 1 shows a simplified diagram of a thermoelectric generator.

The reader is assumed to have a certain amount of knowledge of simple thermoelectric performance modeling. Further background information may be found in many texts on the subject, for example Soo [11]. Interestingly, in the thermoelectric generator (Figure 2), the heat leak is a necessary part of the engine since it is taken to be a result of the thermal conductance of the generator legs themselves and therefore appears in the modeling directly. Moreover, the thermal conductance is inversely proportional to the electrical resistance and therefore the two have conflicting effects.

In most scoping studies of thermoelectric generators the usual practice is to neglect external irreversibilities and to consider only internal effects. Figure 2 shows the schematic model for the usuallly studied, internally irreversible/externally reversible case. Under such conditions, the heat source and heat sink temperatures become identical to the hot junction and cold junction temperatures. The heat input and heat output to and from the generator are given by:

$$Q_1 = SIT_1 - \tfrac{1}{2}I^2R + K(T_1 - T_2) \tag{1}$$

$$Q_2 = SIT_2 + \tfrac{1}{2}I^2R + K(T_1 - T_2) \tag{2}$$

Power is then given by:

$$P = SI(T_1 - T_2) - \tfrac{1}{2}I^2R \tag{3}$$

In the above equations, finite-rate heat transfer to and from the generator is taken as Newtonian. The resulting equation for the efficiency ($\eta = P/Q_1$) is then usually written as:

$$\eta = \eta_C \frac{m}{1 + m - \dfrac{T}{2T_H} + \dfrac{(1+m)^2}{ZT_H}} \tag{4}$$

where $m = R_L/R$ is the "load ratio", $Z = S^2/RK$ the "figure-of merit", R the total electrical resistance, K the total thermal conductance, S the Seebeck coefficient, and η_C the carnot efficiency. This is the equation found in textbooks and the power versus efficiency behaviour is the expected "loop" shape with both effiency and power rising from zero and returning to zero after passing through distinct maximum power and maximum efficiency below the Carnot limit. The loss of power being attributed to heat-leak at open circuit (slow operation; no recovered output) and to joule-heat loss at short circuit (fast operation;friction).

If the generator is made more and more reversible, say by decreasing R (Figure 3(a)) while holding K and S constant, or increasing S (Figure 3(b)) while holding R and K constant, the graph starts to approach the "dome" shape where the maximum efficiency tends toward the Carnot efficiency and the maximum power and maximum efficiency points become increasingly separated. Moreover, From equation (4), with m=1, the efficiency at maximum power is given by:

$$\eta_{mp} = \eta_C \frac{2}{4 - \eta_C + \dfrac{8}{ZT_H}} \tag{5}$$

It is interesting to note that, as the device becomes more reversible (i.e. R decreases and/or S increases), the curves representing the efficiency at maximum power, shown in Figures 3(a) and 3(b), shift to the right and approache a limiting value which is only dependent on the Carnot efficiency. This limiting value is obtainable by requiring $Z \rightarrow \infty$, giving:

$$\lim_{Z \to \infty} \eta_{mp} = \eta_C \frac{2}{4 - \eta_C} \tag{6}$$

The above equation raises an intersting question about the efficiency at maximum power which appears to be limited to 2/3! The maximum efficiency in this case is distinct from the efficiency-at-maximum-power. The former being given by:

$$\eta_{max} = \eta_C \frac{ZT_1 \sqrt{2}\sqrt{(2-\eta_C)ZT_1 + 2}}{(1+ZT_1)\left[4 + \sqrt{2}\sqrt{(2-\eta_C)ZT_1 + 2}\right] - 2\eta_C ZT_1} \tag{7}$$

Of course, in the limit as $Z \rightarrow \infty$, the Carnot limit is approached.

Before proceeding to the irreversible models, it is worth considering the case of the externally irreversible/internally reversible model displayed in Figure 4. Although this model is strictly unrealistic, it nevertheless serves as a link among models. The case requires both R and K to simultaneously vanish (in reality they are inversely proportional) and the resulting energy balances become very simple. The power versus efficiency relation in this case is:

$$P = \frac{K_H K_c \left[(1-\eta)^i T_H^i - T_c^i\right]}{(1-\eta)\left[K_H + (1-\eta)^{(i-1)} K_c\right]} \tag{8}$$

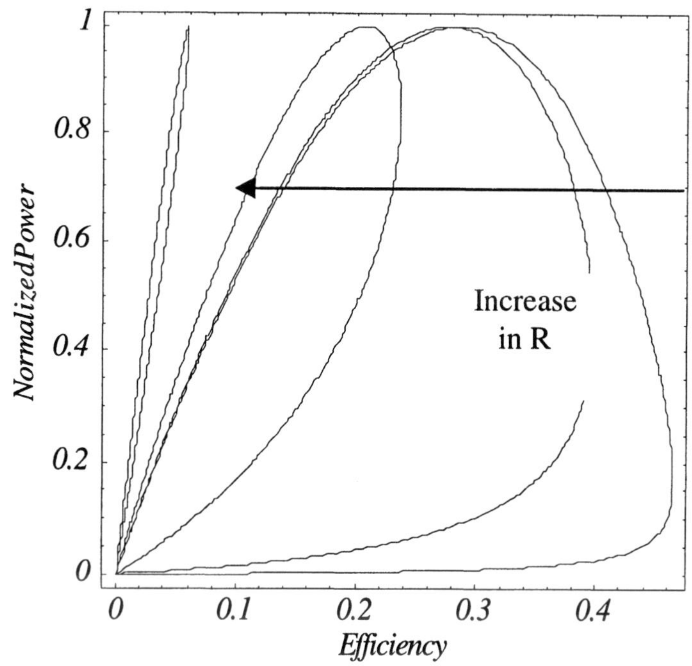

(a)

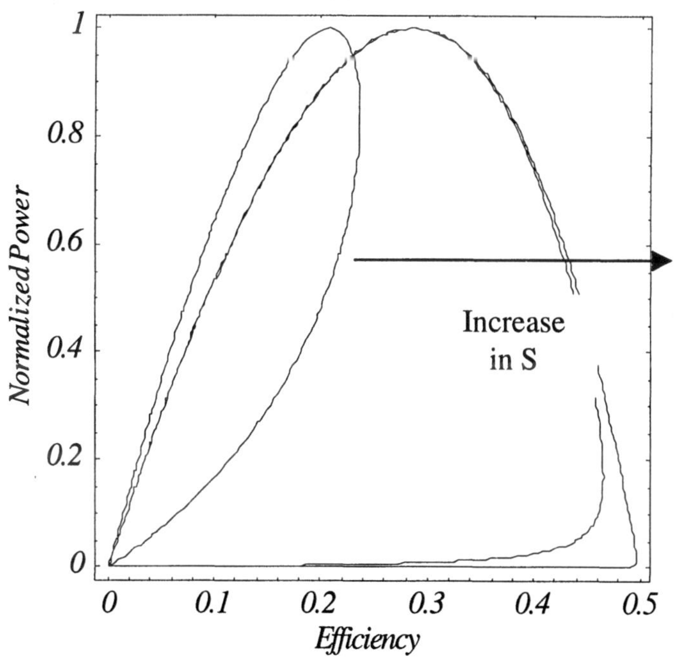

(b)

Figure 3 Variations of Power versus efficiency (a) with R ($R=10^{-9}, 10^{-8}, 10^{-7}, 10^{-6}$, $S=10^{-5}$) and (b) with S ($S=10^{-5}, 10^{-4}, 10^{-3}$, $R=10^{-6}$) in the usually used externally reversible/internally irreversible model. (T_H=600K, η_C=0.5, K = 0.1.)

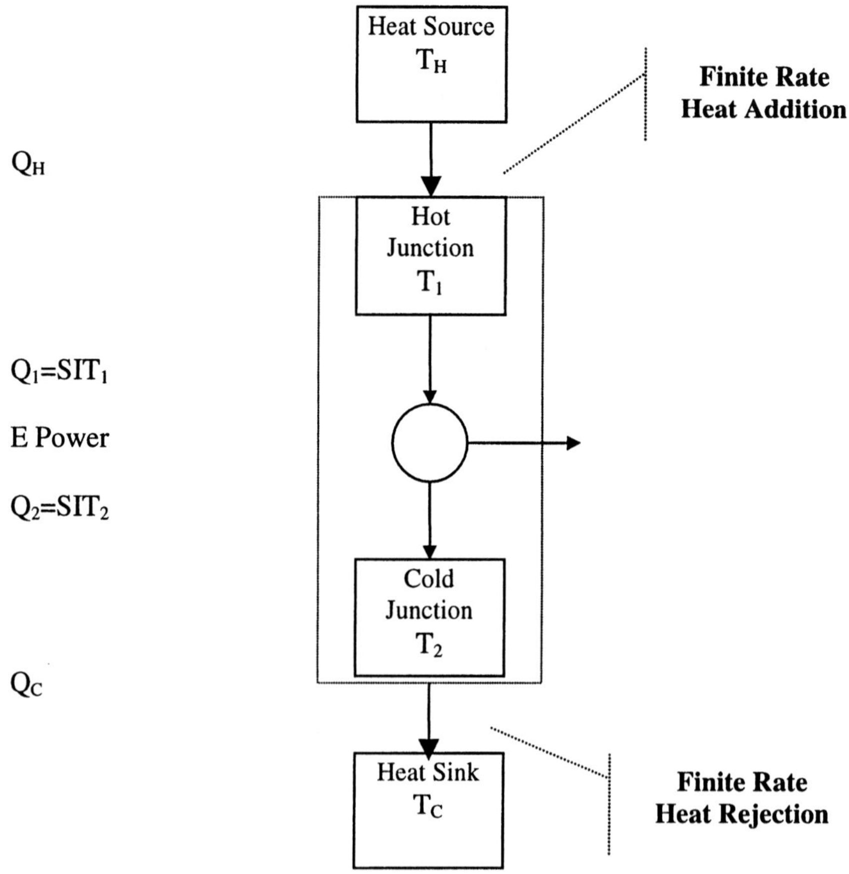

Figure 4 A schematic of an internally reversible/externally irreversible thermoelectric heat engine model.

Where "i" is the heat input/output power mode (unity for conduction/convection or 4 for radiation). Figure 5 shows normalized power $(P^* = P/KT_H^i)$ versus efficiency curves with the Carnot efficiency and the heat transfer mode as parameters. The above relation, being for an internally reversible case, produces the expected "domed" shape with the maximum efficiency terminating at the Carnot limit (by differentiating the above with respect to K and setting to zero one indeed obtains the Carnot limit). The efficiency at maximum power in this "endo reversible" case, obtainable this time by differentiating the above with respect to η and setting to zero, turns out to be the so-called CNCA [3] efficiency given by:

$$\eta_{mp} = 1 - \sqrt{\frac{T_C}{T_H}} \tag{9}$$

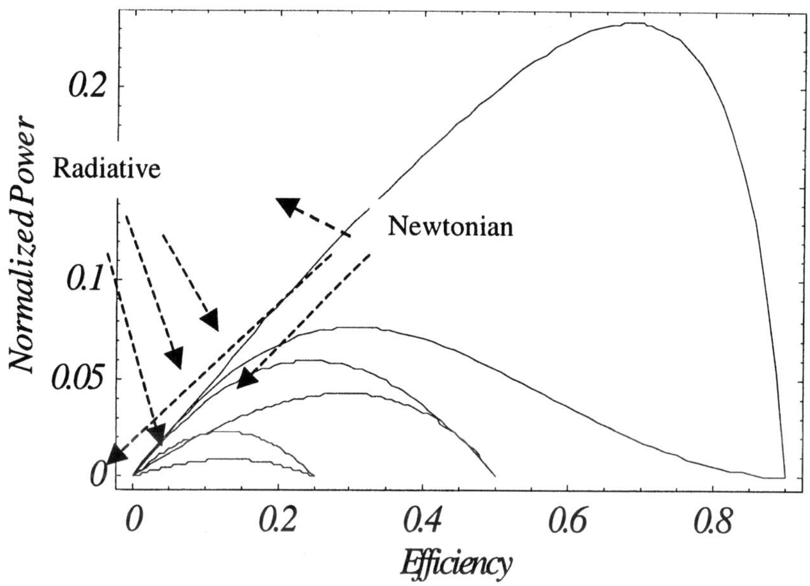

Figure 5 Normalized power versus efficiency curves of an internally reversible/externally irreversible model for three values of η_C=0.25, 0.5, and 0.9 (K_H=K_C=K).

Of course, the limit of this as $T_H \rightarrow \infty$ is unity compared to the limit for the internally irreversible case which is confined to an upper limit of 2/3. Moreover, it turns out that nearly below η_C=0.75, endoreversible (with reversibilities occuring externally) and internally irreversible (with no external irreversibilities) cases produce very similar efficiencies at maximum power as long as Z is not less than about 0.1.

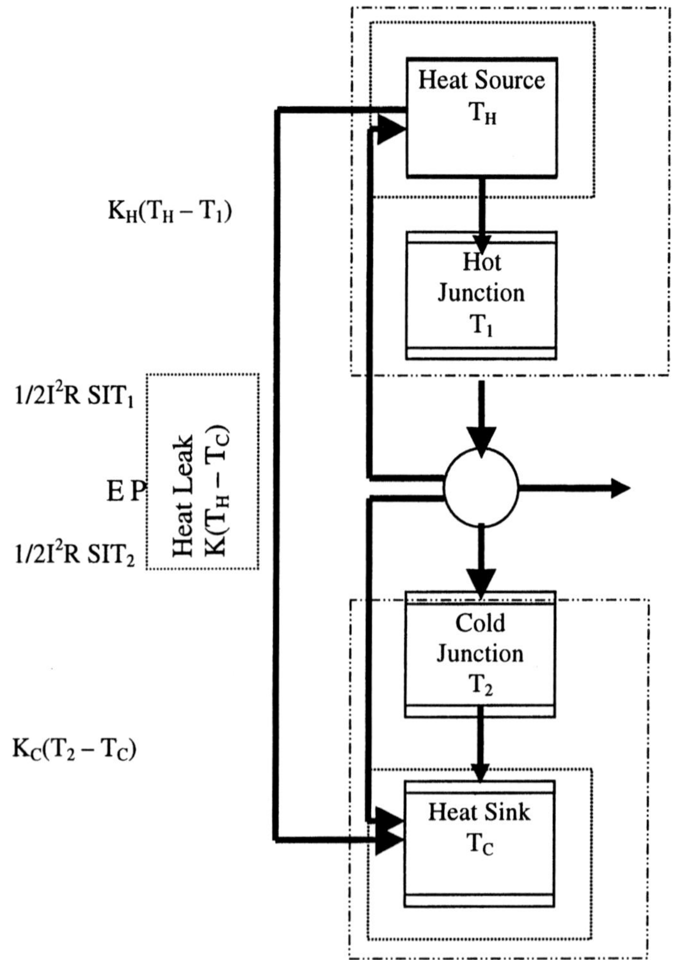

Figure 6 A schematic of the model implied in Gordon's irreversible case.

GORDON'S FINITE-RATE HEAT EXCHANGE THERMOELECTRIC MODEL

In an attempt to make the previous analysis more realistic, Gordon [2] developed a more sophisticated model to investigate the performance of the thermoelectric generator. In his sugested model, Gordon [2] included additional irreversibilities that are associated with heat transfer between the working material and the heat reservoirs. A schematic for the model was not given in Gordon's paper. However, based on the conservation equations used, the perceived schematic of his model is depicted in Figure 6. The power output is given, as usual, by:

$$P = Q_1 - Q_2 \tag{10}$$

Where,

$$Q_1 = SIT_1 - \tfrac{1}{2}I^2R + K(T_H - T_C) \tag{11}$$

and

$$Q_2 = SIT_2 + \tfrac{1}{2}I^2R + K(T_H - T_C) \tag{12}$$

Stating that heat input and output are equal to the reversible Seebeck/Peltier heat, the following are said to hold:

$$K_H(T_H - T_1) = SIT_1 \tag{13}$$

and

$$K_C(T_2 - T_C) = SIT_2 \tag{14}$$

The above conservation equations along with the schematic of the model displayed in Figure 6 indicate that heat leak is assumed to take place between the heat source and heat sink. In reality, it occurs between the hot and cold junctions. Thus, it would have been physically more realistic to consider heat leak to be between the hot and cold junctions. Moreover, the joulean heating is assumed to happen between the heat source or sink and the engine. Since this joulean heating takes place in the legs of the thermoelectric generator (Figure 1), it is more logical to consider it to be between the engine and the hot and cold junctions. With these remarks in mind, the power developed by the generator is given by:

$$P = SI\left[\frac{K_H}{K_H + SI}T_H - \frac{K_C}{K_C - SI}T_C\right] - I^2R \tag{15}$$

It is obvious that the power equation contains no heat-leak factor which indicates that heat leak does not have a direct effect on the power output of the generator!

The efficiency is calculated from

$$\eta = \frac{P}{Q_1} \tag{16}$$

with the relation for Q_H with T_1 eliminated given by

$$Q_1 = \frac{SIK_H T_H}{K_H + SI} - \tfrac{1}{2} I^2 R + K(T_H - T_C) \tag{17}$$

It is obvious that the resulting equations are tedious to solve analytically and should be tackled numerically. The procedure is as follows: Combining the equation defining efficiency with Eq. (14), a new equation in terms of P, I, and η is realized. Eliminating I from the new equation by using Eq. (13), the desired equation relating P and η is obtained. This equation is a very complicated one and may be simplified by introducing the Carnot efficiency and making the assumption that K_H=K_C=K. The final form of this equation is written as:

$$\begin{aligned}
&K^4(\eta_c - 1)T_H\Big[-2K^4\eta^4\eta_c^2T_H^2(-2K^2R^2 + 3K(\eta_c - 2)RT_HS^2 + 27\eta_c^2T_H^2S^4) - \\
&P\Big\{-4K^5(\eta - 2)\eta^3\eta_cR^2T_H - 4(\eta - 2)^2(\eta - 1)P^3S^4 + \\
&2K(\eta - 2)\eta\{22\eta_c + \eta^2(2\eta_c + 1) - \eta(21\eta_c + 2)\}P^2T_HS^4 \\
&- K^3\eta^2T_HS^2\{(\eta - 2)(9\eta\eta_c + 2\eta + 2\eta_c - 4)PR - 18\eta\eta_c^2(4\eta\eta_c + \eta - 9\eta_c)T_H^2S^2\} \\
&- 2K^2\eta^2PS^2\{(\eta - 2)^2PR - 3\eta_c(5\eta^2\eta_c + 2\eta^2 - 28\eta\eta_c - 4\eta + 30\eta_c)T_H^2S^2\} \\
&- K^4\eta^2R\{(\eta - 2)^2PR + 2\eta(4\eta\eta_c^2 + 5\eta_c^2 + 3\eta\eta_c - 10\eta_c + 2\eta)T_H^2S^2\}\Big\}\Big] = 0
\end{aligned} \tag{18}$$

Having found the necessary relation, plots for power versus efficiency are generated and displayed in Figure 7 for a generator resistance (R) ranging from 10^{-4} to 10^{-9} (holding η_c, S, and K constant). The graphs depicted in Figure 7 indicate the existence of four roots giving an acceptable loop shape set of solutions and other solutions that may be rejected since these latter ones are not bounded by the Carnot limit. Additionally, it is seen that as R becomes very small, the non-acceptable solutions disappear. It is not clear whether the unbounded results are due to numerical instabilities or to a physical error in modeling.

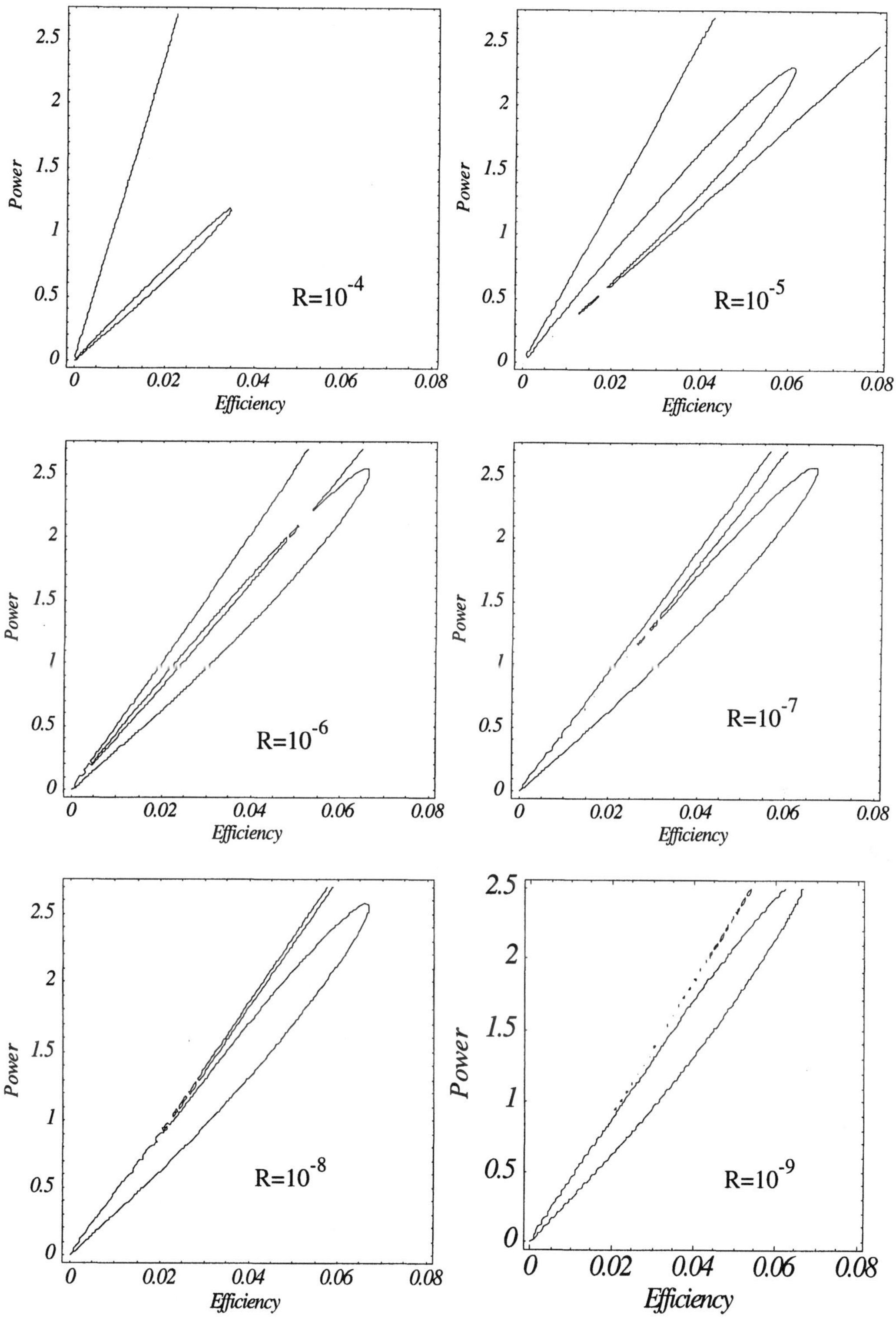

Figure 7 The effect of R on power and efficiency as obtained from Gordon's irreversible model (T_H=600 K, η_C=0.5, K=0.01, S=10^{-6}).

THE NEWLY SUGGESTED THERMOELECTRIC GENERATOR'S MODEL

The general model for Carnot-like engines developed by Moukalled et.al. [1], is modified to account for irreversibilities within the engine and used here to conduct a new analysis of the thermoelectric generator. A schematic of the modified model is depicted in Figure 8. As can be seen, the model allows for finite rate heat exchange with heat transfer obeying Newton's law. Gordon [2] studied the same situation but considered the leakage mechanism to operate between the heat source and the heat sink (T_H and T_C respectively). This enabled him to apply a simple relation between each of T_H and T_1 and T_C and T_2 . With the heat leak, however, more appropriately applied between T_1 and T_2 (Figure 8), the applicable relations become more complex as will be shown. Moreover, Gordon did not explicitly study the effect of heat-leak on the performance of the thermoelectric generator but used it to illustrate certain fundamental points. In this study, the effect of heat leak will be clearly demonstrated. Furthermore, unlike Gordon's model, the joulean heating is considered to occur between the engine and the junctions. Therefore, the new model is physically more sound.

Referring to Figure 8, the heat leaving and entering the hot and cold reservoirs are, respectively, given by:

$$Q_H = K_H(T_H - T_1) \tag{19}$$

$$Q_C = K_C(T_2 - T_C), \tag{20}$$

the heat balance equations at the hot and cold junctions may, respectively, be written as:

$$K_H(T_H - T_1) = SIT_1 - \tfrac{1}{2}I^2R + K(T_1 - T_2) \tag{21}$$

$$K_C(T_2 - T_C) = SIT_2 + \tfrac{1}{2}I^2R + K(T_1 - T_2), \tag{22}$$

the energy quantities entering and leaving the engine are:

$$Q_1 = SIT_1 - \tfrac{1}{2}I^2R \tag{23}$$

$$Q_2 = SIT_2 + \tfrac{1}{2}I^2R, \tag{24}$$

and finally, the expressions for the work and efficiency of the thermoelectric generator are given as:

$$P = Q_1 - Q_2 = SI(T_1 - T_2) - I^2R \tag{25}$$

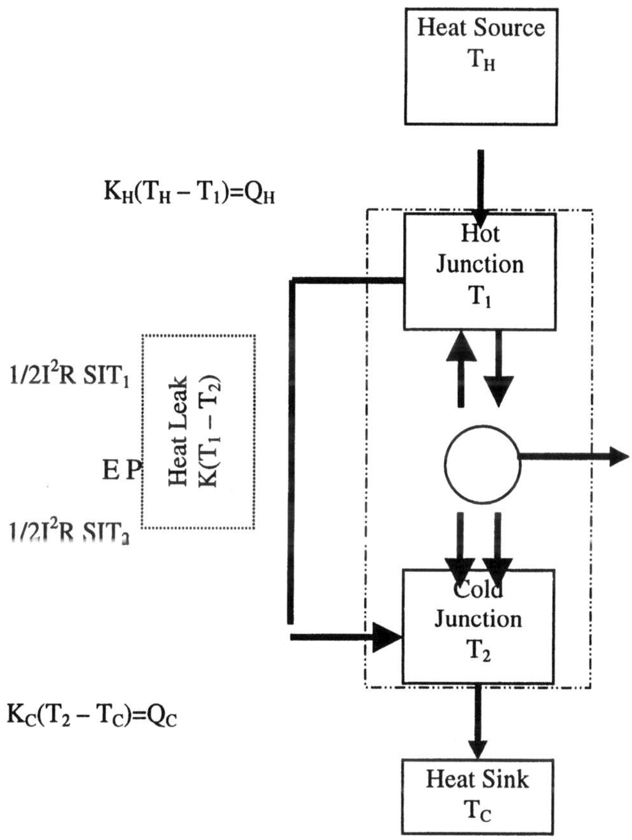

Figure 8 A schematic of the newly proposed model.

$$\eta = \frac{P}{Q_1} = \frac{SI(T_1 - T_2) - I^2R}{SIT_1 - \frac{1}{2}I^2R}, \tag{26}$$

The intention is to derive an equation relating P and η in terms of the constant parameters of the problem. For that purpose, Eqs. (21) and (22) are solved to obtain the following expressions for T_1 and T_2:

$$T_1 = \frac{2I^2KR + I^2K_CR + 2KK_CT_C + 2KK_HT_H + 2K_CK_HT_H - I^3RS - 2IK_HT_HS}{2\left(KK_C + KK_H + K_CK_H + IK_CS - IK_HS - I^2S^2\right)} \tag{27}$$

$$T_2 = \frac{K\left(\frac{1}{2}I^2R + K_HT_H\right) + \left(\frac{1}{2}I^2R + K_CT_C\right)\left(K + K_H + IS\right)}{\left(K + K_H + IS\right)\left(K + K_C - IS\right) - K^2}, \tag{28}$$

The values for T_1 and T_2, as given by the above equations, are substituted in Eq. (25) to yield the following relation between P and I:

$$\begin{aligned} &RS(K_C - K_H)I^3 + 2\left[R(KK_H + K_HK_C + KK_C) + S^2(K_HT_H + K_CT_C - 2P)\right]I^2 - \\ &\quad 2S\left[K_HK_C(T_H - T_C) + P(K_H - K_C)\right]I + 2P(KK_H + K_HK_C + KK_C) = 0 \end{aligned} \tag{29}$$

A similar equation between η and I can be derived by substituting the values of T_1 and T_2 from Eq. (27) into Eq. (26) and is given by:

$$\begin{aligned} &RS\left[(1-\eta)K_H + K_C\right]I^3 + \\ &\quad 2\left\{R\left[KK_C + (1-\eta)KK_H + \left(1 - \frac{\eta}{2}\right)K_HK_C\right] + S^2\left[K_CT_C + (1-\eta)K_HT_H\right]\right\}I^2 \\ &\quad - 2SK_HK_C\left[(1-\eta)T_H - T_C\right]I + 2\eta K_HK_CK(T_H - T_C) = 0 \end{aligned} \tag{30}$$

Combining the previous two equations, the following relation between work and efficiency is obtained:

$$\begin{aligned} &a_2^3d_1^3 + d_1^2\left(a_2b_1b_2^2 - a_1b_2^3 - a_2^2b_2c_1 - 2a_2^2b_1c_2 + 3a_1a_2b_2c_2 - 3a_1a_2^2d_2\right) + \\ &d_1(-a_2b_1b_2c_1c_2 + a_1b_2^2c_1c_2 + a_2^2c_1^2c_2 + a_2b_1^2c_2^2 - a_1b_1b_2c_2^2 - 2a_1a_2c_1c_2^2 + \\ &a_1^2c_2^3 - 2a_2b_1^2b_2d_2 + 2a_1b_1b_2^2d_2 + 3a_2^2b_1c_1d_2 - a_1a_2b_2c_1d_2 + a_1a_2b_1c_2d_2 - \\ &3a_1^2b_2c_2d_2 + 3a_1^2a_2d_2^2) - d_2(-a_2b_1b_2c_1^2 + a_1b_2^2c_1^2 + a_2^2c_1^3 + a_2b_1^2c_1c_2 - \\ &a_1b_1b_2c_1c_2 - 2a_1a_2c_1^2c_2 + a_1^2c_1c_2^2 - a_2b_1^3d_2 + a_1b_1^2b_2d_2 + 3a_1a_2b_1c_1d_2 - \\ &2a_1^2b_2c_1d_2 - a_1^2b_1c_2d_2 + a_1^3d_2^2) = 0 \end{aligned} \tag{31}$$

Where the coefficients are given by:

$$
\begin{aligned}
a_1 &= RS(K_C - K_H) \\
b_1 &= 2\left[R(KK_H + K_H K_C + KK_C) + S^2(K_H T_H + K_C T_C - 2P)\right] \\
c_1 &= -2S\left[K_H K_C (T_H - T_C) + P(K_H - K_C)\right] \\
d_1 &= 2P(KK_H + K_H K_C + KK_C)
\end{aligned}
$$

$$
\begin{aligned}
a_2 &= RS\left[(1-\eta)K_H + K_C\right] \\
b_2 &= 2\left\{R\left[KK_C + (1-\eta)KK_H + \left(1-\frac{\eta}{2}\right)K_H K_C\right] + S^2\left[K_C T_C + (1-\eta)K_H T_H\right]\right\} \\
c_2 &= -2SK_H K_C\left[(1-\eta)T_H - T_C\right] \\
d_2 &= 2\eta K_H K_C K(T_H - T_C)
\end{aligned}
\tag{32}
$$

Using the above-derived equation, the effects of the various parameters on the power and efficiency of the generator are assessed and results are displayed in Figure 9. By augmenting heat leak through increasing the value of the heat transfer coefficient K while holding all other parameters constant (Figure 9(a)), both efficiency and power outpout decrease, as expected, and the efficiency at maximum power and the maximum efficiency of the engine become less distinct. Moreover, as irreversibility increases, the power versus efficiency behaviour change from the "dome" to the "loop" shape. Since the power versus efficiency curves presented in Figure 9 include all possible real roots of Eq. (31), it is obvious that, unlike Gordon's model, the newly suggested one does not predict, in the range of parameters considered, efficiencies above the Carnot limit.

The effects of finite rate heat transfer on variations of power with efficiency are depicted in Figure 9(b). As the resistance to heat transfer increases (i.e. K_H and K_C decrease), the power output, the efficiency at maximum power, and the maximum efficiency of the generator diminish. This behaviour is anticipated because decreasing the thermal resistance to finite rate heat transfer reduces the irreversibilty of the heat engine by forcing the hot and cold junction temperatures to become closer to the heat source and heat sink temperatures (i.e. T_1 and T_2 become closer to T_H and T_C, respectively).

The combined effects of finite rate heat transfer and leakage are presented in Figure 9(c). With decreasing thermal resistance to finite rate heat transfer and leakage by increasing the values of K_H, K_C, and K, an increase in power output and a decrease in efficiency is noticed. The increase in power is caused by the higher rate of heat flow whereas the decrease in efficiencies is due to higher rate of heat leak.

Finally, the power versus efficiency curves for different values of $Z = S^2/RK$ (the figure of merit), holding all other variables constant, are presented in Figure 9(d). As expected, the power and efficiencies increase with Z for the obvious reason that increasing the figure of merit is equivalent to increasing S or decreasing R and/or K.

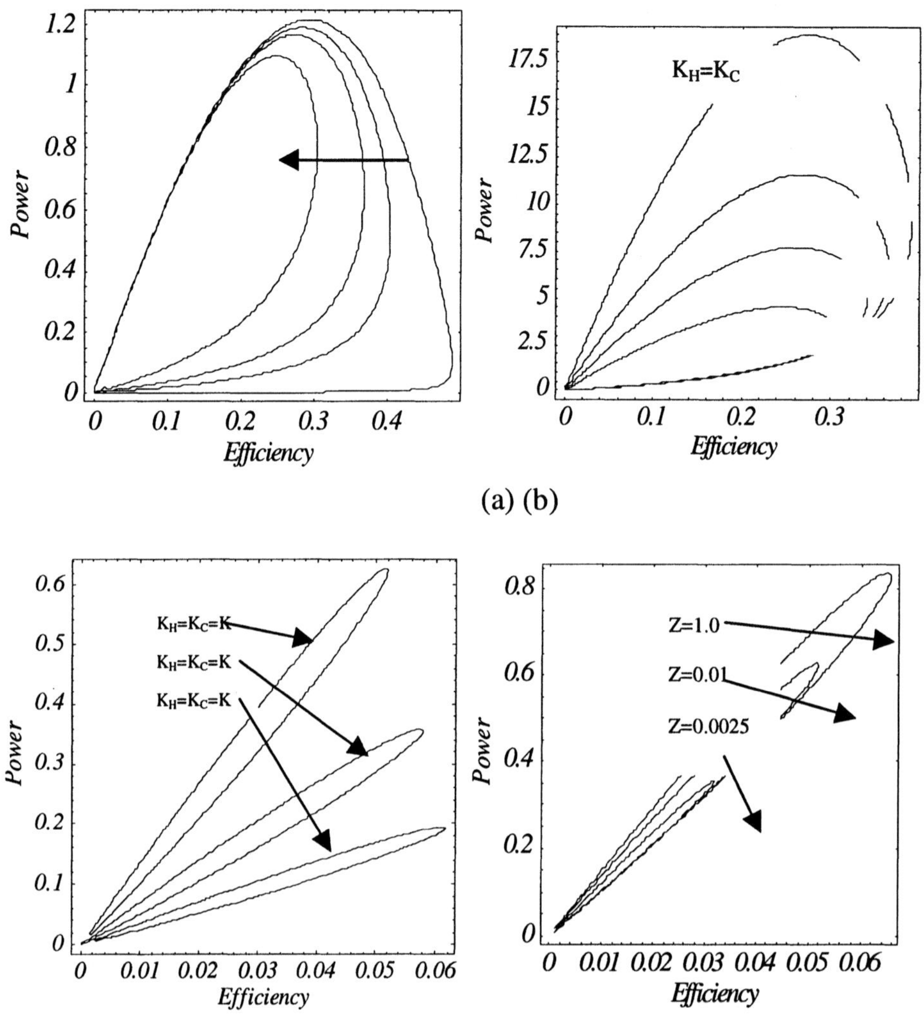

(a) (b)

(c) (d)

Figure 9 (a) Effect of heat leak ($R=10^{-5}$,$S=10^{-4}$,T_H=600 K, T_C=300 K, $K_H=K_C$=0.05, $K=10^{-5}$,0.0005,0.001,0.0025), (b) Effect of finite rate heat transfer (K=0.01, $R=10^{5}$,$S=10^{4}$,T_H=600 K,T_C=300 K), (c) Combined effect of finite rate heat transfer and heat leak ($R=10^{-5}$,$S=10^{-4}$,T_H=600 K, T_C=300 K) and (d) Effect of figure of Merit (T_H=600 K,T_C=300 K, $K=K_H=K_C$=0.1) on the variation of power and efficiency as obtained from the newly developed model.

CONCLUSIONS

The general model developed by Moukalled et. al. [1] for internally reversible/externally irreversible heat engines was modified to account for internal irreversibilities. The modified model, applicable to internally and externally irreversible heat engines, was used to assess the performance of real engines by selecting the thermoelectric generator as a typical example. The effects of heat leak, internal dissipation, and finite rate heat transfer were investigated. The detremental effect of these variables on the performance of the generator was demonstrated. In addition, the performance of the new model was compared against the one developed in [2] and was shown to be superior.

ACKNOWLEDGMENTS

The financial support provided by the University Research Board of the American University of Beirut is gratefully acknowledged. Special thanks are due to Mr. S. Fattal for helping in the numerical solutions of the algebraic equations.

REFERENCES

1. Moukalled, F., Nuwayhid, R., and Fattal, S.,"A Universal Model for Studying the Performance of Carnot-Like Engines at Maximum Power Conditions,", International Journal of Energy Research, vol. 20, pp. 203-214, 1996.
2. Gordon, J.M., "Generalized Power Versus Efficiency Characteristics of Heat Engines:The Thermoelectric Generator as an Instructive Illustration," American Journal of Physics, vol. 59, no. 6, pp. 551-555, 1991.
3. Curzon F. and Ahlborn B., "Efficiency of a Carnot Engine at Maximum Power Output", ," American Journal of Physics, vol. 43, pp. 22-24, 1975.
4. DeVos A., "Efficiency of Some Heat Engines at Maximum Power Conditions," American Journal of Physics, vol. 53, 570-573,1981.
5. Chen L., and Yan Z., 'The Effect of Heat Transfer Law on Performance of a Two Heat-Source Endoreversible Heat Cycle", J. Chem. Phys., vol. 90, no. 7, pp. 3740-3743, 1989
6. Moukalled F., Nuwayhid R., and Noueihed N., "The Efficiency of Endoreversible Heat Engines with Heat-Leak", International Journal of Energy Research, vol. 19, pp. 377-389, 1995.
7. Bejan A., "Entropy generation minimization: The new thermodynamics of finite-size devices and finite finite-time processes", Journal of Applied Physics, vol. 79, No. 3, 1996.
8. Gordon, J.M. and Zarmi, Y.,"Wind Energy as a Solar-Driven Heat Engine: A Thermodynamic Approach", American Journal of Physics, vol. 57, no 11. pp. 995-998, 1989.

9. DeVos, A. and Flater, G.,"The Maximum Efficiency of The Conversion of Solar Energy into Wind Energy", American Journal of Physics, vol. 59, no 8, pp. 751-754, 1991.
10. Nuwayhid, R. and Moukalled, F.,"The Effect of Planet Thermal Conductance on Conversion Efficiency of Solar Energy into Wind Energy", Renewable Energy, vol. 4, No. 1, pp. 53-58, 1994.
11. Soo S.L., "Direct Energy Conversion", pp. 113-149, Prentice-Hall, 1968.

POWER OPTIMIZATIONS OF AN IRREVERSIBLE BRAYTON HEAT ENGINE WITH FINITE HEAT RESERVOIRS

CHING-YANG CHENG
Department of Mechanical Engineering
Nan-Tai Institute of Technology, Yunkan, Taiwan 710, R.O.C.

SHUN-CHING LEE
Department of Mechanical Engineering
National Kaohsiung Institute of Technology, Kaohsiung, Taiwan 807, R.O.C.

and

CHA'O -KUANG CHEN
Department of Mechanical Engineering
National Cheng Kung University, Tainan, Taiwan 701, R.O.C.

Abstract- In this work, finite-time thermodynamics is used to study an irreversible closed Brayton heat engine coupled with heat reservoirs with finite thermal capacitance rates and heat exchangers finite total thermal conductance. The physical model considers two types of irreversibilities: finite thermal conductance between the working fluid and the two finite heat reservoirs and internal irreversibility inside the closed Brayton heat engine. The power output is optimized with respect to the adiabatic temperature ratio, heat exchanger thermal conductance ratio and ratio of working and hot fluid thermal capacitance rates. Effects of total thermal conductances of heat exchangers, ratios of cold and hot fluid thermal capacitance rates, turbine and compressor isentropic efficiencies and reservoir temperature ratios on the maximum power output and its corresponding parameters are studied. The optimum values of adiabatic temperature ratios, heat exchanger thermal conductance ratios and ratios of working and hot fluid thermal capacitance rates could be used in the design of the Brayton heat engine. The thermal capacitance rate of the cold external fluid should be larger than that of the hot external fluid in order to obtain a higher power output. However, increasing the cold fluid capacitance rate tends to lower the maximum-power thermal efficiency and to raise the costs of dealing with the fluids.

Key words: finite-time thermodynamics, heat engine, Brayton cycle, optimization, irreversibility.

INTRODUCTION

Brayton cycles have been widely used in gas-turbine power plants and aircraft propulsion systems. A closed-cycle Brayton heat engine may be built for special purposes, such as space power generation or nuclear power generation. Instead of high-quality fuels, various energy sources can be used to operate a closed Brayton heat engine for generating power. However, this engine needs additional heat transfer equipment. It might be worth studying the influence of the size and allocation of heat transfer equipment and the thermal capacitance rate ratios of fluids on the maximum power output, effectively reducing the cost of heat exchangers and maintaining a greater power output.

The thermal efficiency of a reversible Carnot cycle, given by $\eta_{Carnot} = 1 - (T_L/T_H)$, provides a bound on the optimal operation of heat engines operating between heat reservoirs at temperatures T_H and T_L. However, the power output under reversible condition approaches zero. Novikov [1], El-Wakil [2] and Curzon and Ahlburn [3] stated that the efficiency under the maximum power condition is expressed as $\eta_m = 1 - (T_L/T_H)^{0.5}$, considering the finite heat transfer rate between the heat engine and the infinite heat reservoirs. The maximum-power efficiency obtained was found to be closer to the efficiencies of real engines than the Carnot efficiency. Since then, finite-time thermodynamics has gained much achievement in the past years by many researchers' efforts[4].

There have been a few works about Brayton heat engines. Maximum power of endoreversible Brayton heat engines has been studied by Leff [5], Bejan [6], Wu and Kiang [7], and Ibrahim et. al [8]. Leff [5] showed that the choice of its parameters should be limited for Brayton heat engines to achieve a maximum work. Bejan [6] showed that the power output of an endoreversible Brayton heat engine with infinite heat reservoirs is optimized as the total thermal conductance is equally divided between the two heat reservoirs. Wu and Kiang [7] studied the maximum power output of an endoreversible Brayton heat engine with infinite heat reservoirs, and showed that the power optimization introduces additional constraints on the cycle temperatures such that the power output is a function of the source and the sink temperatures. Ibrahim et. al [8] obtained the formula of the maximum power output of an endoreversible heat engine with infinite or finite heat reservoirs. Wu and Kiang [9] studied the maximum power of an irreversible Brayton heat engine with infinite heat reservoirs, showing the adiabatic efficiencies of turbines and compressors tend to lower the power output and the thermal efficiency. Sahin et. al [10] studied an endoreversible Brayton cycle by optimizing the power density defined as the ratio of power to the maximum volume in the cycle, and found that the maximum-power efficiency is always smaller than the maximum-power-density efficiency. Moreover, Sahin et. al [11] also studied an irreversible Brayton cycle by optimizing the power density, and found that the design parameters at maximum power density lead to smaller and more efficient Brayton heat engines than an engine working at maximum power output conditions. Recently, Cheng and Chen [12] studied the power optimization of an irreversible Brayton heat engine with infinite heat reservoirs, and showed that the optimum ratio of the hot-end heat exchanger conductance to the total conductance of the two heat exchangers is found to be less than 0.5. Here we extend that analysis to incorporate finite thermal capacitance rates of heat reservoirs to increase the degree of realism further.

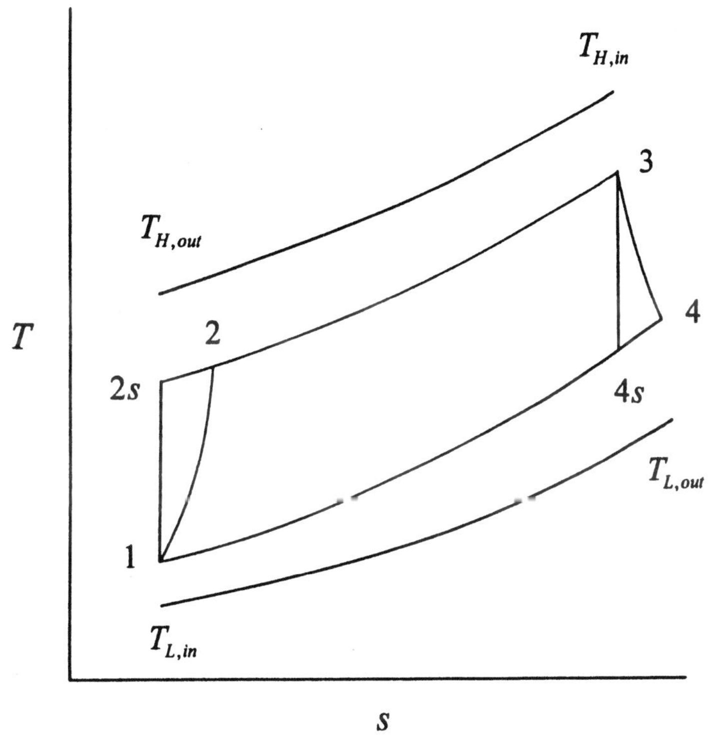

Fig. 1 Temperature-entropy diagram of an irreversible Brayton heat engine.

A steady-flow approach for finite-time thermodynamics is applied in this work to study an irreversible Brayton heat engine with finite heat reservoirs. Here we consider two types of irreversibilities: finite thermal conductance between the working fluid and the two finite heat reservoirs and the irreversibility inside the heat engine. The power output of the Brayton heat engine is optimized with respect to the adiabatic temperature ratio, thermal conductance ratio and ratio of working and hot fluid thermal capacitance rates. The maximum power output and its corresponding parameters are calculated. Moreover, the effects of total thermal conductances of heat exchangers, ratios of cold and hot fluid thermal capacitance rates, turbine and compressor isentropic efficiencies and reservoir temperature ratios on the maximum power output and its corresponding parameters are studied.

ANALYSIS

A steady-flow irreversible Brayton power cycle coupled to two heat reservoirs with finite thermal capacitance rates is sketched in Fig. 1. In the Brayton cycle, first, the working fluid at state 1 undergoes an adiabatic irreversible compression process to state 2, while 1-2s represents the corresponding isentropic process.. Then the working fluid at state 2 undergoes an isobaric heating process to state 3, exchanging heat with a heat reservoir varied from temperature $T_{H,in}$ to temperature $T_{H,out}$. The working fluid produces power during the irreversible process 3-4, with 3-4s representing the corresponding isentropic process. Finally, the working fluid at state 4 undergoes an isobaric cooling process to state 1, exchanging heat with a heat reservoir varied from temperature $T_{L,in}$ to temperature $T_{L,out}$.

Here $\dot{Q}_H$ is the heat rate absorbed from the hot fluid and $\dot{Q}_L$ is the heat rate released to the cold fluid by the working fluid. Moreover, U_H and U_L are respectively the overall heat-transfer coefficients of the hot-end and cold-end heat exchangers. A_H and C_L are respectively the areas of the hot-end and cold-end heat exchangers. Furthermore, $\dot{C}_W$, $\dot{C}_H$ and $\dot{C}_L$ are respectively the thermal capacitance rates of the working fluid, hot fluid and cold fluid. $U_H A_H$ and $U_L A_L$ are respectively the thermal conductances of the hot-end and cold-end heat exchangers. Assume that counter-flow heat exchangers are used, and heat transfer obeys Newton's law of cooling. The heat transfer rates can be expressed as

$$\dot{Q}_H = \dot{C}_{H,\min}\varepsilon_H(T_{H,in} - T_2) = \dot{C}_W(T_3 - T_2) = \dot{C}_H(T_{H,in} - T_{H,out}) \tag{1}$$

$$\dot{Q}_L = \dot{C}_{L,\min}\varepsilon_L(T_4 - T_{L,in}) = \dot{C}_W(T_4 - T_1) = \dot{C}_L(T_{L,out} - T_{L,in}) \tag{2}$$

where the effectivenesses of the hot-end and cold-end heat exchangers ε_H and ε_L are defined respectively as [13]

$$\varepsilon_H = \{1 - \exp[-N_H(1 - R_H)]\} / \{1 - R_H \exp[-N_H(1 - R_H)]\} \tag{3}$$

$$\varepsilon_L = \{1 - \exp[-N_L(1 - R_L)]\} / \{1 - R_L \exp[-N_L(1 - R_L)]\} \tag{4}$$

The numbers of heat transfer units are based on the minimum thermal capacitance rates,

$$N_H = U_H A_H / \dot{C}_{H,\min} \quad N_L = U_L A_L / \dot{C}_{L,\min}$$

and

$$R_H = \dot{C}_{H,\min}/\dot{C}_{H,\max} \quad R_L = \dot{C}_{L,\min}/\dot{C}_{L,\max}$$

$$\dot{C}_{H,\min} = Min(\dot{C}_H, \dot{C}_W) \; \dot{C}_{L,\min} = Min(\dot{C}_L, \dot{C}_W)$$

$$\dot{C}_{H,\max} = Max(\dot{C}_H, \dot{C}_W) \quad \dot{C}_{L,\max} = Max(\dot{C}_L, \dot{C}_W)$$

From the first law of thermodynamics, the power output is

$$\dot{W} = \dot{Q}_H - \dot{Q}_L \tag{5}$$

and the thermal efficiency of a Brayton heat engine is

$$\eta = \frac{\dot{W}}{\dot{Q}_H} = \frac{\dot{Q}_H - \dot{Q}_L}{\dot{Q}_H} \tag{6}$$

Eqs. (1) and (2) give

$$T_3 = (\varepsilon_H R_{HW})T_{H,in} + (1 - \varepsilon_H R_{HW})T_2 \tag{7}$$

$$T_1 = (\varepsilon_L R_{LW})T_{L,in} + (1 - \varepsilon_L R_{LW})T_4 \tag{8}$$

where

$$R_{HW} = \dot{C}_{H,\min}/\dot{C}_W \; R_{LW} = \dot{C}_{L,\min}/\dot{C}_W$$

Moreover, the isentropic efficiencies of gas turbines and compressors are respectively

$$\eta_T = \frac{T_3 - T_4}{T_3 - T_{4s}} \tag{9}$$

$$\eta_C = \frac{T_{2s} - T_1}{T_2 - T_1} \tag{10}$$

Eqs. (9) and (10) yield

$$T_{2s} = \eta_C T_2 + (1 - \eta_C)T_1 \tag{11}$$

$$T_{4s} = (1 - 1/\eta_T)T_3 + (1/\eta_T)T_4 \tag{12}$$

The second law of thermodynamics gives

$$T_{2s}T_{4s} = T_1 T_3 \tag{13}$$

Let

$$x = T_{2s}/T_1 = T_3/T_{4s} \tag{14}$$

where x is defined as the temperature ratio of the adiabatic process. Combining Eqs. (7), (8), (11), (12) and (14) gives

$$T_4 = (\eta_T/x + 1 - \eta_T)(1 - \varepsilon_H R_{HW})T_2 + (\eta_T/x + 1 - \eta_T)(\varepsilon_H R_{HW})T_{H,in} \tag{15}$$

From Eqs. (10) and (14), we get

$$T_1 = \eta_C T_2/(x - 1 + \eta_C) \tag{16}$$

Eqs. (1) and (2) yield

$$T_{H,out} = (1 - \varepsilon_H R_{HH})T_{H,in} + \varepsilon_H R_{HH} T_2 \tag{17}$$

$$T_{L,out} = (1 - \varepsilon_L R_{LL})T_{L,in} + (\eta_T/x - \eta_T + 1)\varepsilon_L \varepsilon_H R_{LL} R_{HW} T_{H,in} + (\eta_T/x - \eta_T + 1)\varepsilon_L R_{LL}(1 - \varepsilon_H R_{HW})T_2 \tag{18}$$

where

$$R_{HH} = \dot{C}_{H,\min}/\dot{C}_H \quad R_{LL} = \dot{C}_{L,\min}/\dot{C}_L$$

From Eqs. (8), (15) and (16), we obtain

$$T_2 = \left(a_1 x^2 + a_2 x + a_3\right)/\left(a_4 x^2 + a_5 x + a_6\right) \tag{19}$$

where

$$a_1 = \varepsilon_H R_{HW}(1 - \varepsilon_L R_{LW})(1 - \eta_T)T_{H,in} + \varepsilon_L R_{LW} T_{L,in}$$

$$a_2 = \varepsilon_H R_{HW}(1 - \varepsilon_L R_{LW})(\eta_C - \eta_T\eta_C + 2\eta_T - 1)T_{H,in} + \varepsilon_L R_{LW}(\eta_C - 1)T_{L,in}$$

$$a_3 = \varepsilon_H R_{HW}(1-\varepsilon_L R_{LW})\eta_T(\eta_C - 1)T_{H,in}$$

$$a_4 = -(1-\varepsilon_H R_{HW})(1-\varepsilon_L R_{LW})(1-\eta_T)$$

$$a_5 = \eta_C - (1-\varepsilon_H R_{HW})(1-\varepsilon_L R_{LW})(\eta_C - \eta_T\eta_C + 2\eta_T - 1)$$

$$a_6 = -(1-\varepsilon_H R_{HW})(1-\varepsilon_L R_{LW})\eta_T(\eta_C - 1)$$

Substitution of Eqs. (1), (2), (7), (15), (16) and (19) into Eq. (5) gives the power output

$$\dot{W} = \left(a_{11}x^3 + a_{12}x^2 + a_{13}x + a_{14}\right)\Big/\left(a_4x^3 + a_5x^2 + a_6x\right) \tag{20}$$

where

$$a_7 = -\dot{C}_{L,\min}\varepsilon_L\eta_T\varepsilon_H R_{HW}T_{H,in}$$

$$a_8 - \dot{C}_{H,\min}\varepsilon_H T_{H,in} - \dot{C}_{L,\min}\varepsilon_L\left[(1-\eta_T)\varepsilon_H R_{HW}T_{H,in} - T_{L,in}\right]$$

$$a_9 = \dot{C}_{L,\min}\varepsilon_L\eta_T(1-\varepsilon_H R_{HW})$$

$$a_{10} = \dot{C}_{H,\min}\varepsilon_H + \dot{C}_{L,\min}\varepsilon_L(1-\eta_T)(1-\varepsilon_H R_{HW})$$

$$a_{11} = a_4a_8 - a_1a_{10}$$

$$a_{12} = a_4a_7 + a_5a_8 - a_1a_9 - a_2a_{10}$$

$$a_{13} = a_5a_7 + a_6a_8 - a_2a_9 - a_3a_{10}$$

$$a_{14} = a_6a_7 - a_3a_9$$

Maximizing the power output $\dot{W}$ with respect to the adiabatic temperature ratio, i. e. setting $\dfrac{\partial\dot{W}}{\partial x} = 0$, gives

$$a_{15}x^4 + a_{16}x^3 + a_{17}x^2 + a_{18}x + a_{19} = 0 \tag{21}$$

where

$$a_{11} = a_4 a_8 - a_1 a_{10}$$

$$a_{12} = a_4 a_7 + a_5 a_8 - a_1 a_9 - a_2 a_{10}$$

$$a_{13} = a_5 a_7 + a_6 a_8 - a_2 a_9 - a_3 a_{10}$$

$$a_{14} = a_6 a_7 - a_3 a_9$$

$$a_{15} = a_5 a_{11} - a_4 a_{12}$$

$$a_{16} = -2a_4 a_{13}$$

$$a_{17} = a_6 a_{12} - a_5 a_{13} - 3a_4 a_{14}$$

$$a_{18} = -2a_5 a_{14}$$

$$a_{19} = -a_6 a_{14}$$

Eq. (21) is then solved to obtain the optimum adiabatic temperature ratio x_{opt} by the secant method [14]. Substitution of this temperature ratio into Eq. (19) gives the optimum cycle temperature $T_{2,opt}$. Then Eqs. (16), (7) and (15) give the optimum temperatures $T_{1,opt}$, $T_{3,opt}$ and $T_{4,opt}$. Substitution of the above-calculated optimum temperatures into Eqs. (1) and (2) yield the heat transfer rates $\dot{Q}_{H,opt}$ and $\dot{Q}_{L,opt}$. Moreover, the maximum power output $\dot{W}_{\max,1}$ is calculated by substituting the optimum adiabatic temperature ratio into Eq. (20).

RESULTS AND DISCUSSION

THE RELATION BETWEEN THE POWER OUTPUT AND THE THERMAL EFFICIENCY

Fig. 2 shows the power output $\dot{W}/(\dot{C}_W T_{H,in})$ as a function of the thermal efficiency η for an irreversible Brayton heat engine. It is shown that the maximum power output is higher than the power output at maximum thermal efficiency. A thermal efficiency corresponds to two different values of the power output, except at the maximum-thermal-efficiency point and the zero-power point. Therefore, as we design an irreversible Brayton heat engine, it would be better to choose the point of higher power output for a given thermal efficiency. Moreover, the operation point had better lie between the maximum-power point and the maximum-efficiency point.

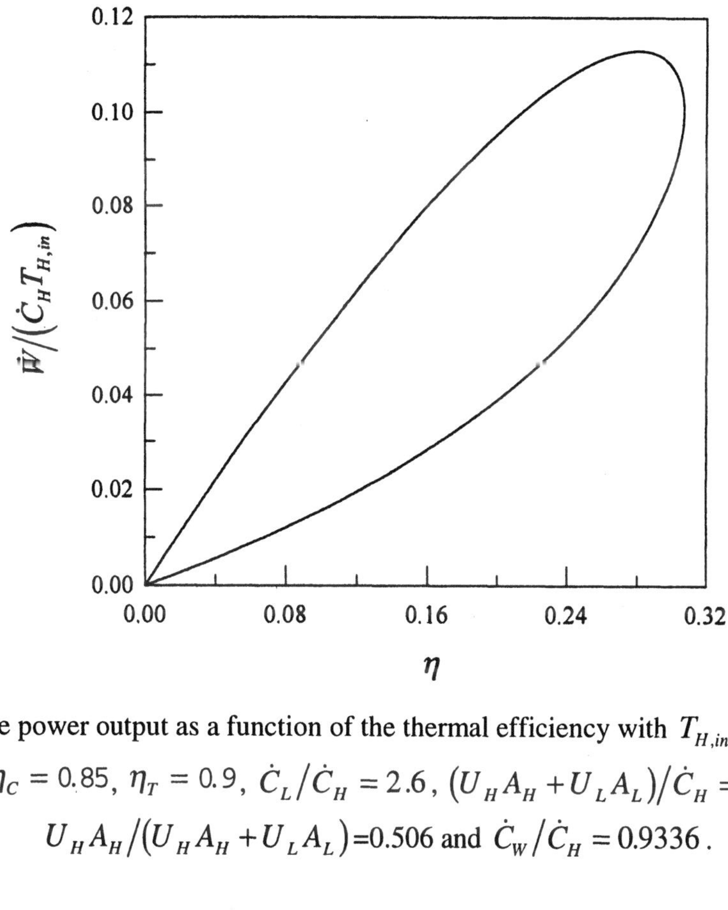

Fig. 2 The power output as a function of the thermal efficiency with $T_{H,in}/T_{L,in} = 4$, $\eta_C = 0.85$, $\eta_T = 0.9$, $\dot{C}_L/\dot{C}_H = 2.6$, $(U_H A_H + U_L A_L)/\dot{C}_H = 8$, $U_H A_H/(U_H A_H + U_L A_L)$=0.506 and $\dot{C}_W/\dot{C}_H = 0.9336$.

THE RELATION OF THE POWER OUTPUT WITH THE THERMAL CONDUCTANCE RATIO

The maximum power output $\dot{W}_{\max,1}/(\dot{C}_W T_{H,in})$ as a function of the thermal conductance ratio $U_H A_H/(U_H A_H + U_L A_L)$ is illustrated in Fig. 3. Here the maximum power output represents the power output that has been maximized with respect to the adiabatic temperature ratio x. The thermal conductance ratio is defined as the ratio of the thermal conductance of the hot-end heat exchanger to the total thermal conductance of the two heat exchangers. The result shows that the once-maximized power output increases, reaches a maximum and then decreases as the thermal conductance ratio increases. It is clearly shown that the power output of an irreversible Brayton heat engine can be further optimized with respect to the thermal conductance ratio. The thermal conductance ratio is found to be an important parameter for the power output of the Brayton heat engine.

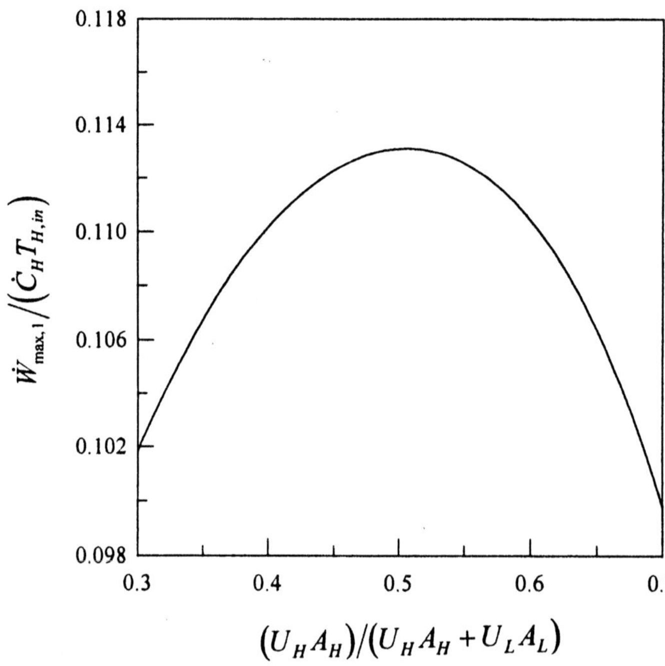

Fig. 3 The maximum power output as a function of the thermal conductance ratio with $T_{H,in}/T_{L,in} = 4$, $\eta_C = 0.85$, $\eta_T = 0.9$, $\dot{C}_L/\dot{C}_H = 2.6$, $(U_H A_H + U_L A_L)/\dot{C}_H = 8$ and $\dot{C}_W/\dot{C}_H = 0.9336$.

THE RELATION OF THE POWER OUTPUT WITH THE RATIO OF WORKING AND HOT FLUID THERMAL CAPACITANCE RATES

The maximum power output $\dot{W}_{max,2}/(\dot{C}_W T_{H,in})$ as a function of the ratio of working and hot fluid thermal capacitance rate $\dot{C}_W/\dot{C}_H$ is plotted in Fig. 4. Here the maximum power output is the power output that has been maximized with respect to the adiabatic temperature ratio x and the thermal conductance ratio $U_H A_H/(U_H A_H + U_L A_L)$. The result shows that the twice-maximized power output increases, reaches a maximum and then decreases as the ratio of working and hot fluid capacitance rates increase. In other words, the power output of an irreversible Brayton heat engine can be further optimized with respect to the ratio of the working and hot fluid capacitance rates.

Here we would like to make a brief conclusion about the power optimization. The power output can be optimized with respect to the adiabatic temperature ratio, and the optimum value of the adiabatic temperature ratio can be calculated from Eq. (21). Moreover, the power output could be further optimized with respect to the thermal conductance ratio, as shown in Fig 3, and the ratio of working and hot fluid thermal capacitance rates, as shown in Fig. 4. However, we cannot perform the power optimization with respect to the thermal conductance ratio and the ratio of working and hot fluid thermal capacitance rates analytically; hence, we calculate these optimum values numerically. Here we present all the results that have been optimized with respect to the adiabatic temperature ratio, the thermal conductance ratio and the ratio of working and hot fluid thermal capacitance rates.

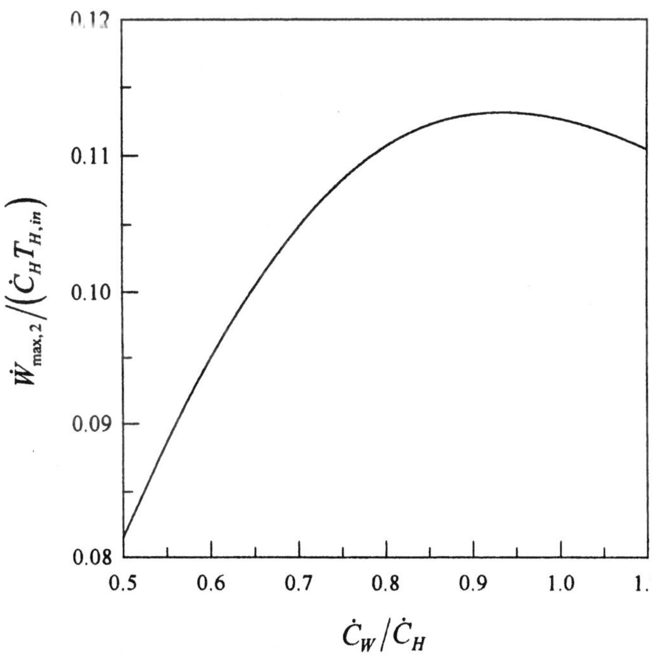

Fig. 4 The maximum power output as a function of the ratio of working and hot fluid thermal capacitance rates with $T_{H,in}/T_{L,in} = 4$, $\eta_C = 0.85$, $\eta_T = 0.9$, $\dot{C}_L/\dot{C}_H = 2.6$ and $(U_H A_H + U_L A_L)/\dot{C}_H = 8$.

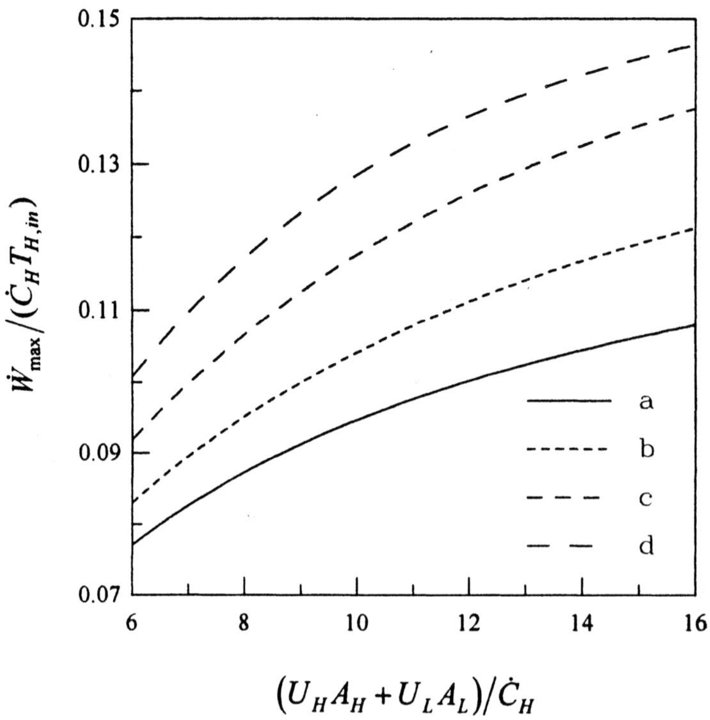

Fig. 5 The effect of the total thermal conductance on the maximum power output with $T_{H,in}/T_{L,in} = 4$, $\eta_C = 0.85$ and $\eta_T = 0.9$: (a) $\dot{C}_L/\dot{C}_H = 0.6$, (b) $\dot{C}_L/\dot{C}_H = 1.0$, (c) $\dot{C}_L/\dot{C}_H = 1.6$ and (d) $\dot{C}_L/\dot{C}_H = 4.6$.

THE EFFECT OF THE TOTAL THERMAL CONDUCTANCE

Fig. 5 shows the effect of the total thermal conductance $(U_H A_H + U_L A_L)/\dot{C}_H$ on the maximum power output $\dot{W}_{max}/(\dot{C}_W T_{H,in})$ for various values of ratios of cold and hot fluid thermal capacitance rates $\dot{C}_L/\dot{C}_H$. As the total thermal conductance of the two heat exchangers increases, the maximum power output tends to increase sharply at low total thermal conductance and then slowly at high total thermal conductance. Therefore, to obtain a reasonable power output (not the greatest that can be attained), we would like to select an appropriate value of the total thermal conductance of the two heat exchangers at a reasonable cost.

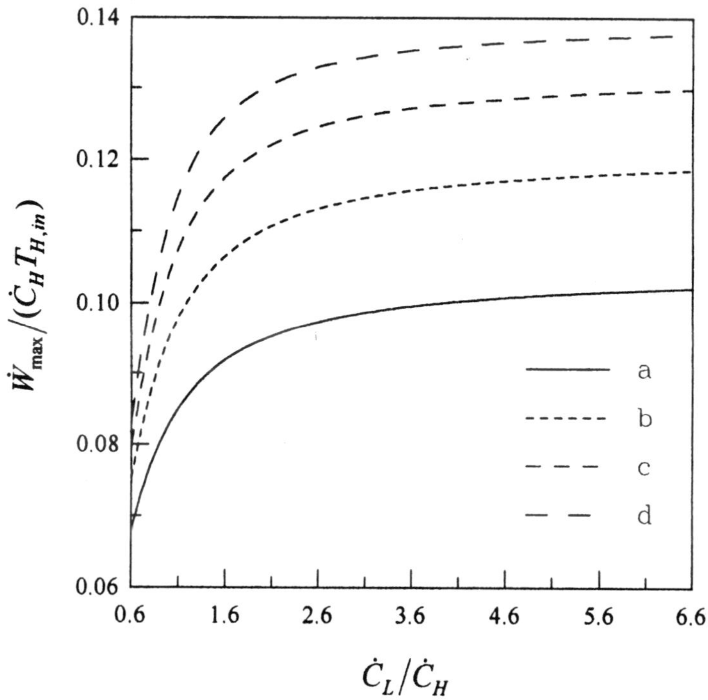

Fig. 6 The effect of the ratio of cold and hot fluid thermal capacitance rates on the maximum power output with $T_{H,in}/T_{L,in} = 4$, $\eta_C = 0.85$ and $\eta_T = 0.9$: (a) $(U_H A_H + U_L A_L)/\dot{C}_H = 6$, (b) $(U_H A_H + U_L A_L)/\dot{C}_H = 8$, (c) $(U_H A_H + U_L A_L)/\dot{C}_H = 10$ and (d) $(U_H A_H + U_L A_L)/\dot{C}_H = 12$.

THE EFFECT OF THE RATIO OF COLD AND HOT FLUID THERMAL CAPACITANCE RATES

Fig. 6 shows the effect of the ratio of cold and hot fluid thermal capacitance rates $\dot{C}_L/\dot{C}_H$ on the maximum power output $\dot{W}_{max}/(\dot{C}_W T_{H,in})$ for various values of the total thermal conductance $(U_H A_H + U_L A_L)/\dot{C}_H$. As the ratio of cold and hot fluid capacitance rates increases, the maximum power output increases rapidly at low capacitance rate ratios and then gradually levels off at high capacitance rate ratios.

Fig. 7 shows the influence of the ratio of cold and hot fluid thermal capacitance rates $\dot{C}_L/\dot{C}_H$ on the thermal efficiency at maximum power output η_{opt} for various values of total thermal conductances $(U_H A_H + U_L A_L)/\dot{C}_H$. As the ratio of cold and hot fluid capacitance rates increases, the thermal efficiency at maximum power output decreases sharply at low capacitance rate ratios and then slowly at high capacitance rate ratios.

Note that the thermal capacitance rate of the cold fluid should always greater than that of the hot fluid in order to obtain a greater power output. However, increasing the cold fluid capacitance rate would lead to a lower maximum-power thermal efficiency as well as higher costs to deal with the fluid. Therefore, we should choose an appropriate value of the ratio of cold and hot fluid thermal capacitance rates by considering three factors: power output, cost and thermal efficiency.

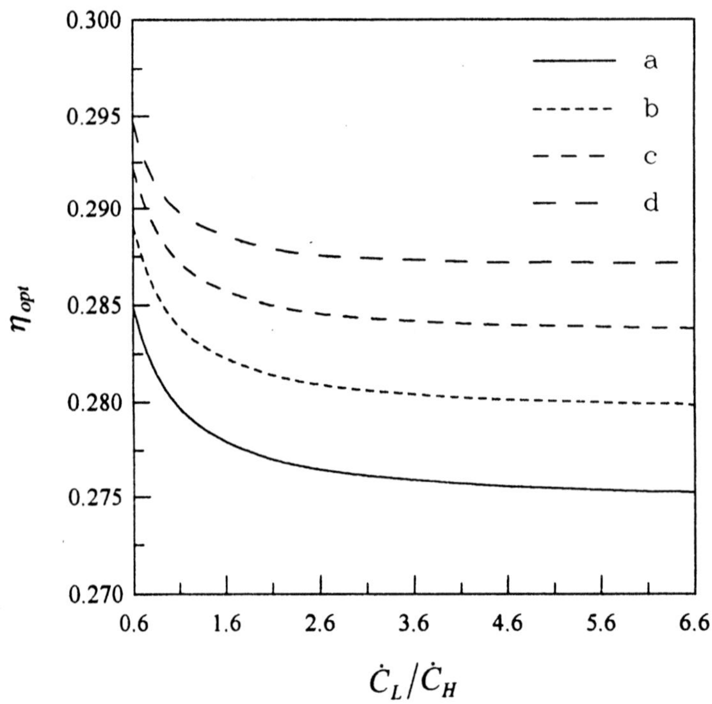

Fig. 7 The effect of the ratio of cold and hot fluid thermal capacitance rates on the thermal efficiency at maximum power output with $T_{H,in}/T_{L,in} = 4$, $\eta_C = 0.85$ and $\eta_T = 0.9$: (a) $(U_H A_H + U_L A_L)/\dot{C}_H = 6$, (b) $(U_H A_H + U_L A_L)/\dot{C}_H = 8$, (c) $(U_H A_H + U_L A_L)/\dot{C}_H = 10$ and (d) $(U_H A_H + U_L A_L)/\dot{C}_H = 12$.

THE EFFECT OF THE RESERVOIR TEMPERATURE RATIO AND THE TURBINE AND COMPRESSOR ISENTROPIC EFFICIENCIES

Fig. 8 shows the effect of the reservoir temperature ratio $T_{H,in}/T_{L,in}$ on the maximum power output $\dot{W}_{\max}/(\dot{C}_W T_{H,in})$ for various isentropic efficiencies of turbines and compressors, η_T and η_C. As shown in Fig. 8, the maximum power output tends to increase almost linearly with increasing reservoir temperature ratios. Moreover, it should be noted that the maximum power output increases with an increase in the isentropic efficiencies of turbines and compressors. Therefore, engineers have always been focusing on the technology both for improving the turbine and compressor isentropic efficiencies and for raising the operating temperatures of turbines and compressors.

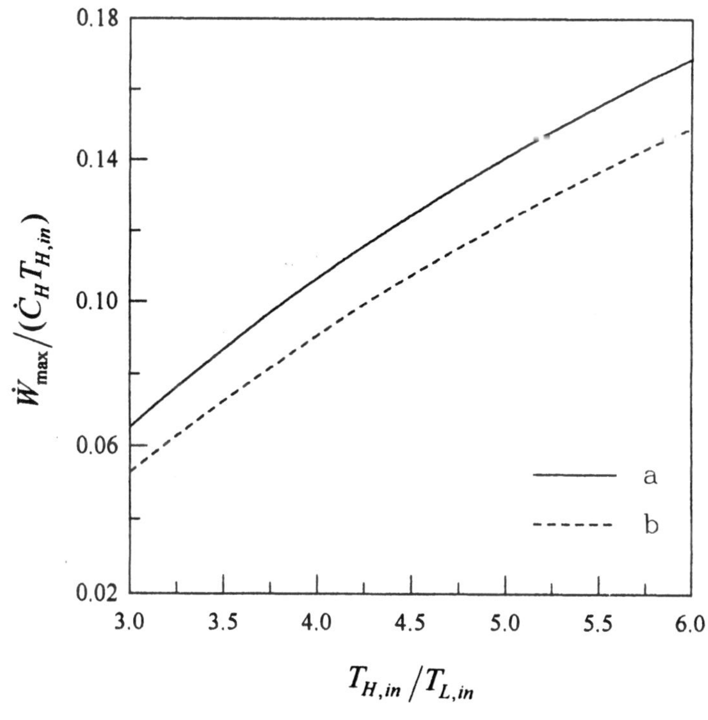

Fig. 8 The effect of the reservoir temperature ratio and the turbine and compressor isentropic efficiencies on the maximum power output with $T_{H,in}/T_{L,in} = 4$, $(U_H A_H + U_L A_L)/\dot{C}_H = 8$ and $\dot{C}_L/\dot{C}_H = 1.6$: (a) $\eta_C = 0.85$ and $\eta_T = 0.9$, and (b) $\eta_C = 0.82$ and $\eta_T = 0.87$.

OPTIMUM ADIABATIC TEMPERATURE RATIO

The variation of the optimum adiabatic temperature ratio x_{opt} with the total thermal conductance $(U_H A_H + U_L A_L)/\dot{C}_H$ and the ratio of cold and hot fluid thermal capacitance rates $\dot{C}_L/\dot{C}_H$ is shown in Fig. 9. The optimum adiabatic temperature ratio increases as the total thermal conductance is increased. Moreover, the optimum adiabatic temperature ratio decreases with an increase in the ratio of cold and hot fluid thermal capacitance rates, when the total thermal conductance is not two great. Fig. 10 illustrates the variation of optimum adiabatic temperature ratio x_{opt} with the reservoir temperature ratio $T_{H,in}/T_{L,in}$ and the turbine and compressor isentropic efficiencies, η_T and η_C. As shown in Fig. 10, the optimum adiabatic temperature ratio increases almost linearly with the reservoir temperature ratio. Moreover, the optimum adiabatic temperature ratio tends to decrease as the isentropic efficiencies of turbines and compressors are increased.

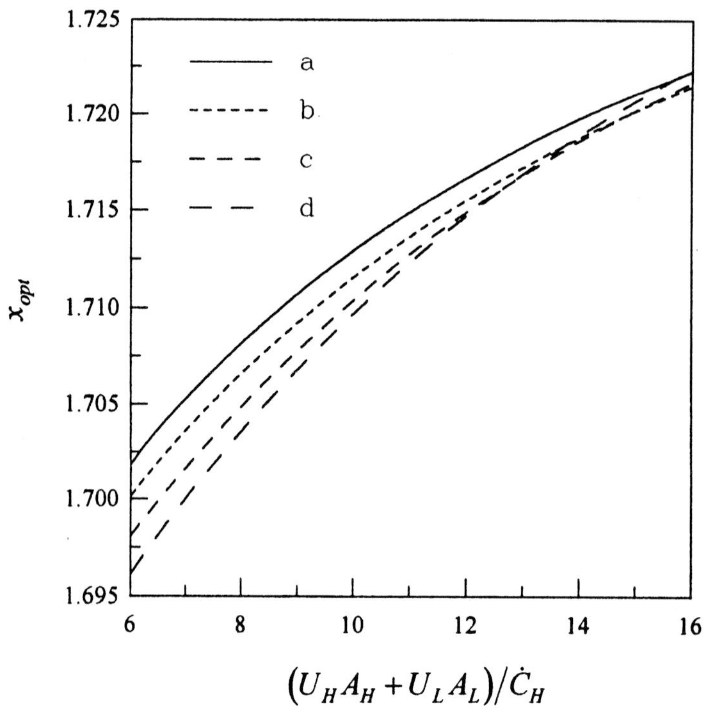

Fig. 9 The variation of the optimum adiabatic temperature ratio with the total thermal conductance and the ratio of cold and hot fluid thermal capacitance rates, $T_{H,in}/T_{L,in} = 4$, $\eta_C = 0.85$ and $\eta_T = 0.9$: (a) $\dot{C}_L/\dot{C}_H = 0.6$, (b) $\dot{C}_L/\dot{C}_H = 1.0$, (c) $\dot{C}_L/\dot{C}_H = 1.6$ and (d) $\dot{C}_L/\dot{C}_H = 4.6$.

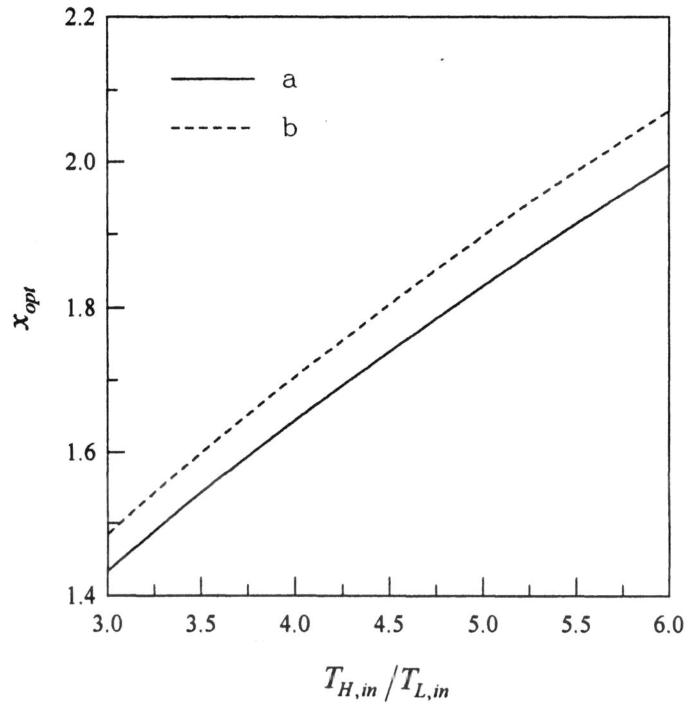

Fig. 10 The variation of the optimum adiabatic temperature ratio with the reservoir temperature ratio and the turbine and compressor isentropic efficiencies with $T_{H,in}/T_{L,in} = 4$, $(U_H A_H + U_L A_L)/\dot{C}_H = 8$, $\dot{C}_L/\dot{C}_H = 1.6$: (a) $\eta_C = 0.85$ and $\eta_T = 0.9$, and (b) $\eta_C = 0.82$ and $\eta_T = 0.87$.

Optimum Thermal Conductance Ratio

Fig. 11 shows the variation of the optimum thermal conductance ratio $[U_H A_H/(U_H A_H + U_L A_L)]_{opt}$ with the total thermal conductance $(U_H A_H + U_L A_L)/\dot{C}_H$ and the ratio of cold and hot fluid thermal capacitance rates $\dot{C}_L/\dot{C}_H$. From curve a in Fig. 11, representing the case of the ratio of cold and hot fluid thermal capacitance rates being smaller than one, it is deduced that the optimum thermal conductance ratio decreases as the total thermal conductance is increased. From curve b in Fig. 12, representing the case of the ratio of cold and hot fluid capacitance rates being equal to one, it is clearly deduced that the optimum thermal conductance ratio is independent of the total thermal conductance. From curves c and d in Fig. 11, representing the case of the ratio of cold and hot fluid thermal capacitance rates being greater than one, it is deduced that the optimum thermal conductance ratio increases as the total thermal conductance is increased.

Fig. 12 illustrates the variation of the optimum thermal conductance ratio $[U_H A_H/(U_H A_H + U_L A_L)]_{opt}$ with the reservoir temperature ratio $T_{H,in}/T_{L,in}$ and the isentropic efficiencies of turbines and compressors, η_T and η_C. It is shown in Fig.12 that the optimum thermal conductance ratio decreases monotonically with the reservoir temperature ratio. Moreover, the optimum thermal conductance ratio tends to decrease as the isentropic efficiencies of turbines and compressors are increased.

It is clear shown in Figs. 11 and 12 that the optimum thermal conductance ratio for an irreversible Brayton heat engine coupled with finite heat reservoirs is not a constant value, and its value may be greater, equal to or less than 0.5. Note that for an irreversible Brayton heat engine coupled with infinite heat reservoirs, this ratio is found to be less than 0.5 by Cheng and Chen [12], and for an endoreversible Brayton heat engine coupled with infinite heat reservoirs, this ratio is found to be equal to 0.5 by Bejan [6].

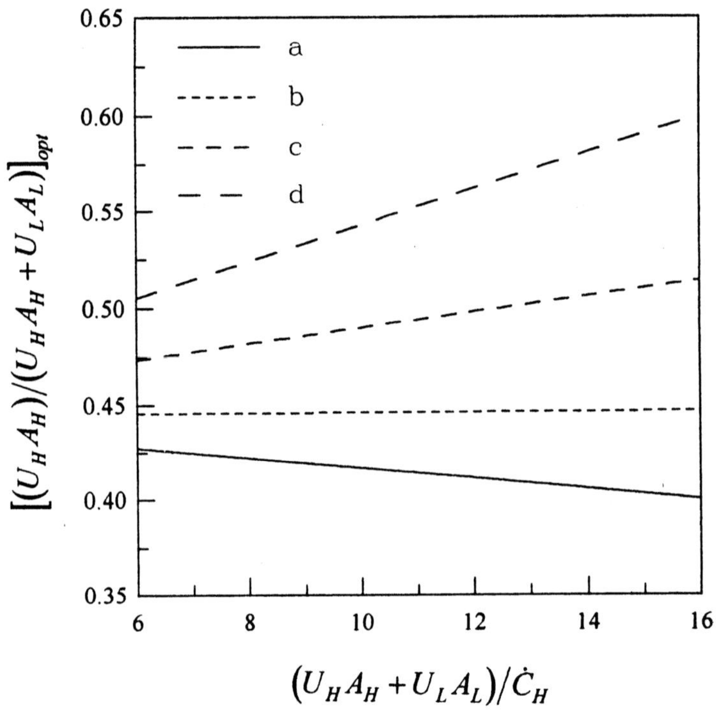

Fig. 11 The variation of the optimum thermal conductance ratio with the total thermal conductance and the ratio of cold and hot fluid thermal capacitance rates with $T_{H,in}/T_{L,in} = 4$, $\eta_C = 0.85$ and $\eta_T = 0.9$: (a) $\dot{C}_L/\dot{C}_H = 0.6$, (b) $\dot{C}_L/\dot{C}_H = 1.0$, (c) $\dot{C}_L/\dot{C}_H = 1.6$ and (d) $\dot{C}_L/\dot{C}_H = 4.6$.

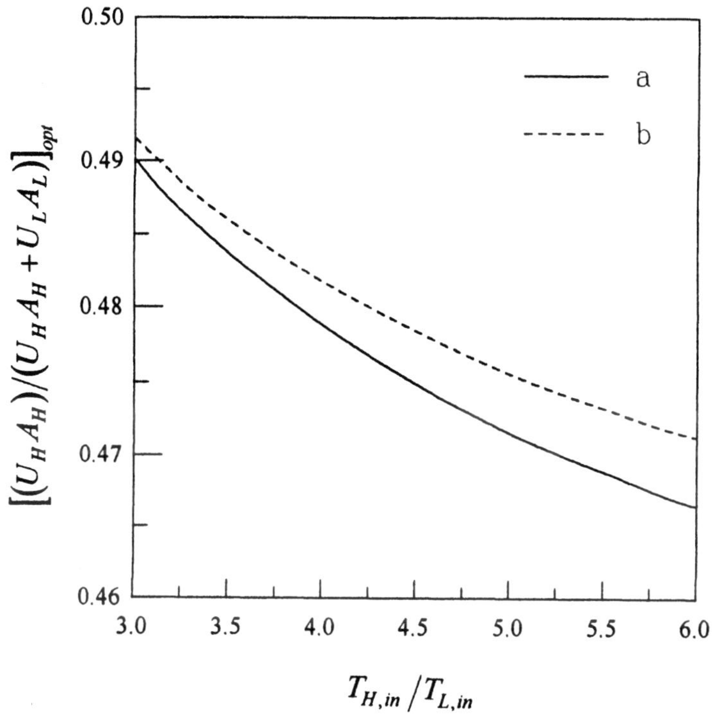

Fig. 12 The variation of the optimum thermal conductance ratio with the reservoir temperature ratio and the turbine and compressor isentropic efficiencie with $T_{H,in}/T_{L,in} = 4$, $(U_H A_H + U_L A_L)/\dot{C}_H = 8$, $\dot{C}_L/\dot{C}_H = 1.6$: (a) $\eta_C = 0.85$ and $\eta_T = 0.9$, and (b) $\eta_C = 0.82$ and $\eta_T = 0.87$.

OPTIMUM RATIO OF WORKING AND HOT FLUID THERMAL CAPACITANCE RATES

Fig. 13 plots the variation of the optimum ratio of working and hot fluid thermal capacitance rates $[\dot{C}_W/\dot{C}_H]_{opt}$ with the total thermal conductance $(U_H A_H + U_L A_L)/\dot{C}_H$ and the ratio of cold and hot fluid thermal capacitance rates $\dot{C}_L/\dot{C}_H$. It is shown in Fig.13 that the optimum ratio of working and hot fluid capacitance rates increases as the total thermal conductance is increased. Moreover, the optimum ratio of working and hot fluid capacitance rates tend to increase as the ratio of cold and hot fluid capacitance rates is increased.

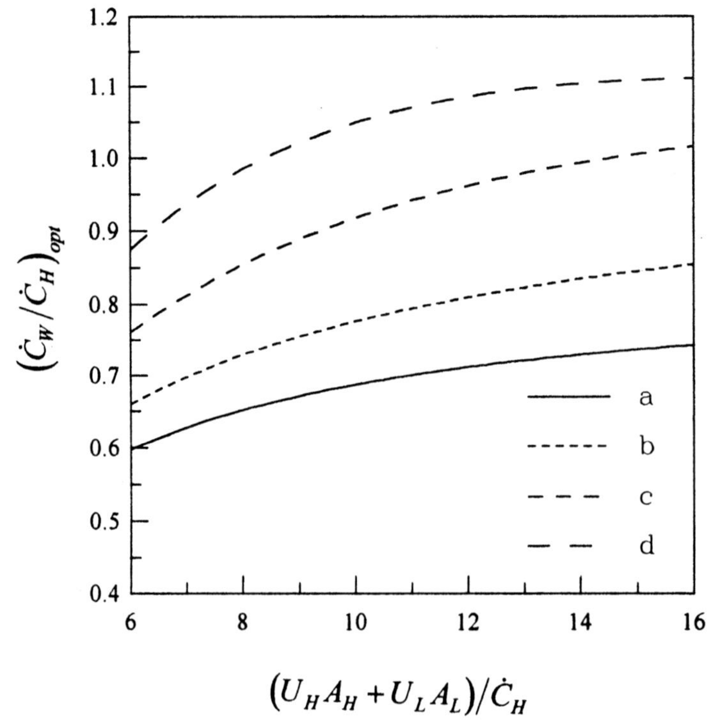

Fig. 13 The variation of the optimum ratio of working and hot fluid thermal capacitance rates with the total thermal conductance and the ratio of cold and hot thermal capacitance rates with $T_{H,in}/T_{L,in} = 4$, $\eta_C = 0.85$ and $\eta_T = 0.9$: (a) $\dot{C}_L/\dot{C}_H = 0.6$, (b) $\dot{C}_L/\dot{C}_H = 1.0$, (c) $\dot{C}_L/\dot{C}_H = 1.6$, and (d) $\dot{C}_L/\dot{C}_H = 4.6$.

Fig. 14 illustrates the relation of the optimum ratio of working and hot fluid thermal capacitance rates $\left[\dot{C}_W/\dot{C}_H\right]_{opt}$ with the reservoir temperature ratio $T_{H,in}/T_{L,in}$ and the isentropic efficiencies of turbines and compressors, η_T and η_C. As shown in Fig. 14, the optimum ratio of working and hot fluid thermal capacitance rates increases with an increase in the reservoir temperature ratio. Moreover, the optimum ratio of working and hot fluid thermal capacitance rates increases as the isentropic efficiencies of turbines and compressors are increased.

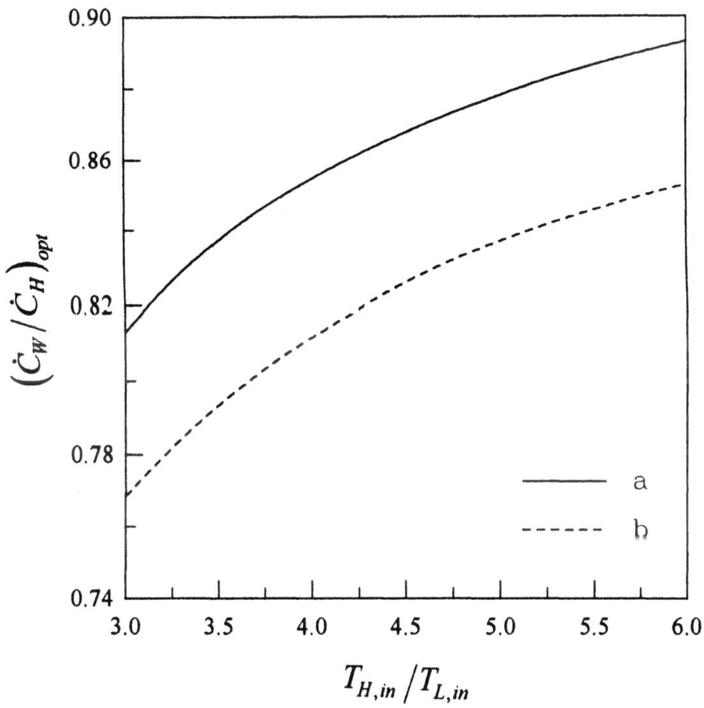

Fig. 14 The variation of the optimum ratio of working and hot thermal fluid capacitance rates with the reservoir temperature ratio and the turbine and compressor isentropic efficiencies with $T_{H,in}/T_{L,in} = 4$, $(U_H A_H + U_L A_L)/\dot{C}_H = 8$ and $\dot{C}_L/\dot{C}_H = 1.6$: (a) $\eta_C = 0.85$ and $\eta_T = 0.9$, and (b) $\eta_C = 0.82$ and $\eta_T = 0.87$.

CONCLUSIONS

Finite-time thermodynamics has been used to study an irreversible Brayton heat engine coupled with heat reservoirs with finite thermal capacitance rates. The power output of the irreversible Brayton heat engine has been optimized with respect to three parameters: adiabatic temperature ratios, heat exchanger thermal conductance ratios, and ratios of working and hot fluid thermal capacitance rates. The effects of total heat exchanger thermal conductances, ratios of cold and hot fluid thermal capacitance rates, reservoir temperature ratios and isentropic efficiencies of turbines and compressors on the maximum power output and its corresponding parameters have been investigated.

To obtain a greater power output, the thermal capacitance rate of the cold fluid should be larger than that of the hot fluid. However, increasing the cold fluid capacitance rate would lead to a lower maximum-power thermal efficiency as well as higher costs to deal with the fluids. Moreover, the optimum values of the important design parameters obtained in this work can be used in the basic design of a closed Brayton heat engine to raise its power output.

NOMENCLATURE

A_H	heat transfer area of the hot-side heat exchanger
A_L	heat transfer area of the cold-side heat exchanger
$\dot{C}_H$	thermal capacitance rate of the hot fluid
$\dot{C}_L$	thermal capacitance rate of the cold fluid
$\dot{C}_W$	thermal capacitance rate of the working fluid
N_H	number of transfer units of the hot-side heat exchanger
N_L	number of transfer units of the cold-side heat exchanger
$\dot{Q}_H$	heat rate absorbed from the hot fluid by the working fluid
$\dot{Q}_L$	heat rate transferred to the cold fluid from the working fluid
$T_{H,in}$	hot fluid inlet temperature
$T_{H,out}$	hot fluid outlet temperature
$T_{L,in}$	cold fluid inlet temperature
$T_{L,out}$	cold fluid outlet temperature
U_H	overall heat transfer coefficient of the hot-side heat exchanger
U_L	overall heat transfer coefficient of the cold-side heat exchanger
$\dot{W}$	net power output
x	adiabatic temperature ratio

Greek letters

ε_H	effectiveness of the hot-side heat exchanger
ε_L	effectiveness of the cold-side heat exchanger
η	thermal efficiency
η_C	compressor isentropic efficiency
η_T	turbine isentropic efficiency

Subscripts

C	compressor
H	hot fluid or hot-side heat exchanger

L	cold fluid or cold-side heat exchanger
max	maximum
opt	optimum
T	gas turbine
W	working fluid

REFERENCES

Novikov, I. I., "The Efficiency of Atomic Power Stations," Journal of Nuclear Energy II, Vol. 7, pp. 125-128, 1958; (translated from Atomnaya Energiya, Vol. 3, No. 11, pp. 409, 1957).

El-Wakil, M. M., Nuclear Power Engineering, pp. 162-165, Mcgraw-Hill, New York, 1962.

Curzon, F. L. and Ahlborn, B., "Efficiency of a Carnot Engine at Maximum Power Output," American Journal of Physics, Vol. 43, No. 1, pp. 22-24, 1975.

Siniutycz, S. and Salamon, P., Finite Time Thermodynamics and Thermoeconomics (Advances in Thermodynamics, Vol. 4), Taylor & Francis, London, U. K., 1990.

Leff, H. S.,"Thermal Efficiency at Maximum Work Output: New Results for Old Engines," American Journal of Physics, Vol. 55, No. 7, pp. 602-610, 1987.

Bejan, A., " Theory of Heat Transfer-Irreversible Power Plants," International Journal of Heat and Mass Transfer, Vol. 31, No. 6, pp.1211-1219, 1988.

Wu, C., and Kiang, R. L. ,"Work and Power Optimization of a Finite-Time Brayton Cycle," International Journal of Ambient Energy, Vol. 11, No. 3, pp. 129-136, 1990.

Ibrahim, O. M., Klein, S. A., and Mitchell, J. W., "Optimum Power Cycles for Specified Boundary Conditions," ASME Journal of Engineering for Gas Turbines and Power, Vol.113, No. 4, pp. 514-521, 1991.

Wu, C., and Kiang, R. L. ,"Power Performance of a Nonisentropic Brayton Cycle," ASME Journal of Engineering for Gas Turbines and Power, Vol. 113, No. 4, pp. 501-504, 1991.

Sahin, B., Kodal, A. and Yavuz, H., "Efficiency of a Joule-Brayton Engine at Maximum Power Density," Journal of Physics D: Applied Physics, Vol. 28, No. 7, pp. 1309-1313, 1995.

Sahin, B., Kodal, A. and Yavuz, H., "Maximum Power Density Analysis of an Irreversible Joule-Brayton Engine," Journal of Physics D: Applied Physics, Vol. 29, No. 5, pp. 1162-1167, 1996.

Cheng, C. Y. and Chen, C. K., "Power Optimization of an Irreversible Brayton Heat Engine," Energy Sources, Vol. 19, No. 5, pp. 461-467, 1997.

Kays, W. M., and London, A. L., Compact Heat Exchangers, 3rd Ed., Chap. 2, McGraw-Hill, New York, 1984.

David, K., and Ward, C., Numerical Analysis, Brooks/Cole, Pacific Grove, Chap. 3, 1991.

TENTATIVE UNIFIED DESCRIPTION OF ENGINES, REFRIGERATORS AND HEAT PUMPS IN CONTACT WITH TWO OR THREE HEAT RESERVOIRS

M. FEIDT

LEMTA - UMR CNRS 7563, Université Henri Poincaré NANCY

2, Avenue de la Forêt de Haye - B.P. 160 - 54504 VANDOEUVRE CEDEX

Abstract- As is well known now, the so called finite time thermodynamics constitute since, approximately twenty years a thermodynamical field of interest.

It appears from the published literature that relatively few efforts have been done to syntethise and normalized the obtained results.

Effectively finite time thermodynamics has been applied to various engines (CARNOT; BRAYTON; STIRLING; ERICSSON; OTTO; DIESEL), and more recently to refrigerators, heat pumps and air conditioning systems.

Since that time numerous papers have been published on two heat reservoirs machines, and on three heat reservoirs one. In that case too, it seems that there is a lack of a general description. The object of this paper is to contribute to a more general model of three heat reservoirs machines, and to propose the corresponding models of two heat reservoirs engines or machines as limits of the preceding one.

INTRODUCTION

As is well known now, the so called finite time thermodynamics constitute since, approximately twenty years a thermodynamical field of interest.

As pointed out by A. BEJAN, V. RADCENCO and us, these considerations are in fact older of the most cited paper of CURZON and AHLBORN [1] due to the study of for example I. NOVIKOV [2] and CHAMBADAL [3]. Now this approach is introduced in teaching books for example [4, 5, 6].

Many publications have been published in the field. The most recent review is probably the one of A. BEJAN [7]. Some research books also have been published recently [8, 9].

It appears from these documents that relatively few efforts has been done to syntethise and normalize the obtained results. It is the goal we try to developp since approximately ten years [10, 11].

Finite time thermodynamics has been successfully applied to various engines (CARNOT ; BRAYTON ; STIRLING ; ERICSON ; OTTO ; DIESEL) [12], and more recently to what we call « inverse cycle machines », to say : refrigerators, heat pump, air conditioning systems, heat transformers. The first paper to be published in that field was the one of BLANCHARD [13].

After what, numerous papers have been published on two reservoirs machines [14], and on three heat reservoirs one [15]. In this case too, it seems that relatively few tentatives have been done, to propose a general description of these inverse cycle machines. We note only a paper of CHEN and YAN, who propose an unified description of endoreversible cycles [16]. The object of this paper is to focus on the more general model of three heat reservoirs machines that can be conceived, owing to the published materials, and to propose the corresponding models of two heat reservoirs engines or inverse cycle machines as limits of the preceding ones.

Comparison with the preceding existing results is considered, and possible extensions are envisaged or proposed.

Relation Between Coefficient of Performance and Useful Effect

Disponible Equations and Hypothesis of the Model

The algebraic form of the first law of thermodynamics, is expressed as:

$$\omega + q_H + q_I + q_C = 0 \qquad (1)$$

ω, mechanical energy (or power) to move the internal (and external) fluid, if necessary.
q_H, heat energy (or power) exchanged at the hot reservoir. This reservoir is supposed isothermal at T_{SH} (thermostat : hypothesis 1). But calculations can be done, with finite thermal capacity [17]; we complete by supposing $T_{SH} \succ T_{SI} \succ T_{SC}$.
q_i, heat energy (or power) exchanged at the intermediate reservoir. The same hypothesis is used : thermostat at T_{SI}.
q_c, heat energy (or power) exchanged at the cold reservoir supposed thermostat at T_{SC}.

Remark 1 : energy are used when we consider, non stationary engines or machines, whereas power are used when we consider stationary engines or machines ; the corresponding cycles are continuous one.

Remark 2 : the sign of each energy quantity (or power) is determined according to the thermodynamics convention : all the quantities entering the thermal system are positive ; all the quantities leaving the system are negative.

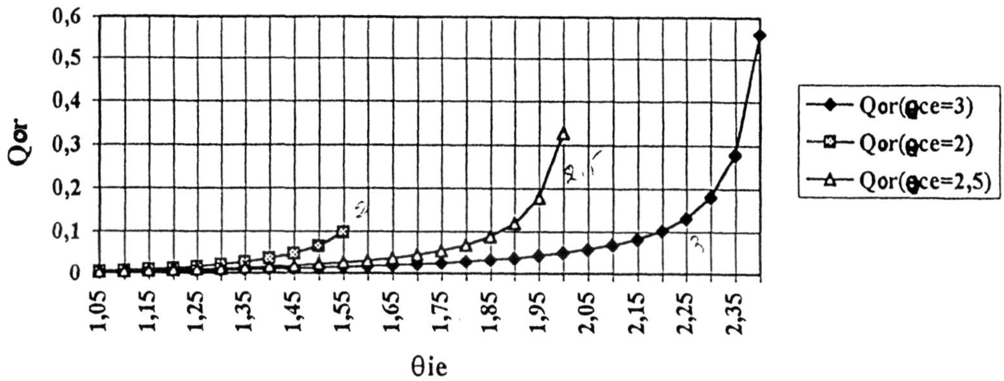

Figure 1 : evolution of q_{or}, optimum heating power at the boiler with $\vartheta_{ie} = {}^{T_{Ie}}\!/\!_{T_{Fe}}$ report of entrance intermediate and cold temperature (finite reservoirs), for three values of $\vartheta_{ce} = {}^{T_{Ce}}\!/\!_{T_{Fe}}$ (T_{Ce} entrance temperature of the external fluid at the boiler), according to [25]

Remark 3 : equation (1) correspond to a cycle (closed loop of thermodynamical transformations, if energy are used ; the same equation is written in terms of fluxes, if a stationary state is corresponding to the studied system.

Similarly, the algebraic forms of the second law of thermodynamics is expressed as

$$\frac{q_H}{T_H} + \frac{q_I}{T_I} + \frac{q_C}{T_C} + S = 0 \qquad (2)$$

T_H, T_I, T_C represents the isothermal temperature (hypothesis 2) of the internal fluid respectively at the hot, intermediate, or cold side.

S represents the internal entropy created in the system during the cycle (respectively the internal entropy flux generated in the stationary system transformations) ; S is a positive quantity for the general case (endo and exo irreversibility) ; S is equal to zero for the endoreversible case, corresponding to the more generally treated form.

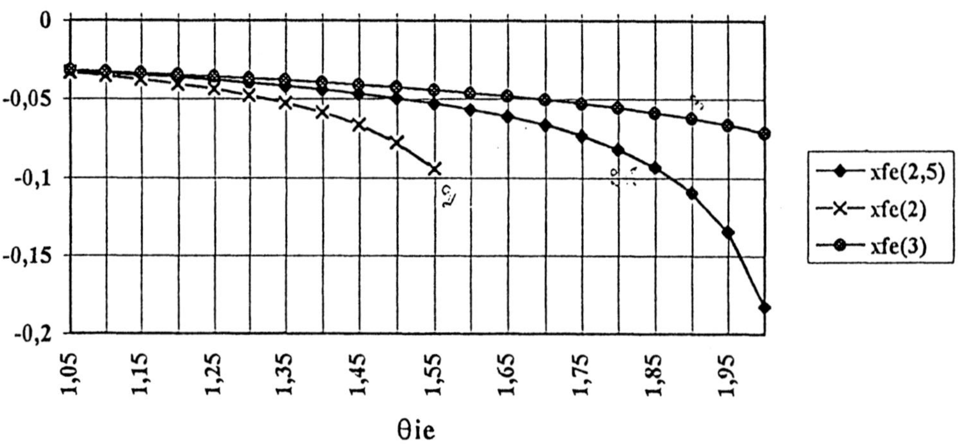

Figure 2 : evolution of x_{fe}, optimal gradient of temperature at entrance of the cold HX of the machine : $x_{fe} = X_{fe}/T_{fe}$; $X_{fe} = T_e - T_{fe}$ according to [25]

Expressions of the Overall Coefficient of Performance C.O.P.

Currently, and whatever is the system, the C.O.P. is adopted as the objective function to optimize. We have to notice, at that time, that it is not the only possible objective function; but it is the one chosen in most cases (see the references), with the constraint of a given useful effect in terms of energy or power.

Case of Real Machines

The C.O.P. is generally defined as the useful effect, divided by the energetic cost. So if we apply the preceding definition, it comes respectively :
for the C.O.P. of a refrigerating machine :

$$\Sigma = {q_C}/{(q_H + \omega)} \qquad (3)$$

for the C.O.P. of a heat pump :

$$\Sigma ' = {- q_I}/{(q_H + \omega)} \qquad (4)$$

for the C.O.P. of a heat transformer :

$$\Sigma '' = {- q_H}/{(qI + \omega)} \qquad (5)$$

Remark 1 : Σ, $\Sigma '$, $\Sigma ''$ are always given as positive quantities.
Remark 2 : relation (1) impose a systematic link between Σ and $\Sigma '$: $\Sigma ' = \Sigma + 1 \succ 1$
Remark 3 : $\Sigma '' = {1}/{\Sigma '} \prec 1$

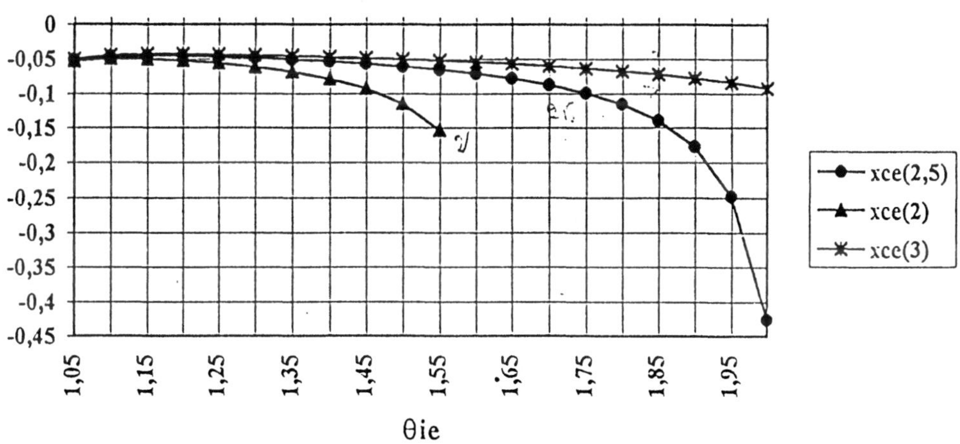

Figure 3 : evolution of x_{ce}, optimal gradient of temperature at the entrance of the hot HX of the machine ; $x_{ce} = {X_{ce}}/{T_{ce}}$; $X_{ce} = T_H - T_{ce}$ according to [25]

Case of Endoreversible Machine

In that case, S = 0. By using equations (1, 2 and 3, 4, 5) for respectively refrigerators, heat pumps, and heat transformers, it comes :

$$\Sigma_e = \frac{1/T_I - 1/T_H}{1/T_C - 1/T_I}\left[1 + \frac{\omega}{\omega + q_H}\,\frac{T_I}{T_H - T_I}\right] \tag{6}$$

$$\Sigma'_e = \frac{1/T_C - 1/T_H}{1/T_C - 1/T_I}\left[1 + \frac{\omega}{\omega + q_H}\,\frac{T_C}{T_H - T_C}\right] \tag{7}$$

$$\Sigma''_e = \frac{1/T_C - 1/T_I}{1/T_C - 1/T_H}\left[1 + \frac{\omega}{\omega + q_I}\,\frac{TC}{T_I - T_C}\right] \tag{8}$$

It is to be noticed that if we neglect the ω contribution, (7-8-9) corresponds to the known relations for three thermal machines generally indicated in the literature, for perfectly endoreversible machines, to say Ψ, Ψ', Ψ''.

For the machines that necessitates a mechanical power (pump), the proposed relations indicate how to correct the results. It appears that corrections depends on the heat injected in the system (q_H, q_I), as well as of the corresponding internal level of temperature (T_H, T_I), but also on the temperature level of rejection of wasted heat (T_I , T_C).

- Remark 1 : the preceding relations can be particularized in the thermostatics limit, for which $T_H \rightarrow T_{SH}$, $T_I \rightarrow T_{SI}$, $T_C \rightarrow T_{SC}$
- Remark 2 : the thermostatics limit allows us to give an estimation of the correction to be done on the current endoreversible optimization of the machines
- Remark 3 : if simultaneous T_H^{-1} and ω tend to zero, we restitute, the values corresponding to two heat reservoirs machines
- Remark 4 : if simultaneously T_I^{-1} and ω tend to zero in (7), we restitute, the value corresponding to a two heat reservoirs engine.

CASE OF EXO AND ENDO IRREVERSIBLE MACHINES

This case is the more general one. It includes $S \neq 0$, the generated entropy that can be in theory expressed using entropy analysis of the studied system [7].

By using (1, 2 and 3, 4, 5), it comes respectively for refrigerators, heat pumps and heat transformer :

$$\Sigma_i = \frac{1/T_I - 1/T_H}{1/T_C - 1/T_C}\left[1 + \frac{\omega - T_H \cdot S}{\omega + q_H}\,\frac{T_I}{T_H - T_I}\right] \tag{9}$$

$$\Sigma'_i = \frac{1/T_C - 1/T_H}{1/T_C - 1/T_I}\left[1 + \frac{\omega - T_H \cdot S}{\omega + q_H}\,\frac{T_C}{T_H - T_C}\right] \tag{10}$$

$$\Sigma''_i = \frac{1/T_C - 1/T_I}{1/T_C - 1/T_H}\left[1 + \frac{\omega - T_I \cdot S}{\omega + q_I}\,\frac{T_C}{T_H - TC}\right] \tag{11}$$

We can see in the three formulae (9-10-11), that the internal irreversibility of the system has opposite action on the correction factors than ω, and that the temperature affected to S differs between refrigerators heat pumps (T_H) and heat transformers (T_I).

We can conclude that the only case, where the generally used definitions of C.O.P. holds for the said endoreversible machines is when $\omega - T.S = 0$, whatever is the machine. Otherwise the C.O.P. can be greater or lower, than the classical value.

VARIOUS EXPRESSIONS OF THE C.O.P. SUBJECT TO CONSTRAINT

As has been demonstrated by some authors, relation between C.O.P., and refrigerating or heating load can be useful in order to perform optimization of the system.

In that sense (9-10-11) constitute a set of equations depending on parameters ω, S but also on q_H or q_i. It means the corresponding C.O.P. are expressed, in fact for given heat input (constraint).

We propose hereafter, the corresponding expressions obtained, when the useful effects are imposed (these cases corresponds to the more usually studied cases in the literature).

for a refrigerating machine, with q_C imposed

$$\frac{1}{\Sigma_i} = \frac{1}{\Psi} - \frac{\omega - T_H S}{q_C}\,\frac{T_I}{T_H - T_I} \tag{12}$$

or, if Σ_i is imposed

$$q_C = \frac{\Psi \Sigma_i}{\Sigma_i - \Psi} \frac{T_I(\omega - T_H S)}{T_H - T_I} \tag{12'}$$

for a heat pump, with q_I imposed

$$\frac{1}{\Sigma'_i} = \frac{1}{\Psi'} - \frac{\omega - T_H S}{q_I} \frac{T_C}{T_H - T_C} \tag{13}$$

or, if Σ'_i is imposed

$$q_i = -\frac{\Psi' \Sigma'_i}{\Sigma'_i - \Psi'} \circ \frac{T_C(\omega - T_H S)}{T_H - T_I} \tag{13'}$$

for a heat transformer, with q_H imposed

$$\frac{1}{\Sigma''_i} = \frac{1}{\Psi''} + \frac{\omega - T_I S}{q_H} \circ \frac{T_I}{T_I - T_C} \tag{14}$$

or, if Σ''_i is imposed

$$q_H = -\frac{\Psi'' \Sigma''_i}{\Sigma''_i - \Psi''} \circ \frac{T_I(\omega - T_I S)}{T_I - T_C} \tag{14'}$$

Remark 1 : it is in an easy way possible also to deduce, three corresponding formulas, characterizing the relation between the heat useful effects, and the heat costs associated with the preceding ones.
Remark 2 : all the preceding expressions are useful tools in order to optimize a given objective function subject to one corresponding constraint.

SENSITIVITY TO THE FORM OF HEAT TRANSFER LAW

INFLUENCE OF THE KIND OF HEAT TRANSFER LAW

- two kinds of heat transfer linear law

The present existing literature shows us that the most used heat transfer law is the one of NEWTON ; that is the linear one.
This law is expressed as :

$$q_X = K_X V_X (T_{SX} - T_X) \qquad (15)$$

In our case, x can be C, I, or H.

If the studied system is a stationary one, this law is imposed in term of heat flux q_X in W ; K_X represents in that case a global heat transfer coefficient $\left(W/_{m^2 \cdot K}\right)$, and V_X a finite size surface $\left(m^2\right)$.

If the system operates on a discontinuous mode, q_X corresponds to a heat quantity (J) ; K_X represents in that case a heat transfer conductance $\left(W/_K\right)$ and V_X the finite time duration of the corresponding transfer (s).

Remark : if V_X corresponds to unity, the two preceding approaches are identical, and the heat transfer law is expressed in function of the heat transfer conductance, which is always a positive physical quantity. So if $T_{si} \succ T_i$, the heat is received by the system ; if $T_i \prec T_{si}$ the heat is furnished by the system.

- generalized radiative heat transfer law

Common radiative heat transfer is characterized by

$$q_X = K_X V_X \left(T_{SX}^{\ 4} - T_X^{\ 4}\right) \qquad (16)$$

This law, said law 1, can be generalized, in the form :

$$q_X = K_X V_X \left(T_{SX}^{\ n} - T_X^{\ n}\right) \qquad (17)$$

In that case K_X is a generalized heat transfer coefficient $\left(W/_{m^2 \cdot K^n}\right)$, or a generalized heat transfer conductance $\left(W/_{K^n}\right)$. The sign convention expressed previously is always valid.

Remark : this law is currently used in the literature, and particularly with n = -1, corresponding to the phenomenological thermodynamics heat transformer law.

- generalize convective heat transfer law

We propose here, to extend the preceding results to a less common heat transfer law ; this law is useful for example for natural convective heat transfer, and for boiling heat transfer also (with respectively n = 0,25 and n = 3).

This law, named law 2, is expressed in the general following form :

$$q_X = K_X V_X (T_{SX} - T_X)^n \tag{18}$$

K_X remains a generalized heat transfer coefficient, or a generalized heat transfer conductance. But we have to take care that (18) is not always valid, due to the sign.

This form is true for the cold part, whatever is the n value. But for the intermediate and hot part of the machine, it is true whatever is n, only for refrigerating machines and heat pumps at the hot side, and for the heat transformer at the intermediate temperatures. If $T_I \succ T_{SI}$, the sign must be changed in (18) for refrigerating machines and heat pumps. If $T_H \succ T_{SH}$, the same holds, at the hot side for a heat transformer.

Remark : the two preceding laws (17, 18) restitute the more common and studied case, in the limit where n = 1 (limit case, corresponding to NEWTON law).

Generalized Description of the Thermal System

It is possible to give a general formal equations set for the three heat transfer quantities, whatever is the law used, (for example (17) or (18) [18] :

$$q_H = K_H V_H f_H (T_{SH}, T_H) \tag{19}$$

$$q_I = K_I V_I f_I (T_{SI}, T_I) \tag{20}$$

$$q_C = K_C V_C f_C (T_{SC}, T_C) \tag{21}$$

It appears in that case, that we have always a constraint of finite size of the system, or of finite duration of the machines cycle so that :

$$V_H + V_I + V_C = V \tag{22}$$

In all cases, V is a given constant.

It is also, to be noticed that (1, 2, 22) give the values of variables V_H, V_I, V_C in function of temperatures and parameters of the problem :

$$V_H = \left\{ VK_I f_I K_C f_C \left(\frac{1}{T_C} - \frac{1}{T_I} \right) + \omega \left(\frac{K_C f_C}{T_C} - \frac{K_i f_i}{T_I} \right) - S(K_C f_C - K_I f_I) \right\} \frac{1}{\Delta} \quad (23)$$

$$V_I = \left\{ VK_C f_C K_H f_H \left(\frac{1}{T_H} - \frac{1}{T_C} \right) + \omega \left(\frac{K_H f_H}{T_H} - \frac{K_C f_C}{T_C} \right) - S(K_H f_H - K_C f_C) \right\} \frac{1}{\Delta} \quad (24)$$

$$V_C = \left\{ VK_H f_H K_I f_I \left(\frac{1}{T_I} - \frac{1}{T_H} \right) + \omega \left(\frac{K_I f_I}{T_I} - \frac{K_H f_H}{T_H} \right) - S(K_I f_I - K_H f_H) \right\} \frac{1}{\Delta} \quad (25)$$

$$\text{with } \Delta = K_I f_I K_C f_C \left(\frac{1}{T_C} - \frac{1}{T_I} \right) + K_C f_C K_H f_H \left(\frac{1}{T_H} - \frac{1}{T_C} \right) + K_H f_H K_I f_I \left(\frac{1}{T_I} - \frac{1}{T_H} \right)$$

$$f_H = f_H(T_{SH}, T_H); f_I = f_I(T_{SI}, T_I); f_C = f_C(T_{SC}, T_C)$$

variables remaining for optimization are T_H, T_I, T_C. Parameters appearing in (23, 24, 25) are $T_{SH}, T_{SI}, T_{SC}, K_H, K_I, K_C, \omega$ and S.

Remark 1 : in order to optimize the machine, it remains to precise the objective function, and eventually a constraint on a $q_X (q_H \; ; \; q_I \; ; \; q_C)$, or eventually the C.O.P.

Remark 2 : the heat transfer laws used are not obligatory the same at the three thermal contacts ; but the hypothesis of similarity of heat transfer law is always used in the literature consulted

Remark 3 : (23, 24, 25) are generalized formulas giving repartition of V_X ; if ω and S equal zero, we obtain particular results, published previously [19]. But it is to remark, that the new relations indicates the influence of ω, and S on the V_X distribution.

Remark 4 : influence of the heat transfer law appears in the relations (23, 24, 25) through the three f_H, f_I, f_C functions previously described

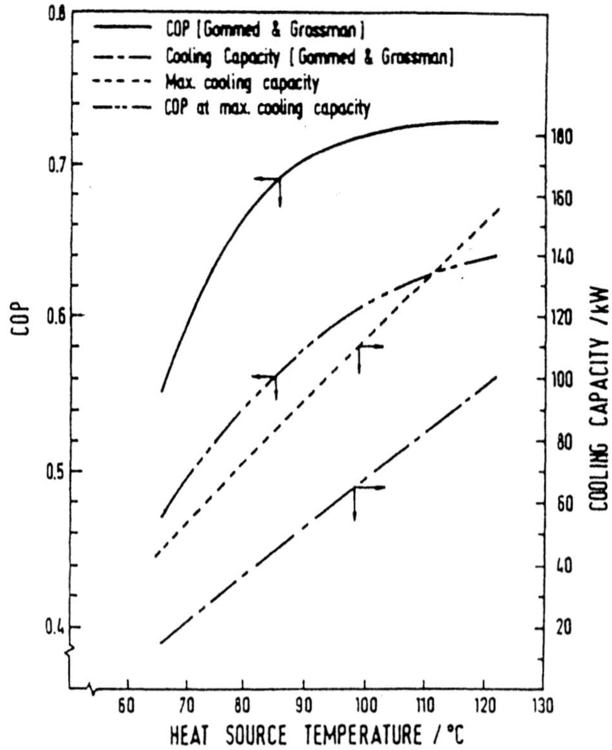

Figure 4 : comparison of maximum cooling capacity and COP with simulated results, according to [30]

Some Particular Configurations

Limit Cases of the Proposed Model

As is proposed in some papers, we can introduced essentially two limit cases, to the preceding generalized three thermal machines :

- two reservoirs refrigerating machine or heat pump

The corresponding relations are in fact deduced from the preceding considerations, but in that case the notion of heat transformer vanishes, as well as ω.

The temperature T_{SH} tends to infinity ; consequently T_H too ; q_H, the corresponding heat becomes, in that case, pure mechanical energy ω_m (or power owing to the case). (19) in that case has to be considered as an undetermined form $0 \times \infty$; effectively $V_H \rightarrow 0$, and simultaneous $K_H.f_H$ tends to infinity in such a way that the product remains the finite value $\omega_m \succ 0$.

These considerations are completely coherent with relation (23).

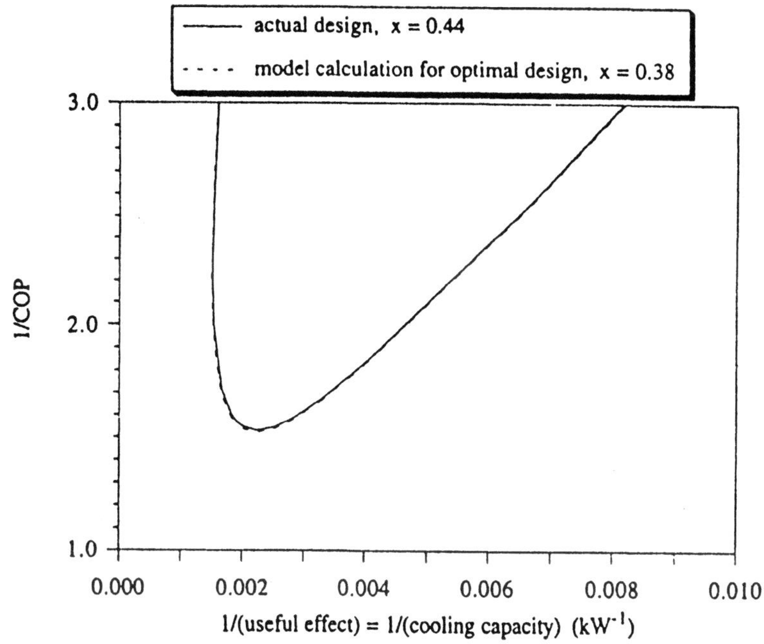

Figure 5 : evolution of $1/_{COP}$ againt $1/_{(cooling\ rate)}$ for a single stage absorption chiller according to [33] ; the control variable x is the division of total heat rejection between condenser and absorber

Remark 1 : limits of relations (9, 10) corresponds to exo-endoirreversible C.O.P. of refrigerating machines or heat pumps ; in the same manner (12, 13) give limits, when q_C is imposed for a refrigerating machine, or when q_I is imposed for a heat pump.
Remark 2 : (12', 13') are used to obtain useful effect, when the C.O.P. is imposed, for the same two reservoirs refrigerating machine or heat pump.
Remark 3 : (24, 25) are the two relations giving V_I, V_C in this limit as :

$$V_I = \frac{\dfrac{VK_C f_C}{T_C} + S}{\dfrac{K_C f_C}{T_C} - \dfrac{K_I f_I}{T_I}} \tag{26}$$

$$V_C = \frac{\dfrac{-VK_I f_I}{T_I} - S}{\dfrac{K_C f_C}{T_C} - \dfrac{K_I f_I}{T_I}} \tag{27}$$

We see, on these results, that the internal irreversibility influences adversely on the two quantities V_I, V_C; generally $\dfrac{K_C f_C}{T_C}$ differs from $\dfrac{K_I f_I}{T_I}$, so that V_I and V_C are not equally distributed (equipartition of V_x does not hold) for this two reservoirs generalized refrigerating machine or heat pump.

- two reservoirs engine

This case corresponds to the limit of a heat pump ; so the notion of refrigerating machine or heat transformer vanishes, as well as ω.

The temperature T_{SI} tends to infinity ; consequently T_I too ; q_I the corresponding heat becomes, pure mechanical energy ω_m (or power owing to the case). (20) in that case has to be considered as an undetermined form $0 \times \infty$; effectively $V_I \to 0$, and simultaneously $K_I f_I$ tends to infinity in such a way, that the product remains the finite value $\omega_m \prec 0$. These considerations are completely coherent with relation (24).

Remark 1 : limit of relation (10) corresponds to exo-endo irreversible efficiency of the engine :

$$\Im_i = \frac{T_H - T_C}{T_H} - \frac{T_C . S}{q_H} \tag{28}$$

we see from (28), that the presence of internal irreversibility reduces the corresponding efficiency proportionally to the report T_C / q_H. So the level of the cold internal temperature has to be kept low, and the heat exchanged high (size of the engine) at a given dissipation, in order to obtain an efficiency as closed as possible of the endoreversible one.

Remark 2 : (23, 25) are the two relations giving V_H, V_C, in this limit as :

$$V_H = \frac{\frac{V K_C f_C}{T_C} + S}{\frac{K_C f_C}{T_C} - \frac{K_H f_H}{T_H}} \qquad (29)$$

$$V_C = \frac{\frac{-V K_H f_H}{T_H} - S}{\frac{K_C f_C}{T_C} - \frac{K_H f_H}{T_H}} \qquad (30)$$

Again, it appears, that the internal irreversibilities influes adversely on the two quantities V_H, V_C ; generally $\frac{K_C f_C}{T_C}$ differs from $\frac{K_I f_I}{T_I}$, so that V_H and V_C are not equally distributed (equipartitions of V_x does not hold for this generalized engine).

MORE SOPHISTICATED MACHINES

- heat pumps moved by a thermal engine

This particular configuration is being studied and developed particularly by M. SASSO and his co-workers [20]. Results relative to this case can be found in the corresponding papers or communications.

We will only notice here, that on a formal point of view, many authors of the literature develop their model of three heat reservoirs machines on the basis of a schematic decomposition of the machine combining a two heat reservoirs engine powering a two heat reservoirs heat pump [21].

- solar three heat reservoir machines

In that case the hot side of the machine corresponds to a solar collector. The essential particularity of this system comes from the heat losses associated with the solar collector. These losses occur between the temperature T_{SH} of the isothermal collector and the ambient, proportionally to a heat loss conductance, or to a global heat transfer loss coefficient.

So it is well known that the solar collector efficiency is currently a linearly decreasing function of T_{SH}.

Correspondingly, the thermostatics efficiency of three heat reservoirs refrigerators or heat pumps is increasing function of the same temperature. So it exists an optimum of overall efficiency of the system, for a given particular temperature T_{SH}. This case has been considered in the literature and extended to endoreversible machines [22, 23] and endoirreversible machines [24].

The considerations exposed in the present paper allows us to extend the range of validity of these preceding models.

- finite thermal capacities of the machines

As can be seen in the preceding solar system, the fluid heated in the solar collector is generally not subject to phase change ; so he cannot be considered as isothermal (hypothesis : $T_H =_C te$). This fact has been considered in one paper [16], to take account of finite size thermal capacity of a system under transient conditions.

We have adopted the same demarch, for a permanent system [17]. In that case we take account of the finite mass fluxes of external fluids. Corresponding results are disponible in recent papers [25].

Some figures illustrating the preceding results are proposed hereafter (figures 1, 2, 3).

POSSIBLE EXTENSIONS OF THE PRECEDING ANALYSIS

DETAILED EXPRESSION OF THE DISSIPATION FUNCTION

Relation (2) introduces S, internal entropy created in the system, but on a lumped (global) form. This is interesting to see the sensitivity of the optimization performed, to the global dissipation of the machine ; also when experiments are done, this lumped value is more easily accessible on a bench. These considerations are useful for global design, or on site evaluation or maintenance.

But for a more powerful and precise model, S as to be analyzed in details ; in that case entropy analysis could introduce more realistic considerations. But probably, the corresponding complexity of the model, implies more variables and parameters, as well as indispensable and efficient computer means (with longer calculation time and cost).

EXERGETIC ANALYSIS

Exergy analysis is well known in industry ; it has been used in some works [26]. In these, the only reference used is the ambient temperature.

Exergetic efficiency is a very interesting potential objective function, because it corresponds to the efficiency of the system, relative to the second law of the thermodynamics. So it is a criterion of the quality of the system in term of irreversibility.

We think also that it introduces a kind of nondimensionality. We try in all the work we do, to propose nondimensionnal form for the results, as some other authors [27]. The paper of A. BEJAN et al [28] seems to be more judicious in term of reference conditions, to our opinion.

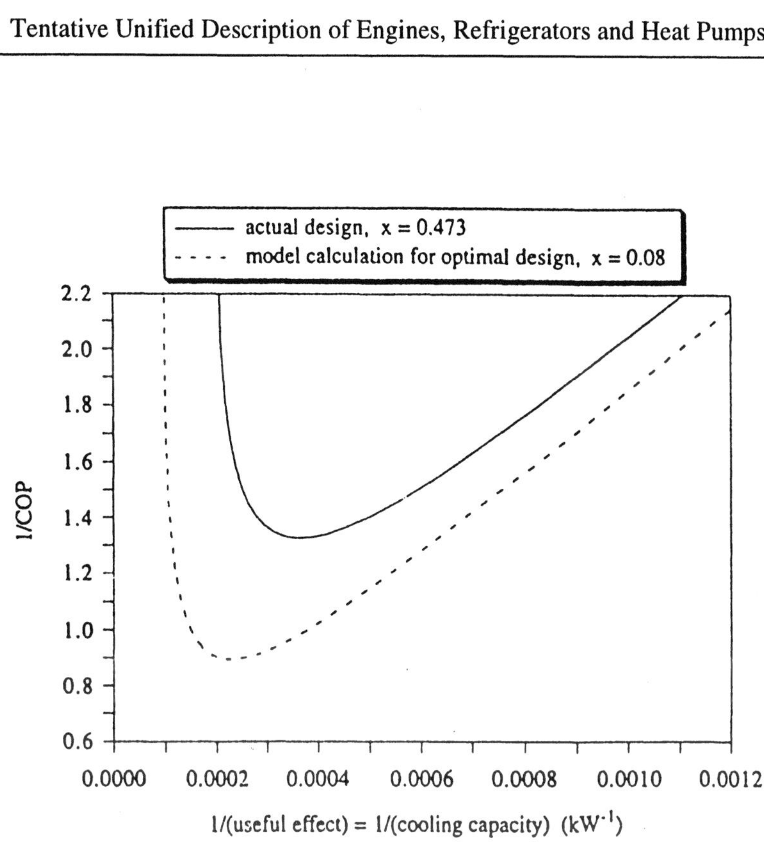

Figure 6 : representation as in figure 5, but computer simulation (rather than experimental) results for a single stage absorption chiller according to [33].

THERMOECONOMIC CONSIDERATIONS

Finally industry and applications necessitate to consider economic and financial aspects. This seems actually poorly developed. Some papers [26, 28] have introduced this aspect. These tentatives introduces respectively the prices of heat input and output relative to energy or exergy. It results from these papers existence of a maximum profit depending on the two preceding prices. But it seems that these thermoeconomic models remain relatively crude, and can be developed in the future [29].

VALIDATION OF MODELS

- Comparison with detailed numerical calculations

Until recently, relatively few tentatives have been reported, in order to compare the results of the lumped thermodynamics models proposed here, to more classical and detailed physical ones.

Recently a paper reports such a comparative studies [30, 31]. The corresponding results seems promising. The figure 4 illustrates the present state of the art.

Future works will probably give a new and more complete insight in the possibilities and workness of the new thermodynamics approach.

- Comparison with performance of industrial systems

Since some years, a very interesting work has been initiated by GORDON and coworkers [32]. These authors propose to use existing performance characteristics of disponible apparatus, to compare with what is expected from the corresponding model. They propose a characteristic curve of systems and particularly chillers :

$$\frac{1}{C.O.P.} = f\left(\frac{1}{q_{evap}}\right)$$

This comparison seems pertinent, if we remember formula (12) previously established in this paper.

These authors relate the consistency of the existing practical data with the fact that absorption chillers appear to have been designed toward maximizing their C.O.P. ; this one is achieved around their maximum cooling capacity, as was seen previously for vapour compression chillers.

The other major results of the comparisons done is that performance of three heat reservoirs machines is dominated by heat transfer irreversibilities, and internal entropy generation ; the effect of heat leaks is marginal for these authors.

So the proposed generalised insight developed here seems able to precise the relative importance and sensitivity to heat transfer irreversibilities, and internal entropy generation.

Internal entropy generation that contribute for GORDON, to half of the efficiency, has to be analysed in detail. Entropy analysis of heat and mass transfer in the heat exchangers, internal regeneration, and throtthing, is now being developed in our research group, and probably by others, in order to ameliorate one more time the optimization results.

Some comparison with laboratory experiments are also intended [33] (figures 5, 6).

CONCLUSIONS

The present research subject developed and synthetised here is now quickly going, evolving one. The proposed approach is completely complementary of detailed physical calculations ; this lumped irreversible approach is more powerful than the one relative to thermostatics ; it constitutes the only true thermodynamics.

It allows a quick diagnosis of the operating characteristics or of the design for a relatively complex system, via a lumped representation, but obviously more realistic description than the ancient thermostatic one.

ACKNOWLEDGEMENTS

I thank, Professor C. WU, for the honour to invite me for this submission.

In memory of my father, who left us just one year before: without him, this paper has never been proposed for publication.

M. FEIDT
08.03.1998

REFERENCES

1. CURZON F.L., AHLBORN B., « Efficiency of a Carnot engine at maximum power conditions », Am. J. Phys., vol. 53, p. 570-573, (1975)
2. NOVIKOV I.I., « The efficiency of atomic stations (a review) », Journal Nuclear Energy, vol. 7, p. 125-128, (1958)
3. CHAMBADAL P., « Les centrales nucléaires », Editeur Armand Colin, Paris, (1957), « Evolutions et applications du concept d'Entropie », Dunod, Paris, p. 84, (1963)
4. FEIDT M., « Thermodynamique et Optimisation energétique des systèmes et procédés », Technique et Documentation, Paris, (1987)
5. BEJAN A., « Advanced engineering Thermodynamics », Wihy Interscience, New York, (1988)
6. RADCENCO V.T., « Generalized thermodynamics », Editura tehnica, Bucharest, (1994)
7. BEJAN A., « Entropy generation minimization : the new thermodynamics of finite time size devices and finite time processes », J. Appl. Phys., vol. 79, n° 3, p. 1191-1204, (Feb. 1996)
8. SIENIUTYCZ S., SALAMON P., « Finite time Thermodynamics and thermoeconomics », adv. in Thermodynamics, vol. 4, Taylor and Francis, New York, (1990)
9. DE VOS A., « Endoreversible Thermodynamics of solar energy conversion », Oxford science publications, Oxford, (1992)
10. FEIDT M., PHILIPPI I., « Basis of a general approach for Finite Time Thermodynamics applied to two heat reservoirs machines », communication ECOS'92, Zaragoza, Spain, (June 1992)
11. FEIDT M., RAMANY BALA P., BENELMIR R., « Synthesis, unification, and generalization of preceding finite time Thermodynamics studies devoted to

CARNOT cycles », communication proceedings p. 429-441, FLOWERS'94, Firenze, Italia, (July 1994)

12. FEIDT M., « Sur une systématique des cycles imparfaits », ENTROPIE, n° 205, p. 53-61, (1997)
13. BLANCHARD C.H., « Coefficient of performance for finite speed heat pump », J. Appl. Phys., vol. 51, p. 2471-2472, (May 1980)
14. FEIDT M., BENELMIR R., « Thermodynamic optimization of irreversible inverse cycle : case of a vapour compression machine », communication 19th congress IIR-IIF, La Haye, The Netherlands, (Août 1995), proceedings vol. IV.a, p. 178-186
15. DINY M., BOUSSEHAIN R., FEIDT M., « Optimisation thermodynamique des machines à froid trithermes », communication, 19th congress IIR-IIF, La Haye, The Netherlands, (Août 1995), proceedings vol. IV.a, p. 147-154
16. CHEN J., YAN Z., « unified description of endoreversible cycles », Phys. rev. A., gen. phys., vol. 39, n° 9, p. 4140-4147, (Apr. 1989)
17. FEIDT M., « Finite time Thermodynamics applied to optimization of heat pump and refrigerating machines cycles », communication IMACS congress, Paris, (July 1988)
18. PHILIPPI I., FEIDT M., « Finite time Thermodynamics applied to inverse cycle machine », Montreal, Canada, (August 1991)
19. FEIDT M., BENELMIR R., « Première approche de l'optimisation des échangeurs de machines trithermiques », communication journée d'étude S.F.T., Paris, France, (8.04.1992)
20. DENTICE D'ACCADIA M., et al, « Design consideration for irreversible heat engine driven heat pumps », communication ECOS'96, Stockholm, Sweden, (June 1996), proceedings p. 135-140
21. CHEN J., YAN Z., « Equivalent combined systems with three heat sources heat pumps », J. chem; Phys., vol. 90, n° 9, p. 4951-4955, (May 1989)
22. WU C., CHEN L., SUN F., « Optimization of solar absorption refrigerator », Applied thermal Engineering, vol. 17, n° 2, p. 203-208, (1997)
23. YAN Z., CHEN J., « The maximum overall coefficient of performance of a solar driven heat pump system », J. Appl. Phys., vol. 76, n° 12, p. 8129-8134, (1994)
24. GUOXING L., YAN Z., « The optimal performance of an irrreversible solar driven heat pump system », ECOS'96 congress, Stockholm, Sweden, 1996, proceedings, p. 141-146
25. DINY M., FEIDT M., BOUSSEHAIN R., « Optimal performance of three heat sources machines », submitted for publication in Cryogenics
26. CHEN L., WU C., SUN F., « Maximum profit performance of an absorption refrigerator », I.J.E.E.E., vol. 4, n° 1, p. 1-7, (1996)
27. BOUSSEHAIN R., FEIDT M., « Cycle optimization for refrigeration system with three heat reservoirs », Mathematische Modellierung, Kassel, Germany, (1984), proceedings Mc GRAW-Hill, p. 46-56, (1986)
28. BEJAN A., VARGAS J.V.C., « Optimal allocation of a heat exchanger inventory in heat driven refrigerators », Int. J. Heat Mass Transfer, vol. 38, n° 16, p. 2997-3004, (1995)

29. GROSU L., BENELMIR R., FEIDT M., « Contribution to exergo economy of refrigerating machine », communication accepted for ECOS'98, Nancy, France, (July 1998)
30. WISEYSUNDERA N.E., « Analysis of the ideal absorption cycle with external heat transfer irreversibilities », Energy, vol. 20, n° 2, p. 123-130, (1995)
31. WISEYSUNDERA N.E., « Performance limits of absorption cycles with external heat transfer irreversibilities », Applied Thermal Engineering, vol. 16, n° 2, p. 175-181, (1996)
32. GORDON J.M., NG K.C., « A general thermodynamic model for absorption chillers ; theory and experiment », Heat Recovery Syst. CHP, vol. 15, n° 1, p. 73-83, (01.1995)
33. CHUA H.T., GORDON J.M., NG K.C., HAN Q., « Entropy production analysis and experimental confirmation of absorption systems », Int. J. Refrig., vol. 20, n° 3, p. 179-190, (1997)

COMPARISON OF OPTIMUM THERMAL CONDUCTANCE DISTRIBUTIONS FOR AN IRREVERSIBLE CARNOT CYCLE UNDER DIFFERENT OPTIMIZATION CRITERIA

A. CALVO HERNÁNDEZ, A. MEDINA, J. M. M. ROCO, AND S. VELASCO
Departamento de Física Aplicada,
Universidad de Salamanca, 37008 Salamanca, Spain

Abstract- A Carnot-like irreversible power cycle accounting for finite-rate heat transfers between the working fluid and the external heat reservoirs, heat leak between heat reservoirs, and internal dissipations of the working fluid is considered. For this model we present results of the optimum distribution of the external and internal thermal heat conductances in terms of a dimensionless price parameter, the engine temperature ratio τ and an internal irreversibility factor. We use three different optimization criteria: one based on the maximization of the power output and two more based on the maximization of the per-unit-time efficiency. The last ones show different behaviors with the engine temperature ratio, but they become identical to the τ-independent values predicted by the optimization criterion that maximizes the power output when $\tau \to 1$.

Key words: Carnot engine, Irreversible Performance, Thermal Conductance Distribution, Optimization Criteria

1. INTRODUCTION

As it has been pointed out by Berry [1] one fundamental characteristic of Finite-Time Thermodynamics is the freedom to elect the objective function to be optimized and, as a consequence, the flexibility in the comparison of the performance of systems optimized under different criteria. In this work we shall try to apply this idea to a Carnot-like irreversible power cycle, making special emphasis in the different behavior of the involved thermal conductance distributions, a subject of technical and economical interest. Before go on we summarize some previous and related results.

It is now widely recognized that the three major sources of irreversibilities in Carnot-like engines are those coming from the finite-rate heat transfer between the external heat reservoirs and the working fluid, the dissipative processes inside the working fluid (as friction, turbulences, inertia phenomena, and so on), and heat leak between heat reservoirs. The consideration of these irreversibilities reproduces the key features of real heat engines: loop-shaped power versus efficiency curves, relative proximity of maximum power and maximum efficiency points, and a maximum efficiency far below the Carnot value [2-5].

Another interesting point is the one concerning the analysis of the optimum distribution of the heat-exchanger equipment and the associated investment cost. In this regard we mention the recent works by Ait-Ali [6], Bejan [7], and Cheng and Chen [8]. Ait-Ali [6] optimized an irreversible Carnot heat engine for maximum power output and maximum thermal efficiency where, besides external finite-rate heat transfer irreversibility, a source of internal irreversibility produces entropy at a rate proportional to the external heat conductance and the engine temperature ratio, τ. This author found that the cold-end (condenser) heat conductance is always higher than that of the hot-end (boiler), although they present a quite different behavior with τ under maximum power and maximum efficiency solutions. The limit value of the Curzon-Ahlborn [9] maximum power cycle (overall heat conductance equally distributed between the hot-end and cold-end heat exchangers), is recovered by this author when $\tau \rightarrow 1$. Cheng and Chen [8], using a very similar model to account for the internal irreversibilities but optimizing the irreversible Carnot heat engine under the ecological criterion [10] (best compromise between power output and entropy generation), also found that the thermal conductance of the hot-end heat exchanger should be smaller than that of the cold-end heat exchanger in order to achieve better ecological performances. Bejan [7], besides of the two external finite-size heat exchangers considered in the abovementioned models, introduces the existence of an internal heat leak closely related to the thermal insulation of the power plant. By optimizing the efficiency under maximum power conditions when the total resources available for the internal thermal resistance and the external thermal conductance are fixed, this author found an optimum investment allocation ratio between the heat exchanger thermal conductances and the internal thermal insulation.

Our purpose in this work is to compare three different optimization criteria which give rise to different behaviors of the external and internal thermal conductance distributions. While the first criterion essentially follows Bejan's method [7] the other ones are based in the consideration of the per-unit-time efficiency as the objective function, a novel optimization criterion recently proposed [11] and applied in the analysis of the real low-temperature refrigerators [12], and power cycles [13]. To get this we use a quite standard model which incorporates the three abovementioned irreversibilities (Sec. 1). In Sec. 2 we briefly re-derive the analytical results by Bejan [7] with minor changes in order to introduce an internal irreversibility parameter not explicitly considered by this author in his model. In Secs. 3.2 and 3.3 we present the two per-unit-time solutions which must be numerically obtained. Moreover, we shall proof that Bejan's results (Sec. 3.1) are the limit values that the two per-unit-time solutions reach when $\tau \rightarrow 1$. Finally, in Sec. 4, we present a brief summary and the main conclusions of this work.

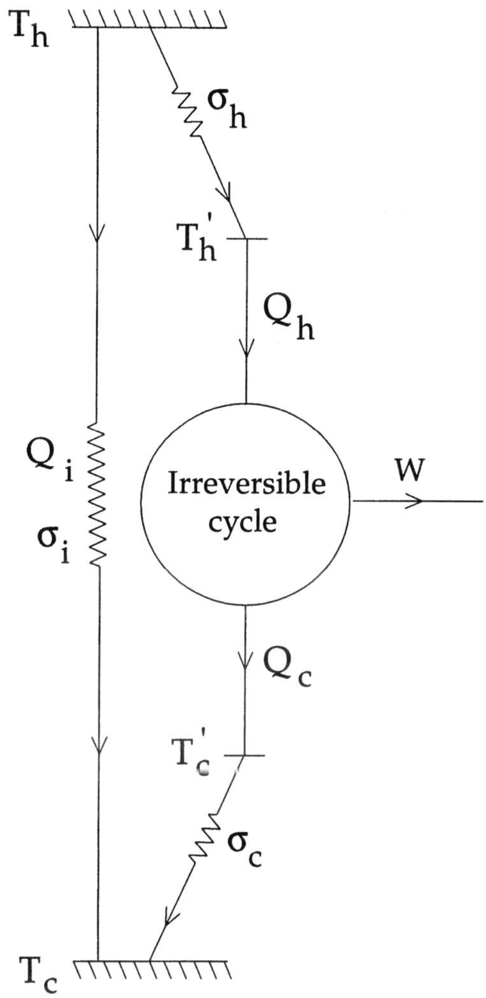

Fig.1 Irreversible Carnot cycle model with bypass heat leak, two finite-size heat exchangers and internal irreversibilities for the working fluid.

2. THEORETICAL MODEL

One of the most usual theoretical modelizations [2,4-7] for Carnot power plants is the so-called steady-flow or continuous irreversible Carnot power cycle. It is schematized on Fig. 1, and it essentially consists of two internal reservoirs at temperatures T_h' and T_c' coupled to another two external reservoirs at temperatures T_h and T_c. It is considered the existence of two types of irreversibilities: external and internal. The external ones come from the heat transfer across the finite temperature gaps: $|Q_h|$, the heat supplied per cycle by the hot reservoir at temperature T_h and $|Q_c|$, the heat absorbed per cycle by the heat sink at temperature T_c. Assuming linear heat transfer laws and denoting by t the overall cycle time:

$$|Q_h| = \sigma_h (T_h - T_h')t = \sigma_h T_h \left(1 - \frac{1}{a_h}\right)t \tag{1}$$

$$|Q_c| = \sigma_c (T_c - T_c')t = \sigma_c T_c (a_c - 1)t \tag{2}$$

where $a_h = T_h / T_h' \geq 1$, $a_c = T_c' / T_c \geq 1$, and σ_h, σ_c are respectively the hot-end and cold-end thermal conductances.

Within the internal sources of irreversibility we distinguish two contributions: internal irreversibilities of the working fluid (turbulences, frictions, viscosity, etc.) and bypass heat leak (heat transfer through the engine). Those of the working fluid can be quantified by means of the Clausius inequality [3-5]:

$$\frac{|Q_h|}{T_h'} = I \frac{|Q_c|}{T_c'} \tag{3}$$

where I is a parameter in the interval $0 < I \leq 1$ that quantifies this kind of irreversibilities. If we assume again a linear law for the heat transfer per cycle through the machine, Q_i [2,3,7]:

$$|Q_i| = \sigma_i (T_h - T_c)t = \sigma_i T_h (1 - \tau)t \tag{4}$$

where σ_i is the internal thermal conductance and $\tau = T_c / T_h$. As we shall discuss later on, the heat transfer (4) is essentially associated to the thermal isolation of the plant, a fundamental point of interest from an economic viewpoint when designing this kind of plants. The particular limit case when $I = 1$ and $\sigma_i = 0$ represents the absence of internal irreversibilities, thus the heat engine becomes endoreversible [9].

It is easy to obtain the net power output, P, and the efficiency, η, for this irreversible Carnot cycle by substituting in Eq. (3) the Eqs. (1) and (2) for Q_h and Q_c. From that we get:

$$a_c = \frac{I}{I - \sigma_{hc}(a_h - 1)} \tag{5}$$

$$P = \frac{|Q_h| - |Q_c|}{t} = \sigma_h T_h \frac{I(a_h - 1) - \sigma_{hc}(a_h - 1)^2 - \tau(a_h^2 - a_h)}{a_h(I + \sigma_{hc}) - \sigma_{hc} a_h^2} \tag{6}$$

$$\eta = \frac{|Q_h| - |Q_c|}{|Q_h| + |Q_i|} = \left[1 - \frac{a_h \tau}{I - \sigma_{hc}(a_h - 1)}\right]\left[\frac{a_h - 1}{a_h - 1 + \sigma_{ih} a_h (1 - \tau)}\right] \tag{7}$$

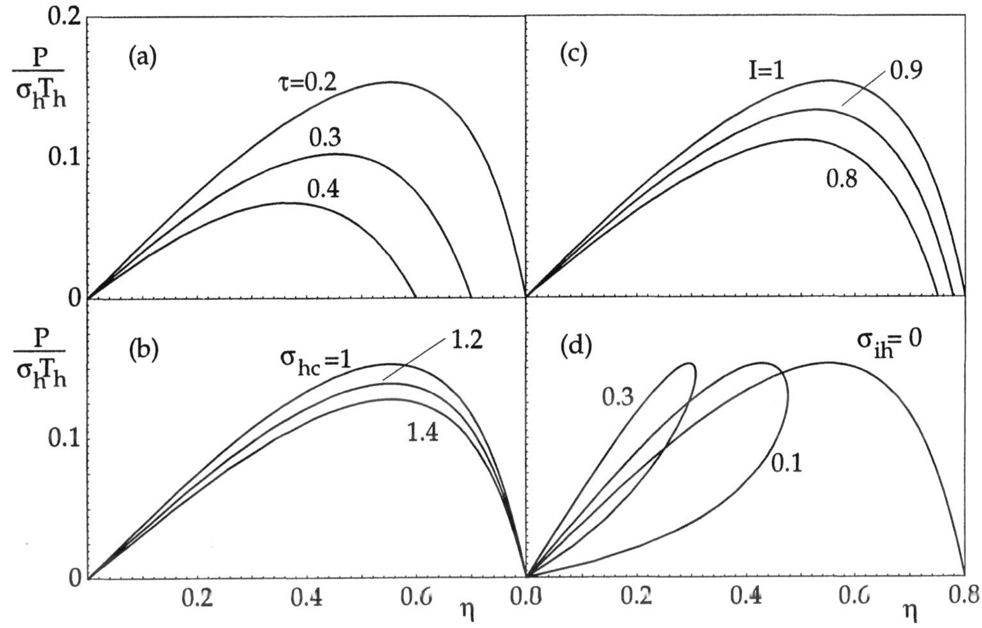

Fig.2 Reduced power vs efficiency curves for the considered irreversible Carnot cycle. (a) Influence of the temperature ratio, τ ($I = 1, \sigma_{hc} = 1, \sigma_{ih} = 0$); (b) influence of the external heat conductances ratio, σ_{hc} ($I = 1, \sigma_{ih} = 0, \tau = 0.2$); (c) influence of the internal irreversibilities of the working fluid ($\sigma_{hc} = 1, \sigma_{ih} = 0, \tau = 0.2$); (d) influence of the heat leak between external reservoirs ($I = 1, \sigma_{hc} = 1, \tau = 0.2$).

where $\sigma_{hc} = \sigma_h / \sigma_c$ and $\sigma_{ih} = \sigma_i / \sigma_h$. Then, the net power output is a function of six parameters, $P = P(\sigma_h, T_h, a_h, \sigma_{hc}, \tau, I)$ and the efficiency is a five-parameter function, $\eta = \eta(a_h, \sigma_{hc}, \sigma_{ih}, \tau, I)$. Taking into account that $|Q_h| \propto T_h'$ and $|Q_c| \propto T_c'$, we found the following condition for the overall cycle time:

$$t \propto \frac{1}{\sigma_h (a_h - 1)} = \frac{a_c}{\sigma_c (a_c - 1)} \tag{8}$$

and thus, we can define the per-unit-time efficiency, $\eta / t = \dot{\eta}$, as:

$$\dot{\eta} \propto \left[1-\frac{a_h\tau}{I-\sigma_{hc}(a_h-1)}\right]\left[\frac{a_h-1}{a_h-1+\sigma_{ih}a_h(1-\tau)}\right]\sigma_h(a_h-1) \tag{9}$$

To illustrate the behavior of the net power output (6) and the efficiency (7) obtained for this model, we show in Fig. 2 the reduced power $P/(\sigma_h T_h)$ versus η for several values of the parameters σ_{hc}, σ_{ih}, τ, and I. These curves have been obtained by eliminating the parameter a_h between eqs. (6) and (7). Fig. 2a shows the $P-\eta$ curves for $I=1$, $\sigma_{hc}=1$, $\sigma_{ih}=0$ and several values of τ (0.2, 0.3, and 0.4). When the engine is endoreversible, an increase in the temperature ratio, τ, provokes a decrease both in the power output and in the efficiency (this one always varies between $\eta=0$ and the corresponding Carnot limit, $\eta_C=1-\tau$). The maximum value of P corresponds to the Curzon-Ahlborn limit of η, i.e., the efficiency of an endoreversible Carnot cycle optimized respect to maximum power conditions, $\eta_{CA}=1-\sqrt{\tau}$ [9]. Figure 2b shows the evolution of the $P-\eta$ curves for $I=1$, $\sigma_{ih}=0$, $\tau=0.2$. In this case, P decreases when σ_{hc} increases, but the maximum efficiency is unaltered.

We have represented in Figs. 2c and 2d the influence of internal irreversibilities. In figure 2c we fixed $\sigma_{hc}=1$, $\sigma_{ih}=0$, $\tau=0.2$ to study the evolution of the $P-\eta$ curves with I. From these curves, it is evident that a decrease of I below 1 causes a clear decrease of the maximum power output and the efficiency. But, in any case, the qualitative behavior of the case $I=1$ is unchanged. On the contrary, from Fig. 2d, we can conclude a different behavior when the parameter I is fixed and the heat leak between external reservoirs, σ_{ih}, is changed. Now, the inclusion of a parameter σ_{ih} different from zero converts to loops the otherwise parabolic $P-\eta$ curves, that is, when there exists internal heat leak, maximum power output and maximum efficiency are very close one to another. Moreover, an increase of σ_{ih} causes a decrease of the efficiency. The power is unaffected by a change in σ_{ih}, because as it can be observed in Eq. (6) it does not depend on that kind of internal irreversibilities. In summary, the behavior of the proposed continuous Carnot cycle globally is capable to reproduce the main characteristics observed in real power plants, at least from a qualitative point of view [2,3,5,6].

Another possible representation for these functions, power output, efficiency and per-unit-time efficiency, can give valuable information on the trade-off between heat exhanger inventory and thermal insulation. This alternative scheme is due to Bejan [7]. If we denote $\sigma=\sigma_h+\sigma_c$ for the overall external heat conductance and by $R_i=1/\sigma_i$, the internal thermal resistance between heat reservoirs, the investment in the plant can be balanced in investment associated to R_i and σ. If the total resources available for these items are fixed ($P_\sigma\sigma+P_iR_i=constant$) we can define a constraint as:

$$\sigma + \frac{P_i}{P_\sigma} R_i = C \tag{10}$$

where C is the total thermal conductance that could be built with the resources available both for σ and R_i . P_i, P_σ represent the unit cost associated with, respectively, adding more internal thermal insulation and more external thermal conductance. (Note that their units are different: $\$K/W$ for P_σ and $\$W/K$ for P_i). As suggested by Bejan, it is convenient to define an investment allocation ratio $z = \sigma/C = 1 - (P_i R_i)/(CP_\sigma)$. With this definition, the external hot-end thermal conductance σ_h, becomes a function of z, C, and σ_{hc} in this way:

$$\sigma_h = \frac{\sigma\sigma_{hc}}{1+\sigma_{hc}} = \frac{zC\sigma_{hc}}{1+\sigma_{hc}} \tag{11}$$

and σ_{ih} is now a function of z, σ_{hc} and a dimensionless price parameter, $\Pi = P_i/(P_\sigma C^2)$ as:

$$\sigma_{ih} = \sigma_{ih}(\sigma_{ih}, z, \Pi) = \frac{P_i}{P_\sigma C(1-z)} \frac{1+\sigma_{hc}}{Cz\sigma_{hc}} = \Pi \frac{1+\sigma_{hc}}{\sigma_{hc} z(1-z)} \tag{12}$$

In this new representation, P, η and $\dot{\eta}$ are functions of the following parameters:

$$P = P(\sigma, T_h, a_h, \sigma_{hc}, z, \tau, I) \tag{13}$$

$$\eta = \eta(a_h, \sigma_{hc}, z, \Pi, \tau, I) \tag{14}$$

$$\dot{\eta} = \dot{\eta}(a_h, \sigma_{hc}, z, \Pi, \tau, I) \tag{15}$$

Respect to the definition of the price parameter, Π, it is noteworthy to stress that fixed the unitary costs of the external heat conductance and of the thermal insulation, P_σ and P_i, a decrease of Π means an increase of C, i.e., an increase of the available thermal conductance, σ, and/or the available thermal insulation, R_i (see Eq. (10)). In particular, when the parameter Π approaches zero, for fixed values of P_σ, P_i and σ, the available thermal insulation, R_i, becomes infinite and, as a consequence the power plant behaves as endoreversible.

3. Different Optimization Procedures

3.1. Maximum power conditions

Probably, the simplest optimization procedure (widely analyzed in the literature because it can be solved analytically [3,5,7]) for the model we have outlined consists in the calculation of the optimum heat exchanger temperature differences by maximizing the power output. For the model outlined here this is given by the condition:

$$\left(\frac{\partial P}{\partial a_h}\right)_{a_h=\bar{a}_{h,1}} = 0 \tag{16}$$

that yields to an optimum value for a_h,

$$\bar{a}_{h,1} = \frac{I+\sigma_{hc}}{\sigma_{hc}+\sqrt{\tau I}} \tag{17}$$

When substituted in eq. (6) for P, we obtain the maximum power, $\bar{P}_1$:

$$\bar{P}_1 = \sigma T_h \frac{\sigma_{hc}}{1+\sigma_{hc}} \frac{\left(\sqrt{I}-\sqrt{\tau}\right)^2}{I+\sigma_{hc}} \tag{18}$$

where we have used the relation, $\sigma_h = (\sigma\sigma_{hc})/(1+\sigma_{hc})$. Note that $\bar{P}_1$ does not depend on the internal heat leak conductance ratio, σ_{ih}.

The efficiency at maximum power conditions can be obtained from eq. (7) by substituting a_h by its optimum value $\bar{a}_{h,1}$:

$$\bar{\eta}_1 = \frac{1-\sqrt{\tau/I}}{1+\sigma_{ih}(1-\tau)\left(\dfrac{I+\sigma_{hc}}{I-\sqrt{\tau I}}\right)} \tag{19}$$

The endoreversible limit for the power output, $\bar{P}_{e,1}$, and the efficiency at maximum power conditions, $\bar{\eta}_{e,1}$, are obtained by setting $I=1$ and $\sigma_{ih}=0$ in eqs. (18) and (19):

$$\bar{P}_{e,1} = \sigma T_h \frac{\sigma_{hc}}{(1+\sigma_{hc})^2}\left(1-\sqrt{\tau}\right)^2 \tag{20}$$

$$\bar{\eta}_{e,1} = 1-\sqrt{\tau} = \eta_{CA} \tag{21}$$

Thus, as expected, the optimization respect to the power output in endoreversible conditions yields to the Curzon-Ahlborn efficiency [9].

From a technical point of view, there are two points of great interest when designing power plants [7]: the optimum allocation of the heat exchanger inventory and thermal insulation. Information on the optimum values for σ_{hc} can be obtained in an easy way by considering the total thermal conductance $\sigma = \sigma_h + \sigma_c$ as a fixed constraint and by performing a second optimization of the power output, in this case of the maximum power, $\overline{P}_1$. So, the new optimization condition is:

$$\left(\frac{\partial \overline{P}_1}{\partial \sigma_{hc}}\right)_{\sigma_{hc}=\overline{\sigma}_{hc,1}} = 0 \tag{22}$$

This condition is also analytically solvable, and gives an optimum value $\overline{\sigma}_{hc,1} = \sqrt{I}$. So, the optimum allocation of the heat exchanger inventory depends only on the internal irreversibilities of the working fluid. In the endoreversible limit, $I = 1$, thus the hot-end and cold-end thermal conductances are identical $\sigma_h = \sigma_c$, but in other cases, when $I < 1$, σ_c will be always greater than σ_h. By substituting $\overline{\sigma}_{hc,1}$ in eq. (18), and defining a reduced power function as $P^* = P/(\sigma T_h)$, we can get an expression for the twice optimized reduced power output in terms only of two parameter, τ and I:

$$\overline{P}^*_1 = \overline{P}^*_1\left(\overline{\sigma}_{hc,1}, \tau, I\right) = \frac{\left(\sqrt{I} - \sqrt{\tau}\right)^2}{\left(1 + \sqrt{I}\right)^2} \tag{23}$$

On the contrary, practical information on the trade-off between heat exchanger inventory and thermal insulation can not be obtained from the power, because it never depends on σ_{ih}. So, another function should be elected. We shall perform this third optimization on the efficiency, at maximum power conditions, $\overline{\eta}_1$. Moreover, we can directly get information on costs by introducing the investment allocation ratio, z, and the price parameter, Π, as discussed in Sec. 2. Thus by using eq. (12) and eq. (19), $\overline{\eta}_1$ can be expressed as:

$$\overline{\eta}_1 = \eta\left(\overline{a}_{h,1}, \overline{\sigma}_{hc,1}, \sigma_{ih}\left(\overline{\sigma}_{hc,1}, z, \Pi\right), \tau, I\right) = \frac{1 - \sqrt{\tau/I}}{1 + \Pi \dfrac{\left(1 + 1/\sqrt{I}\right)}{z(1-z)} \left(\dfrac{I + \sqrt{I}}{I - \sqrt{\tau I}}\right)(1-\tau)} = \overline{\eta}_1(z, \Pi, \tau, I) \tag{24}$$

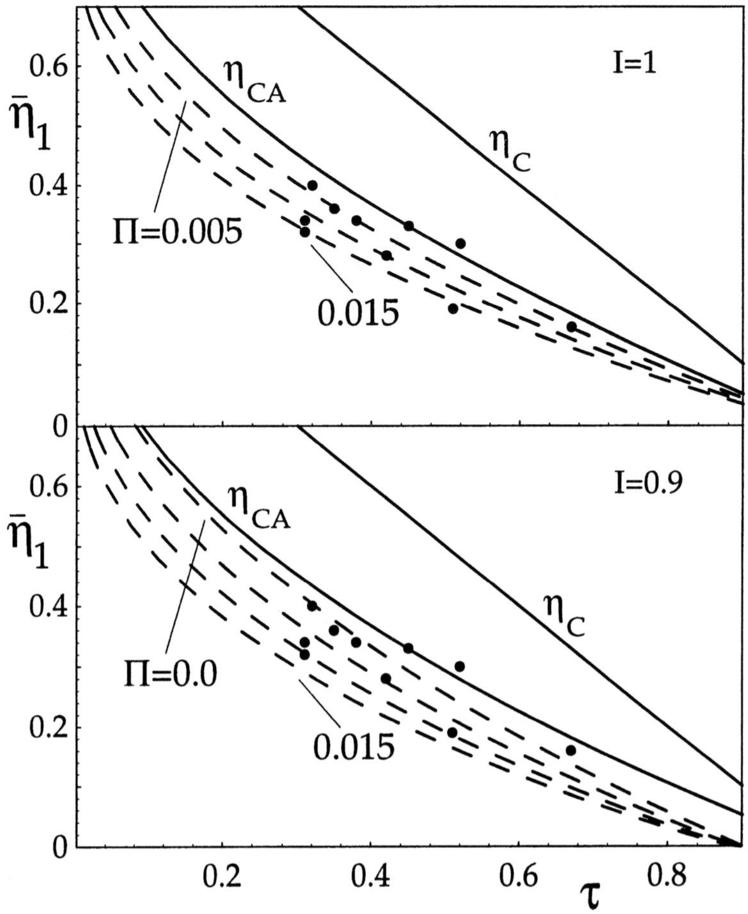

Fig.3 Optimized efficiency, $\bar{\eta}_1$, as given by eq. (26) for several values of the price parameter, Π (0.0, 0.005, 0.01, and 0.015) (dashed lines). In the figures are also shown the Carnot efficiency, η_C, the Curzon-Ahlborn efficiency, η_{CA}, and some numerical values of the efficiency for real power plants as reported by Bejan (dots). Note that for $I=1$ and $\Pi=0.0$, $\bar{\eta}_1=\eta_{CA}$.

Now, it is straightforward to get the optimum investment allocation ratio by solving:

$$\left(\frac{\partial\bar{\eta}_1}{\partial z}\right)_{z=\bar{z}_1}=0 \tag{25}$$

The solution of this equation is again analytical and gives $\bar{z}_1=1/2$. This means that $P_iR_i=P_\sigma\sigma$ (see Eq. (10)), i.e., the total investment in the thermal insulation of the plant equals the investment in the total external conductance.

By substituting $\overline{z}_1 = 1/2$ in eq. (24), we can write an expression for the three times optimized efficiency:

$$\overline{\eta}_1(\overline{z}_1, \Pi, \tau, I) = \frac{1 - \sqrt{\tau/I}}{1 + 4\Pi\left(1 + \frac{1}{\sqrt{I}}\right)\left(\frac{I + \sqrt{I}}{I - \sqrt{\tau I}}\right)(1 - \tau)} \tag{26}$$

In summary, this optimization procedure gives optimum values for the temperature ratio a_h by maximizing the power output, optimum values for the ratio between external conductances, by maximizing the maximum power output ($\overline{\sigma}_{hc,1} = \sqrt{I}$ and $\overline{\sigma}_{ih} = 4\Pi \frac{\left(1 + \sqrt{I}\right)}{\sqrt{I}}$, see eq.(12)) and optimum values for the investment allocation ratio, $\overline{z}_1 = 1/2$, by maximizing the efficiency at maximum power conditions. Results by Bejan in Ref. [7] are recovered from those in this subsection when $I = 1$.

In Fig. 3 we represent $\overline{\eta}_1$ as given by Eq. (26) versus τ for several values of the price parameter Π in the cases $I = 1$ (upper graph) and $I = 0.9$ (lower graph). We also plot the Carnot and Curzon-Ahlborn efficiencies and numerical values of observed effciencies for real power plants as reported by Bejan [14]. For the case $I = 1$, it is observed that the price parameter Π that yields to efficiency values similar to the real ones are in the interval $\Pi = 0.0$, $\Pi = 0.015$. For $\Pi = 0.0$, (equivalent to fixing $\sigma_{ih} = 0$, as we commented before), the Curzon-Ahlborn efficiency is exactly reproduced and greater Π values yield to lower efficiencies. The mentioned interval for Π is in good concordance with results of Bejan [7] for a similar model. When I<1, the efficiencies decrease, so the values that takes the price parameter, Π, in order to reproduce experimental data have to be slightly diminished, i.e., C, that globally represents the total cost of the plant (for fixed P_i and P_σ, see eq. (10)) must be increased when internal irreversibilities increase. In any case, the inclusion of internal irreversibilities ($I < 1$) does not produce significant changes on the behavior of $\overline{\eta}_1$.

3.2. Per-Unit-Time Efficiency Optimization

An alternative way to calculate the optimum temperature ratio, a_h, from the model we are analyzing is by maximizing the per-unit-time efficiency. Its physical meaning was suggested by Ma [15]: in order to reach its maximum efficiency, a heat engine must be infinitely slow (reversible limit), but to increase the operation speed, the efficiency has to be partially yielded. Thus, the optimization of the per-unit-time efficiency provides a quantification of the best compromise between operation speed and efficiency.

Consequently, we can begin now from eq. (9) that gives the per-unit-time efficiency as a function, $\dot{\eta} = \dot{\eta}(a_h, \sigma_{hc}, \sigma_{ih}, \tau, I)$, and solve:

$$\left(\frac{\partial \dot{\eta}}{\partial a_h}\right)_{a_h=\bar{a}_{h,2}} = 0 \qquad (27)$$

to obtain the optimum temperature ratio in conditions of maximum per-unit-time efficiency. Unfortunately, this equation has to be solved numerically. There is only a particular case where the optimum ratio $\bar{a}_h$ can be expressed analytically. That is the endoreversible case, $I = 1$ and $\sigma_{ih} = 0$. In these conditions, it is easy to find:

$$\bar{a}_{h,2} = 1 + \frac{1}{\sigma_{hc}} - \frac{1}{\sigma_{hc}}\sqrt{\frac{(1+\sigma_{hc})\tau}{\tau + \sigma_{hc}}} \qquad (28)$$

which substituted in (7) leads to:

$$\bar{\eta}_{e,2}\left(\bar{a}_{h,2}, \sigma_{hc}, \sigma_{ih} = 0, \tau, I = 1\right) = 1 + \frac{\tau}{\sigma_{hc}} - \frac{\sqrt{(1+\sigma_{hc})(\sigma_{hc} + \tau)\tau}}{\sigma_{hc}} \qquad (29)$$

It is easy to proof that this endoreversible efficiency has an upper bound given by the Curzon-Ahlborn efficiency,

$$\bar{\eta}_{e,2}\left(\bar{a}_{h,2}, \sigma_{hc}, \sigma_{ih} = 0, \tau, I = 1\right) < 1 - \sqrt{\tau} \qquad (30)$$

In general, for $I < 1$ and $\sigma_{ih} \neq 0$, eq. (27) does not yield to analytical solutions, so $\bar{a}_{h,2}$ has to be investigated numerically. By substituting $\bar{a}_{h,2}$ in eqs. (6) and (7) we get the efficiency, $\bar{\eta}_2 = \eta\left(\bar{a}_{h,2}, \sigma_{hc}, \sigma_{ih}, \tau, I\right)$, and the power, $\bar{P}_2 = P\left(\sigma_h, T_h, \bar{a}_{h,2}, \sigma_{hc}, \tau, I\right)$, in conditions of maximum per-unit-time efficiency. From these equations is possible to get information on the optimum allocation of the heat exchanger inventory and thermal insulation by making use of the representation given by eqs. (13) and (14) and solving:

$$\left(\frac{\partial \bar{P}_2}{\partial \sigma_{hc}}\right)_{\sigma_{hc}=\bar{\sigma}_{hc,2}} = 0 \qquad \text{and} \qquad \left(\frac{\partial \bar{\eta}_2}{\partial z}\right)_{z=\bar{z}_2} = 0 \qquad (31)$$

In particular, for the optimum investment allocation ratio, $\bar{z}_2$, numerical solutions of (31) give a value $\bar{z}_2 = 1/2$, i.e., the same value that the one obtained with the maximization of the power output to get $\bar{a}_{h,1}$. Thus, $\bar{z}_1 = \bar{z}_2$ and again $P_i R_i = P_\sigma \sigma$.

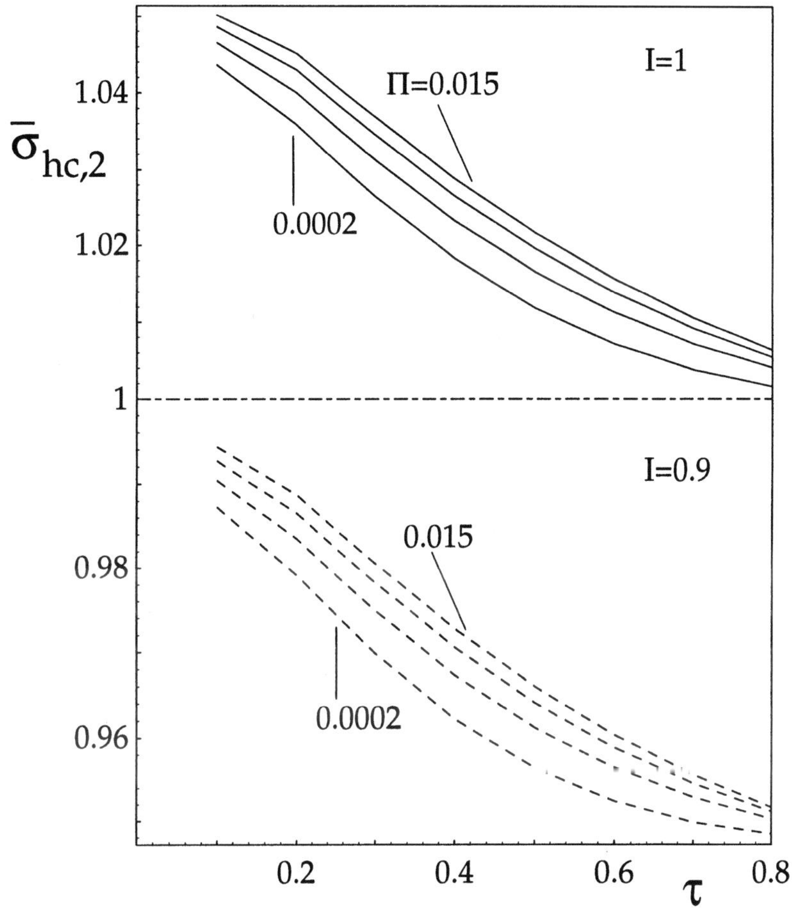

Fig. 4 Optimum external heat conductance ratio, $\bar{\sigma}_{hc,2}$, obtained by solving eq. (31) for $\Pi = 0.0002$, 0.005, 0.01, and 0.015. Solid lines, $I = 1$; dashed lines, $I = 0.9$.

Numerical solutions of the optimum $\bar{\sigma}_{hc,2}$ are shown in Fig. 4 for several values of the parameters Π and I. First, it is noticeable that $\bar{\sigma}_{hc,2}$ for $I = 1$ and any Π-value, is always greater than unity, so the thermal conductance of the hot-end exchanger is greater than that of the condenser. Reversely, when internal irreversibilities of the working fluid are considered ($I < 1$), $\bar{\sigma}_{hc,2} < 1$ for any value of Π. For both I values, $\bar{\sigma}_{hc,2}$ decreases with τ for a fixed price parameter. For $I = 1$, $\bar{\sigma}_{hc,2} \to 1$ for high τ values, but for $I = 0.9$, $\bar{\sigma}_{hc,2}$ approaches the value $\sqrt{I}$, i.e., the optimum thermal conductances ratio obtained by getting $\bar{a}_h$ with an optimization of the power output. Respect to the behavior with the price parameter Π, we can say that, with or without internal dissipations for the working fluid, $\bar{\sigma}_{hc,2}$ increases as Π increases.

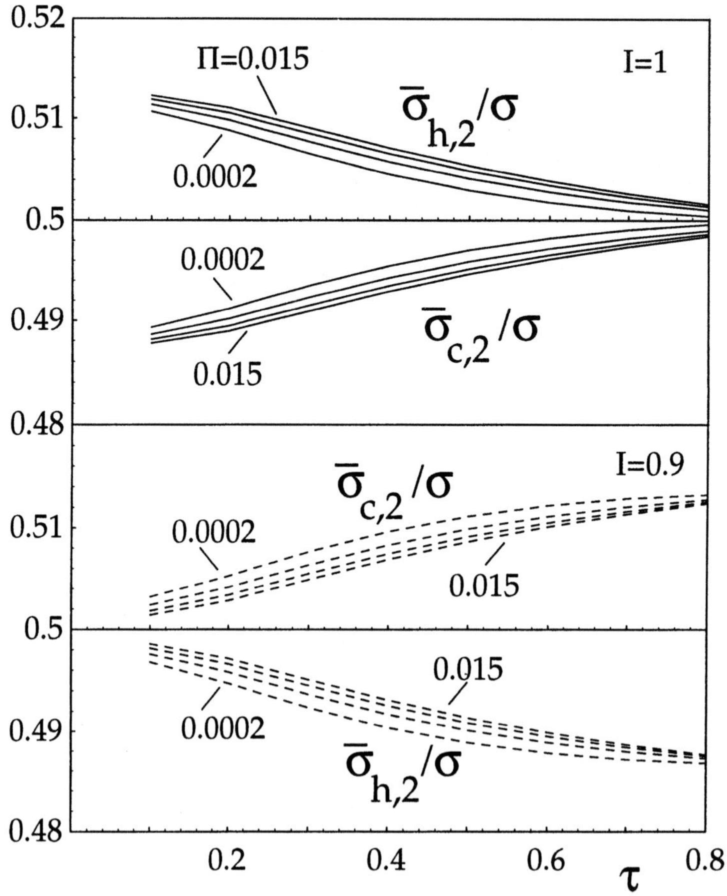

Fig.5 Optimum external heat conductance distributions, $\bar{\sigma}_{h,2}/\sigma$ and $\bar{\sigma}_{c,2}/\sigma$, for Π= 0.0002, 0.005, 0.01, and 0.015. Solid lines, $I = 1$; dashed lines, $I = 0.9$.

We plot in Fig. 5 the optimum external heat conductance distributions $\bar{\sigma}_{h,2}/\sigma = \bar{\sigma}_{hc,2}/\left(\bar{\sigma}_{hc,2}+1\right)$ and $\bar{\sigma}_{c,2}/\sigma = 1-\bar{\sigma}_{h,2}/\sigma$ versus τ for some values of I and Π. It is very clear from the figure that for $I = 1$, $\bar{\sigma}_{h,2} > \bar{\sigma}_{c,2}$ and $\bar{\sigma}_{h,2} \approx \bar{\sigma}_{c,2}$ for high τ values. For $I = 0.9$, $\bar{\sigma}_{c,2} > \bar{\sigma}_{h,2}$ and only for small τ values $\bar{\sigma}_{h,2} \approx \bar{\sigma}_{c,2}$. We note that the progressive splitting of $\bar{\sigma}_{c,2}$ and $\bar{\sigma}_{h,2}$ as τ increases (observed in the lower graph of Fig. 5) agrees with results reported by Ait-Ali for $\tau > 0.5$ and obtained under maximum efficiency conditions for an irreversible model without heat leaks between the external reservoirs (see Fig. 6 in Ref. [6]).

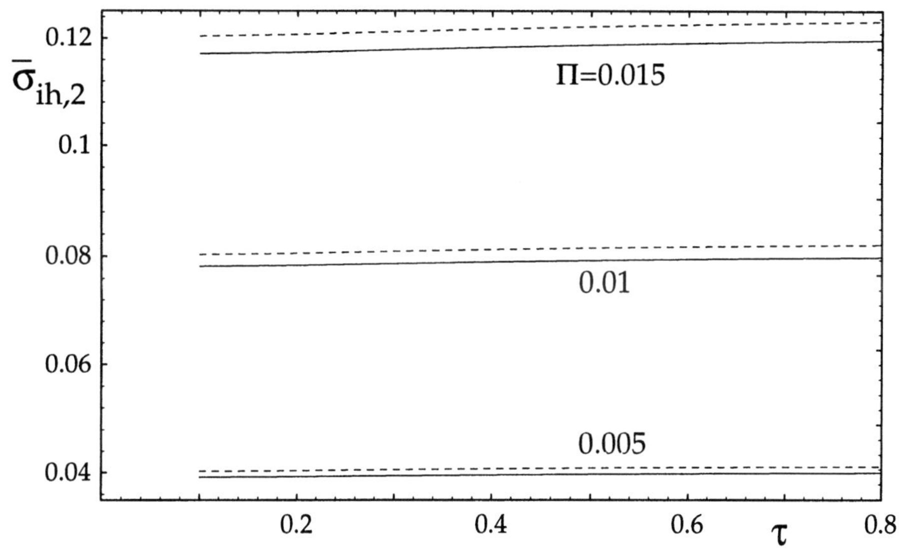

Fig.6 Optimum heat conductance distribution $\bar{\sigma}_{ih,2}$, for several Π-values. Solid lines, $I = 1$; dashed lines, $I = 0.9$.

The ratio between the optimum internal heat conductance and the hot-end external heat conductance, $\bar{\sigma}_{ih,2} = 4\Pi\left(1+\bar{\sigma}_{hc,2}\right)/\bar{\sigma}_{hc,2}$ (Eq. (12) with $\bar{z}_2 = 1/2$) is represented in Fig. 6 versus τ. We can conclude from this figure that the qualitative behavior of $\bar{\sigma}_{ih,2}$ does not depend greatly on either I or τ for a fixed Π value. Nevertheless the changing behavior of $\bar{\sigma}_{ih,2}$ with Π is appreciable: $\bar{\sigma}_{ih,2}$ increases with increasing Π values. For instance, for $\tau = 0.4$ (a representative value in real power plants) and $\Pi = 0.005$, $\bar{\sigma}_{ih,2}$ amounts approximately 4%, but for $\Pi = 0.015$ it amounts 12% with a small difference for the cases $I = 1$ or $I = 0.9$.

Although we do not show explicitly the figure, the optimized efficiencies at maximum per-unit-time efficiency conditions are very similar to those shown in Fig. 3 when the power output is maximized. The Π-values reproducing observed efficiencies for real plants are very similar (always in the interval $\Pi = 0.0 - 0.015$) and efficiencies always decrease as Π increases. The only difference when comparing with figure 3 would be that in the endoreversible limit ($I = 1, \Pi = 0.0$), $\bar{\eta}_2$ is not exactly η_{CA}, but it is always true that $\bar{\eta}_2 < \eta_{CA}$.

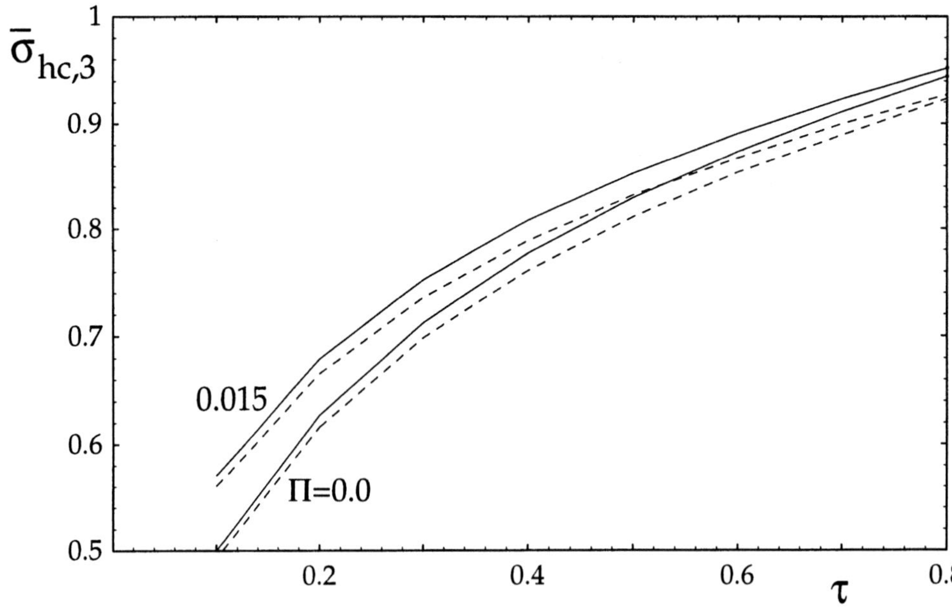

Fig. 7 Optimum external heat conductance ratio, $\bar{\sigma}_{hc,3}$, obtained by solving eq. (33) for Π= 0.0 and 0.015. Solid lines, $I = 1$; dashed lines, $I = 0.9$.

3.3. Full Optimization of the Per-Unit-Time Efficiency

As we noted in Sec. 2, the per-unit-time efficiency, in the parametrical representation that permits to investigate the optimum allocation of the heat exchanger inventory and thermal insulation, is a six-parameter function of this form: $\dot{\eta} = \dot{\eta}(a_h, \sigma_{hc}, z, \Pi, \tau, I)$. Then, in principle, it is feasible to get optimized values for a_h, σ_{hc} and z by maximizing this function only. This procedure will give us a final efficiency in conditions of fully-optimized per-unit-time efficiency depending on three parameters: Π, τ, and I. As in Sec. 3.2 this optimization must be solved numerically.

If we proceed step by step, the first equation to solve in order to get the optimum distribution of temperatures is:

$$\left(\frac{\partial \dot{\eta}}{\partial a_h}\right)_{a_h = \bar{a}_{h,3}} = 0 \qquad (32)$$

This equation is identical to eq. (27), so the solution $\bar{a}_{h,3}$ verifies:
$\bar{a}_{h,3} = \bar{a}_{h,2}$. The design parameters σ_{hc} and z can be obtained by maximizing $\dot{\eta}$, within the representation given by eq. (15). The equations to solve now are:

$$\left(\frac{\partial \bar{\dot{\eta}}_3}{\partial \sigma_{hc}}\right)_{\sigma_{hc}=\bar{\sigma}_{hc,3}} = 0 \qquad \text{and} \qquad \left(\frac{\partial \bar{\dot{\eta}}_3}{\partial z}\right)_{z=\bar{z}_3} = 0 \tag{33}$$

where $\bar{\dot{\eta}}_3 = \dot{\eta}(\bar{a}_{h,3}, \sigma_{hc}, z, \Pi, \tau, I)$. The solutions of these equations show that the optimum allocation ratio is, as it happens for the preceding optimization procedures, constant, and with numerical value $\bar{z}_3 = 1/2$.

The solutions of eq. (33) (see fig. (7)) show that the optimum values for $\bar{\sigma}_{hc,3}$ are always below the unity, independently of the existence or not of internal irreversibilities on the working fluid. $\bar{\sigma}_{hc,3}$ increases as it does τ, and in the high-τ regime it is easy to realize that $\bar{\sigma}_{hc,3} \to \sqrt{I}$.

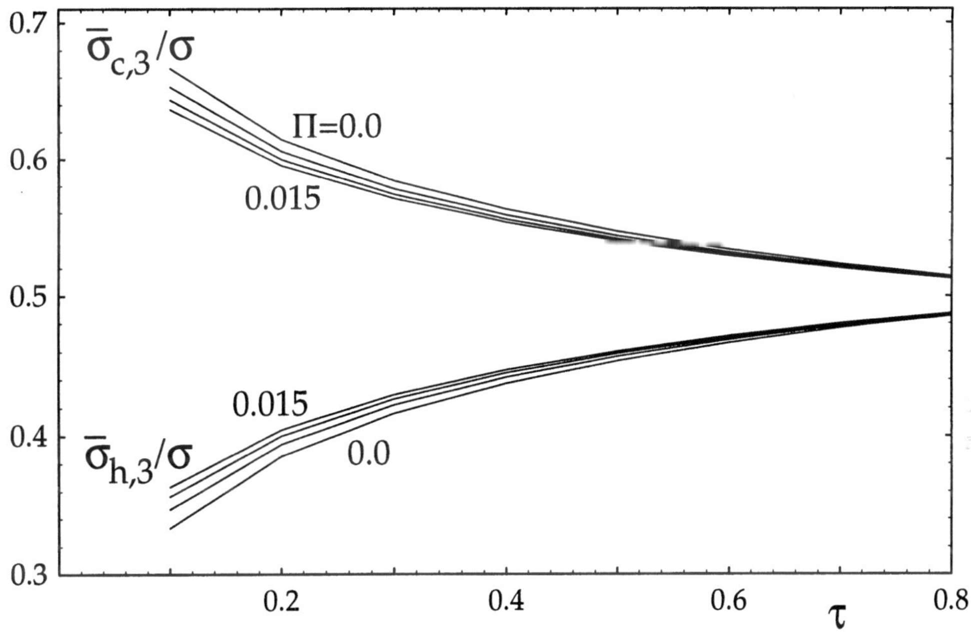

Fig.8 Optimum external heat conductance distributions, $\bar{\sigma}_{h,3}/\sigma$ and $\bar{\sigma}_{c,3}/\sigma$, for $I=1$ and Π-values: 0.0, 0.005, 0.01, and 0.015. For $I<1$ (not shown in the figure), only small numerical differences are appreciated.

We have plotted in Fig. 8 the evolution of $\bar{\sigma}_{c,3}/\sigma$ and $\bar{\sigma}_{h,3}/\sigma$ versus τ for $I=1$ and some values for the cost parameter between $\Pi = 0.0$ and $\Pi = 0.015$. The inclusion of internal irreversibilities ($I<1$) does not change the behavior of the curves in this figure, there appear only small numerical variations. It is observed in the plot that $\bar{\sigma}_{c,3}$ is always greater than $\bar{\sigma}_{h,3}$ and that $\bar{\sigma}_{c,3}/\sigma$ decreases as Π increases. For $\bar{\sigma}_{h,3}/\sigma$ the evolution with Π is opposite. For a fixed Π–value, the behavior of $\overline{\sigma}_{h,3}$ and $\overline{\sigma}_{c,3}$ with τ in Fig. 8 is very similar to the one reported by Ait-Ali under maximum power conditions (see Fig. 5 in [6]).

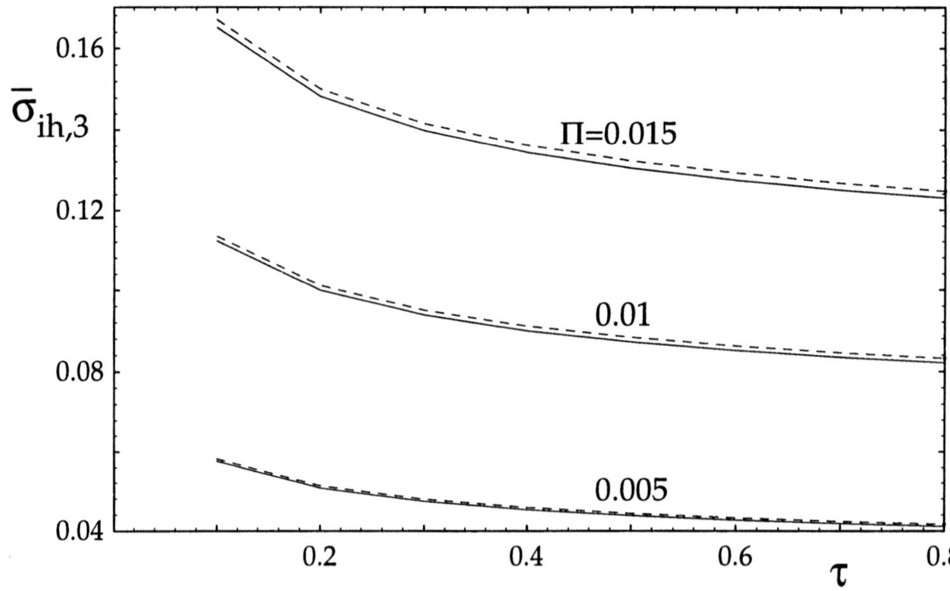

Fig. 9 Optimum heat conductance distributions, $\overline{\sigma}_{ih,3}$, for several Π-values. Solid lines, $I=1$; dashed lines, $I=0.9$.

The optimum value for the ratio $\overline{\sigma}_{ih,3}$ is shown in Fig. 9. As in Sec. 3.2 the influence of I on this optimum ratio is small, but on the contrary, within this optimization procedure, it is observed a clear decrease of $\overline{\sigma}_{ih,3}$ with τ as it increases. So, the sensitivity of $\overline{\sigma}_{ih}$ is much more important when the optimization is relative only to the per-unit-time efficiency. As it happened with $\overline{\sigma}_{ih,2}$, $\overline{\sigma}_{ih,3}$ decreases markedly as Π increases. It is easy to proof that for high τ , the corresponding limit values of $\overline{\sigma}_{ih,2}$ and $\overline{\sigma}_{ih,3}$ coincide for any value of Π and I.

We do not show the figures, but the efficiencies optimized only with respect to the per-unit-time efficiency, $\overline{\eta}_3 = \eta_3\left(\overline{a}_{h,3}, \overline{\sigma}_{hc,3}, \overline{z}_3, \tau, \Pi, I\right)$ are very similar to the efficiencies obtained with the previously considered optimization criteria, $\overline{\eta}_1$ and $\overline{\eta}_2$. They are always below η_{CA}, decrease as Π increases and the adequate Π-values to reproduce real efficiencies are in the interval $\Pi = 0.0 - 0.015$.

4. Summary and Conclusions

We have presented an irreversible, three-parametric Carnot-like cycle model which captures the main features of the real power plants in terms of the engine temperature ratio, τ, a parameter accounting for internal irreversibilities in the working fluid, I, and a dimensionless price parameter, Π, that takes into account the investment on both the external thermal conductance (the one associated with the hot-end and cold-end heat exchangers) and the internal thermal conductance (the one associated with the thermal insulation of the power plant). For this model we have performed three different optimization procedures in order to get information about the optimum thermal conductance distributions $\overline{\sigma}_{hc}$ and $\overline{\sigma}_{ih}$. The three criteria we have used are: a (double) maximization of the power output with a_h and σ_{hc} and a third maximization of the efficiency under maximum power conditions respect to z (criterion 1); maximization of the per-unit-time efficiency respect to a_h, and later maximizations of the power, respect to σ_{hc} and efficiency, respect to z (criterion 2); and full maximization of the per-unit-time efficiency (criterion 3).

At sight of the results presented along the work we summarize the following conclusions:

a. When $\overline{\sigma}_{hc}$ and $\overline{\sigma}_{ih}$ are calculated from criterion 1 we obtain τ -independent expressions given by $\overline{\sigma}_{hc,1} = \sqrt{I}$ and $\overline{\sigma}_{ih,1} = 4\Pi\left(\sqrt{I}+1\right)/\sqrt{I}$. In the endoreversible limit ($I = 1, \Pi = 0$), these expressions become the well-known results $\overline{\sigma}_{hc} = 1$ and $\overline{\sigma}_{ih} = 0$, i.e., the hot-end and cold-end thermal conductances are equally distributed and the thermal insulation is infinite. If there exist internal irreversibilities ($I < 1$, $\Pi > 0$) the cold-end thermal conductance must be greater than the hot-end thermal conductance, and $\overline{\sigma}_{ih}$ increases as Π increases, with minor numerical differences as I decreases. This criterion 1 recovers the corresponding results by Bejan setting $I = 1$ [7].

b. When $\overline{\sigma}_{hc}$ and $\overline{\sigma}_{ih}$ are calculated from criterion 2 we obtain the τ-dependent behaviors plotted in figures 4, 5, and 6. In absence of internal irreversibilities in the working fluid, $I = 1$, the hot-end thermal conductance is greater than the cold-end thermal conductance but the differences progressively disappear as τ increases (see upper parts in figures 4 and 5). Minor numerical changes are observed by the consideration of non-zero Π values. If internal irreversibilities of the working fluid are considered $I < 1$ the opposite behavior is observed. The cold-end thermal conductance is now greater (as it happened with criterion 1) and the differences progressively disappear as τ decreases (see lower parts in figures 4 and 5). The internal thermal conductance distribution (see figure 6) shows a slightly increase with τ but the Π-dependence remains as in criterion 1.

c. When $\overline{\sigma}_{hc}$ and $\overline{\sigma}_{ih}$ are calculated from criterion 3 we obtain the τ-dependent behaviors plotted in figures 7, 8, and 9. This criterion, independently of the I-

values, predicts that the cold-end thermal conductances are greater than the hot-end ones and that the differences between them disappear as the temperature ratio increases (see figures 7 and 8). The internal thermal conductance distribution (see figure 9) shows a very clear decrease with the temperature ratio τ.

Finally, we stress that both per-unit-time efficiency optimization criteria under (real) internal irreversible conditions ($I < 1,\ \Pi > 0$) predicts that the optimum cold-end heat conductance is greater than the optimum hot-end heat conductance. The differences between them are due to the different τ -behavior. While within criterion 2 the cold-end (hot-end) thermal conductance increases (decreases) as τ increases, the opposite behavior is obtained with criterion 3. The τ -independent results obtained with criterion 1 are recovered in the limit $\tau \rightarrow 1$ for the criteria 2 and 3.

ACKNOWLEDGMENTS

We thank financial support from DGICYT of Spain under Grant PB95-0934 and from Junta de Castilla y León (Spain) under Grant SA78/96.

REFERENCES

1. R. S. Berry, in *Finite-Time Thermodynamics and Thermoeconomics*, edited by S. Sieniutycz and P. Salamon (Taylor and Francis, 1990), p. vii.
2. J. M. Gordon and M. Huleihil, J. Appl. Phys. **72**, 829 (1992).
3. J. Chen, J. Phys. D: Appl. Phys. **27**, 1144 (1994).
4. S. Ozkaynak, S. Goktun, and H. Yavuz, J. Phys. D: Appl. Phys. **27**, 1139 (1994).
5. L. Chen, F. Sun, and C. Wu, Energy Convers. Mgmt. **38**, 1501 (1997).
6. M. A. Ait-Ali, J. Appl. Phys. **78**, 4313 (1995).
7. A. Bejan, *Entropy Generation Minimization* (CRC Press, Boca Raton, 1996) Ch. 8.1.
8. C.-Y. Cheng and C.-K. Chen, J. Phys. D: Appl. Phys. **30**, 1602 (1997).
9. F. Curzon and B. Ahlborn, Am. J. Phys. **43**, 22 (1975).
10. F. Angulo-Brown, J. Appl. Phys. **69**, 7465 (1991).
11. S. Velasco, J. M. M. Roco, A. Medina and A. Calvo Hernández, Phys. Rev. Lett. **78**, 3241 (1997).
12. S. Velasco, J. M. M. Roco, A. Medina and A. Calvo Hernández, Appl. Phys. Lett. **71**, 1130 (1997).
13. A. Calvo Hernández, A. Medina, J. M. M. Roco, and S. Velasco, Appl. Phys. Lett. Submitted.
14. A. Bejan, *Advanced Engineering Thermodynamics* (Wiley, 1988), p. 408.
15. S. K. Ma, *Statistical Mechanics* (World Scientific, Singapore, 1985), pp. 24-28.

AN IRREVERSIBLE OTTO CYCLE MODEL INCLUDING CHEMICAL REACTIONS

F. ANGULO-BROWN,
Departamento de Física, Escuela Superior de Física y Matemáticas, Instituto Politécnico Nacional, Edificio N# 9 U.P. Zacatenco, 07738 México D.F., México

T.D. NAVARRETE-GONZÁLEZ AND J.A. ROCHA-MARTÍNEZ
Area de Física, Departamento de Ciencias Básicas, Universidad Autónoma Metropolitana Azcapotzalco, Avenida San Pablo 180, 02200 México D.F., México

Abstract- In this work we present a simplified Otto cycle model that includes power losses due to a kind of lumped friction and it also considers the combustion reaction at the end of the adiabatic compression. The cycle with chemical reaction is analyzed for the cases of methane and isooctane with and without excess of air, which is measured through a parameter α¤ 1. We calculate the power output (P) and the efficiency (η) of the cycle as functions of the compression ratio, r, and α. We find convex curves of P and η in terms of r and loop-shaped curves of P vs. η. Our results suggest that the cycle's performance is mainly dominated by the caloric properties (heat capacities) of the fuels.

INTRODUCTION

Models for internal combustion engines are discussed in many textbooks of thermodynamics [1-5]. The most common models are the reversible ones. As it is well known, the reversible models are only useful as limit cases and they serve for putting upper (or lower) bounds to several process variables, such as efficiency and work, for example. All reversible models of a thermal engine has null power output. Thus, the proposal of thermal engine models with power different from zero may be useful for a better understanding of the thermodynamics of the engine performance. In recent years, several authors [6-9] have proposed models for the Otto cycle with the property of having power and efficiency curves with a convex behavior with respect to the compression ratio (maximum volume divided by minimum volume). Although, these models are oversimplified in comparison with a real gasoline engine, they have the virtue of recovering in a reasonable manner typical real curves for power (P) and efficiency (η) of actual engines. Besides, these models lead to loop-shaped power-versus-efficiency curves as is common to almost all real heat engines [10]. The models considered in Refs. 6 to 9, have not taken into account explicitly the chemical reactions occurring inside the working gas of the thermal engine. In this paper, we consider the combustion reactions occurring at the end of the compression stroke, and we calculate the temperature T_3 (see Fig. 1) at the state where the power stroke begins, by means of the estimation of the heat of reaction released by the combustion of fuels such as methane and isooctane. For these two fuels, we calculate the power and efficiency of an Otto cycle as functions of the compression ratio. In these cases, we again obtain convex curves for P and η, indicating that there exists an optimum compression ratio if one wishes to work that cycle in the maximum power or maximum efficiency regimes. We also made an analysis of the Otto cycle-performance in terms of a parameter α ¤ 1 which measures the excess of air in the reactive mixture. We find, that the better values for P and η correspond for the case α=1. In summary, in this paper, we show that with the simple techniques of the finite-time thermodynamics, it is possible to reach some insights over the performance of internal combustion engines.

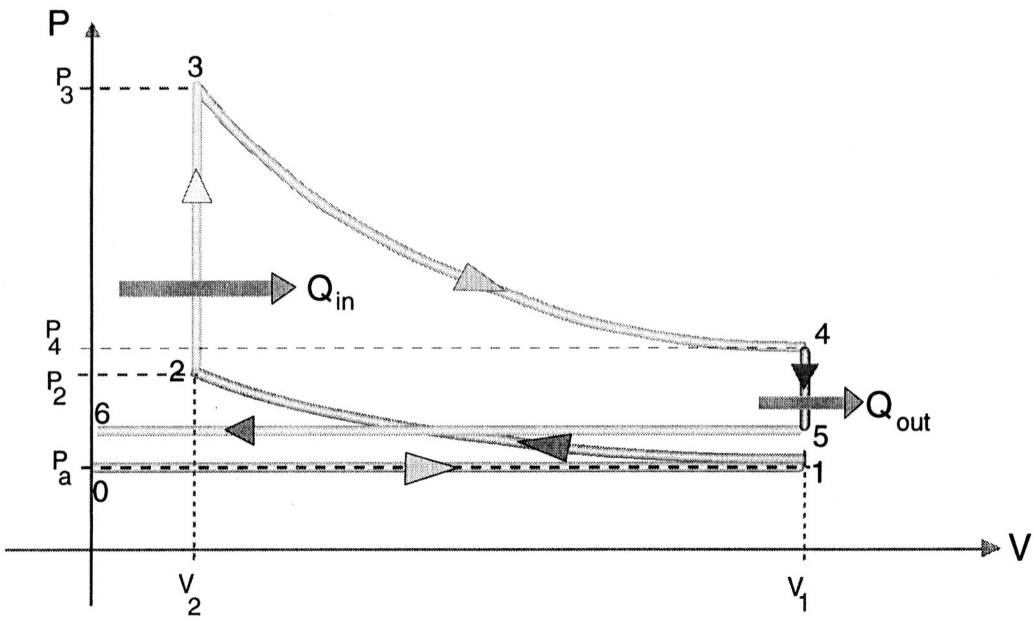

Fig.1.- Pressure - volume diagram of the Otto cycle

2.- COMBUSTION REACTION AND THERMAL CYCLE

In a real gasoline engine the chemical reaction between the constituients of the reactive mixture, that is, between air plus fuels is a very complex reaction, which leads to more than one hundred products. In our approach we only consider a typical textbook combustion reaction as the following [2],

$$C_aH_b + \alpha\left(a+\frac{b}{4}\right)(O_2 + 3.773N_2) = aCO_2 + \frac{b}{2}H_2O + \alpha\left(a+\frac{b}{4}\right)(3.773N_2) + \left(a+\frac{b}{4}\right)(\alpha-1)O_2 \tag{1}$$

where C_aH_b is a hydrocarbon fuel, and $\alpha \geq 1$ is a parameter that measures the excess of air in the unburned mixture. Both, the reactive mixture (left hand of Eq. 1) and the product mixture (right hand of Eq. 1) will be considered as ideal gas mixtures. In what follows we call gas R to the reactive mixture and gas P to the product mixture. In Fig. 1, we depict an Otto cycle model with seven corner states (0,1,...,6), which we describe as follows: in state 0, begins the admission process of gas R inside the cylinder at constant pressure (atmospheric pressure) until state 1 (maximum volume, V_1). In this state, the gas R has a number of moles given by

$$n_1 = \frac{p_1V_1}{\tilde{R}T_1} = n_R \tag{2}$$

where $\tilde{R} = 8.31451 \frac{J}{mol\ K}$ is the universal constant gas; $p_1 = 1.013\times10^5 \frac{N}{m^2}$ is the atmospheric pressure, $T_1 = 350\ K$ and we take as a typical value $V_1 = 4\times10^{-4}\ m^3$. Thus, $n_R = 0.013924\ mol$. In state 1 initiates an adiabatic compression ending in state 2 (minimum volume, V_2) . The compression ratio is

$$r = \frac{V_1}{V_2} \tag{3}$$

and the ideal adiabatic equation is $T\,V^{\gamma_1} = cte.$, where $\gamma_1 = \frac{C_{pR}}{C_{vR}}$, being C_{pR} and C_{vR} the constant-pressure and the constant-volume heat capacities of gas R, respectively. In our approach we take C_{pR} , C_{vR} and γ_1 as functions of the temperature [see next section]. By means of the adiabatic equation written in terms of the compression ratio

$$T_2 = T_1\, r^{\gamma_1 - 1} \tag{4}$$

and the ideal gas equation

$$p_2 = \frac{n_2\, \tilde{R}\, T_2}{V_2} \tag{5}$$

where $n_2 = n_R$ due to mass conservation, the state 2 turns out to be determined. During the process $2 \rightarrow 3$, the isochoric combustion takes place. Thus, in certain part of this process the conversion of gas R into gas P is carried out. In this process a quantity of heat Q_{in} is released within the working gas. We take Q_{in} given by,

$$Q_{in} = C_{V1}\left(T_3 - T_2\right) = H_P\left(T_3\right) - H_R\left(T_2\right) \tag{6}$$

where C_{V1} is an average value of C_V for gas R and gas P (see next section), and $H_{P,R}$ are the enthalpies of gas P at T_3 and gas R at T_2 respectively. By means of known polynomials [2] for the enthalpy of all gases involved in the chemical reaction [see next section] , Eq. (6) permits to obtain T_3 as a real solution of a 5th-degree algebraic equation. Thus, we calculate p_3 by means of

$$p_3 = \frac{n_3 \tilde{R} T_3}{V_2} \tag{7}$$

where n_3 is the number of moles of gas P, that is, the product mixture (see next section).

Starting in state 3, gas P undergoes an adiabatic expansion (the power stroke) ending at the volume V_1 . In this state 4, the temperature is calculated by means of the ideal adiabatic equation

$$T_4\, V_1^{\gamma_2 - 1} = T_3\, V_2^{\gamma_2 - 1}$$

which become

$$T_4 = T_3\, r^{1 - \gamma_2} \tag{8}$$

where $\gamma_2 = \frac{C_{pP}}{C_{VP}}$, being C_{pP} and C_{VP} the heat capacities of gas P, which are taken as functions of the temperature (see next section).

From state 4 to sate 5, gas P is cooled isochorically releasing an amount of heat Q_{out}, given by $Q_{out} = C_{V2}(T_4 - T_5)$, where C_{v2} is an average value of C_V for the gas P between states 4 and 5. Starting in state 5 gas P is expelled (see Fig. 1). We estimate T_5 by means of the relation $T_5 = q\ T_4$, where q is a number smaller than one, which we adjust in such a manner that the cycle's efficiency for isooctane with r = 8 be approximately 0.28 [6,7]. Thus, state 5 is finally determined by,

$$p_5 = \frac{n_5 \tilde{R} T_5}{V_1}. \tag{9}$$

where $n_5 = n_3$ due to mass conservation.

3.- Power and Efficiency Calculations

With the procedure described in section 2, we determine all the states necessary for the power and efficiency calculations. In Refs. 6 and 7, convex curves with one maximum point were obtained for power and efficiency by means of a kind of lumped friction embracing global losses. This dissipative term was taken proportional to the mean piston velocity, leading to a "loss power" of the form [6,7],

$$P_\mu = -b(r-1)^2 \tag{10}$$

where $b = \frac{\mu X_2^2}{(\Delta t_{12})^2}$, being X_2 the piston position at minimum volume; μ a coefficient of friction (which takes into account the global losses); and Δt_{12} the time spent in the power stroke. In Refs. 6 and 7, Δt_{12} was taken as $\Delta t_{12} \approx \tau/2$, being τ the cycle's period taken as fixed. This approximation was based in the hypothesis that adiabats can be treated as instantaneous processes [12]. If one uses non-instantaneous adiabats [13], assuming for example, that they evolve with the sound velocity [14], $u = (\gamma \tilde{R} T / M)^{\frac{1}{2}}$, one finds that contribution of adiabats to cycle's time is must smaller than that of non-adiabatic processes (in fact, the adiabatic time is of the order of 5% of cycle's period). Nevertheless, in the present paper, we consider that the compression stroke also contributes to the friction losses. This is accomplished by the introduction of a factor 2/R in Eq. (10), where R is the so-called parameter of non-endoreversibility given by $\mathcal{R} = \frac{C_{V1}}{C_{V2}}$ [7], and C_{V1} and C_{V2} were previously described. Thus the "loss power" turns out to be

$$P_\mu = -2b\frac{C_{V2}}{C_{V1}}(r-1)^2 \tag{11}$$

With the previous considerations, the useful power becomes

$$P = \frac{C_{V1}(T_3 - T_2) - C_{V2}(T_4 - T_5)}{\tau} - 2b\frac{C_{V2}}{C_{V1}}(r-1)^2, \tag{12}$$

Where the first term in the right hand is the "reversible power" [7]. Therefore, the cycle's thermal efficiency is given by

$$\eta = \frac{P}{Q_{in}/\tau} = \frac{P\tau}{C_{V1}(T_3 - T_2)}$$

or

$$\eta = 1 - \left[\frac{C_{V2}(T_4 - T_5)}{C_{V1}(T_3 - T_2)} + \frac{2bC_{V2}(r-1)^2}{C_{V1}{}^2(T_3 - T_2)}\right] \tag{13}$$

Now we present the further steps for the numerical procedure for calculating P and η. By means of Eq. (2) we calculate the number of moles in sates 1 and 2 for gas R. As we saw in section 1, gas R is the reactive mixture given by

$$gas\ R = C_a H_b + \alpha\left[a + \frac{b}{4}\right](O_2 + 3.7773\, N_2),$$

which has a total molar number, n_R=0.013924 mol. For calculating the contribution of each constituent of gas R to n_R, we use the following relation

$$\frac{n_i}{n_R} = \frac{N_i}{N_R},$$

where n_i is the molar number of species i in sates 1 and 2; N_i is the molar number of species i in the left hand of chemical equation (1); and N_R is the total molar number in the left hand of the chemical equation. Thus,

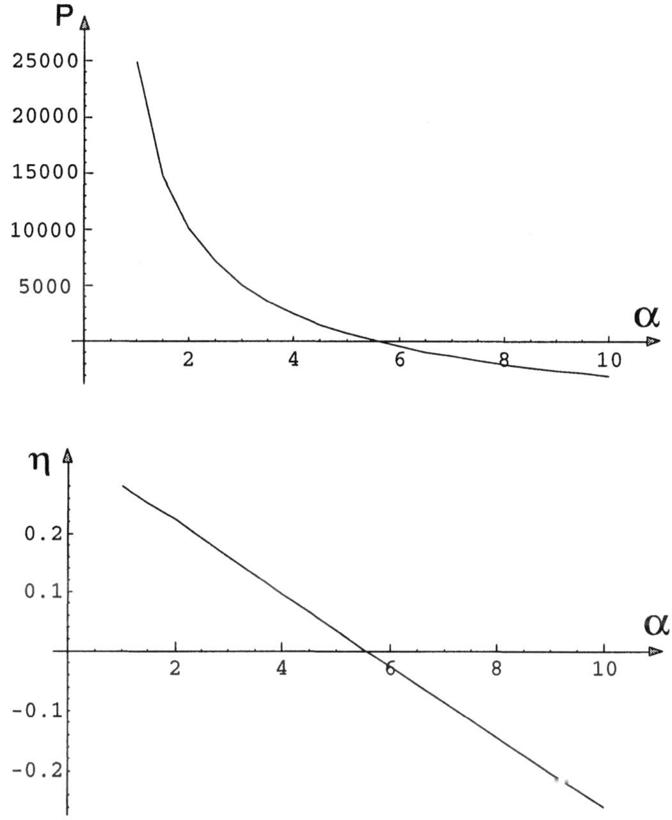

Fig.2.- P and η behavior in terms of the parameter of excess of air, α.

$$N_{C_aH_b} = 1; \qquad N_{O_2} = \left(a + \frac{b}{4}\right)\alpha; \qquad N_{N_2} = 3.773\left(a + \frac{b}{4}\right)\alpha$$

and

$$N_R = 1 + (4.773)\left(a + \frac{b}{4}\right)\alpha,$$

and for example, for n_{O_2} we have

$$n_{O_2} = \frac{(0.013924)\left(a + \frac{b}{4}\right)\alpha}{1 + 4.7731\left(a + \frac{b}{4}\right)\alpha} \tag{14}$$

For gas P (product mixture), we proceed in the same way.

For the calculation of heat capacities (and therefore for γ_1 and γ_2), we use the well known relations for gas mixtures on a molar basis [2],

$$\tilde{c}_v = \sum \tilde{x}_i \tilde{c}_{v\,i} \quad \text{and} \quad \tilde{c}_p = \sum \tilde{x}_i \tilde{c}_{p\,i} \tag{15}$$

where $\tilde{x}_i$ is the molar fraction $\frac{n_i}{n_R}$ (or $\frac{n_i}{n_P}$) and $\tilde{c}_{v\,i}$ and $\tilde{c}_{p\,i}$ are the molar heat capacities of species i.

For calculating $\tilde{c}_{p\,i}$ we use the JANAF table thermodynamic data [5] in the following manner, for the constituents of gas R we take an average between typical temperatures in the adiabatic compression, that is,

$$\left\langle \tilde{c}_p \right\rangle_i = \frac{1}{2}\left[\tilde{c}_{p\,i}(298\ K) + \tilde{c}_{p\,i}(1000\ K)\right]$$

and for constituents of gas P, we also take an average between two typical temperatures

$$\left\langle \tilde{c}_p \right\rangle_i = \frac{1}{2}\left[\tilde{c}_{p\,i}(2000\ K) + \tilde{c}_{p\,i}(3500\ K)\right].$$

The next step is to use Eqs.(15) by means of the averages values, that is

$$\tilde{c}_p = \sum_i \tilde{x}_i \left\langle \tilde{c}_p \right\rangle_i = \frac{1}{n_{(R,P)}} \sum_i n_i \left\langle \tilde{c}_p \right\rangle_i$$

therefore, the heat capacity is

$$C_{P_{(R,P)}} = n_{(R,P)} \tilde{c}_p = \sum_i n_i \left\langle \tilde{c}_p \right\rangle_i \tag{16.a}$$

in both cases, (gases R and P).

For calculating C_V we use the well known relation

$$C_{V\,R} = C_{p\,R} - n_R \tilde{R}$$

and

$$C_{V\,P} = C_{p\,P} - n_P \tilde{R}\,. \tag{16.b}$$

By using Eqs. (15) and (16), we immediately obtain

$$\gamma_1 = \frac{C_{p\,R}}{C_{V\,R}}$$

and

$$\gamma_2 = \frac{C_{p\,P}}{C_{V\,P}}. \tag{17}$$

Evidently, $C_{p\,P}$, $C_{V\,P}$, $C_{p\,R}$, $C_{V\,R}$, γ_1 and γ_2 are all functions of the excess parameter α, through Eqs. as(14), and (16). In the process $2 \rightarrow 3$ (see Fig. 1), we estimate C_{V1} as the average between C_V of gas R in state 2 and C_V of gas P in state 3, that is,

$$C_{V\,1} \approx \frac{1}{2}\left[C_{V\,R} + C_{V\,P}\right]$$

and for the process $4 \rightarrow 5$, we take $C_{V\,2} = C_{V\,P}$, that is, the heat capacity of the products of the combustion reaction. For the same reasons mentioned above, also $C_{V\,1}$ and $C_{V\,2}$ are functions of α.

For calculating T_3 after the chemical reaction we proceed as follows, first we take Q_{in} given by

$$Q_{in} = C_{V1}\left(T_3 - T_2\right) = H_P\left(T_3\right) - H_R\left(T_2\right), \tag{18}$$

where H_p and H_R are the enthalpies of gases P and R in states 3 and 2 respectively. For the calculation of these enthalpies we use the empirical relationships for molar enthalpies given in Ref. [2], which are,

$$\tilde{h}_i(T) = \tilde{R}T\left[a_{i1} + \frac{a_{i2}}{2}T + \frac{a_{i3}}{3}T^2 + \frac{a_{i4}}{4}T^3 + \frac{a_{i5}}{5}T^4 + \frac{a_{i6}}{T}\right] \tag{19}$$

for O_2, N_2, CO_2, and H_2O, and

$$\tilde{h}_f(T) = A_{f1}t + \frac{A_{f2}}{2}t^2 + \frac{A_{f3}}{3}t^3 + \frac{A_{f4}}{4}t^4 - \frac{A_{f5}}{t} + A_{f6} + A_{f8} \tag{20}$$

with $t = \frac{T\,(K)}{1000}$, for the hydrocarbons of the type C_aH_b. All coefficients a_i and A_f are given in tables 4.10 and 4.11 of Ref. [2], respectively.

By using Eqs. (19), (20) and $\tilde{h} = \sum_i \tilde{x}_i \tilde{h}_i$, we calculate the enthalpy of gas R and gas P as,

$$H_R(T) = n_R \, \tilde{h}_R(T) \tag{21}$$

and

$$H_P(T) = n_P \, \tilde{h}_P(T) .$$

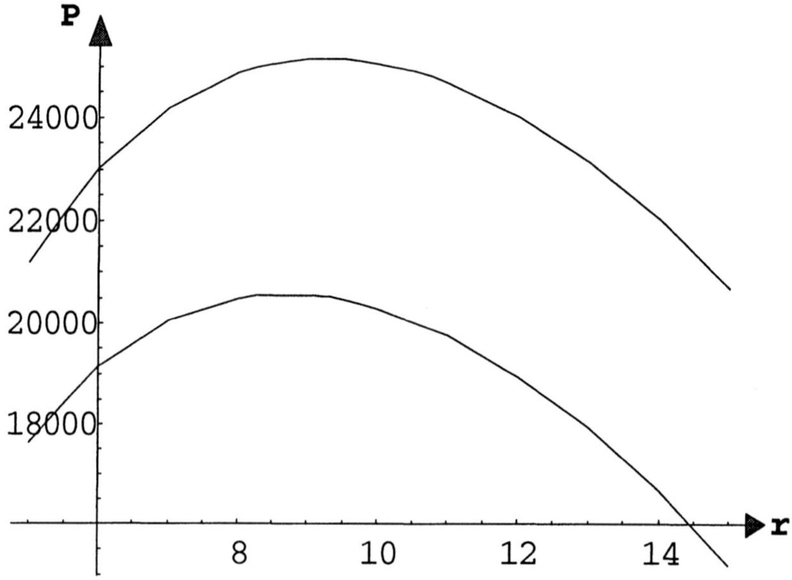

Fig.3.- Comparison of the power as function of r (α=1) for isooctane (upper curve) and methane (lower curve).

When we substitute Eqs. (21) in Eq. (18), we obtain a polynomial of fifth degree in the variable T_3. We numerically solve this equation for determining T_3. Thus, taking $b = 32.5 \ W$ [6,7], now we have all data for the calculation of the power P [Eq. (12)] and efficiency η [Eq. (13)]. The parameter q mentioned at the end of section 2 results q=0.1. In the next section we show numerical calculations of P and η as functions of r and α for two hydrocarbons, CH_4 (methane) and C_8H_{18} (isooctane).

4.- NUMERICAL RESULTS AND CONCLUSIONS

As an example of the procedure described in the previous section, we show in Tables I and II the performance data of isooctane (C_8H_{18}) and methane (CH_4) for r=8 in both cases. In Fig. 2, we see (for isooctane) that both P and η decrease with increasing α, decaying both to zero values at $\alpha \approx 5.8$. This same behavior is observed for the methane case and for any r in both cases. In Fig. 3 we depict P vs r-curves for both fuels (α=1) and it can be seen that the power values for C_8H_{18} are above of the corresponding values of CH_4. This same behavior occurs for η vs r-curves (Fig. 4). In Fig. 5, we see that loop-shaped P vs. η-curves result in our model, as it is common in real engines [10]. In Fig. 3, the compression ratios where P reach a maximum value are $r_1^* = 8.75$ for methane and $r_2^* = 9.25$ for isooctane. In Fig. 4, the corresponding values of r for maximum efficiency are $r_1^+ = 9$ for methane and $r_2^+ = 9.5$ for isooctane. This fact permits the obtention of loop-shaped curves as that of Fig. 5. In general, for all α and r in their physical ranges (P>0 and η >0), P and η for isooctane are above of the corresponding values for methane. A remarkable fact is that $T_1, T_2, \ldots, T_5$ for both fuels have very small differences (below 5 percent, see Tables I and II), however their power and efficiency (Fig. 3 and 4) have noticeable differences (above 10 percent for P and above 7 percent for η). In our calculations we find that T_3, that is, the temperature after the combustion reaction is practically insensitive to the chemical differences between CH_4 and C_8H_{18}, nevertheless , as we mentioned P and η are sensitive to the different fuels. In our opinion, this fact is due to the differences in the heat capacities of both substances. That is, seemingly, the engine-performance is dominated for the caloric properties of the fuels, such as we suggested in a previous paper [7].

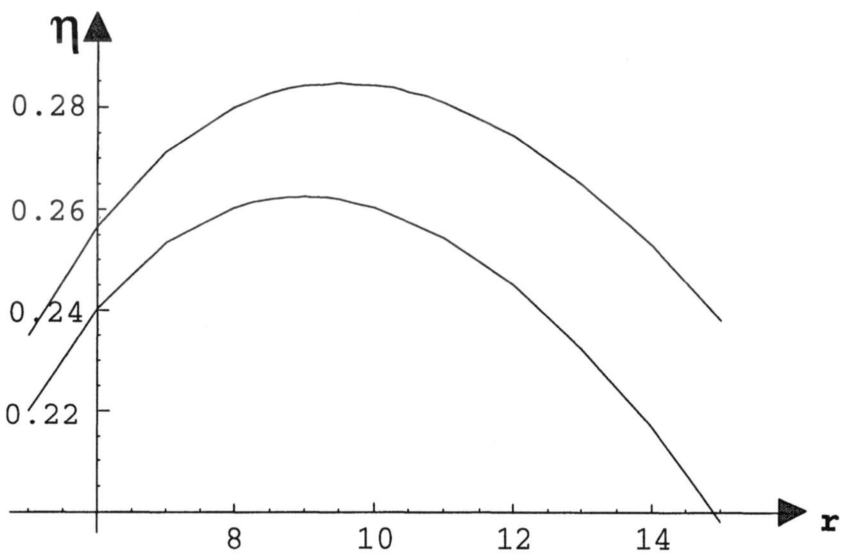

Fig.4.- Comparison of the efficiency as function of r (α=1) for isooctane (upper curve) and methane (lower curve).

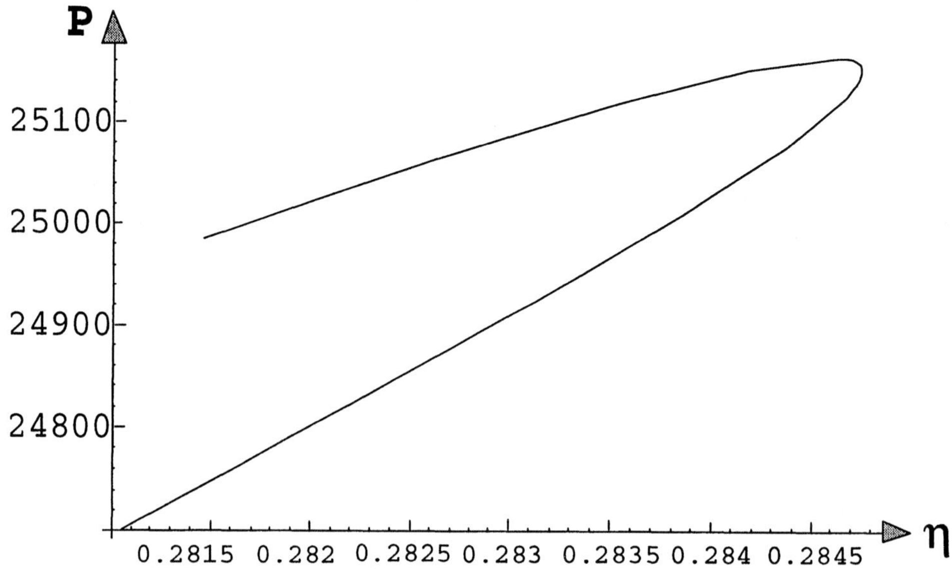

Fig.5.- Part of a loop-shaped P vs. η curve for isooctane (α=1).

a	T_2	G_{v1}	T_3	g_1	T_4	g_2	G_{v2}	T_5	P	h
1	663.691	0.437705	7364.61	1.30772	4422.02	1.2453	0.499185	431.694	24886.4	0.280001
1.5	686.842	0.412203	5382.73	1.32421	3152.25	1.25732	0.467316	307.735	14764.	0.251704
2.	699.82	0.399346	4465.46	1.33321	2578.88	1.26402	0.451249	251.761	10148.9	0.222713
2.5	708.121	0.391598	3873.74	1.33888	2217.36	1.26829	0.441567	216.467	7200.18	0.191672
3.	713.887	0.386418	3456.92	1.34278	1966.61	1.27126	0.435093	191.988	5135.95	0.159899
3.5	718.124	0.382711	3146.61	1.34563	1782.	1.27343	0.430461	173.965	3605.86	0.128031
4.	721.371	0.379927	2906.01	1.3478	1640.05	1.2751	0.426982	160.108	2423.36	0.0963502
4.5	723.937	0.377759	2713.54	1.3495	1527.25	1.27641	0.424273	149.096	1479.75	0.0649713
5.	726.017	0.376024	2555.71	1.35088	1435.23	1.27748	0.422104	140.113	707.498	0.0339349
5.5	727.736	0.374603	2423.68	1.35202	1358.6	1.27836	0.420328	132.632	62.4813	0.0032455
6.	729.182	0.373418	2311.4	1.35297	1293.67	1.2791	0.418847	126.293	-485.343	-0.0271083
6.5	730.414	0.372415	2214.59	1.35379	1237.86	1.27973	0.417594	120.845	-957.145	-0.0571452
7.	731.476	0.371555	2130.15	1.35449	1189.32	1.28027	0.416519	116.106	-1368.28	-0.0868861
7.5	732.402	0.370809	2055.77	1.35509	1146.66	1.28075	0.415587	111.942	-1730.18	-0.116352
8.	733.216	0.370156	1989.68	1.35563	1108.84	1.28116	0.414772	108.249	-2051.52	-0.145564
8.5	733.937	0.36958	1930.52	1.3561	1075.04	1.28153	0.414052	104.95	-2339.01	-0.17454
9.	734.58	0.369068	1877.21	1.35652	1044.64	1.28186	0.413412	101.982	-2597.95	-0.203298
9.5	735.158	0.36861	1828.88	1.3569	1017.12	1.28216	0.412839	99.2955	-2832.54	-0.231854
10.	735.679	0.368197	1784.85	1.35724	992.087	1.28242	0.412324	96.8514	-3046.21	-0.260223

Table I.- Thermodynamic and performance data for isooctane (r=8)

a	T_2	G_{v1}	T_3	g_1	T_4	g_2	G_{v2}	T_5	P	h
1	699.106	0.41075	7027.34	1.33272	4226.74	1.24448	0.473544	422.674	20507.8	0.260359
1.5	712.482	0.395045	5339.09	1.34183	3132.25	1.25647	0.451412	313.225	13184.1	0.238043
2.	719.909	0.386802	4463.02	1.34682	2581.67	1.26324	0.439795	258.167	9287.11	0.211676
2.5	724.633	0.381723	3888.4	1.34997	2229.	1.26759	0.432638	222.9	6686.24	0.182702
3.	727.903	0.37828	3479.47	1.35213	1982.03	1.27063	0.427786	198.203	4815.36	0.152668
3.5	730.3	0.375793	3172.78	1.35371	1798.95	1.27287	0.42428	179.895	3402.08	0.122315
4.	732.133	0.373911	2933.7	1.35492	1657.46	1.27458	0.421629	165.746	2294.56	0.0919845
4.5	733.579	0.372438	2741.64	1.35587	1544.58	1.27594	0.419553	154.458	1401.38	0.0618355
5.	734.75	0.371253	2583.65	1.35663	1452.24	1.27704	0.417884	145.224	664.29	0.0319366
5.5	735.718	0.37028	2451.14	1.35727	1375.15	1.27795	0.416512	137.515	44.5168	0.00231279
6.	736.53	0.369467	2338.22	1.3578	1309.71	1.27872	0.415365	130.971	-484.762	-0.0270326
6.5	737.222	0.368776	2240.71	1.35825	1253.38	1.27938	0.414392	125.338	-942.685	-0.0561072
7.	737.819	0.368183	2155.54	1.35864	1204.32	1.27994	0.413556	120.432	-1343.28	-0.0849235
7.5	738.338	0.367667	2080.42	1.35898	1161.16	1.28043	0.41283	116.116	-1697.08	-0.113496
8.	738.795	0.367216	2013.61	1.35927	1122.86	1.28087	0.412193	112.286	-2012.13	-0.141841
8.5	739.199	0.366816	1953.75	1.35954	1088.61	1.28125	0.411631	108.861	-2294.71	-0.169972
9.	739.56	0.366461	1899.77	1.35977	1057.78	1.28159	0.41113	105.778	-2549.79	-0.197905
9.5	739.884	0.366143	1850.8	1.35998	1029.86	1.2819	0.410681	102.986	-2781.36	-0.225652
10.	740.176	0.365856	1806.16	1.36017	1004.44	1.28218	0.410277	100.444	-2992.65	-0.253225

Table II.- Thermodynamic and performance data for methane (r=8)

ACKNOWLEDGMENT

This work was partially supported by COFAA-IPN and Instituto Mexicano del Petróleo.

REFERENCES

1.- Zemansky M.W. and Dittman R.H., 1987, Heat and Thermodynamics (New York, Mc Graw-Hill)

2.- Heywood J.B., 1988, Internal Combustion Engine Fundamentals (New York, Mc Graw-Hill)

3.- Taylor C.F. 1968, The Internal Combustion Engine in Theory and Practice, Vol. 1 (Cambridge Mass , MIT Press)

4.- Faires V.G., 1962, Thermodynamics (London, Mc Millan)

5.- Decher Reiner, Energy Conversion Oxford University Press New York 1994.

6.- Angulo-Brown F. Fernández-Betanzos J. and Díaz-Pico C.A., 1994, Eur. J. Phys. 15,38

7.- Angulo-Brown F., Rocha-Martínez J.A., and Navarrete-González T.D., 1996, J. Phys. D: Appl. Phys. 29, 80

8.- Calvo Hernández A., Medina A., Roco J.M.M. and Velasco S., 1995, Eur. J Phys. 16,73

9.- Calvo Hernández A., Roco J.M.M., Medina A. and Velasco S., 1996, Eur. J. Phys. 17,11

10.- Gordon J.M. and Huleihil M., 1992, J. Appl. Phys. 72,829;Chen L., Sun F. and Wu C. 1997, Energy Convers. Mgmt., 38,1501

11.- Mozurkewich M. and Berry R.S., 1982, J. Appl. Phys. 53,34

12.- Rubin M.H., 1979, Phys. Rev. A 19,1272

13.- Agrawal D.C., Gordon J.M. and Huleihil M.,1994, Indian J. Eng. and Mat. Sci. 1,195
14.- Callen H.J. 1988, Thermodynamics and an Introduction to Thermostatics (New York, Wiley).

FINITE-TIME CHARACTERIZATION OF IRREVERSIBLE THERMALLY-DRIVEN HEAT PUMPS

M. DENTICE D'ACCADIA, M. SASSO AND S. SIBILIO
DETEC - Università degli Studi di Napoli "Federico II"
P. le Tecchio, 80 - 80125 Napoli, Italy

ABSTRACT

The coupling of a heat-driven prime mover with a heat pump is studied with a finite-time thermodynamic approach. Recovery of engine waste heat is taken into account. An endoreversible model is developed, that permits evaluation of the influence of design variables, such as thermal resistance in heat exchangers, on system performance. The search for optimum design conditions is included, from both thermodynamic and thermoeconomic points of view.

Key words: endoreversible thermodynamics, heat engine-driven heat pump, optimum design, thermoeconomics.

INTRODUCTION

In the field of finite-time thermodynamics, FTT, considerable research efforts have been spent on finding more realistic limits on the optimal operation of heat machines in finite time. From a FTT point of view, optimum conditions are those in which the maximum energy output in a finite time is supplied. Great emphasis has therefore been put on the problem of obtaining the maximum work flow rate from a heat engine. The most common thermodynamic cycles, such as Otto, Brayton, cascade, combined cycles, have been modelled thermodynamically assuming a finite rate of heat interaction [1 - 12]. The optimal performance of endoreversible refrigerators and heat pumps has been investigated, too [13 - 20].

In this paper, the coupling of a prime mover, PM, with a vapour-compression heat pump, HP, is considered (heat engine-driven heat pump). If only mechanical link between PM and HP is taken into account, without considering the use of engine waste heat, the optimal performance of the whole plant (maximum cooling or heating capacity) corresponds to that of the PM (maximum shaft work). However, one of the most interesting features of a heat engine-driven heat pump is the availability of waste heat from engine coolant and exhaust gas, for heating purposes and/or hot-water production. When engine waste heat is recovered, optimal performance conditions are different from those of the simple mechanical link case, and are influenced by several variables.

In this paper, referring to a heat engine-driven heat pump recovering waste engine heat, and assuming that irreversibility is only due to heat transfer under finite temperature differences (endoreversible model), according to the finite-time thermodynamic approach, performance of the plant and its main subsystems, Prime Mover and Heat Pump, has been analyzed as a function of heat exchangers thermal resistances and temperature levels of surrounding heat reservoirs [21 - 23].

This approach leads to interesting design consideration, such as the optimal allocation of the heat exchanger inventory in the plant as a whole and in each of its main subsystems.

Furthermore, the irreversible model developed in this paper opens the way to thermoeconomic considerations about the optimal trade-off between capital cost, proportional to the size, and therefore to the thermal conductance, and operating cost, that, at fixed desired thermal output, decreases when thermal conductance, and therefore efficiency, rise.

ENDOREVERSIBLE THERMODYNAMIC MODEL

In figure 1 a schematic of a heat transformer, HT, is shown. The system is made up by two main subsystem, a Prime Mover, PM, and a Heat Pump, HP, driven by the heat engine. In the heating mode, the HT takes heat from high-level and low-level heat sources, and delivers the useful thermal flow rates, $\dot{Q}_M^{PM}$ and $\dot{Q}_M^{HP}$, to the medium temperature sinks, at temperatures T_M^W and T_M^S, respectively. In the cooling season, the useful thermal flow rate, $\dot{Q}_L^{HP}$, is drawn from the low-level heat source and the engine waste heat, $\dot{Q}_M^{PM}$, is recovered to supply hot water (heat reservoir at temperature T_M^W). Usually, the low temperature reservoir simulates the external environment during the heating mode, and the indoor ambient during cooling mode.

An example of real machine that operates similarly to the system in figure 1 is the Gas engine driven Heat Pump, GHP, shown in figure 2. In this case, since in the engine an internal combustion is realized, some problem occurs in defining the temperature T_H of the high level reservoir. However, if we imagine that the combustion process is substituted by a heat transfer to the PM working fluid, plausible values of T_H may be obtained starting from the adiabatic flame temperature related to the fuel used in the engine.

When the system supplies heat for two different purposes, the Heat Transformer operates among four different heat reservoirs (*quadrithermal system*). In many applications, the heat output of the vapour compression loop (condenser) and the engine waste heat are both used for heating the same space; consequently, the system operates with only one heat sink at medium temperature, $T_M^S = T_M^W = T_M$ (*trithermal system*).

Both cases (quadrithermal and trithermal Heat Transformers) will be considered in the following, with the finite-time thermodynamic approach.

Heat transfers will be considered as governed by linear equations (Newton law). This permits to take into account the relation between heat transfer rate and temperature difference and to overcame the limit of Classical Thermodynamics, that considers thermal equilibrium during heat transfer interactions, no realistic in actual applications. The heat flow rates in the four heat exchangers will be therefore expressed as follows:

$$\dot{Q}_H = C_H\left(T_H - \overline{T}_H\right) \tag{1}$$

$$\dot{Q}_M^{PM} = C_M^{PM}\left(\overline{T}_M^{PM} - T_M^W\right) \tag{2}$$

$$\dot{Q}_M^{HP} = C_M^{HP}\left(\overline{T}_M^{HP} - T_M^S\right) \tag{3}$$

$$\dot{Q}_L = C_L\left(T_L - \overline{T}_L\right) \tag{4}$$

where C_H, C_M^{PM}, C_M^{HP} and C_L are the thermal conductances in the four heat exchangers, defined as the product of the heat transfer area by the overall heat-transfer coefficient based on that area.

In order to determine the optimum allocation of the global conductance, C_G, sum of the conductances of all heat exchangers, among PM and HP, and then the best distribution among the four heat exchangers, the following allocation factors may be introduced:

- z among PM and HP ($z = C^{PM}/C^{HP}$);
- x among PM heat exchangers ($x = C_M^{PM} / C_H$);
- y among HP heat exchangers ($y = C_L / C_M^{HP}$).

Furthermore, the temperature ratios τ and τ_C may be introduced [4]:

$$\tau = \frac{T_H}{T_M^W}\ , \qquad \tau_c = \frac{\bar{T}_H}{\bar{T}_M^{PM}}$$

By applying first and second law of thermodynamics to the plant as a whole and to its main subsystems, under the hypothesis of endoreversible cycles for both Prime Mover and Heat Pump, that work according to Carnot cycles, it may be derived that all thermal and work flow rates in figure 1 are a function of the following variables [22, 23]:

$$T_M^W, T_M^S, T_L, C_G, x, y, z, \tau, \tau_c$$

While in the case of power plant the maximum of the desired output $(\dot{W})$ is independent from the conductances [1, 4], in this case the maximum of the overall energy output $\dot{Q}_M^{HT} = \dot{Q}_M^{PM} + \dot{Q}_M^{HP}$ is strongly influenced by their values, so that the analysis of the problem is quite complicated. The choice of the values to be assigned to x and y and z must be guided by the aim of maximizing $\dot{Q}_M^{HT}$. It must be observed that the Prime Mover contributes to the desired effect in two ways: directly, by means of its waste heat, $\dot{Q}_M^{PM}$, and indirectly, by means of the shaft work supplied to the Heat Pump.

Since $\dot{Q}_M^{HP}$ is a monotone increasing function of $\dot{W}$, the best case occurs if $\dot{W}$ and $\dot{Q}_M^{PM}$ achieved their maximum for the same value of x. This situation is verified for $x = x_{opt}=1/2$, according to [4]. Referring to the Heat Pump, the maximum of $\dot{Q}_M^{HP}$, for a fixed $\dot{W}$, is achieved for $y = y_{opt}=1/2$, again.

Therefore, for x=1/2 and y=1/2, the aim of optimizing the total heat flow rate $\dot{Q}_M^{HT}$ is satisfied. From the previous analysis, for fixed x and y, we have:

$$\dot{Q}_M^{HT}, \dot{Q}_M^{PM}, \dot{Q}_M^{HP}, \dot{W} = f(T_M^W, T_M^S, T_L, C_G, z, \tau, \tau_c) \tag{5}$$

Since all work and heat flow rates vary linearly with the global conductance C_G, the corresponding quantities per unit C_G may be introduced if one wants to reduce the number of independent variables to be considered. Some example will be provided in the following, referring to both quadrithermal and trithermal systems.

For a thorough examination of the analytical model assumed to represent the Heat Transformer in figure 1, the complete set of equations (5) is reported in references [22, 23], together with all equations and mathematical passages needed to derive them.

Quadrithermal Thermodynamic System

In this case, referring to the heating mode, the Heat Transformer supplies heat to two different heat reservoirs. The first one, at temperature T_M^W, uses heat recovered from the engine, the second one, at temperature T_M^S, is served by the Heat-Pump condenser In figure 3, assuming $x=x_{opt}$, $y=y_{opt}$, z=0.9, τ=1000/294, T_M^S=294 K, T_M^W=310 K, T_L=280 K, the overall heat flow rate per unit C_G supplied by the HT is shown as a function of τ_c, together with the desegregated contributes of HP condenser and PM waste heat. In the same figure the COP, calculated as the ratio of $\dot{Q}_M^{HT}$ to $\dot{Q}_H^{PM}$, is shown too.

It is interesting to stress the contradictions between the choice of optimizing heat output (Finite-Time path) and that of finding the better COP conditions (Classical path). For $\tau_C = 1$ the PM doesn't supply any work to the HP. The entire heat flow rate supplied to the medium temperature is therefore due to the waste engine heat, that equals the HT input, i.e. the PM just works as a thermal resistance between two heat reservoirs. Consequently, the COP reaches its minimum, COP = 1.

When τ_C increases, the temperature differences in the PM heat exchangers tend to zero; then, the working fluids tend to a perfect thermal equilibrium with each temperature reservoir. From a finite-time point of view, system performance, that is: heat and work flow rates actually supplied, decrease, but COP increases up to its higher limit (for $\tau_C = \tau$). Optimum HT performance (maximum heat flow rate supplied) is achieved for a $\tau_C = \tau_{opt} = 1.4$; unfortunately, the corresponding COP is very low. The results provided by the model, for $\tau_C = \tau_{opt}$, compare well with performance of actual GHP models designed to supply heat for both space heating and hot water production. For example, the ratio of the HP thermal output to the PM thermal output is about 1.66, instead of the actual average value of 1.7, and the COP value provided by the model is about 1.90, while it is about 2.00 for actual GHPs.

In figure 3 the well known Novikov, Curzon and Ahlborn optimum value, for $\tau_C = \tau_{NCA}$ [1, 2], is shown too: this is the operating condition that maximizes the work output of the Prime Mover and therefore the useful output of the Heat Pump. It may be noted that τ_{NCA} value is quite different by τ_{opt}, as a consequence of the fact that also engine waste heat is used for heating purposes.

In the example developed for the quadrithermal system, a fixed value of parameter z has been assumed. The dependance of Heat Transformer performance on z will be analyzed in the case of trithermal system.

Trithermal Thermodynamic System

The simplest type of Heat Transformer, is considered, steadily operating among three different heat sources and/or heat sinks ($T_M^W = T_M^S = T_M$).

In this case, the influence of the conductance distribution, z, on the maximum output provided to the end user may be investigated. In figure 4 the influence of z and the temperature ratio τ_C on HT performance is shown. The overall thermal output per unit C_G and the COP^{HT} are both referred to their upper limits, that are, respectively:

- $\dot{Q}^{HT}_{M,max}$, attained for z=0 and τ_C=1;
- the maximum Carnot COP, COP_C^{HT}, attained for $\tau_C = \tau$.

Both these limit conditions are unattainable for real plants. For z=0 the thermal resistances in the HP heat exchangers tend to the infinity and consequently only the PM can work. For $\tau_C = \tau$, as previously stated, in spite of the fact that the overall COP reaches its maximum limit of 15.1, the heat output is zero.

Another limit condition occurs when z = 1: the PM thermal resistances tend to ihe infinity, so that neither PM nor HP can operate.

The comparison between figures 3 and 4 is interesting: as previously noted, the region of high thermal output is opposite to the region of high COP, so, a high value of the overall heat flow rate may be achieved only with low efficiency.

In order to find a reasonable range of variation for the independent variables of figure 4, we have to take into account that:

- as shown in figure 3, the optimum operation condition occurs, for each value of z, at low value of τ_C; in [23] it has been evaluated that the HT usually achieves the desired optimum output in a range of τ_C from 1 to 1.53;
- as a consequence of the high temperature difference in the PM heat exchangers, the global conductance is mainly allocated in the HP, so that a realistic range of variation for z is 0.8 to 0.95.

In this way it is possible to locate on the surfaces of figure 4 two small fields corresponding to possible operation conditions for actual plants. In these fields, the heat delivered by the Heat Transformer is less than 30% of $\dot{Q}^{HT}_{M,max}$ and the efficiency is less than 20% of COP_C.

THERMOECONOMIC OPTIMIZATION OF THE ENDOREVERSIBLE MODEL

As an example of further application, thermoeconomic optimization of the endoreversible Heat Transformer of figure 1 is dealt with, referring to the case of trithermal system operating in the heating mode.

The overall cost of the thermal output per unit energy, π_{out}, may be expressed as:

$$\pi_{out} = \frac{\pi_{in}}{COP^{HT}} + \frac{\varsigma \times \Phi}{E_t} \qquad (6)$$

where π_{out} is the unit cost of the input energy, Φ the overall cost of the plant, ζ the capital recovery factor and E_t the thermal energy supplied every year by the equipment.

Other costs, such as maintenance cost, are assumed not to be affected by system design: they will be therefore neglected.

We assume that the optimization only aims at choosing the best value of the global conductance C_G, for a fixed value of the maximum thermal output $\dot{Q}_M^{HT}$. Therefore, we will assume that the Heat Transformer cost, Φ, only depends on C_G, according to the following relation:

$$\Phi = \Phi_0 \times (C_G / C_{G,0})^n \tag{7}$$

where Φ_0, $C_{G,0}$ and n are constant.

If $\dot{Q}_M^{HT}$ is fixed, when C_G increases both COP^{HT} and Φ rise. Therefore, π_{out} has a minimum for a given thermal conductance, that corresponds to an optimum condition from a thermoeconomic point of view.

As an example, we consider a Heat Transformer for which the following hypotheses are assumed: $\dot{Q}_M^{HT}$ = 100 kW, x = y = 1/2, T_H = 1000 K, T_M = 294 K, T_L = 280 K, E_t = 360 GJ/year, Φ_0 = 20,000 \$, $C_{G,0}$ = 1.0 kW/K and n = 0.75. Furthermore, a capital recovery factor ζ = 0.0814 is assumed, that corresponds to a ten-year life cycle and to a discount rate of 5.0 %. Referring to these assumptions, figure 5 shows the maximum COP^{HT} for various C_G, starting from the minimum value needed for a thermal output of 100 kW (about 0.62 kW/K). These maximum values have been obtained from the mathematical model symbolically expressed by equations (5), by choosing, for each C_G, the value of z which maximizes the HT Coefficient Of Performance.

In figure 6 the dependence of π_{out} on the thermal conductance C_G is shown, for different values of π_{in}. As previously stated, for a fixed unit cost of the energy input, an optimal value of the global conductance C_G may be determined, which minimizes the unit cost of thermal output. For example, if π_{in} is 20 \$/GJ, we get the minimum π_{out} for $C_G \cong 2.5$ kW/K, which corresponds to a maximum COP^{HT} of about 1.9.

CONCLUSIONS

Attention has been paid to the search for the optimum operation conditions of a heat engine-driven heat pump with engine waste heat recovery. To get closer to actual machines, the influence of irreversible heat transfer in the Prime Mover and Heat Pump heat exchangers have been considered to develop an endoreversible finite-time model. The results provided by the model compare well with performances of actual Gas engine driven Heat pumps, showing that the finite-time approach provides a reference more realistic than the traditional analysis.

With the help of the above mentioned model, design considerations have been made about the influence of thermal resistances and their distribution on plant performance. As a further example of application, thermoeconomic optimization of an endoreversible system has been investigated, too.

NOMENCLATURE

C conductance of heat exchanger, kW/Km^2

COP Coefficient Of Performance

E Energy, GJ/year

$\dot{Q}$ thermal flow rate, kW

$\dot{W}$ work transfer rate, kW

T	absolute temperature, K
$\overline{T}$	thermodynamic mean temperature, K
x	Prime Mover conductance (C^{PM}) allocation ratio
y	Heat Pump conductance (C^{HP}) allocation ratio
z	global conductance (C_G) allocation ratio

Greek symbols

ζ	capital recovery factor
π	unit economic cost, \$/GJ or \$/kWh
τ	temperature ratio, $\frac{T_H}{T_M^W}$
τ_C	prime mover Carnot temperature ratio, $\frac{\overline{T}_H}{\overline{T}_M^{PM}}$

Subscripts

C	referred to Carnot cycle
G	referred to the global conductance
H	referred to high temperature thermal reservoir or to high level heat exchanger
in	input
L	referred to low temperature thermal reservoir or to low level heat exchanger
M	referred to medium temp. thermal reservoir or to medium level heat exchanger
max	referred to the maximimum heat output
0	reference or environmental value
opt	optimum
out	output
t	thermal

Superscripts

HP referred to the vapour compression Heat Pump

HT referred to the Heat Transformer

PM referred to the Prime Mover

S referred to the hot water

W referred to the room (heating) or to the environment (cooling)

REFERENCES

[1] I.I. Novikov, *Atomnaya Engergiya*, **3**, (11), 409 (1957), English translation in *J. Nuclear Energy II*, **7**, 125 (1958).

[2] F.L. Curzon, B. Ahlborn, *American Journal Phys.* **43**, 22 (1975).

[3] P. Salamon , Y.B. Band, O. Kafri, *Journal Applied Phys.* **53**, 197 (1982)

[4] A. Bejan, *International Journal of Heat & Mass Transfer,* **31**, 1211(1988)

[5] G. Grazzini, Maximum work with isothermal exchange irreversible heat engine. In *Proc. Florence World Energy Research Symposium FLOWERS '90,* (Edited by S.S. Stecco, M.J. Moran), 473-484, Pergamon Press, New York (1990)

[6] C. Wu, *Energy Convers. Mgmt* **30**, 261 (1990)

[7] C. Wu, *Energy Convers. Mgmt* **32**, 249 (1991)

[8] C. Wu, *Energy Convers. Mgmt* **34**, 1239 (1993)

[9] C. Wu, D.A. Blank, *Energy Convers. Mgmt* **34**,1255 (1993)

[10] A. De Vos, *Energy Convers. Mgmt* **36**, 1 (1995)

[11] C. Wu, W.H. Schulden, *Heat Recovery Systems & CHP* **15**, 13 (1995)

[12] C.Wu, *Heat Recovery Systems & CHP* **15**, 351 (1995)

[13] Z. Yan, J. Chen, Coefficient of performance for an irreversible Carnot heat pump, In *Proc. Florence World Energy Research Symposium FLOWERS '90,* (Edited by S.S. Stecco, M.J. Moran), 193-200, Pergamon Press, New York (1990).

[14] C. Wu, *Int. J. of Ambient Energy* **13**, 133 (1992)

[15] C. Wu, *Heat Recovery Systems & CHP* **13**, 161 (1993)

[16] S.A. Klein, *Int. J. Refrigeration* **15**, 181 (1992)

[17] G. Grazzini, *Int. J. Refrigeration* **16**, 101 (1993)

[18] C. Wu, *Energy Convers. Mgmt* **34**, 1249 (1993)

[19] C. Wu, *Energy Convers. Mgmt* **36**, 7 (1995)

[20] C. Wu., J. Chen, *Int. J. of Ambient Energy* **16,** 17 (1995)

[21] M. Dentice d'Accadia, M. Sasso, S. Sibilio, *Thermodynamics and the Design, analysis, and Improvement of Energy Systems,* (Edited by R.J. Krane), 263, AES, **33**, ASME (1994)

[22] M. Dentice d'Accadia, M. Sasso, S. Sibilio, In *Proc. ECOS'96 Symposium Efficiency, Cost, Optimization, Simulation and environmental aspects of energy systems,* (Edited by P. Alvfors, L. Eidensten, G. Svedberg, J. Yan), 135 (1996)

[23] M. Dentice d'Accadia, M. Sasso, S. Sibilio, *Energy Conversion and Management* **38**, 401 (1997)

[24] E.P. Gyftopoulos, G.P. Beretta, *Thermodynamics Foundations and Applications,* Mac Millan C., 586-613 (1991)

[25] S. Kawasaki et al., *IEA Heat Pump Centre Newsletter,* 8 (3), 11 (1990).

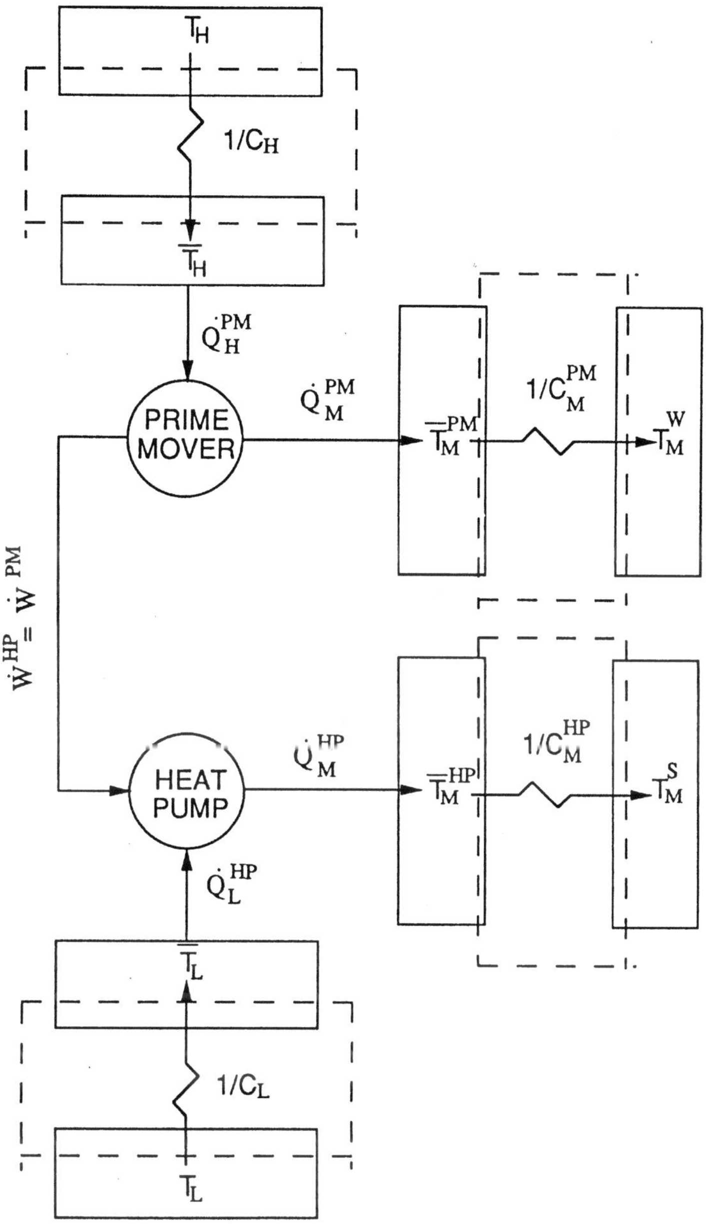

Figure 1 - Schematic of a Heat Pump moved by a thermally-driven Prime Mover.

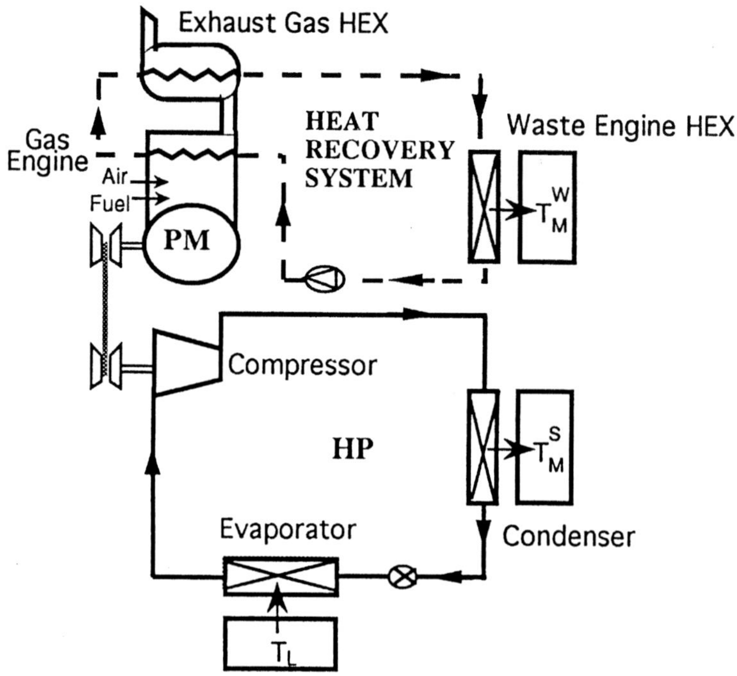

Figure 2 - Schematic of a Gas engine driven Heat Pump, GHP.

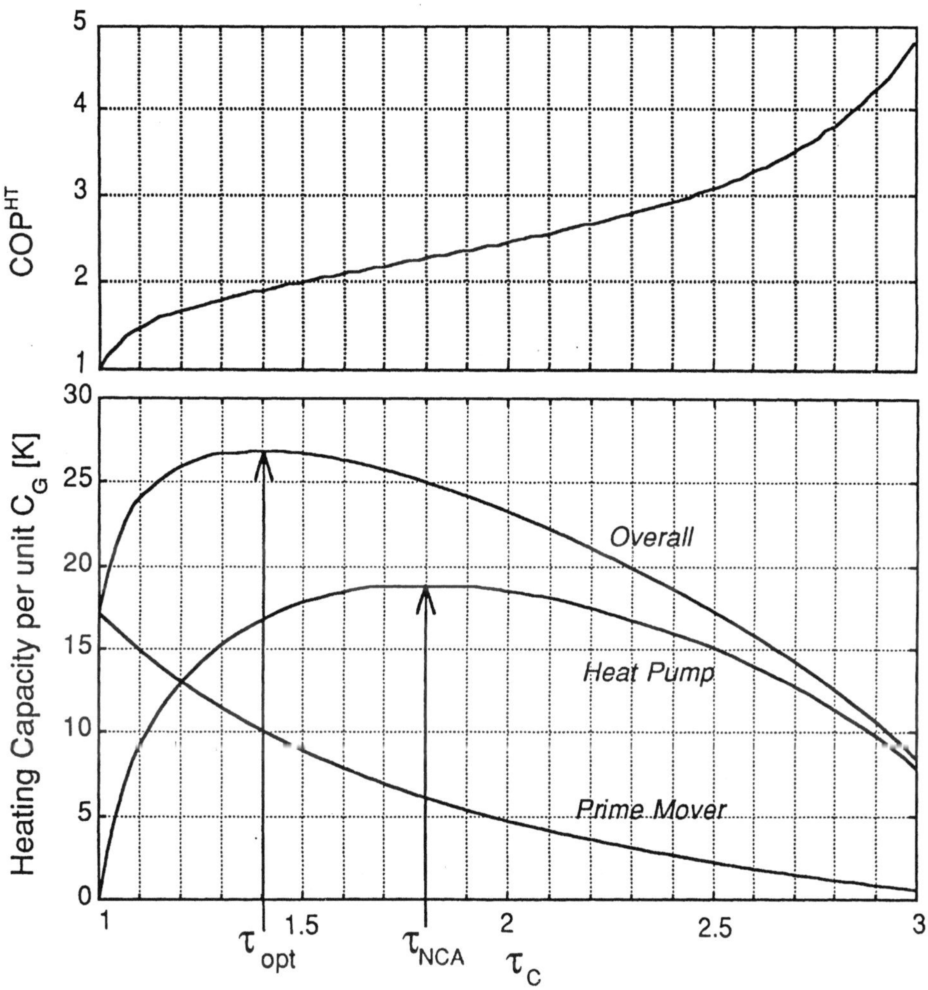

Figure 3 - Heat flow rates per unit C_G and COP of a quadrithermal system vs. τ_C ($x=x_{opt}$, $y=y_{opt}$, $z=0.9$, $\tau=1000/294$, $T_M^S=294$ K, $T_M^W=310$ K, $T_L=280$ K).

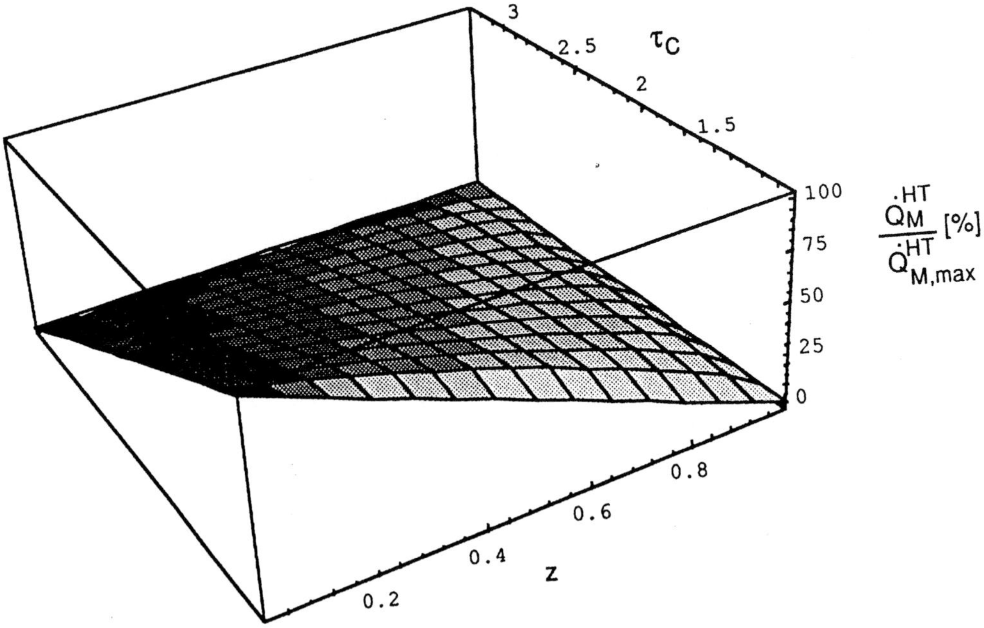

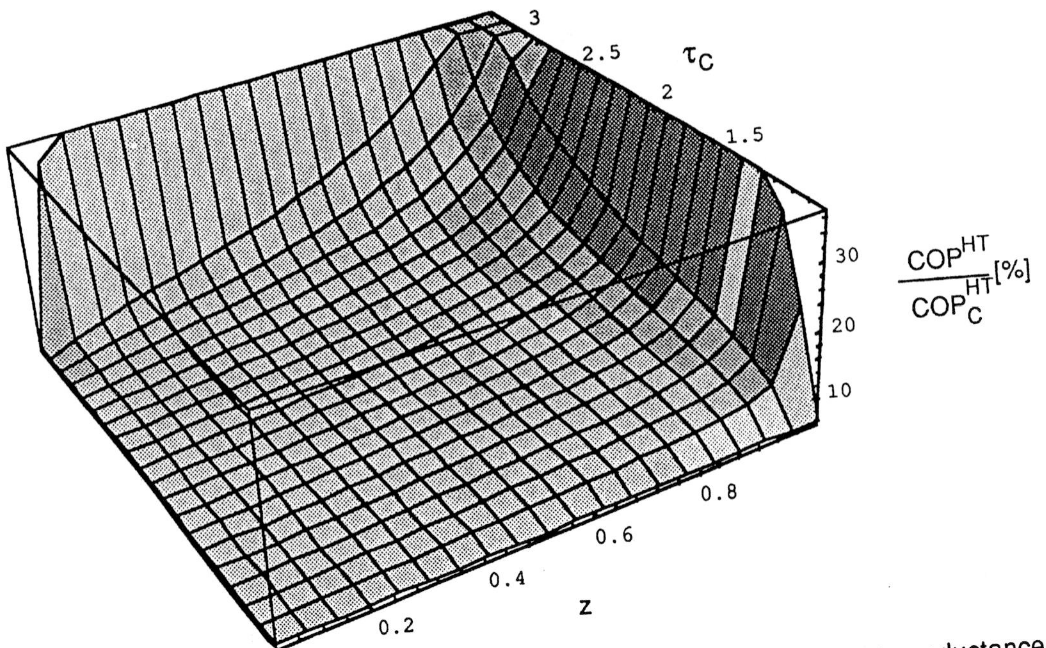

Figure 4 - Dimensionless HT heat output and COP vs. τ_C and global conductance allocation ratio values, z for a thrithermal system.

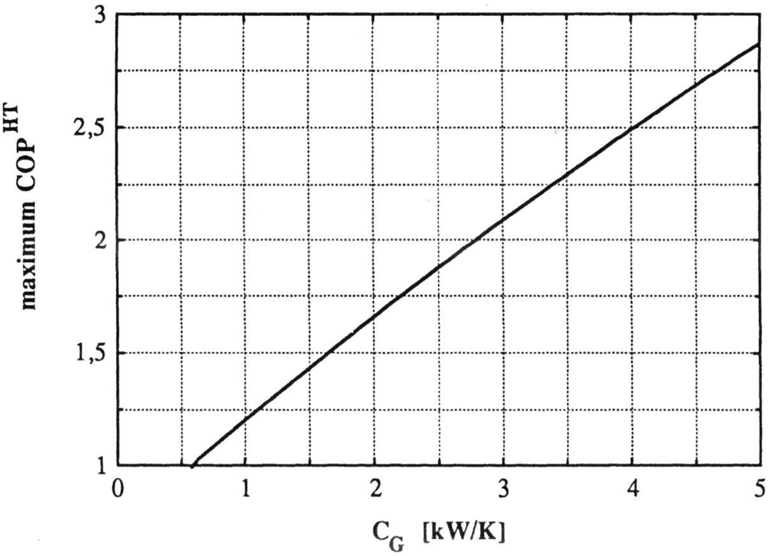

Figure 5 - Maximum COP^{HT} vs C_G. $\dot{Q}_M^{HT}$ = 100 kW, x = y = 1/2, T_H = 1000 K, T_M = 294 K, T_L = 280 K.

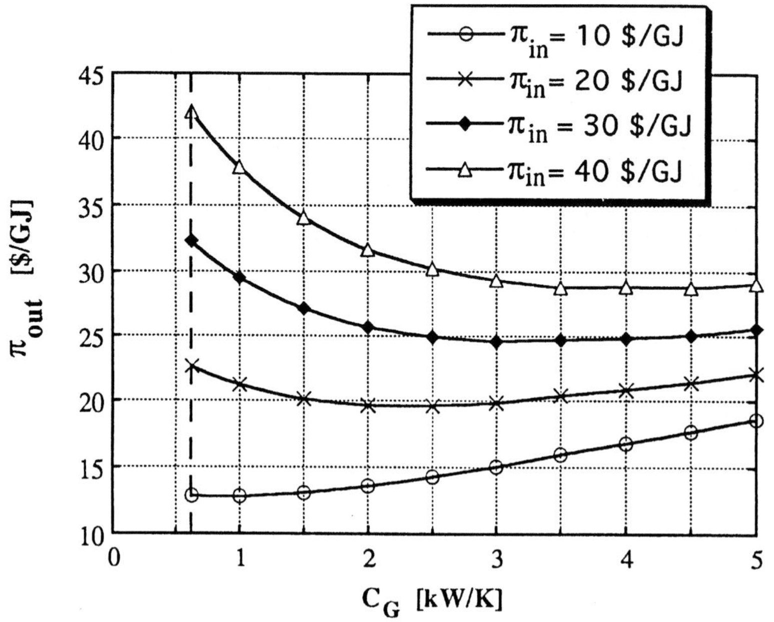

Figure 6 - Unit cost of the thermal output vs C_G. $\dot{Q}_M^{HT}$ = 100 kW, x = y = 1/2, T_H = 1000 K, T_M = 294 K, T_L = 280 K, E_t = 360 GJ/year, Φ_0 = 20,000 \$, $C_{G,0}$ = 1.0 kW/K, n = 0.75, ζ = 0.0814.

A GENERALIZED MODEL OF A COMBINED HEAT PUMP CYCLE AND ITS PERFORMANCE

LINGEN CHEN
Faculty 306, Naval Academy of Engineering, Wuhan 430033,
People's Republic of China

YUEHONG BI
Thermal Energy Engineering Department, Beijing Polytechnic University, Beijing 100022, People's Republic of China

and

CHIH WU
Mechanical Engineering Department, U.S. Naval Academy,
Annapolis, Md 21402, USA

Abstract-A steady flow combined heat pump cycle model with heat resistance, heat leakage and internal irreversibility is built in this paper. The optimal performance of the model is studied. The relation between optimal heating load and coefficient of performance (COP), as well as the maximum COP and the corresponding heating load are derived.

Keywords: combined heat pump cycle, performance, optimization

NOMENCLATURE

A a function defined in Eq.(42)
B a function defined in Eq.(31)
B_1 a function defined in Eq.(43)
B_2 a function defined in Eq.(48)
B_3 a function defined in Eq.(49)
C_i internal heat conductance (kW/K)

F_T total heat transfer surface area (m^2)
F_1 heat transfer surface area of high-temperature side heat exchanger (m^2)
F_2 heat transfer surface area of intermediate-temperature heat exchanger (m^2)
F_3 heat transfer surface area of low-temperature side heat exchanger (m^2)
K_i overall heat transfer coefficient of i th stage heat exchanger [$kW/(km^2)$]
K_1 overall heat transfer coefficient of high-temperature heat exchanger [$kW/(km^2)$]
K_2 overall heat transfer coefficient of intermediate-temperature heat exchanger [$kW/(km^2)$]
K_3 overall heat transfer coefficient of low-temperature side heat exchanger [$kW/(km^2)$
M a function defined in Eq.(27)
M_{min} minimum value of M
n counting number defined in Eq.(43)
P total power input (kW)
P_1 power input to topping cycle (kW)
P_2 power input to bottom cycle (kW)
Q_H rate of heat transfer to heat sink (heating load) (kW)
q rate of heat leak (kW)
Q_L rate of heat removed from heat source (kW)
Q_1 rate of heat transfer from topping cycle to heat sink in the high-temperature side heat exchanger(kW)
Q_2 rate of heat transfer from bottom cycle to topping cycle (kW)
Q_2' rate of heat transfer from bottom cycle to topping endoreversible cycle (kW)
Q_3 rate of heat transfer from heat source to bottom cycle in the low-temperature side heat exchanger (kW)
Q_3' rate of heat transfer from heat source to bottom endoreversible cycle in the low-temperate side heat exchanger (kW)
T_H temperature of heat sink (K)
T_L temperature of heat source (K)
T_1 temperature of working fluid in the warm-side of topping cycle (K)
T_2 temperature of working fluid in the cold-side of topping cycle (K)
T_3 temperature of working fluid in the warm-side of bottom cycle (K)
T_4 temperature of working fluid in the cold-side of bottom cycle (K)
U_T total heat exchanger inventory (kW/K)
U_1 heat exchanger inventory of high-temperature side heat exchanger (kW/K)
U_2 heat exchanger inventory of intermediate-temperature heat exchanger (kW/K)
U_3 heat exchanger inventory of low-temperature side heat exchanger (kW/K)
x heat transfer surface area ratio defined in Eq.(25)
x_{opt} optimum value of x
y heat transfer surface area ratio defined in Eq.(25)
y_{opt} optimum value of y

z heat transfer surface area ratio defined in Eq.(25)
z_{opt} optimum value of z
β coefficient of performance (COP)
β_{max} maximum COP
π heating load (kW)
π_m heating load at the maximum COP point (kW)
π_{max} maximum heating load (kW)
π_{min} minimum heating load (kW)
Φ_i degree of internal dissipation of i th stage refrigeration cycle
Φ_T total degree of internal dissipation of combined cycle
Φ_1 degree of internal dissipation of topping cycle
Φ_2 degree of internal dissipation of bottom cycle

INTRODUCTION

According to classical thermodynamics, the performance of a combined heat pump cycle is identical to that of a single-cycle operating in the same temperature range. For example, when a reversible combined cycle heat pump operates between hot and cold reservoirs at temperatures T_H and T_L, respectively, its COP is equal to the COP of a reversible Carnot heat pump operating in the same temperature range. However, the use of reversible processes as standards of performances in industry is not desirable because a reversible process must be carried out at an infinitesimally slow pace. The high performance criterion of an ideal reversible heat pump is seldom approached.

The consequence of incorporating finite-time processes into an otherwise ideal thermodynamic cycle was elegantly demonstrated by Novikov (1957) [1], Chambadal (1957) [2] and Curzon and Ahlborn (1975) [3]. They considered the case of finite rates of heat transfer to and from a Carnot heat engine. After maximizing the power output, they derived a simple expression for the efficiency that was different from the well-known Carnot efficiency. Since finite-time thermodynamics or entropy generation minimization was advanced, many authors have studied the effect of irreversibilities on the performance of thermodynamic processes and cycles. Some detailed literature surreys were given in several books by Andresen [4], Tsirlin [5], Feidt [6], Sieniutycz and Salamon [7], De Vos [8], Radcenco [9], Bejan [10], and Bejan, Tsatsaronis and Moran [11] and some review articles by Andresen etal [12-14], Orlov etal [15], Wu [16,17], Chen etal [18,19],Sieniutycz and Shiner [20] and Bejan[21-24]. Several researchers including Chen and Wu [25] and Chen etal [26-28] have investigated effects of finite rate of heat transfer, together with other major irreversibilities on the performance of combined heat pump cycles using endoreversible (heat resistance) model [25,26,27] and heat resistance and heat leak model [28]. There exist different characteristic for various irreversibilities effects on the performance of single-cycle heat pumps [29-37] and combined-cycle heat pump [28]. Recently, Chen etal [38-40] presented a generalized irreversible Carnot heat pump model by taking additional account of several internal irreversibilities of the heat pumps. This was done by means of a constant parameter and a constant coefficient, together with loss of heat resistance. It is the purpose of this paper to present a more complete combined heat pump model, including both internal and external irreversibilities of the combined plant by extending the Chen' s work [38-40] to the combined-cycle heat pumps, and to analyze the optimal performance of a combined heat pump system affected by the irreversibility of finite rate heat transfer, internal dissipation of the working fluid and the bypass heat leakage.

MODEL OF THE COMBINED HEAT PUMP PLANT

The real combined heat pump cycle formed by two steady irreversible cycles and its surrounding heat reservoirs are shown in Fig. 1. The combined cycle operates between two reservoirs at temperatures T_H and T_L ($T_H > T_L$). The temperatures of the working fluid in the top irreversible cycle are T_1 and T_2, when heat leaves and enters the working fluid. In the bottom irreversible cycle, these temperatures are T_3 and T_4 .The following temperature relation hold:

$$T_1 > T_H > T_3 > T_2 > T_L > T_4 \quad (1)$$

Heat exchange between the two irreversible cycles is carried out directly through the same heat transfer surface area so that the quantity of heat absorbed by the working fluid in the top cycle is equal to that rejected by the working fluid in the bottom cycle. The irreversibilities are constant rate of bypass heat leak q which flows from the heat sink at temperature T_H to heat source at temperature T_L, the internal dissipation of the working fluids in the top and bottom cycles, and the three heat transfers caused by the three finite temperature differences ($T_1 - T_H$), ($T_3 - T_2$) and ($T_L - T_4$) required by the heat rejection and absorption rates.

The first law of thermodynamics requires

$$Q_1 = Q_H + q = \pi + q \quad (2)$$

$$Q_3 = Q_L + q \quad (3)$$

where Q_1 and Q_3 are the rates of heat flow from the top cycle to the heat sink in the high-temperature side heat exchanger and from the heat source to the bottom cycle in the low-temperature side heat exchanger of the combined cycle respectively, Q_H and Q_L are the heat transfer rate to the heat sink (heating load, π) and the heat removed rate from the heat source , respectively.

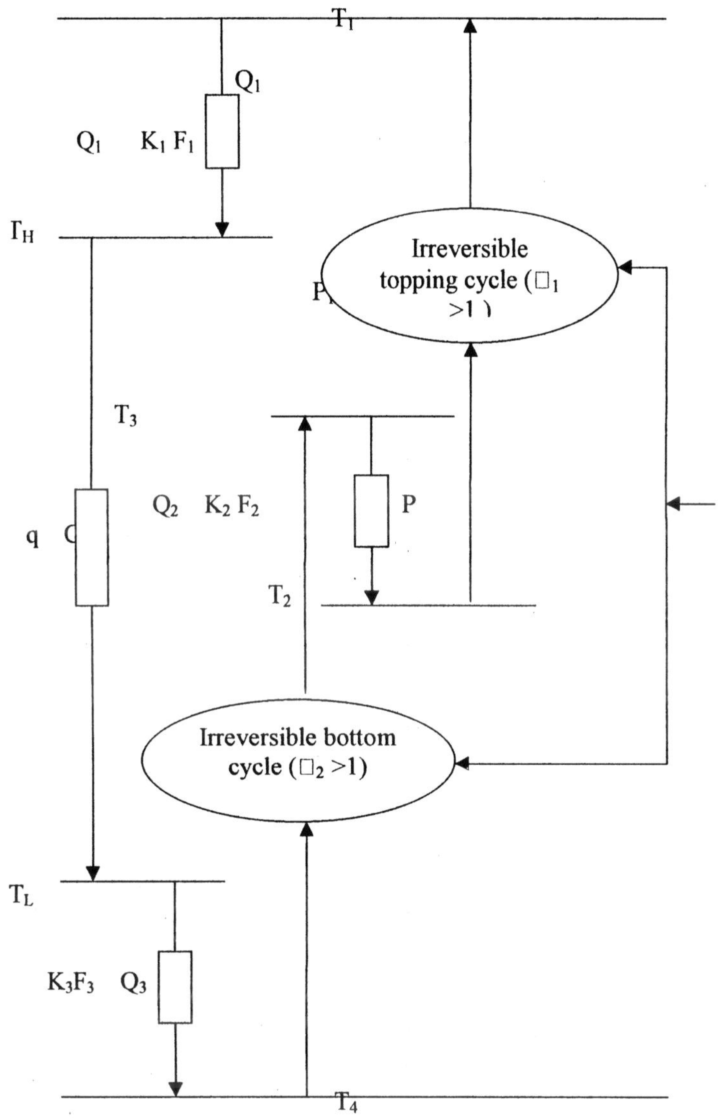

Figure 1. Generalized combined heat pump cycle with all irreversibilities

The internal dissipation including miscellaneous factors such as friction, turbulence and non-equilibrium activities inside the two cycles are characterized by two constant coefficients [38-40]

$$\Phi_1 = \frac{Q_2^{'}}{Q_2} \geq 1 \tag{4}$$

$$\Phi_2 = \frac{Q_3^{'}}{Q_3} \geq 1 \tag{5}$$

where Q_2 is the rate of heat exchange between the top and bottom irreversible cycles in the intermediate heat exchanger, $Q_2^{'}$ is the rate of heat absorbed from the bottom cycle to the top endoreversible cycle in the intermediate heat exchanger, and $Q_3^{'}$ is the rate of heat absorbed from the heat sink to the bottom endoreversible cycle in the low-temperature side heat exchanger. Because the power input required by the real heat pump is larger than that of the endoreversible one with the same heating load, the rate of heat absorbed from the heat source to the cold working fluid for the real cycle is less than that for the endoreversible one. Thus, $Q_2^{'} > Q_2$ and $Q_3^{'} > Q_3$ hold.

For the top endoreversible heat pump, the second law of thermodynamics requires that (no matter whether the bottom heat pump is endoreversible)

$$\frac{Q_2^{'}}{Q_1} = \frac{T_2}{T_1} \tag{6}$$

For the bottom endoreversible heat pump, the second law of thermodynamics requires that (no matter whether the topping heat pump is endoreversible)

$$\frac{Q_3^{'}}{Q_2} = \frac{T_4}{T_3} \tag{7}$$

Combining Eqs.(4), (5), (6) and (7) gives

$$\frac{Q_2}{Q_1} = \frac{T_2}{\Phi_1 T_1} \tag{8}$$

$$\frac{Q_3}{Q_2} = \frac{T_4}{\Phi_2 T_3} \tag{9}$$

The power input, P, to the combined heat pump is

$$P=P_1+P_2=Q_H-Q_L=Q_1-Q_3 \tag{10}$$

where P_1, P_2 and P are the power input to the topping and bottom cycles and the total power input to the combined heat pump.

The COP of the combined heat pump plant is

$$\beta=\frac{Q_H}{P}=\frac{\pi}{P}=\frac{1-q/Q_1}{1-Q_3/Q_1} \tag{11}$$

Assuming that the rates of heat flow in the heat exchangers follow Newton's law, then

$$Q_1=(T_1-T_H)K_1F_1 \tag{12}$$

$$Q_2=(T_3-T_2)K_2F_2 \tag{13}$$

$$Q_3=(T_L-T_4)K_3F_3 \tag{14}$$

where K_1 is the overall heat transfer coefficient and F_1 is the heat transfer surface area of the high-temperature (hot) side heat exchanger between the topping cycle and the heat sink, K_2 is the overall heat transfer coefficient and F_2 is the heat transfer surface area of the intermediate heat exchanger between the topping and bottom cycle, and K_3 is the overall heat transfer coefficient and F_3 is the heat transfer surface area of the low-temperature (cold) side heat exchanger between the bottom cycle and the heat source.

The total heat transfer surface area (F_T) of the combined cycle is assumed to be a constant

$$F_T=F_1+F_2+F_3 \tag{15}$$

ANALYSIS AND OPTIMIZATION

Solving Eq.(12) for T_1

$$T_1=T_H+\frac{Q_1}{K_1F_1} \tag{16}$$

Substituting Eqs.(13) and (16) into Eq.(8) yields

$$T_2 = \frac{T_3[\Phi_1 T_H + Q_1/(K_1F_1/\Phi_1)]}{\Phi_1 T_H + Q_1[(K_2F_2)^{-1} + (K_1F_1/\ \Phi_1)^{-1}]} \tag{17}$$

Combining Eqs.(13), (14) and (9) gives

$$T_4 = \frac{K_3F_3T_LT_3}{T_3(K_3F_3 + K_2F_2/\Phi_2) - T_2K_2F_2/\Phi_2} \tag{18}$$

Substituting Eq.(17) into Eq.(18) yields

$$T_4 = \frac{T_L\{\Phi_T T_H + Q_1[(K_1F_1/\Phi_T)^{-1} + (K_2F_2/\Phi_2)^{-1}]\}}{\Phi_T T_H + Q_1[(K_1F_1/\Phi_T)^{-1} + (K_2F_2/\Phi_2)^{-1} + (K_3F_3)^{-1}]} \tag{19}$$

where Φ_T is the total degree of internal dissipation in the combined system.

$$\Phi_T = \Phi_1\Phi_2 \tag{20}$$

Solving Eq.(14) for T_4

$$T_4 = T_L - \frac{Q_3}{K_3F_3} \tag{21}$$

Combining Eqs.(19) and (21) gives

$$\frac{Q_1}{Q_3} = \frac{\Phi_T T_H + Q_1[(K_3F_3)^{-1} + (K_2F_2/\ \Phi_2)^{-1} + (K_1F_1/\ \Phi_T)^{-1}]}{T_L} \tag{22}$$

Substituting Eq.(22) into Eqs. (11) yields

$$\beta = \frac{\pi}{\pi + q}\left\{1 - \frac{T_L}{\Phi_T T_H + (\pi + q)[(K_3F_3)^{-1} + (K_2F_2/\Phi_2)^{-1} + (K_1F_1/\Phi_T)^{-1}])}\right\}^{-1} \tag{23}$$

Equation (23) provides a general relation between the COP and heating load of the combined plant. For given values of T_H , T_L, Φ_1, Φ_2, q, K_1, K_2, K_3, F_T and • , • is a function of the distribution among F_1, F_2 and F_3. Equation (23) indicates that the maximization of • is equivalent to the minimization of M, as defined by Eq.(24)

$$M = (K_3F_3)^{-1} + (K_2F_2/\ \Phi_2)^{-1} + (K_1F_1/\ \Phi_T)^{-1} \tag{24}$$

Define three heat transfer surface area ratios x, y and z,

$$x = \frac{F_1}{F_T}, y = \frac{F_2}{F_T}, z = \frac{F_3}{F_T} \tag{25}$$

thus

$$F_1 = xF_T, F_2 = yF_T, F_3 = zF_T, z = 1 - x - y \tag{26}$$

and

$$M = \frac{\Phi_T}{xK_1F_T} + \frac{\Phi_2}{yK_2F_T} + \frac{\Phi_1}{(1-x-y)K_3F_T} \tag{27}$$

To find the minimum M from Eq.(27), taking the partial derivative of M with respect to x and y and setting them equal to zero ($\partial M/ \partial x = 0$ and $\partial M/ \partial y = 0$) gives the optimal heat transfer surface area distribution (χ_{opt}, y_{opt} and z_{opt}) and the minimum M (M_{min}) as follows:

$$\chi_{opt} = \left(\frac{F_1}{F_T}\right)_{opt} = \left(\frac{\Phi_T}{K_1}\right)^{\frac{1}{2}} \left[\frac{1}{(K_3)^{\frac{1}{2}}} + \left(\frac{\Phi_2}{K_2}\right)^{\frac{1}{2}} + \left(\frac{\Phi_T}{K_1}\right)^{\frac{1}{2}}\right]^{-1} \tag{28}$$

$$y_{opt} = \left(\frac{F_2}{F_T}\right)_{opt} = \left(\frac{\Phi_2}{K_2}\right)^{\frac{1}{2}} \left[\frac{1}{(K_3)^{\frac{1}{2}}} + \left(\frac{\Phi_2}{K_2}\right)^{\frac{1}{2}} + \left(\frac{\Phi_T}{K_1}\right)^{\frac{1}{2}}\right]^{-1} \tag{29}$$

$$z_{opt} = \left(\frac{F_3}{F_T}\right)_{opt} = \left(\frac{1}{K_3}\right)^{\frac{1}{2}} \left[\frac{1}{(K_3)^{\frac{1}{2}}} + \left(\frac{\Phi_2}{K_2}\right)^{\frac{1}{2}} + \left(\frac{\Phi_T}{K_1}\right)^{\frac{1}{2}}\right]^{-1} \tag{30}$$

and

$$M_{\min} = \left[\left(\frac{1}{K_3}\right)^{\frac{1}{2}} + \left(\frac{\Phi_2}{K_2}\right)^{\frac{1}{2}} + \left(\frac{\Phi_T}{K_1}\right)^{\frac{1}{2}}\right]^2 \frac{1}{F_T} = B \tag{31}$$

Substituting Eq.(31) into Eq.(23) yields

$$\beta = \frac{(\Phi_T T_H + qB)\pi + B\pi^2}{B\pi^2 + (\Phi_T T_H - T_L + 2qB)\pi + (\Phi_T T_H - T_L + qB)q} \tag{32}$$

Equation (32) is the optimal performance characteristics of the generalized model of real combined heat pump plants with irreversibilities of heat resistance, heat leakage and internal dissipation. This equation determines the optimal COP for the given π. Equation (32) also indicates that $\beta=1$ when $\pi=\pi_{max}$ and $\pi=\pi_{min}$, where

$$\pi_{\min} = \frac{\Phi_T q T_H}{T_L - qB} - q \tag{33}$$

There exists a maximum COP point. To find the maximum COP from Eq.(32), taking the derivative of with respect to π and setting it equal to zero ($d\beta/d\pi=0$) gives the maximum COP (β_{max}) and the corresponding heating load (π_m) as follows:

$$\beta_{\max} = \left\{ \frac{\Phi_T T_H + qB}{(qBT_L)^{1/2} + [\Phi_T T_H (\Phi_T T_H - T_L + qB)]^{1/2}} \right\}^2 \tag{34}$$

and

$$\pi_m = \frac{[\Phi_T T_H T_L (\Phi_T T_H - T_L + qB) qB]^{1/2} + (\Phi_T T_H - T_L + qB) qB}{B(\Phi_T T_H + qB)} \tag{35}$$

Combining Eqs. (11) and (32) gives the optimal heating load and the optimal COP for a given total power input (P), and the minimum total power input for a given heating load in the following forms:

$$\pi = \frac{\{[\Phi_T T_H - T_L - B(P-2q)]^2 + 4B[T_L q + (\Phi_T T_H + qB)(P-q)]\}^{1/2} - [\Phi_T T_H - T_L - B(P-2q)]}{2B} \tag{36}$$

$$\varepsilon = \frac{\{[\Phi_T T_H - T_L - B(P-2q)]^2 + 4B[T_L q + (\Phi_T T_H + qB)(P-q)]\}^{1/2} - [\Phi_T T_H - T_L - B(P-2q)]}{2BP} \tag{37}$$

and

$$P = \frac{B\pi^2 + (\Phi_T T_H - T_L + 2qB)\pi + (\Phi_T T_H - T_L + qB)q}{\Phi_T T_H + qB + B\pi} \tag{38}$$

When the combined system obtains the optimal COP for a given π, the temperatures of the working fluid in the isothermal processes must satisfy some optimal relations. Substituting Eqs. (28) - (31) into Eqs. (16), (17) and (19) yields

$$T_1 = T_H + (\pi + q)A[1/(\Phi_T K_1)]^{1/2} \tag{39}$$

$$\frac{T_3}{T_2} = \frac{\Phi_T T_H + (\pi + q)A[(\Phi_2/K_2)^{1/2} + (\Phi_T/K_1)^{1/2}}{\Phi_T T_H + (\pi + q)A(\Phi_T/K_1)^{1/2}} \tag{40}$$

$$T_4 = T_L \frac{\Phi_T T_H + (\pi + q)A[(\Phi_2/K_2)^{1/2} + (\Phi_T/K_1)^{1/2}]}{\Phi_T T_H + (\pi + q)B} \tag{41}$$

where

$$A = \frac{1}{F_T}\left[\left(\frac{\Phi_T}{K_1}\right)^{1/2} + \left(\frac{\Phi_2}{K_2}\right)^{1/2} + \left(\frac{1}{K_3}\right)^{1/2}\right] \tag{42}$$

The COP versus heating load of the generalized irreversible combined heat pump plant is a parabolic one. This qualitative behavior of COP versus heating load characteristics appears to be common for real heat pumps [41-43].

DISCUSSION

For a combined heat pump cycle, which is formed by several irreversible heat pump cycles, with bypass heat leakage, the optimal performance characteristics have the same expressions as Eqs.(32)- (42), simply replace B by B_1 from i=1 to i=n+1

$$B_1 = \frac{1}{F_T}\left\{\sum_{i=1}^{n+1}\left(\prod_{j=i}^{n+1}\Phi_j/\ K_i\right)^{\frac{1}{2}}\right\}^2 \tag{43}$$

where n is the number of stages of the combined cycle, K_i is the heat transfer coefficient for the ith stage heat exchanger between working fluids or between working fluid and the reservoirs, and Φ_j is the internal dissipation for the jth stage heat pump cycle and $\Phi_{n+1}=1$ is taken.

For the combined cycle, the optimization problem can be extended by distributing the total heat exchanger inventory U_T [27,28] among heat exchanger sizes: $U_1 = K_1F_1$, $U_2 = K_2 F_2$, and $U_3 = K_3 F_3$.

The total heat exchanger inventory (U_T) of the combined cycle is assumed to be a constant

$$U_T = U_1 + U_2 + U_3 \tag{44}$$

Using the above constraint equation, the optimal distribution of the heat exchanger inventory is obtained in the following expressions:

$$(\frac{U_1}{U_T})_{opt} = \frac{(\Phi_T)^{\frac{1}{2}}}{1+(\Phi_2)^{\frac{1}{2}}+(\Phi_T)^{\frac{1}{2}}} \tag{45}$$

$$(\frac{U_2}{U_T})_{opt} = \frac{(\Phi_2)^{\frac{1}{2}}}{1+(\Phi_2)^{\frac{1}{2}}+(\Phi_T)^{\frac{1}{2}}} \tag{46}$$

and

$$(\frac{U_3}{U_T})_{opt} = \frac{1}{1+(\Phi_2)^{\frac{1}{2}}+(\Phi_T)^{\frac{1}{2}}} \tag{47}$$

The optimal performance characteristics have the same expressions as Eqs.(32) – (42) again, simply replace B by B_2

$$B_2 = \frac{1}{U_T}[1+(\Phi_2)^{\frac{1}{2}}+(\Phi_T)^{\frac{1}{2}}]^2 \tag{48}$$

For a combined cycle formed by n stages of irreversible heat pump cycles, B should be replaced by B_3

$$B_3 = \frac{1}{U_T}\left\{\sum_{i=1}^{n+1}[\prod_{j=i}^{n+1}\Phi_j]^{\frac{1}{2}}\right\}^2 \tag{49}$$

CONCLUSION

In the design of combined heat pump cycle systems, knowledge of the upper bound of the COP is highly desirable. This paper presents a generalized irreversible combined heat pump cycle model by using heat transfers, heat leakage and internal dissipation to simulate the energy exchange and irreversibilities. An analytical investigation of the performance of the model is undertaken. The COP versus heating characteristics, and the maximum COP and the corresponding heating load are derived. The optimal distributions of the heat transfer surface area or the heat transfer inventory for the heat exchangers of the combined heat pump plant are also obtained. It provides new theoretical limits for designing a real combined heat pump plant and for performance comparison between existing combined plants.

REFERENCES

1. Novikov I I. The efficiency of atomic power stations (A review). Atommaya Energiya 3, 1957(11): 409
2. Chambadal P. Les Centrales Nucleaires. Paris: Armand Colin, 1957:41-58
3. Curzon F L and Ahlborn B. Efficiency of a Carnot engine at maximum power output. Am. J. Phys., 1975, 43(1):22-24
4. Andresen B. Finite –Time Thermodynamics. Physics Laboratory II, University of Copenhagen, 1983
5. Tsirlin A M. Optimal Cycles and Cycle Regimes. Moscaw: Energomizdat, 1985 (in Russian)
6. Feidt M. Thermodynamique et Optimisation Energetique des Systems et Procedes. Paris : Technique et Documentation, Lavoisier, 1987(in French)
7. Sieniutycz S and Salamon P (eds.). Finite-Time Thermodynamics and Thermoeconomics. New York: Taylor and Francis, 1990
8. De Vos A. Endoreversible Thermodynamics of Solar Energy Conversion. Oxford: Oxford University, 1992
9. Radcenco V. Generalized Thermodynamics. Bucharest: Editura Techica, 1994 (in English)
10. Bejan A. Entropy Generation Minimization. Boca Raton FL: CRC Press, 1996
11. Bejan A, Tsatsaronis G and Moran M. Thermal Design & Optimization. New York: Wiley, 1996
12. Andresen B, Berry R S, Ondrechen M J and Salamon P. Thermodynamics for processes in finite time. Acc. Chem. Res. 1984, 17(4):266-271
13. Andresen B, Salamon P and Berry R S. Thermodynamics in finite time. Phys. Today, 1984(Sept.):62-70
14. Andresen B. Finite-time thermodynamics and thermodynamic length. Rev. Gen.Therm., 1996, 35(418/419):647-650.
15. Orlov V N and Rudenko A V. Optimal control in problems of extremal of irreversible thermodynamic processes. Avtomatika i Telemekhanika, 1985(5):7-41
16. Wu C. Power optimization of a finite-time Carnot heat engine. Energy, The Int. J., 1988, 13(9):681-687
17. Wu C, Kiang R L, Lopardo V J and Karpouzian G N. Finite-time thermodynamics and endoreversible heat engines. Int. J. Mech. Engng. Edu., 1993, 21(4):337-346
18. Chen L, Sun F and Chen W. New developments of finite time thermodynamics. Nature J., 1992, 15(4):249-253

19. Chen L, Sun F and Chen W. The present state and the trend of finite time thermodynamics for energy systems. Adv. Mechanics, 1992, 22(4):479-488
20. Sieniutycz S and Shiner J S. Thermodynamics of irreversible processes and its relation to chemical engineering: Second law analyses and finite time thermodynamics. J. Non-Equilib. Thermodyn., 1994, 19(4):303-348
21. Bejan A. Power generation and refrigeration models with heat transfer irreversibilities. J. Heat Transfer Soc. Japan, 1994, 33(128):68-75
22. Bejan A. Entropy generation minimization: The new thermodynamics of finite-size devices and finite-time processes. J. Appl. Phys., 1996, 79(3):1191-1218
23. Bejan A. Method of entropy generation minimization, or modeling and optimization based on combined heat transfer and thermodynamics. Rev. Gen. Therm., 1996, 35(418/419):637-646
24. Bejan A. Fundamental optima in thermal Science. Int. J. Mech. Engng. Edu., 1997, 25(1):33-47
25. Chen J and Wu C. Performance of a cascade combined heat pump system. J. Institute Energy, 1995, 68 (476): 137-141
26. Chen L, Sun F and Chen W. Finite time thermodynamic analysis for combined heat pump cycles. J. Engng. Thermophys., 1996, 17 (Suppl.):13-15
27. Chen L, Wu C and Sun F. Optimization of specific rate of heat pumping in combined heat pump cycles. Energy Convers. Mgmt., 1998, 39 (1/2): 113-116
28. Chen L and Sun F. Effects of heat leak on the performance of steady flow combined heat pump cycles. Pow. Sys. Engng., 1997, 13 (2):1-4
29. Chen L, Sun F and Chen W. Finite-time thermodynamic performance of heat pumps with the consideration of heat leak. J. Engng. Thermal Energy Pow., 1994, 9 (2): 121-125
30. Chen L, Sun F and Chen W. Finite-time thermodynamic performance of a class of irreversible two-heat-reservoir forward and reverse cycle. Sci. Tech. Bullet., 1995, 11 (2):126
31. Chen L, Wu C and Sun F. Heat transfer effect on the specific heating load of heat pumps. Appl. Thermal Engng., 1997, 17(11): 103-110
32. Chen L, Wu C and Sun F. Heat pump performance with internal heat leak. Int. J. Ambient Energy, 1997, 18 (3):129-134
33. Wu C, Chen L and Sun F. Finite time thermodynamic performance for a class of irreversible heat pumps. Presented at ASME ASIA'97 Conf., 30, Sept. –2, Oct., 1997, Singapore.
34. Sun F, Chen W, Chen L and Wu C. Optimal performance of an endoreversible Carnot heat pump. Energy Convers. Mgmt., 1997, 38 (14): 1439-1443
35. Blanchard C H. Coefficient of performance for finite-speed heat pump. J. Appl. Phys., 1980, 51 (5): 2471-2472
36. Goth Y and Feidt M. Optimum COP for endoreversible heat pump or refrigerating machine. C. R. Acad. Sc. Paris, 1986, 303 (1): 19-24
37. Ait-Ali M M. The maximum coefficient of performance of internally irreversible refrigerators and heat pumps. J. phys. D: Appl. Phys., 1996, 29(4): 975-980

38. Chen L and Sun F. A new model of irreversible heat pumps and its performance optimization. J. Naval Acad. Engng., 1995 (4): 49-58
39. Cheng C and Chen C. Performance optimization of an irreversible heat pump. J. Phys. D: Appl. Phys.,1995, 28 (12): 2451-2454
40. Chen L and Sun F. The effect of heat leak, heat resistance and internal irreversibility on the optimal performance of Carnot heat pumps. J. Engng. Thermophys., 1997, 18 (1): 25-27
41. Chen L, Ni N and Sun F. Effects of heat resistance and internal irreversibility on the characteristics of air heat pump cycles. Acta Energiae Solaris Sinica, 1997, 18 (1): 86-91
42. Chen L, Ni N, Sun F and Wu C. Performance of real regenerated air heat pumps. Proc. IASTED Int. Conf., High Technology in the Power Industry, Oct. 27-30, 1997, Orlando, Florida, USA. Editor: A. Domigan, Jr., IASTED/Acta Press. pp. 387-392
43. Sun F, Chen L and Xue M. Heat transfer surface area minimization for naval air-conditioning plants. Ship Engng., 1997(3):33-34

A CONTRAST BETWEEN THE PHYSICAL AND THE ENGINEERING APPROACHES TO FINITE-TIME THERMODYNAMIC MODELS AND OPTIMIZATIONS

PETER SALAMON
Department of Mathematical and Computer Sciences
San Diego State University
San Diego, CA 92182-7720

1. INTRODUCTION

Finite-time thermodynamics [1] is an extension of traditional thermodynamics that seeks to characterize in-principle limits to the performance of thermodynamic processes given the constraint that such processes take place in a finite time. This constraint forces any process that involves transport through a finite conductance to produce entropy and thus leads to strengthened versions of the second law and provides positive lower bounds on the associated entropy production.

The present paper differentiates between the physics and the engineering perspectives regarding the value of such studies and gives some of the highlights of what can be gained from physical (as opposed to engineering) studies of such processes. The difference can be characterized as follows: physical studies seek in-principle limits to what is possible while engineering studies seek to characterize the best possible process design given current technology and current prices.

2. BOUNDS AND THE ROLE OF OPTIMIZATION

One important reason for studying thermodynamic processes in finite time is to pursue the quest for understanding the limits of what can be achieved in such processes quite aside from any direct implications for engineering. The simplest analyses of this kind reveal how different sources of irreversibility individually influence the limits to power generation and entropy production. In such analyses there is no pretense of counting **all** sources of irreversibility -- one merely counts those sources one is focusing on and thereby obtains **bounds** on what can happen in real processes.

The underlying mathematical fact that allows this is a basic lemma of optimization theory: solving an optimization problem with fewer constraints (additional controls) bounds the value of the objective for the original problem. Analyzing the optimal performance of a system when we improve the control that we are able to exercise, can only improve the optimal value of our objective and thereby gives a bound on what is physically possible.

One important example of this lemma in finite-time thermodynamics is the theorem of Orlov and Berry [2] which is striking in its simplicity and power. They begin by writing down the full dynamical equations for the working fluid, the heat baths, and the power source coupled to a heat engine complete with partial differential equations for the distributed subsystems. The heat transfer to the working fluid is proportional to the temperature difference across a surface. They complete the careful formulation of their control problem by specifying an objective function f. They then switch to a problem in which the temperature at each position inside the working fluid and the heat exchangers is controllable. This discards the partial differential equation constraints describing the dynamics of the temperature in the regions *inside* the engine and heat exchangers, thereby changing these temperatures from state variables to control variables. The optimum in the simplified problem bounds the optimum in the real problem since the simplified problem results by effecting better controls. Ignoring a great number of the dynamical equations converts state variables to control variables and enlarges the set of allowed processes. Of course this must be done carefully respecting the laws of thermodynamics -- these are constraints that we must not ignore.

Physically it is easy to ensure that the laws of thermodynamics are not violated. We simply take the additional (fictitious) controls on our process to be reversible, i.e. effected by reversible processes using additional work and heat sources as needed. Thus we assume reversible control over all but a few of the degrees of freedom in our system. We then ask what is the least entropy production, or maximum power output, or the optimal value of various other possible objectives, when we control the rest of the universe in an ideal way -- namely reversibly. This focuses the analysis on the limitations to energy conversion inherent in that part of the system or device which we did **not** assume to be reversibly controllable. The approach is not limited to heat engines and the results bound the possible values of the objective in a more meaningful way than the bounds provided by classical thermodynamics. While the additional controls may be unrealistic from an engineering point of view, they do not violate physical laws and can help us extract information regarding the limitations to real processes imposed by specific modes of irreversibility which are the focus of a specific study. Since typical applications of the Orlov-Berry theorem require the controls to maintain uniform intensities inside a system, they can even be realized by real processes in which the transport coefficients inside a system are large compared to the transport coefficients across the boundary. This is usually referred to as endoreversibility which is often useful in simplifying the analysis of limitations to the behavior of thermodynamic systems[1]. Note that this is not always

[1] Prior to the Orlov-Berry theorem, the argument required an assumption of the separability of the time scales for the transport processes within a system and across system boundaries[3].

needed for our analysis but only when the dissipation inside a system is not part of the irreversible modes which are the focus of our study. Note further that the term "endo" is slightly misleading since this argument can be equally well applied to subsystems such as thermal reservoirs which form part of the "environment" and are thus more aptly labeled "exo".

A second theorem due to A. Tsirlin [4] serves to simplify finite-time thermodynamic studies even further. This theorem states that provided the constraints and objective of a problem depend only on net (or average) process quantities, i.e. on integrals over time, the optimal way to control the system is to take our controls to be piecewise constant in time. These two theorems make simple models interesting and able to give useful, rigorous bounds regarding what can be achieved during the control of real processes. The simplified problems often lead to solutions which tell us to run the process infinitely fast or infinitesimally slow. The interesting domain is in between: well posed problems which ignore enough irreversibility to be tractable and yet count enough irreversibilities to keep the solution finite. The value of the objective obtained in the simpler optimization bounds the value of the objective in real processes in a more realistic way than a reversible bound, although it is still often very far from what can be achieved in reality.

The results must be used with great care however. Other features of the optimal solution obtained cannot be directly compared with similar features of real processes. A simple and familiar example is provided by the well known equation of Curzon-Ahlborn-Novikov (CAN) [5,6] for the efficiency at maximum power

$$\eta = 1 - \sqrt{\frac{T_c}{T_h}} . \tag{1}$$

The striking simplicity of this equation, as well as its similarity to the well known efficiency of Carnot, has brought it rapid acceptance. Nonetheless, as pointed out in [8], this efficiency is surpassed in real power plants! The explanation is that this equation does **not** represent the maximum efficiency but rather the efficiency at maximum power. The associated maximum power

$$P = \frac{K}{4}\left(\sqrt{T_h} - \sqrt{T_c}\right)^2 \tag{2}$$

must be more than the power delivered in any process which corresponds to the CAN calculation augmented by additional dynamical equations. This much is mathematical fact. Note that there is **no** implication regarding a comparison between the efficiency obtained at this power and efficiencies in real processes -- even when the real processes are also designed to maximize power, albeit with additional constraints!

3. Some Misunderstandings

One of my goals in the present paper is to respond to certain criticisms of what I call the physics approach to the analysis of thermodynamic processes [7-9]. The discussion above regarding the CAN efficiency is one such issue. A second issue also concerns unwarranted conclusions based on the above formalism. Finite-time thermodynamics makes no claims to the effect that *every* process approaches reversible as the rate is slowed down. Thus Professor Gytopoulos's examples [8] concerning 1000 spark plugs and batteries attack a non-existent claim. Certainly, unless we are able to exercise a great deal of control over the process, slowing it down does not help. The strongest correct statement along these lines which one can make is as follows: Finite process time is a very real constraint and forces any process to produce entropy provided some transport is required and conductivities are finite. When the process time constraint is increased, it is *possible* to use more time although it is *not necessary* to do so unless we assume further restrictions on our controls. Thus we are free to use the same optimal control that was allowed for constrained time τ even after this is relaxed to a larger constrained time $\tau' > \tau$. It thus follows that increasing the allowed process time cannot make the optimal process perform worse by any measure since all we have done is allow additional possibilities. Thus the issue is whether one understands time as a constraint for a control problem or as a traditional physical variable.

Whether or not additional time actually helps improve a process depends on many factors. Transport processes through a finite conductance can proceed closer to reversibly, provided we have adequate control which can take advantage of the longer time. Systems which have an exergy leak, [8] such as electrochemical cells, certainly do not do well when operated very slowly. This is an essential feature of a finite-time thermodynamic analysis for these systems [10].

Another issue related to repairing the road of communications between science and engineering concerns the idea that there can only be one "freely variable reservoir". Bejan, foremost among authors, has advanced this unfortunate point of view. I begin with a quote from his referee report on one of my papers [2].

> "In my 1996 Am. J. Phys. paper I [Bejan] explained why the Curzon & Ahlborn constant is unrealistic, and why related claims are unfounded. The reason is that the optimization performed by Curzon & Ahlborn is based on the unrealistic assumption that the heat input (amount of fuel) to the engine is a true degree of freedom in the design, i.e., a quantity that varies as freely as the rejected heat. This assumption would hold only in a world with two environments, or two atmospheres ..., certainly not in today's power plants on earth.
> My 1996 paper [9] was a challenge to modelers such as Curzon and Ahlborn to explain specifically (or graphically) where their free input comes from. This challenge was sidestepped by the present authors."

[2] He chose to disclose his identity in the report.

Let me not sidestep the issue any further. As I hope was made clear in the discussion above, this challenge comes from a mistaken conception of the *physical* subject of finite-time thermodynamics which by its nature seeks bounds through the use of very limited counting. This is also the answer to Bejan's similar challenge issued in this same 1996 paper [9]

> "Whether the entropy generation estimate of Salamon et al. accounts for all the irreversibilities of the power plant is a topic I would like to open for discussion."

Let me reiterate; there is no attempt to count *all* the irreversibility but simply to *bound* the performance by counting only *some* of the irreversibility in order to understand the intrinsic limitations set by the effects of these modes.

Bejan's idea is that we need to count the complete process from the burning of the fossil fuel to the equilibrated combustion gases at atmospheric temperature. According to Bejan "Any other 'entropy generation rate' calculation is incomplete, i.e. irrelevant and useless." I now argue that this statement is completely untenable from a physical, an engineering, and an economic point of view.

Let us begin with the physics. Here the debate is perhaps moot since Bejan has admitted in print that if one is talking about a heat engine operating between the atmospheres of earth and another planet, one can consider both reservoirs to be freely varying. Thus the idea of a single freely variable reservoir is terracentric at best. To focus our analysis on the effects of a particular mode of irreversible interaction, it helps to ignore the irreversibilities of the other processes. This is the essence of the program I outlined above. Thus we are free to treat the combustion process and the residual cooling of the exhaust gases as reversible. Is there something physically sacred about combustion and exhaust (as opposed to, say, the inside of the engine compartment) that we are allowed to take one as reversible and the other not? We can focus our irreversibility analyzer on any mode that we choose and there is no physical grounds for limiting what irreversibilities one chooses to count. Recall that the physics we are talking about understanding is used for the analysis of processes from the molecular to the cosmic scales. There are many more thermodynamic systems than the ones commonly met in mechanical engineering and thermodynamics applies to them all. In fact, some of the finest accomplishments of finite-time thermodynamics have been at the molecular level [11]. The idea of only one freely variable reservoir is irrelevant for a beam of molecules undergoing interactions with a buffer gas and a laser beam.

Are there good engineering grounds for insisting that one must count the irreversibilities in the combustion of the fossil fuels and the exergy thrown away with the exhaust gases? After all, both of these represent very significant exergy costs in most real power plants. In the analyses of these plants for engineering purposes I certainly agree that it is reasonable, nay even crucial, to count the irreversibility of the discarded exhaust. This is not the same as discounting calculations that do not count this irreversibility as "irrelevant and useless". This would discard the engineering study of solar and geothermal power plants not to mention devices such as James Senft's icewater powered heat engines. This is an unaffordably narrow view of engineering. Are we to be forced to count the fossil fuel burned in producing the ice for the icewater engine? Of what relevance is this for the *engineering* problem of how best to design an engine that is to be powered by the temperature difference between a glass of icewater and the atmosphere? The icewater in this engineering problem *should* be counted as a "freely variable" reservoir.

Bejan dismissed the geothermal power plant example by saying that "in the end the hot stream becomes a very expensive per-unit commodity, not a freely varying flow rate as assumed in ...". Let us therefore examine a possible economic basis to the idea of a single freely variable reservoir. Once we let prices become thermodynamic variables we enter the domain of what has been called thermoeconomics [1]. A careful examination then reveals that no reservoir is free. Pumping river water to cool process streams or building cooling towers is not free but has a very real price associated with amortizing the capital costs of the heat exchangers, pumps, etc.. Thus the atmosphere is not exactly freely varying either. Is it a useful approximation to take heat exchange with the atmosphere as free? Sometimes. In a more careful economic analysis, heat from each reservoir has a shadow price and these prices will be of both signs and none of them will equal zero.

4. Some Classic Results from Finite Time Thermodynamics

In the domain of well posed problems there is a veritable plethora of possible optimizations and analyses by varying the system, the constraints, or the objective. This fact has brought some indignation from various editors flooded with papers working out these various and sundry cases. This argues all the more in favor of a physics which could serve to unify these sundry results. I now cite a few of the steps in the direction of such a physical theory.

Above, I have already cited the classic theorems of Orlov-Berry and Tsirlin. These two theorems enable us to reduce most analyses to the finite dimensional level where simple calculus suffices. Another classic result is a folk theorem due to Gouy and Stodola [12, 13] which was rediscovered several times by various authors [14]. This theorem guarantees that entropy production is always just the exergy loss divided by the atmospheric temperature; thus minimizing loss of exergy is always equivalent to minimizing entropy production. The atmosphere is certainly a unique heat bath in defining this constant of proportionality.

A similar result regarding the equivalence of minimum entropy production and maximum power also holds provided one specifies the initial and final states of all participating systems except the atmosphere and a work reservoir. Once one has committed to a certain change in the exergy of all the participating systems except one (note that any change in thermal energy of the atmosphere carries exergy zero), any uncaptured work is perforce released into the atmosphere and so maximizing power is equivalent to minimizing entropy production. If one includes the possibility of saving some of the exergy for later use, e.g., burn less fuel, use waste heat for cogeneration, etc., the final states can be allowed to vary and minimum entropy generation and maximum power become opposing objectives; minimum entropy production generally means opting for conservation while maximum power means opting for rapid consumption. For example, in a heat engine working between two heat reservoirs with a given entropy excursion[3], $\int |ds|$, where s is the entropy of the working fluid, there is a range of interesting optima with extremes at maximum power and at minimum entropy production [6]. There is interesting new physics at either end and thermoeconomics in between.

For linear heat exchange, simple results abound. The minimum entropy production ΔS^u_{min} goes as the square of the entropy excursion of the working fluid, i.e., is given by

$$\Delta S^u_{min} = \left(\int_0^\tau |ds| \right)^2 / \kappa\tau \tag{3}$$

where κ is the conductance to the working fluid and τ is the cycle time [3]. The maximum power from a working fluid in a periodically changing environment $T^{ex}(t)$ is given by

$$P_{max} = \kappa \operatorname{var}\left(\sqrt{T_{ex}(t)}\right), \tag{4}$$

where var indicates the variance in time [15]. If the conductance to the working fluid can vary between the two heat exchange branches, the optimal allocation of area is inversely proportional to the square root of the conductivity [16].

These early results have been worked and reworked, extended and applied many times to many different systems [1, 17, 18]. So far the early results have served most importantly as a paradigm for how to calculate and reason. As argued above, the reasoning involved is delicate. For building a physical theory of such processes, reworking and reformulation is a necessary step toward coming up with a useful and general formalism. Such reworking brings the proper mathematical structure to the surface and allows us to give all of these problems a synthesis within one framework. Such a synthesis is hinted at in a number of recent efforts [19, 20].

5. The Horse-Carrot Process and Related Theorems

One particularly important and general problem concerns controlling a system to follow a prespecified sequence of states. This has been termed a horse-carrot process by analogy since it requires coaxing a system (the horse) to move toward equilibrium with a

---

[3] For a simple cycle, the entropy excursion of the working fluid is just twice the width of the cycle in a T,S diagram. See references [3].

controlled sequence of baths (the carrots). Here the coupling between the system and its environment is the mode of irreversibility considered -- the environment with which the system interacts is assumed reversibly controllable. For the process in a (sufficiently large) constrained time, the optimal solution is to control the system so the rate of entropy production is constant [21, 22]. The resulting entropy production is bounded by

$$\Delta S^u \geq L^2 \varepsilon / \tau \qquad (5)$$

where L is the thermodynamic distance traversed in the state space of the system and ε is a mean relaxation time [23].

The theorem is perhaps more interesting for a staged process in which the system proceeds by a sequence of equilibrium stages. For this case, the horse-carrot inequality takes the form

$$\Delta S^u \geq L^2 / 2k \qquad (6)$$

where k is the number of stages. For minimum entropy production, these stages should be equally spaced in thermodynamic distance $\Delta L = L/k = \text{constant}$ [24].

The version of the theorem for staged processes has been extended to steady flow processes including fractional distillation [25]. In this analysis, the irreversibility comes from the equilibration of the material that moves to the next tray with the material on that next tray, the latter acting as the bath. This result has very real implications for column design: *Distribute the heating so the temperature profile along the column is in steps of equal thermodynamic distance.* Economic use of the design requires a heat pump which can work efficiently between many temperature stages.

6. Mathematical Structure

As the above discussion illustrates, understanding the physics is different from understanding the engineering. Physicists are not directly interested in how to design or operate a specific device or process but rather in understanding what limitations are set for the process by the thermodynamic laws. In mathematical physics, such understanding usually comes by studying the mathematical structure of the set of possibilities. This is not a problem that engineers have been concerned with, even when very similar optimizations were performed for very similar systems. It is a problem worth studying.

Gibbs' reformulation of thermodynamics focusing on the geometric structure of state space is a prime example of the type of mathematical structure I have in mind here. So is the Horse-Carrot theorem described above. The theorem has served to elucidate the Riemannian geometric structure on the state space[4] of a thermodynamic system -- a

[4] The Riemannian metric is given by the second derivative matrix of the entropy of the system as a function of the extensive variables.

structure which turns out to extend from the quantum mechanical to the macroscopic level and which has given us an improved fluctuation theory accurate to larger fluctuations [26].

7. Further Difficulties of Communication

There exist nontrivial barriers to communication between physicists and engineers working in this area even at the level of how we use the English language. To make this point, I cite an anecdote regarding the use of the phrase "industrial example". In theorist-speak, this phrase means that *some* industry might care about it *some* day. In the referee report cited above, Adrian Bejan, dismissed geothermal power plants as "hardly an industrial example" since their "share of the global energy industry is meaningless."[5]

As a second example I consider the word "applied". In a mathematics department, this word is used to mean that the arguments and results relate to something outside the sphere of mathematics, even when this "use" is extremely esoteric, e.g. characterizing certain cross sections of a fiber bundle over a seventeen dimensional space-time. Note that this use is very far from what "applied" typically means in a physics department which is still dramatically different from what "applied" means in an engineering department.

Other examples of words that mean very different things to engineers, physicists, and mathematicians include "realistic", "general", "optimal" and even "understanding".

8. Conclusions

In our modern multicultural and multidisciplinary world it is important to adopt a wider perspective. While the type of physical studies outlined above may be of interest to engineering design only in its earliest stages, there is a long tradition of physics *ultimately* leading to principles important for engineering. It is important to accept the fact that our field is interdisciplinary and to try to move beyond the perspectives offered by our own, narrowly defined disciplines.

References

1. Sieniutycz, S. and Salamon, P. (1990) *Finite–Time Thermodynamics and Thermoeconomics*, Taylor and Francis, New York.
2. Orlov, V. and Berry, R. S. (1990) Power output from an irreversible heat engine with a nonuniform working fluid, *Physical Review* A **42**, 7230-7235.
3. Salamon, P., Nitzan, A., Andresen, B. and Berry, R. S. (1980) Minimum entropy generation and the optimization of heat engines, *Physical Review A* **21**, 2115-2129.

[5] We remark that in Iceland, the overwhelming majority of the energy industry is geothermal. Thus the perspectives offered by different cultural and ethnic affiliations also create barriers to communications.

4. Tsirlin, A. M. (1986) *Optimal Control of Technological Processes*, Energoatomizdat, Moscow.
5. Novikov, I. I. (1958) The Efficiency of Nuclear Power Stations, *Journal of Nuclear Energy II* (U.S.S.R.) **7**, 125-128.
6. Curzon, F. L. and Ahlborn, B. (1975) Efficiency of a Carnot engine at maximum power output, *American Journal of Physics* **43**, 22-24.
7. Bejan, A. (1996) "Entropy Generation Minimization: The New Thermodynamics of Finite-size Devices and Finite-time Processes", *Journal of Applied Physics* **79**, 1191-1218.
8. Gyftopoulos, E. (1997) "Fundamentals of Analysis of Processes", *Energy Conversion and Management* **38**, 1525-1533.
9. Bejan, A. (1996) "Models of Power Plants that Generate Minimum Entropy while Operating at Maximum Power", *American Journal of Physics* **64**, 1054-1059.
10. Watowich, S. J. and Berry, R. S. (1986) "Optimal Current Paths for Model Electrochemical Systems", *Journal of Physical Chemistry* **90**, 4624-4631.
11. Geva, E. and Kosloff, R. (1992) "On the classical limit of quantum thermodynamics in finite time", *Journal of Chemical Physics* **97**, 4398-4412.
12. Gouy, M. (1889) "Sur L'Energie Utilizable", *Journal de Physique* **8**, 501.
13. Stodola, A. (1910) *Steam and Gas Turbines*, McGraw-Hill, New York.
14. Tolman, R. C. and Fine, P. C. (1948) "On the irreversible production of entropy", *Reviews of Modern Physics* **20**, 51.
15. Salamon, P., Band, Y. B. and Kafri, O. (1982) "Maximum power from a cycling working fluid", *Journal of Applied Physics* **53**, 197-202.
16. Goth, Y. and Feidt, M. (1986), "Conditions optimales de fonctionnement des pompes a chaleur ou machines a froid associees a un cycle de Carnot endoreversible", 303, 19-24.
17. Bejan, A. (1996) *Entropy generation minimization : the method of thermodynamic optimization of finite-size systems and finite-time processes*, CRC Press.
18. Feidt, M. (1996) *Thermodynamique et optimisation energetique*, Tec & Doc-Lavoisier.
19. Nulton, J.D. and Salamon, P. (1998) "Finite-time thermodynamic analysis of controlled heat integration", *Proceedings of ECOS98*, ed. M. Feidt.
20. Feidt, M. (1998) "Thermodynamics and optimization of reverse cycle machines: refrigeration, heat pumps, air conditioning, cryogenics", *Proceedings of NATO ASI on Thermodynamics and the optimization of complex energy systems*, ed. A. Bejan and E. Mamut.
21. Spirkl, W. and Ries, H. (1995) "Optimal finite-time endoreversible processes", *Physical Review E* **52**, 3485-3489.
22. L. Diosi, K. Kulacsy, B. Lukacs, and A. Racz, *Journal of Chemical Physics* **105**, 11220 (1996).
23. Salamon, P. and Berry, R. S. (1983) "Thermodynamic Length and Dissipated Availability", *Physical Review Letters* **51**, 1127 - 1130.
24. Nulton, J., Salamon, P., Andresen, B. and Anmin, Q. (1985), "Quasistatic Processes as Step Equilibrations", *Journal of Chemical Physics* **83**, 334-338.

25. Salamon, P. and Nulton, J.D. (1998), The Geometry of Separation Processes: A Horse-Carrot Theorem for Steady Flow Systems, *Europhysics Letters* **42**, 571-576.
26. G. Ruppeiner (1995) *Reviews of Modern Physics* **67**, 605.

INDEX

A

B

C

D

E

I

J

K

L

V

W